Basic Building Data

Third Edition

Don Graf

Revised by S. Blackwell Duncan

VAN NOSTRAND REINHOLD COMPANY
New York

Copyright ©1985, by Van Nostrand Reinhold Company Inc.
Library of Congress Catalog Card Number 84-2343
ISBN 0-442-22738-8

All rights reserved. No part of this work covered by the
copyright hereon may be reproduced or used in any form
or by any means—graphic, electronic, or mechanical,
including photocopying, recording, taping, or information
storage and retrieval systems—without written permission
of the publisher.

Printed in the United States of America

Published by Van Nostrand Reinhold Company Inc.
135 West 50th Street
New York, New York 10020

Van Nostrand Reinhold Company Limited
Molly Millars Lane
Wokingham, Berkshire RG11 2PY, England

Van Nostrand Reinhold
480 La Trobe Street
Melbourne, Victoria 3000, Australia

Macmillan of Canada
Division of Canada Publishing Corporation
164 Commander Boulevard
Agincourt, Ontario M1S 3C7, Canada

16 15 14 13 12 11 10 9 8 7 6 5 4 3 2 1

Library of Congress Cataloging in Publication Data

Graf, Don.
  Basic building data.

  Includes index.
  1. Building—Handbooks, manuals, etc. I. Duncan,
Stuart B. II. Title.
TH151.G68 1984   690   84-2343
ISBN 0-442-22738-8

PREFACE

This book began as a collection of loose-leaf sheets about manufactured products and basic design data to help the author in his active architectural practice. From 1932 to 1942, these sheets were published and augmented with the help of numerous manufacturers and associations, many of which are still represented in the third edition.

The first bound book was issued and copyrighted by Reinhold Publishing Corporation in 1944; in 1949, it underwent a complete revision. The present edition, once again totally revised to reflect the many changes that have occurred over the past decades, is issued and copyrighted by Van Nostrand Reinhold Company.

Because the material contained herein is basic, references to specific products have not been included. In using the data, bear in mind that much of it—with the exception of such purely factual material as mantissas, mathematical formulae, conversion factors—is intended simply to be representative of the types of specific building situations, products, and materials that will actually be encountered and to suggest guidelines for analogous circumstances. In most instances, once a particular idea has evolved or a preliminary design has been worked out, the reader should turn to appropriate trade associations for further guidance, to particular manufacturers' literature for information on applications, specifications, availability, and design or product changes, and to model and local building codes to determine permissibility and practicability.

The reader must also realize that over the past two decades or so there has been a veritable explosion of knowledge, materials, techniques, and products in the building trades, as well as a substantial expansion of building codes and similar regulations, product standards, federal specifications, and the like. Thus, there are many practices, techniques, products, specifications, and designs available to people involved with the building trades today in addition to those discussed here. Numerous alternatives to much of what *is* discussed are also available. Neither this book nor any other could hope to cover the thousands of possible details found in modern-day construction. Nonetheless, this volume will serve as a handy and valuable reference for much of the basic information required in general building construction, and as a preliminary guide in the formulation of ideas and designs; better still, the user will find contained herein definitive answers to a good many puzzling questions.

To every manufacturer and association that cooperated with this revision, thanks are due. Where the text or detailing varies with their recommendations, it reflects the judgment of the author (and/or the reviser) that such was necessary for the purpose of this book. The outstanding associations and manufacturers who have assisted in the assembly of the material are listed, for your convenience, on the following pages.

DON GRAF

RFD 2, Ossining, New York
1949

S. BLACKWELL DUNCAN

Gateway Road, Snowmass, Colorado
1985

ACKNOWLEDGMENTS

AMATEUR ATHLETIC UNION OF U.S., Indianapolis, IN

AMERICAN CONCRETE INSTITUTE, Detroit, MI

AMERICAN HARDBOARD ASSOCIATION, Palatine, IL

AMERICAN INSTITUTE OF ARCHITECTS, Washington, DC

AMERICAN INSURANCE ASSOCIATION, New York, NY

AMERICAN LAUNDRY MACHINERY, INC., Cincinnati, OH

AMERICAN NATIONAL STANDARDS INSTITUTE, New York, NY

AMERICAN PLYWOOD ASSOCIATION, Tacoma, WA

AMERICAN SOCIETY FOR TESTING AND MATERIALS, Philadelphia, PA

AMERICAN TELEPHONE AND TELEGRAPH CO., New York, NY

ARMCO, INC., Middletown, OH...... *Roofing, siding, and flashings*

BARROWS CORP., Cincinnati, OH. *Porcelain enamel letters for signs*

BRICK INSTITUTE OF AMERICA, McLean, VA

BRUNSWICK CORP., Skokie, IL...... *Billiard tables, bowling alleys*

CAST IRON SOIL PIPE INSTITUTE, McLean, VA

EASTMAN KODAK CO., Rochester, NY.
Darkrooms, home movie equipment

FAIRBANKS-MORSE PUMP DIV., COLT INDUSTRIES, INC., Kansas City, KS. .... *Water systems, well pumps*

GENERAL ELECTRIC CO., Schenectady, NY... *Underwater lighting*

HEYWOOD-WAKEFIELD CO., Gardner, MA...... *Theater seating*

HOFFMAN SPECIALTY ITT, Indianapolis, IN... *Heating specialties*

INDEPENDENT PROTECTION CO., Goshen, IN.
Lightning protection equipment

INLAND STEEL CO., Chicago, IL..... *Stair treads, hatchway covers*

KOHLER CO., Kohler, WI................... *Plumbing fixtures*

KOPPERS CO., Pittsburgh, PA............ *Coal tar pitch products*

LIBBEY-OWENS-FORD CO., Toledo, OH........ *Multiple glazing*

METAL LATH/STEEL FRAMING ASSOCIATION, Chicago, IL

MONTGOMERY ELEVATOR CO., Moline, IL.... *Elevator systems*

MUTSCHLER BROS. CO., Nappanee, IN......... *Kitchen cabinets*

NATIONAL FIRE PROTECTION ASSOCIATION, Quincy, MA

NATIONAL GYPSUM CO., Dallas, TX.......... *Plaster partitions*

NATIONAL PAINT AND COATINGS ASSOCIATION,
Washington, DC

NATIONAL TERRAZZO AND MOSAIC ASSOCIATION,
Des Plaines, IL

PECORA CHEMICAL CORP., Garland, TX. . . *Mastics and caulking*

THE PEELE CO., Bay Shore, NY.................... *Fire doors*

PITTSBURGH-CORNING CORP., Pittsburgh, PA...... *Glass block*

PORTLAND CEMENT ASSOCIATION, Skokie, IL

SCOTT PAPER CO., Philadelphia, PA....... *Washroom equipment*

SOUTHERN FOREST PRODUCTS ASSOCIATION,
New Orleans, LA

SOUTHERN PINE INSPECTION BUREAU, Pensacola, FL

THE STANLEY WORKS, New Britain, CT............. *Hardware*

THOMPSON LIGHTNING PROTECTION, INC., St. Paul, MN

UNDERWRITER'S LABORATORIES, INC., Chicago, IL

U.S. DEPARTMENT OF AGRICULTURE, Washington, DC

U.S. DEPARTMENT OF COMMERCE, Washington, DC

U.S. ENVIRONMENTAL PROTECTION AGENCY, Washington, DC

U.S. PLYWOOD DIV., CHAMPION INTERNATIONAL CORP.,
Stamford, CT

U.S. PUBLIC HEALTH SERVICE, Washington, DC

UNIVERSAL ATLAS CEMENT DIV., UNITED STATES STEEL
CORP., Pittsburgh, PA

WESTERN WOOD MOULDING AND MILLWORK PRODUCERS,
Portland, OR

BRICK BONDS AND
MORTAR JOINTS

RUNNING

COMMON

CHECKER BOARD

RUNNING HEADER

ENGLISH

DUTCH, DUTCH CROSS
OR ENGLISH CROSS

GARDEN WALL

GARDEN WALL CROSS

FLEMISH

FLEMISH CROSS

SPIRAL STRETCHER

FLEMISH SPIRAL

STRUCK
WEATHERED
FLUSH
RAKED
STRIPPED
V-JOINT
CONCAVE
BEADED

HEIGHTS FOR 2½"
BRICK COURSES

BASED ON STANDARD BRICK 2¼" + ¼" JOINT

No. of Courses	Vertical Height	No. of Courses	Vertical Height	No. of Courses	Vertical Height
		50	10'- 5"	100	20'-10"
1	2½"	51	10'- 7½"	101	21'- 0½"
2	5"	52	10'-10"	102	21'- 3"
3	7½"	53	11'- 0½"	103	21'- 5½"
4	10"	54	11'- 3"	104	21'- 8"
5	1'- 0½"	55	11'- 5½"	105	21'-10½"
6	1'- 3"	56	11'- 8"	106	22'- 1"
7	1'- 5½"	57	11'-10½"	107	22'- 3½"
8	1'- 8"	58	12'- 1"	108	22'- 6"
9	1'-10½"	59	12'- 3½"	109	22'- 8½"
10	2'- 1"	60	12'- 6"	110	22'-11"
11	2'- 3½"	61	12'- 8½"	111	23'- 1½"
12	2'- 6"	62	12'-11"	112	23'- 4"
13	2'- 8½"	63	13'- 1½"	113	23'- 6½"
14	2'-11"	64	13'- 4"	114	23'- 9"
15	3'- 1½"	65	13'- 6½"	115	23'-11½"
16	3'- 4"	66	13'- 9"	116	24'- 2"
17	3'- 6½"	67	13'-11½"	117	24'- 4½"
18	3'- 9"	68	14'- 2"	118	24'- 7"
19	3'-11½"	69	14'- 4½"	119	24'- 9½"
20	4'- 2"	70	14'- 7"	120	25'- 0"
21	4'- 4½"	71	14'- 9½"	121	25'- 2½"
22	4'- 7"	72	15'- 0"	122	25'- 5"
23	4'- 9½"	73	15'- 2½"	123	25'- 7½"
24	5'- 0"	74	15'- 5"	124	25'-10"
25	5'- 2½"	75	15'- 7½"	125	26'- 0½"
26	5'- 5"	76	15'-10"	126	26'- 3"
27	5'- 7½"	77	16'- 0½"	127	26'- 5½"
28	5'-10"	78	16'- 3"	128	26'- 8"
29	6'- 0½"	79	16'- 5½"	129	26'-10½"
30	6'- 3"	80	16'- 8"	130	27'- 1"
31	6'- 5½"	81	16'-10½"	131	27'- 3½"
32	6'- 8"	82	17'- 1"	132	27'- 6"
33	6'-10½"	83	17'- 3½"	133	27'- 8½"
34	7'- 1"	84	17'- 6"	134	27'-11"
35	7'- 3½"	85	17'- 8½"	135	28'- 1½"
36	7'- 6"	86	17'-11"	136	28'- 4"
37	7'- 8½"	87	18'- 1½"	137	28'- 6½"
38	7'-11"	88	18'- 4"	138	28'- 9"
39	8'- 1½"	89	18'- 6½"	139	23'-11½"
40	8'- 4"	90	18'- 9"	140	29'- 2"
41	8'- 6½"	91	18'-11½"	141	29'- 4½"
42	8'- 9"	92	19'- 2"	142	29'- 7"
43	8'-11½"	93	19'- 4½"	143	29'- 9½"
44	9'- 2"	94	19'- 7"	144	30'- 0"
45	9'- 4½"	95	19'- 9½"	145	30'- 2½"
46	9'- 7"	96	20'- 0"	146	30'- 5"
47	9'- 9½"	97	20'- 2½"	147	30'- 7½"
48	10'- 0"	98	20'- 5"	148	30'-10"
49	10'- 2½"	99	20'- 7½"	149	31'- 0½"

HEIGHTS FOR 2⅝"
BRICK COURSES

BASED ON STANDARD BRICK 2¼" + ⅜" JOINT

No. of Courses	Vertical Height	No. of Courses	Vertical Height	No. of Courses	Vertical Height
		50	10'-11¼"	100	21'-10½"
1	2⅝"	51	11'- 1⅞"	101	22'- 1⅛"
2	5¼"	52	11'- 4½"	102	22'- 3¾"
3	7⅞"	53	11'- 7⅛"	103	22'- 6⅜"
4	10½"	54	11'- 9¾"	104	22'- 9"
5	1'- 1⅛"	55	12'- 0⅜"	105	22'-11⅝"
6	1'- 3¾"	56	12'- 3"	106	23'- 2¼"
7	1'- 6⅜"	57	12'- 5⅝"	107	23'- 4⅞"
8	1'- 9"	58	12'- 8¼"	108	23'- 7½"
9	1'-11⅝"	59	12'-10⅞"	109	23'-10⅛"
10	2'- 2¼"	60	13'- 1½"	110	24'- 0¾"
11	2'- 4⅞"	61	13'- 4⅛"	111	24'- 3⅜"
12	2'- 7½"	62	13'- 6¾"	112	24'- 6"
13	2'-10⅛"	63	13'- 9⅜"	113	24'- 8⅝"
14	3'- 0¾"	64	14'- 0"	114	24'-11¼"
15	3'- 3⅜"	65	14'- 2⅝"	115	25'- 1⅞"
16	3'- 6"	66	14'- 5¼"	116	25'- 4½"
17	3'- 8⅝"	67	14'- 7⅞"	117	25'- 7⅛"
18	3'-11¼"	68	14'-10½"	118	25'- 9¾"
19	4'- 1⅞"	69	15'- 1⅛"	119	26'- 0⅜"
20	4'- 4½"	70	15'- 3¾"	120	26'- 3"
21	4'- 7⅛"	71	15'- 6⅜"	121	26'- 5⅝"
22	4'- 9¾"	72	15'- 9"	122	26'- 8¼"
23	5'- 0⅜"	73	15'-11⅝"	123	26'-10⅞"
24	5'- 3"	74	16'- 2¼"	124	27'- 1½"
25	5'- 5⅝"	75	16'- 4⅞"	125	27'- 4⅛"
26	5'- 8¼"	76	16'- 7½"	126	27'- 6¾"
27	5'-10⅞"	77	16'-10⅛"	127	27'- 9⅜"
28	6'- 1½"	78	17'- 0¾"	128	28'- 0"
29	6'- 4⅛"	79	17'- 3⅜"	129	28'- 2⅝"
30	6'- 6¾"	80	17'- 6"	130	28'- 5¼"
31	6'- 9⅜"	81	17'- 8⅝"	131	28'- 7⅞"
32	7'- 0"	82	17'-11¼"	132	28'-10½"
33	7'- 2⅝"	83	18'- 1⅞"	133	29'- 1⅛"
34	7'- 5¼"	84	18'- 4½"	134	29'- 3¾"
35	7'- 7⅞"	85	18'- 7⅛"	135	29'- 6⅜"
36	7'-10½"	86	18'- 9¾"	136	29'- 9"
37	8'- 1⅛"	87	19'- 0⅜"	137	29'-11⅝"
38	8'- 3¾"	88	19'- 3"	138	30'- 2¼"
39	8'- 6⅜"	89	19'- 5⅝"	139	30'- 4⅞"
40	8'- 9"	90	19'- 8¼"	140	30'- 7½"
41	8'-11⅝"	91	19'-10⅞"	141	30'-10⅛"
42	9'- 2¼"	92	20'- 1½"	142	31'- 0¾"
43	9'- 4⅞"	93	20'- 4⅛"	143	31'- 3⅜"
44	9'- 7½"	94	20'- 6¾"	144	31'- 6"
45	9'-10⅛"	95	20'- 9⅜"	145	31'- 8⅝"
46	10'- 0¾"	96	21'- 0"	146	31'-11¼"
47	10'- 3⅜"	97	21'- 2⅝"	147	32'- 1⅞"
48	10'- 6"	98	21'- 5¼"	148	32'- 4½"
49	10'- 8⅝"	99	21'- 7⅞"	149	32'- 7⅛"

HEIGHTS FOR 2¾"
BRICK COURSES

BASED ON STANDARD BRICK 2¼" + ½" JOINT

No. of Courses	Vertical Height	No. of Courses	Vertical Height	No. of Courses	Vertical Height
1	2¾"	50	11'- 5½"	100	22'-11"
2	5½"	51	11'- 8¼"	101	23'- 1¾"
3	8¼"	52	11'-11"	102	23'- 4½"
4	11"	53	12'- 1¾"	103	23'- 7¼"
5	1'- 1¾"	54	12'- 4½"	104	23'-10"
6	1'- 4½"	55	12'- 7¼"	105	24'- 0¾"
7	1'- 7¼"	56	12'-10"	106	24'- 3½"
8	1'-10"	57	13'- 0¾"	107	24'- 6¼"
9	2'- 0¾"	58	13'- 3½"	108	24'- 9"
		59	13'- 6¼"	109	24'-11¾"
10	2'- 3½"				
11	2'- 6¼"	60	13'- 9"	110	25'- 2½"
12	2'- 9"	61	13'-11¾"	111	25'- 5¼"
13	2'-11¾"	62	14'- 2½"	112	25'- 8"
14	3'- 2½"	63	14'- 5¼"	113	25'-10¾"
15	3'- 5¼"	64	14'- 8"	114	26'- 1½"
16	3'- 8"	65	14'-10¾"	115	26'- 4¼"
17	3'-10¾"	66	15'- 1½"	116	26'- 7"
18	4'- 1½"	67	15'- 4¼"	117	26'- 9¾"
19	4'- 4¼"	68	15'- 7"	118	27'- 0½"
		69	15'- 9¾"	119	27'- 3¼"
20	4'- 7"				
21	4'- 9¾"	70	16'- 0½"	120	27'- 6"
22	5'- 0½"	71	16'- 3¼"	121	27'- 8¾"
23	5'- 3¼"	72	16'- 6"	122	27'-11½"
24	5'- 6"	73	16'- 8¾"	123	28'- 2¼"
25	5'- 8¾"	74	16'-11½"	124	28'- 5"
26	5'-11½"	75	17'- 2¼"	125	28'- 7¾"
27	6'- 2¼"	76	17'- 5"	126	28'-10½"
28	6'-'5"	77	17'- 7¾"	127	29'- 1¼"
29	6'- 7¾"	78	17'-10½"	128	29'- 4"
		79	18'- 1¼"	129	29'- 6¾"
30	6'-10½"				
31	7'- 1¼"	80	18'- 4"	130	29'- 9½"
32	7'- 4"	81	18'- 6¾"	131	30'- 0¼"
33	7'- 6¾"	82	18'- 9½"	132	30'- 3"
34	7'- 9½"	83	19'- 0¼"	133	30'- 5¾"
35	8'- 0¼"	84	19'- 3"	134	30'- 8½"
36	8'- 3"	85	19'- 5¾"	135	30'-11¼"
37	8'- 5¾"	86	19'- 8½"	136	31'- 2"
38	8'- 8½"	87	19'-11¼"	137	31'- 4¾"
39	8'-11¼"	88	20'- 2"	138	31'- 7½"
		89	20'- 4¾"	139	31'-10¼"
40	9'- 2"				
41	9'- 4¾"	90	20'- 7½"	140	32'- 1"
42	9'- 7½"	91	20'-10¼"	141	32'- 3¾"
43	9'-10¼"	92	21'- 1"	142	32'- 6½"
44	10'- 1"	93	21'- 3¾"	143	32'- 9¼"
45	10'- 3¾"	94	21'- 6½"	144	33'- 0"
46	10'- 6½"	95	21'- 9¼"	145	33'- 2¾"
47	10'- 9¼"	96	22'- 0"	146	33'- 5½"
48	11'- 0"	97	22'- 2¾"	147	33'- 8¼"
49	11'- 2¾"	98	22'- 5½"	148	33'-11"
		99	22'- 8¼"	149	34'- 1¾"

HEIGHTS FOR 2⅞" BRICK COURSES

BASED ON STANDARD BRICK 2¼" + ⅝" JOINT

No. of Courses	*Vertical Height*	*No. of Courses*	*Vertical Height*	*No. of Courses*	*Vertical Height*
		50	11'-11¾"	100	23'-11½"
1	2⅞"	51	12'- 2⅝"	101	24'- 2⅜"
2	5¾"	52	12'- 5½"	102	24'- 5¼"
3	8⅝"	53	12'- 8⅜"	103	24'- 8⅛"
4	11½"	54	12'-11¼"	104	24'-11"
5	1'- 2⅜"	55	13'- 2⅛"	105	25'- 1⅞"
6	1'- 5¼"	56	13'- 5"	106	25'- 4¾"
7	1'- 8⅛"	57	13'- 7⅞"	107	25'- 7⅝"
8	1'-11"	58	13'-10¾"	108	25'-10½"
9	2'- 1⅞"	59	14'- 1⅝"	109	26'- 1⅜"
10	2'- 4¾"	60	14'- 4½"	110	26'- 4¼"
11	2'- 7⅝"	61	14'- 7⅜"	111	26'- 7⅛"
12	2'-10½"	62	14'-10¼"	112	26'-10"
13	3'- 1⅜"	63	15'- 1⅛"	113	27'- 0⅞"
14	3'- 4¼"	64	15'- 4"	114	27'- 3¾"
15	3'- 7⅛"	65	15'- 6⅞"	115	27'- 6⅝"
16	3'-10"	66	15'- 9¾"	116	27'- 9½"
17	4'- 0⅞"	67	16'- 0⅝"	117	28'- 0⅜"
18	4'- 3¾"	68	16'- 3½"	118	28'- 3¼"
19	4'- 6⅝"	69	16'- 6⅜"	119	28'- 6⅛"
20	4'- 9½"	70	16'- 9¼"	120	28'- 9"
21	5'- 0⅜"	71	17'- 0⅛"	121	28'-11⅞"
22	5'- 3¼"	72	17'- 3"	122	29'- 2¾"
23	5'- 6⅛"	73	17'- 5⅞"	123	29'- 5⅝"
24	5'- 9"	74	17'- 8¾"	124	29'- 8½"
25	5'-11⅞"	75	17'-11⅝"	125	29'-11⅜"
26	6'- 2¾"	76	18'- 2½"	126	30'- 2¼"
27	6'- 5⅝"	77	18'- 5⅜"	127	30'- 5⅛"
28	6'- 8½"	78	18'- 8¼"	128	30'- 8"
29	6'-11⅜"	79	18'-11⅛"	129	30'-10⅞"
30	7'- 2¼"	80	19'- 2"	130	31'- 1¾"
31	7'- 5⅛"	81	19'- 4⅞"	131	31'- 4⅝"
32	7'- 8"	82	19'- 7¾"	132	31'- 7½"
33	7'-10⅞"	83	19'-10⅝"	133	31'-10⅜"
34	8'- 1¾"	84	20'- 1½"	134	32'- 1¼"
35	8'- 4⅝"	85	20'- 4⅜"	135	32'- 4⅛"
36	8'- 7½"	86	20'- 7¼"	136	32'- 7"
37	8'-10⅜"	87	20'-10⅛"	137	32'- 9⅞"
38	9'- 1¼"	88	21'- 1"	138	33'- 0¾"
39	9'- 4⅛"	89	21'- 3⅞"	139	33'- 3⅝"
40	9'- 7"	90	21'- 6¾"	140	33'- 6½"
41	9'- 9⅞"	91	21'- 9⅝"	141	33'- 9⅜"
42	10'- 0¾"	92	22'- 0½"	142	34'- 0¼"
43	10'- 3⅝"	93	22'- 3⅜"	143	34'- 3⅛"
44	10'- 6½"	94	22'- 6¼"	144	34'- 6"
45	10'- 9⅜"	95	22'- 9⅛"	145	34'- 8⅞"
46	11'- 0¼"	96	23'- 0"	146	34'-11¾"
47	11'- 3⅛"	97	23'- 2⅞"	147	35'- 2⅝"
48	11'- 6"	98	23'- 5¾"	148	35'- 5½"
49	11'- 8⅞"	99	23'- 8⅝"	149	35'- 8⅜"

HEIGHTS FOR 3″
BRICK COURSES

BASED ON STANDARD BRICK 2¼″ + ¾″ JOINT

No. of Courses	Vertical Height	No. of Courses	Vertical Height	No. of Courses	Vertical Height
1	3″	50	12′-6″	100	25′-0″
2	6″	51	12′-9″	101	25′-3″
3	9″	52	13′-0″	102	25′-6″
4	1′-0″	53	13′-3″	103	25′-9″
5	1′-3″	54	13′-6″	104	26′-0″
6	1′-6″	55	13′-9″	105	26′-3″
7	1′-9″	56	14′-0″	106	26′-6″
8	2′-0″	57	14′-3″	107	26′-9″
9	2′-3″	58	14′-6″	108	27′-0″
		59	14′-9″	109	27′-3″
10	2′-6″				
11	2′-9″	60	15′-0″	110	27′-6″
12	3′-0″	61	15′-3″	111	27′-9″
13	3′-3″	62	15′-6″	112	28′-0″
14	3′-6″	63	15′-9″	113	28′-3″
15	3′-9″	64	16′-0″	114	28′-6″
16	4′-0″	65	16′-3″	115	28′-9″
17	4′-3″	66	16′-6″	116	29′-0″
18	4′-6″	67	16′-9″	117	29′-3″
19	4′-9″	68	17′-0″	118	29′-6″
		69	17′-3″	119	29′-9″
20	5′-0″				
21	5′-3″	70	17′-6″	120	30′-0″
22	5′-6″	71	17′-9″	121	30′-3″
23	5′-9″	72	18′-0″	122	30′-6″
24	6′-0″	73	18′-3″	123	30′-9″
25	6′-3″	74	18′-6″	124	31′-0″
26	6′-6″	75	18′-9″	125	31′-3″
27	6′-9″	76	19′-0″	126	31′-6″
28	7′-0″	77	19′-3″	127	31′-9″
29	7′-3″	78	19′-6″	128	32′-0″
		79	19′-9″	129	32′-3″
30	7′-6″				
31	7′-9″	80	20′-0″	130	32′-6″
32	8′-0″	81	20′-3″	131	32′-9″
33	8′-3″	82	20′-6″	132	33′-0″
34	8′-6″	83	20′-9″	133	33′-3″
35	8′-9″	84	21′-0″	134	33′-6″
36	9′-0″	85	21′-3″	135	33′-9″
37	9′-3″	86	21′-6″	136	34′-0″
38	9′-6″	87	21′-9″	137	34′-3″
39	9′-9″	88	22′-0″	138	34′-6″
		89	22′-3″	139	34′-9″
40	10′-0″				
41	10′-3″	90	22′-6″	140	35′-0″
42	10′-6″	91	22′-9″	141	35′-3″
43	10′-9″	92	23′-0″	142	35′-6″
44	11′-0″	93	23′-3″	143	35′-9″
45	11′-3″	94	23′-6″	144	36′-0″
46	11′-6″	95	23′-9″	145	36′-3″
47	11′-9″	96	24′-0″	146	36′-6″
48	12′-0″	97	24′-3″	147	36′-9″
49	12′-3″	98	24′-6″	148	37′-0″
		99	24′-9″	149	37′-3″

HEIGHTS FOR 3⅛″
BRICK COURSES

BASED ON STANDARD BRICK 2¼″ + ⅞″ JOINT

No. of Courses	Vertical Height	No. of Courses	Vertical Height	No. of Courses	Vertical Height
		50	13′- 0¼″	100	26′- 0½″
1	3⅛″	51	13′- 3⅜″	101	26′- 3⅝″
2	6¼″	52	13′- 6½″	102	26′- 6¾″
3	9⅜″	53	13′- 9⅝″	103	26′- 9⅞″
4	1′- 0½″	54	14′- 0¾″	104	27′- 1″
5	1′- 3⅝″	55	14′- 3⅞″	105	27′- 4⅛″
6	1′- 6¾″	56	14′- 7″	106	27′- 7¼″
7	1′- 9⅞″	57	14′-10⅛″	107	27′-10⅜″
8	2′- 1″	58	15′- 1¼″	108	28′- 1½″
9	2′- 4⅛″	59	15′- 4⅜″	109	28′- 4⅝″
10	2′- 7¼″	60	15′- 7½″	110	28′- 7¾″
11	2′-10⅜″	61	15′-10⅝″	111	28′-10⅞″
12	3′- 1½″	62	16′- 1¾″	112	29′- 2″
13	3′- 4⅝″	63	16′- 4⅞″	113	29′- 5⅛″
14	3′- 7¾″	64	16′- 8″	114	29′- 8¼″
15	3′-10⅞″	65	16′-11⅛″	115	29′-11⅜″
16	4′- 2″	66	17′- 2¼″	116	30′- 2½″
17	4′- 5⅛″	67	17′- 5⅜″	117	30′- 5⅝″
18	4′- 8¼″	68	17′- 8½″	118	30′- 8¾″
19	4′-11⅜″	69	17′-11⅝″	119	30′-11⅞″
20	5′- 2½″	70	18′- 2¾″	120	31′- 3″
21	5′- 5⅝″	71	18′- 5⅞″	121	31′- 6⅛″
22	5′- 8¾″	72	18′- 9″	122	31′- 9¼″
23	5′-11⅞″	73	19′- 0⅛″	123	32′- 0⅜″
24	6′- 3″	74	19′- 3¼″	124	32′- 3½″
25	6′- 6⅛″	75	19′- 6⅜″	125	32′- 6⅝″
26	6′- 9¼″	76	19′- 9½″	126	32′- 9¾″
27	7′- 0⅜″	77	20′- 0⅝″	127	33′- 0⅞″
28	7′- 3½″	78	20′- 3¾″	128	33′- 4″
29	7′- 6⅝″	79	20′- 6⅞″	129	33′- 7⅛″
30	7′- 9¾″	80	20′-10″	130	33′-10¼″
31	8′- 0⅞″	81	21′- 1⅛″	131	34′- 1⅜″
32	8′- 4″	82	21′- 4¼″	132	34′- 4½″
33	8′- 7⅛″	83	21′- 7⅜″	133	34′- 7⅝″
34	8′-10¼″	84	21′-10½″	134	34′-10¾″
35	9′- 1⅜″	85	22′- 1⅝″	135	35′- 1⅞″
36	9′- 4½″	86	22′- 4¾″	136	35′- 5″
37	9′- 7⅝″	87	22′- 7⅞″	137	35′- 8⅛″
38	9′-10¾″	88	22′-11″	138	35′-11¼″
39	10′- 1⅞″	89	23′- 2⅛″	139	36′- 2⅜″
40	10′- 5″	90	23′- 5¼″	140	36′- 5½″
41	10′- 8⅛″	91	23′- 8⅜″	141	36′- 8⅝″
42	10′-11¼″	92	23′-11½″	142	36′-11¾″
43	11′- 2⅜″	93	24′- 2⅝″	143	37′- 2⅞″
44	11′- 5½″	94	24′- 5¾″	144	37′- 6″
45	11′- 8⅝″	95	24′- 8⅞″	145	37′- 9⅛″
46	11′-11¾″	96	25′- 0″	146	38′- 0¼″
47	12′- 2⅞″	97	25′- 3⅛″	147	38′- 3⅜″
48	12′- 6″	98	25′- 6¼″	148	38′- 6½″
49	12′- 9⅛″	99	25′- 9⅜″	149	38′- 9⅝″

HEIGHTS FOR 3¼"
BRICK COURSES

BASED ON STANDARD BRICK 2¼" + 1" JOINT

No. of Courses	Vertical Height	No. of Courses	Vertical Height	No. of Courses	Vertical Height
		50	13'- 6½"	100	27'- 1"
1	3¼"	51	13'- 9¾"	101	27'- 4¼"
2	6½"	52	14'- 1"	102	27'- 7½"
3	9¾"	53	14'- 4¼"	103	27'-10¾"
4	1'- 1"	54	14'- 7½"	104	28'- 2"
5	1'- 4¼"	55	14'-10¾"	105	28'- 5¼"
6	1'- 7½"	56	15'- 2"	106	28'- 8½"
7	1'-10¾"	57	15'- 5¼"	107	28'-11¾"
8	2'- 2"	58	15'- 8½"	108	29'- 3"
9	2'- 5¼"	59	15'-11¾"	109	29'- 6¼"
10	2'- 8½"	60	16'- 3"	110	29'- 9½"
11	2'-11¾"	61	16'- 6¼"	111	30'- 0¾"
12	3'- 3"	62	16'- 9½"	112	30'- 4"
13	3'- 6¼"	63	17'- 0¾"	113	30'- 7¼"
14	3'- 9½"	64	17'- 4"	114	30'-10½"
15	4'- 0¾"	65	17'- 7¼"	115	31'- 1¾"
16	4'- 4"	66	17'-10½"	116	31'- 5"
17	4'- 7¼"	67	18'- 1¾"	117	31'- 8¼"
18	4'-10½"	68	18'- 5"	118	31'-11½"
19	5'- 1¾"	69	18'- 8¼"	119	32'- 2¾"
20	5'- 5"	70	18'-11½"	120	32'- 6"
21	5'- 8¼"	71	19'- 2¾"	121	32'- 9¼"
22	5'-11½"	72	19'- 6"	122	33'- 0½"
23	6'- 2¾"	73	19'- 9¼"	123	33'- 3¾"
24	6'- 6"	74	20'- 0½"	124	33'- 7"
25	6'- 9¼"	75	20'- 3¾"	125	33'-10¼"
26	7'- 0½"	76	20'- 7"	126	34'- 1½"
27	7'- 3¾"	77	20'-10¼"	127	34'- 4¾"
28	7'- 7"	78	21'- 1½"	128	34'- 8"
29	7'-10¼"	79	21'- 4¾"	129	34'-11¼"
30	8'- 1½"	80	21'- 8"	130	35'- 2½"
31	8'- 4¾"	81	21'-11¼"	131	35'- 5¾"
32	8'- 8"	82	22'- 2½"	132	35'- 9"
33	8'-11¼"	83	22'- 5¾"	133	36'- 0¼"
34	9'- 2½"	84	22'- 9"	134	36'- 3½"
35	9'- 5¾"	85	23'- 0¼"	135	36'- 6¾"
36	9'- 9"	86	23'- 3½"	136	36'-10"
37	10'- 0¼"	87	23'- 6¾"	137	37'- 1¼"
38	10'- 3½"	88	23'-10"	138	37'- 4½"
39	10'- 6¾"	89	24'- 1¼"	139	37'- 7¾"
40	10'-10"	90	24'- 4½"	140	37'-11"
41	11'- 1¼"	91	24'- 7¾"	141	38'- 2¼"
42	11'- 4½"	92	24'-11"	142	38'- 5½"
43	11'- 7¾"	93	25'- 2¼"	143	38'- 8¾"
44	11'-11"	94	25'- 5½"	144	39'- 0"
45	12'- 2¼"	95	25'- 8¾"	145	39'- 3¼"
46	12'- 5½"	96	26'- 0"	146	39'- 6½"
47	12'- 8¾"	97	26'- 3¼"	147	39'- 9¾"
48	13'- 0"	98	26'- 6½"	148	40'- 1"
49	13'- 3¼"	99	26'- 9¾"	149	40'- 4¼"

WIDTHS OF
BRICK PIERS

ONE STRETCHER

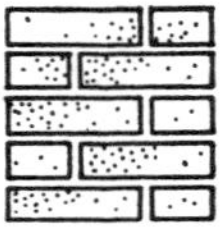

STRETCHER + HEADER

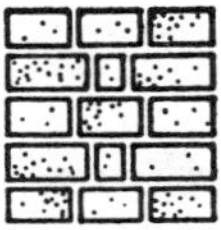

THREE HEADERS

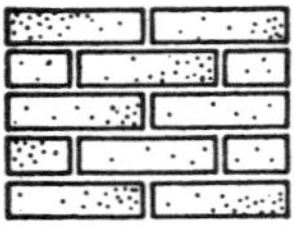

TWO STRETCHERS

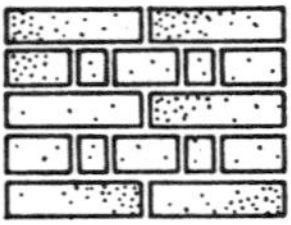

TWO STRETCHERS

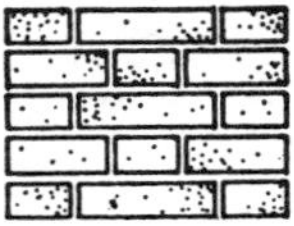

STRETCHER + TWO HEADERS

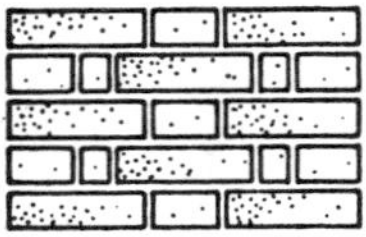

HEADER AND TWO STRETCHERS

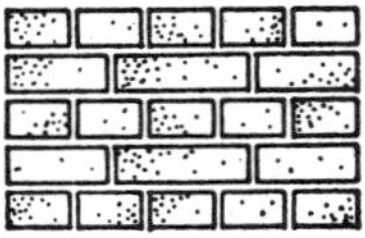

FIVE HEADERS

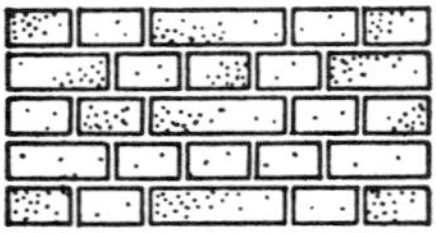

STRETCHER AND FOUR HEADERS

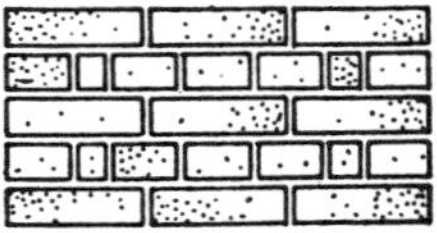

THREE STRETCHERS

Width of pier* determined by	Thickness of vertical mortar joints						
	1/4"	3/8"	1/2"	5/8"	3/4"	7/8"	1"
1 Stretcher	No joints—7½						
2 Headers	7¼	7⅜	7½	7⅝	7¾	7⅞	8
1 Stretcher & 1 Header	11¼	11⅜	11½	11⅝	11¾	11⅞	12
3 Headers	11	11¼	11½	11¾	12	12¼	12½
2 Stretchers	15¼	15⅜	15½	15⅝	15¾	15⅞	16
1 Stretcher & 2 Headers	15	15¼	15½	15¾	16	16¼	16½
1 Header & 2 Stretchers	19	19¼	19½	19¾	20 ·	20¼	20½
5 Headers	18½	19	19½	20	20½	21	21½
1 Stretcher & 4 Headers	22½	23	23½	24	24½	25	25½
3 Stretchers	23	23¼	23½	23¾	24	24¼	24½

*Based on standard common brick length of 7½", width of 3½"

9

LAYING OUT
PATTERN BRICKWORK

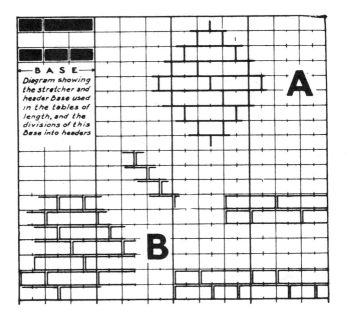

The first step in laying out ornamental patterns in brickwork is to construct a bond diagram, as shown above.

As an example, suppose the brick selected is 2¼″ thick, 3¾″ wide and 8″ long. The joints are to be ½″. The proper vertical scale would be 2¾″, which is equal to the height of one brick plus one joint. The proper horizontal scale would be 12¾″, which equals 8″ plus 3¾″ plus two ½″ joints.

The basis of most ornamental joints is the shifting of the vertical joints in successive courses by one-quarter brick (or more), as indicated in the diagram above. To make the diagram, the base should be laid off using the proper horizontal scale, with lines one-half brick apart. The vertical divisions are drawn by using the proper vertical scale.

In making *small scale* diagrams, it is sufficient to indicate the mortar joints by solid lines, as indicated at "A." It is most convenient in making *large scale* drawings to use the guide lines of the diagram as the bottom and right-hand edge of the brick itself, as shown in several examples at "B."

All diagonal brickwork patterns require an *odd* number of vertical courses to make the pattern come out right.

HORIZONTAL DIMENSIONS FOR BRICKWORK

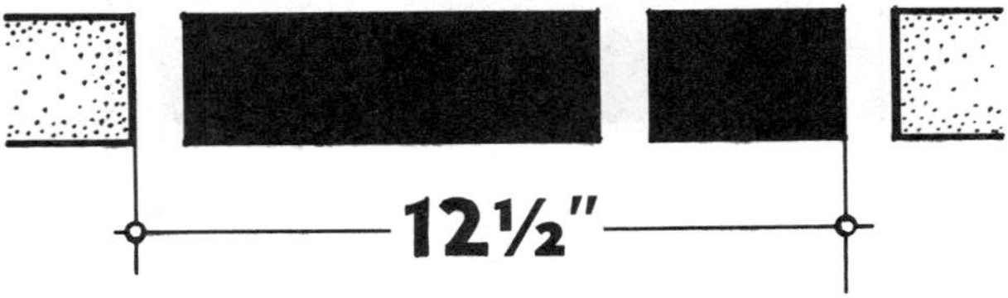

BASE: 1½ Bricks + 2 Vertical Joints

Number of Half Bricks	*Width*	*Number of Half Bricks*	*Width*	*Number of Half Bricks*	*Width*
1	4⅙"	25	8'- 8⅙"	49	17'- 0⅙"
2	8⅓"	26	9'- 0⅓"	50	17'- 4⅓"
3	1'- 0½"	27	9'- 4½"	51	17'- 8½"
4	1'- 4⅔"	28	9'- 8⅔"	52	18'- 0⅔"
5	1'- 8⅚"	29	10'- 0⅚"	53	18'- 4⅚"
6	2'- 1"	30	10'- 5"	54	18'- 9"
7	2'- 5⅙"	31	10'- 9⅙"	55	19'- 1⅙"
8	2'- 9⅓"	32	11'- 1⅓"	56	19'- 5⅓"
9	3'- 1½"	33	11'- 5½"	57	19'- 9½"
10	3'- 5⅔"	34	11'- 9⅔"	58	20'- 1⅔"
11	3'- 9⅚"	35	12'- 1⅚"	59	20'- 5⅚"
12	4'- 2"	36	12'- 6"	60	20'-10"
13	4'- 6⅙"	37	12'-10⅙"	61	21'- 2⅙"
14	4'-10⅓"	38	13'- 2⅓"	62	21'- 6⅓"
15	5'- 2½"	39	13'- 6½"	63	21'-10½"
16	5'- 6⅔"	40	13'-10⅔"	64	22'- 2⅔"
17	5'-10⅚"	41	14'- 2⅚"	65	22'- 6⅚"
18	6'- 3"	42	14'- 7"	66	22'-11"
19	6'- 7⅙"	43	14'-11⅙"	67	23'- 3⅙"
20	6'-11⅓"	44	15'- 3⅓"	68	23'- 7⅓"
21	7'- 3½"	45	15'- 7½"	69	23'-11½"
22	7'- 7⅔"	46	15'-11⅔"	70	24'- 3⅔"
23	7'-11⅚"	47	16'- 3⅚"	71	24'- 7⅚"
24	8'- 4"	48	16'- 8"	72	25'- 0"

HORIZONTAL DIMENSIONS
FOR BRICKWORK

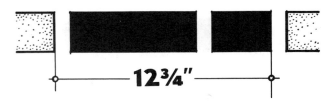

12¾"

BASE: 1½ Bricks + 2 Vertical Joints

Number of Half Bricks	Width		Number of Half Bricks	Width		Number of Half Bricks	Width
1	4¼"		25	8'-10¼"		49	17'- 4¼"
2	8½"		26	9'- 2½"		50	17'- 8½"
3	1'- 0¾"		27	9'- 6¾"		51	18'- 0¾"
4	1'- 5"		28	9'-11"		52	18'- 5"
5	1'- 9¼"		29	10'- 3¼"		53	18'- 9¼"
6	2'- 1½"		30	10'- 7½"		54	19'- 1½"
7	2'- 5¾"		31	10'-11¾"		55	19'- 5¾"
8	2'-10"		32	11'- 4"		56	19'-10"
9	3'- 2¼"		33	11'- 8¼"		57	20'- 2¼"
10	3'- 6½"		34	12'- 0½"		58	20'- 6½"
11	3'-10¾"		35	12'- 4¾"		59	20'-10¾"
12	4'- 3"		36	12'- 9"		60	21'- 3"
13	4'- 7¼"		37	13'- 1¼"		61	21'- 7¼"
14	4'-11½"		38	13'- 5½"		62	21'-11½"
15	5'- 3¾"		39	13'- 9¾"		63	22'- 3¾"
16	5'- 8"		40	14'- 2"		64	22'- 8"
17	6'- 0¼"		41	14'- 6¼"		65	23'- 0¼"
18	6'- 4½"		42	14'-10½"		66	23'- 4½"
19	6'- 8¾"		43	15'- 2¾"		67	23'- 8¾"
20	7'- 1"		44	15'- 7"		68	24'- 1"
21	7'- 5¼"		45	15'-11¼"		69	24'- 5¼"
22	7'- 9½"		46	16'- 3½"		70	24'- 9½"
23	8'- 1¾"		47	16'- 7¾"		71	25'- 1¾"
24	8'- 6"		48	17'- 0"		72	25'- 6"

HORIZONTAL DIMENSIONS FOR BRICKWORK

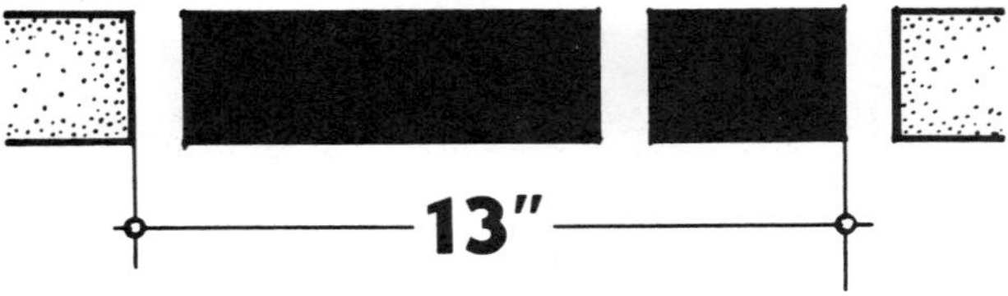

BASE: 1½ Bricks + 2 Vertical Joints

Number of Half Bricks	*Width*	*Number of Half Bricks*	*Width*	*Number of Half Bricks*	*Width*
1	4⅓"	25	9'- 0⅓"	49	17'- 8⅓"
2	8⅔"	26	9'- 4⅔"	50	18'- 0⅔"
3	1'- 1"	27	9'- 9"	51	18'- 5"
4	1'- 5⅓"	28	10'- 1⅓"	52	18'- 9⅓"
5	1'- 9⅔"	29	10'- 5⅔"	53	19'- 1⅔"
6	2'- 2"	30	10'-10"	54	19'- 6"
7	2'- 6⅓"	31	11'- 2⅓"	55	19'-10⅓"
8	2'-10⅔"	32	11'- 6⅔"	56	20'- 2⅔"
9	3'- 3"	33	11'-11"	57	20'- 7"
10	3'- 7⅓"	34	12'- 3⅓"	58	20'-11⅓"
11	3'-11⅔"	35	12'- 7⅔"	59	21'- 3⅔"
12	4'- 4"	36	13'- 0"	60	21'- 8"
13	4'- 8⅓"	37	13'- 4⅓"	61	22'- 0⅓"
14	5'- 0⅔"	38	13'- 8⅔"	62	22'- 4⅔"
15	5'- 5"	39	14'- 1"	63	22'- 9"
16	5'- 9⅓"	40	14'- 5⅓"	64	23'- 1⅓"
17	6'- 1⅔"	41	14'- 9⅔"	65	23'- 5⅔"
18	6'- 6"	42	15'- 2"	66	23'-10"
19	6'-10⅓"	43	15'- 6⅓"	67	24'- 2⅓"
20	7'- 2⅔"	44	15'-10⅔"	68	24'- 6⅔"
21	7'- 7"	45	16'- 3"	69	24'-11"
22	7'-11⅓"	46	16'- 7⅓"	70	25'- 3⅓"
23	8'- 3⅔"	47	16'-11⅔"	71	25'- 7⅔"
24	8'- 8"	48	17'- 4"	72	26'- 0"

HORIZONTAL DIMENSIONS
FOR BRICKWORK

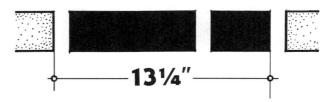

13¼"

BASE: 1½ Bricks + 2 Vertical Joints

Number of Half Bricks	Width		Number of Half Bricks	Width		Number of Half Bricks	Width
1	4⁵⁄₁₂"		25	9'- 2⁵⁄₁₂"		49	18'- 0⁵⁄₁₂"
2	8⁵⁄₆"		26	9'- 6⁵⁄₆"		50	18'- 4⁵⁄₆"
3	1'- 1¼"		27	9'-11¼"		51	18'- 9¼"
4	1'- 5⅔"		28	10'- 3⅔"		52	19'- 1⅔"
5	1'-10¹⁄₁₂"		29	10'- 8¹⁄₁₂"		53	19'- 6¹⁄₁₂"
6	2'- 2½"		30	11'- 0½"		54	19'-10½"
7	2'- 6¹¹⁄₁₂"		31	11'- 4¹¹⁄₁₂"		55	20'- 2¹¹⁄₁₂"
8	2'-11⅓"		32	11'- 9⅓"		56	20'- 7⅓"
9	3'- 3¾"		33	12'- 1¾"		57	20'-11¾"
10	3'- 8⅙"		34	12'- 6⅙"		58	21'- 4⅙"
11	4'- 0⁷⁄₁₂"		35	12'-10⁷⁄₁₂"		59	21'- 8⁷⁄₁₂"
12	4'- 5"		36	13'- 3"		60	22'- 1"
13	4'- 9⁵⁄₁₂"		37	13'- 7⁵⁄₁₂"		61	22'- 5⁵⁄₁₂"
14	5'- 1⅚"		38	13'-11⅚"		62	22'- 9⅚"
15	5'- 6¼"		39	14'- 4¼"		63	23'- 2¼"
16	5'-10⅔"		40	14'- 8⅔"		64	23'- 6⅔"
17	6'- 3¹⁄₁₂"		41	15'- 1¹⁄₁₂"		65	23'-11¹⁄₁₂"
18	6'- 7½"		42	15'- 5½"		66	24'- 3½"
19	6'-11¹¹⁄₁₂"		43	15'- 9¹¹⁄₁₂"		67	24'- 7¹¹⁄₁₂"
20	7'- 4⅓"		44	16'- 2⅓"		68	25'- 0⅓"
21	7'- 8¾"		45	16'- 6¾"		69	25'- 4¾"
22	8'- 1⅙"		46	16'-11⅙"		70	25'- 9⅙"
23	8'- 5⁷⁄₁₂"		47	17'- 3⁷⁄₁₂"		71	26'- 1⁷⁄₁₂"
24	8'-10"		48	17'- 8"		72	26'- 6"

HORIZONTAL DIMENSIONS
FOR BRICKWORK

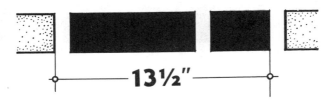

13½"

BASE: 1½ Bricks + 2 Vertical Joints

Number of Half Bricks	Width
1	4½"
2	9"
3	1'- 1½"
4	1'- 6"
5	1'-10½"
6	2'- 3"
7	2'- 7½"
8	3'- 0"
9	3'- 4½"
10	3'- 9"
11	4'- 1½"
12	4'- 6"
13	4'-10½"
14	5'- 3"
15	5'- 7½"
16	6'- 0"
17	6'- 4½"
18	6'- 9"
19	7'- 1½"
20	7'- 6"
21	7'-10½"
22	8'- 3"
23	8'- 7½"
24	9'- 0"

Number of Half Bricks	Width
25	9'- 4½"
26	9'- 9"
27	10'- 1½"
28	10'- 6"
29	10'-10½"
30	11'- 3"
31	11'- 7½"
32	12'- 0"
33	12'- 4½"
34	12'- 9"
35	13'- 1½"
36	13'- 6"
37	13'-10½"
38	14'- 3"
39	14'- 7½"
40	15'- 0"
41	15'- 4½"
42	15'- 9"
43	16'- 1½"
44	16'- 6"
45	16'-10½"
46	17'- 3"
47	17'- 7½"
48	18'- 0"

Number of Half Bricks	Width
49	18'- 4½"
50	18'- 9"
51	19'- 1½"
52	19'- 6"
53	19'-10½"
54	20'- 3"
55	20'- 7½"
56	21'- 0"
57	21'- 4½"
58	21'- 9"
59	22'- 1½"
60	22'- 6"
61	22'-10½"
62	23'- 3"
63	23'- 7½"
64	24'- 0"
65	24'- 4½"
66	24'- 9"
67	25'- 1½"
68	25'- 6"
69	25'-10½"
70	26'- 3"
71	26'- 7½"
72	27'- 0"

HORIZONTAL DIMENSIONS
FOR BRICKWORK

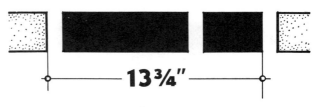

13¾"

BASE: 1½ Bricks + 2 Vertical Joints

Number of Half Bricks	Width	Number of Half Bricks	Width	Number of Half Bricks	Width
1	4⁷⁄₁₂"	25	9'- 6⁷⁄₁₂"	49	18'- 8⁷⁄₁₂"
2	9⅙"	26	9'-11⅙"	50	19'- 1⅙"
3	1'- 1¾"	27	10'- 3¾"	51	19'- 5¾"
4	1'- 6⅓"	28	10'- 8⅓"	52	19'-10⅓"
5	1'-10¹¹⁄₁₂"	29	11'- 0¹¹⁄₁₂"	53	20'- 2¹¹⁄₁₂"
6	2'- 3½"	30	11'- 5½"	54	20'- 7½"
7	2'- 8¹⁄₁₂"	31	11'-10¹⁄₁₂"	55	21'- 0¹⁄₁₂"
8	3'- 0⅔"	32	12'- 2⅔"	56	21'- 4⅔"
9	3'- 5¼"	33	12'- 7¼"	57	21'- 9¼"
10	3'- 9⅚"	34	12'-11⅚"	58	22'- 1⅚"
11	4'- 2⁵⁄₁₂"	35	13'- 4⁵⁄₁₂"	59	22'- 6⁵⁄₁₂"
12	4'- 7"	36	13'- 9"	60	22'-11"
13	4'-11⁷⁄₁₂"	37	14'- 1⁷⁄₁₂"	61	23'- 3⁷⁄₁₂"
14	5'- 4⅙"	38	14'- 6⅙"	62	23'- 8⅙"
15	5'- 8¾"	39	14'-10¾"	63	24'- 0¾"
16	6'- 1⅓"	40	15'- 3⅓"	64	24'- 5⅓"
17	6'- 5¹¹⁄₁₂"	41	15'- 7¹¹⁄₁₂"	65	24'- 9¹¹⁄₁₂"
18	6'-10½"	42	16'- 0½"	66	25'- 2½"
19	7'- 3¹⁄₁₂"	43	16'- 5¹⁄₁₂"	67	25'- 7¹⁄₁₂"
20	7'- 7⅔"	44	16'- 9⅔"	68	25'-11⅔"
21	8'- 0¼"	45	17'- 2¼"	69	26'- 4¼"
22	8'- 4⅚"	46	17'- 6⅚"	70	26'- 8⅚"
23	8'- 9⁵⁄₁₂"	47	17'-11⁵⁄₁₂"	71	27'- 1⁵⁄₁₂"
24	9'- 2"	48	18'- 4"	72	27'- 6"

HORIZONTAL DIMENSIONS
FOR BRICKWORK

14"

BASE: 1½ Bricks + 2 Vertical Joints

Number of Half Bricks	Width	Number of Half Bricks	Width	Number of Half Bricks	Width
1	4⅔"	49	9'- 8⅔"	25	19'- 0⅔"
2	9⅓"	50	10'- 1⅓"	26	19'- 5⅓"
3	1'- 2"	51	10'- 6"	27	19'-10"
4	1'- 6⅔"	52	10'-10⅔"	28	20'- 2⅔"
5	1'-11⅓"	53	11'- 3⅓"	29	20'- 7⅓"
6	2'- 4"	54	11'- 8"	30	21'- 0"
7	2'- 8⅔"	55	12'- 0⅔"	31	21'- 4⅔"
8	3'- 1⅓"	56	12'- 5⅓"	32	21'- 9⅓"
9	3'- 6"	57	12'-10"	33	22'- 2"
10	3'-10⅔"	58	13'- 2⅔"	34	22'- 6⅔"
11	4'- 3⅓"	59	13'- 7⅓"	35	22'-11⅓"
12	4'- 8"	60	14'- 0"	36	23'- 4"
13	5'- 0⅔"	61	14'- 4⅔"	37	23'- 8⅔"
14	5'- 5⅓"	62	14'- 9⅓"	38	24'- 1⅓"
15	5'-10"	63	15'- 2"	39	24'- 6"
16	6'- 2⅔"	64	15'- 6⅔"	40	24'-10⅔"
17	6'- 7⅓"	65	15'-11⅓"	41	25'- 3⅓"
18	7'- 0"	66	16'- 4"	42	25'- 8"
19	7'- 4⅔"	67	16'- 8⅔"	43	26'- 0⅔"
20	7'- 9⅓"	68	17'- 1⅓"	44	26'- 5⅓"
21	8'- 2"	69	17'- 6"	45	26'-10"
22	8'- 6⅔"	70	17'-10⅔"	46	27'- 2⅔"
23	8'-11⅓"	71	18'- 3⅓"	47	27'- 7⅓"
24	9'- 4"	72	18'- 8"	48	28'- 0"

SAFE LOADS ON LIMESTONE LINTELS

SAFE SUPERIMPOSED UNIFORM LOAD PER FOOT OF SPAN FOR SIMPLY SUPPORTED LINTELS 1″ THICK.

Height of Lintel	Span in Feet — Coefficient of Deflection in italics.									
	4 *.014*	5 *.021*	6 *.031*	7 *.042*	8 *.054*	9 *.069*	10 *.085*	12 *.123*	14 *.167*	16 *.218*
6″	25	14	8	4	2	0				
8″	48	28	17	10	6	3	1	0		
10″	77	46	29	18	12	7	4	0		
1′- 0″	114	68	44	29	19	13	8	2	0	
1′- 2″	157	96	62	42	29	20	13	5	0	
1′- 4″	208	128	74	57	40	28	20	9	1	0
1′- 6″	264	163	108	74	53	38	27	13	5	0
1′- 8″	330	204	136	94	67	49	36	19	8	2
1′-10″	400	248	166	116	84	62	46	25	12	4
2′- 0″	480	298	200	140	102	75	56	32	17	7
2′- 2″	565	352	238	167	122	91	69	40	22	11
2′- 4″	656	410	280	196	144	107	81	48	28	15
2′- 6″	757	472	320	227	167	126	96	57	34	19
2′- 8″	860	542	368	261	192	145	111	63	41	24
2′-10″	980	614	416	296	220	166	128	71	49	29
3′- 0″		692	470	334	249	188	145	90	57	36
3′- 2″		770	526	374	278	212	163	103	65	40
3′- 4″		858	584	418	310	236	184	115	74	47
3′- 6″		946	645	466	344	263	204	129	84	54
3′- 8″			712	510	380	291	227	144	94	62
3′-10″			776	555	416	320	250	160	105	70

Table is based on the following conditions:
Extreme fibre stress = 125 lbs. per □ inch.
Unit shear = 150 lbs. per □ inch.
Modulus of elasticity = 4 400 000.
Factor of safety = 8 to 10.
Weight of the lintel itself has been deducted.
Weight of limestone taken as 144 lbs. per cubic foot.

The deflection of the lintel in inches when loaded with the superimposed loads shown in the table may be found by dividing the deflection coefficient by the height of the lintel in inches.

Formulae used in calculating table values:
Superimposed bending load = 14 d² / L² — d.
Superimposed shear load = 300 d / L — d.
Maximum deflection = L² / 1173 d.
d = height of lintel in inches.
L = span of lintel in feet.

UNCOURSED
STONEWORK

Stone dressed to permit laying with uniformly thick horizontal joints of ½" or less is called *ashlar*. Stone roughly dressed to permit laying with uniformly thick horizontal joints of over ½" is called *squared stone,* and is adapted to the same bonds as ashlar. Natural stone which does not permit laying with uniformly thick joints, or dressed stone not permitting horizontal joints, is classed as *rubble*.

Stone laid without continuous horizontal joints is called *uncoursed,* or *random.* (Note particularly the distinction between random masonry and random coursed masonry.)

"THREE UNIT"-ie stones of three heights have been used.

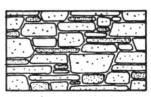

"BROKEN END"- cut beds with angular broken ends.

RANDOM   ASHLAR

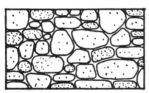

Undressed stone resulting in joints of varying thickness

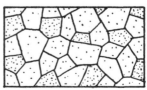

Stratified undressed stone resulting in fairly level beds

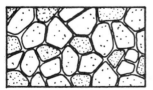

"MOSAIC"-Stone roughly shaped, joints of varying thickness.

"POLYGONAL"-Stone accurately dressed to result in uniform joints.

RANDOM   RUBBLE

COURSED
STONEWORK

Stone dressed to permit laying with uniformly thick horizontal joints of ½" or less is called *ashlar*. Stone roughly squared to permit laying with uniformly thick joints of only greater than ½" is called *squared stone*, and is adapted to the same bonds as ashlar. Regular coursed square-stone masonry is occasionally termed *block in course* masonry or as *hammer dressed ashlar*.

It should be evident that undressed natural stone is not adapted to the bonds shown on this sheet, on account of its inherent variety of thicknesses and unevenness.

If the stones are coursed and of equal lengths with the vertical joints over the center of the preceding course, the masonry is said to be laid in *plumb bond*.

Stone laid with continuous horizontal joints is called *coursed*, or *range work*.

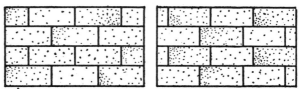

All stones same length and height. Equal course heights may also be jointed as at (6), lower right illustration below.

Irregular lengths, all stones are the same height.

Same as at left except that small stones spots are added

REGULAR COURSED ASHLAR

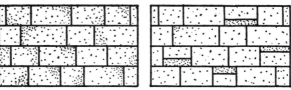

Two heights of stone alternating in a regular arrangement.

ALTERNATING COURSED ASHLAR

COURSED
STONEWORK

Stone dressed to permit laying with uniformly thick horizontal joints of ½″ or less is called *ashlar*. Stone roughly dressed to permit laying with uniformly thick horizontal joints only greater than ½″ is called *squared stone*, and is adapted to the same bonds as ashlar. Natural stone that does not permit laying with uniformly thick joints, or dressed stone that does not permit horizontal joints, is classed as *rubble*.

Stone laid with continuous horizontal joints is called *coursed*, or *range work*. If the heights of the courses are in no regularly recurring arrangement, the stone is called *random coursed*. (Note the distinction between random coursed masonry and random masonry.) If the horizontal courses are continuous for short distances only, the stone is called *broken range*.

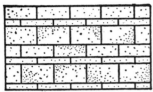

Stones of various heights but all the same lengths.

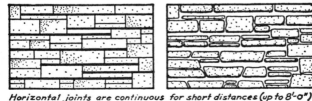

Stones of various heights and lengths.

RANDOM COURSED ASHLAR

Same as (2) above except that small stone spots are added.

RANDOM COURSED ASHLAR

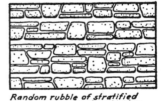

Random rubble of stratified stones brought to level beds at varying vertical intervals.

RANDOM COURSED RANDOM RUBBLE

Horizontal joints are continuous

BROKEN RANGE RANDOM ASHLAR

for short distances (up to 8′-0″)

BROKEN RANGE RANDOM RUBBLE

21

ATTACHMENT METHODS
FOR CLAY TILE

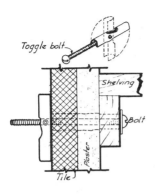

Toggle bolt
Shelving
Bolt
Plaster
Tile

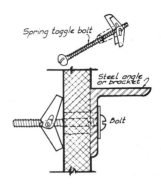

Spring toggle bolt
Steel angle or bracket
Bolt

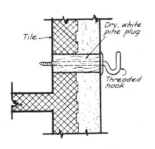

Tile
Dry, white pine plug
Threaded hook

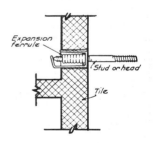

Expansion ferrule
Stud or head
Tile

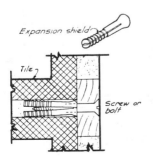

Expansion shield
Tile
Screw or bolt

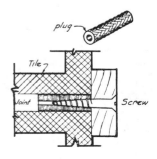

plug
Tile
Joint
Screw

DECAY RESISTANCE
OF WOODS

The natural decay resistance of all common native species of wood lies in the heartwood. When untreated, the sapwood of substantially all species has low resistance to decay and usually has a short life under decay-producing conditions. The decay resistance or durability of heartwood in service is greatly affected by differences in the character of the wood, the attacking fungus, and the conditions of exposure.

The following grouping arranges common species into four classes according to the resistance of each species to heartwood decay. The classification is based upon service records, when they are available, and on general experience.

Heartwood exceptionally durable even under conditions that favor decay	Locust, black
	Mulberry, red
	Osage-orange
Heartwood of high durability	Baldcypress (old)
	Catalpas (all)
	Cedars (all)
	Cherry, black
	Chestnut
	Cypress, Arizona
	Junipers (all)
	Oak, bur
	Oak, chestnut
	Oak, Gambel
	Oak, Oregon white
	Oak, post
	Oak, white
	Redwood
	Walnut, black
	Yew, Pacific
Heartwood of moderate durability	Baldcypress (young)
	Douglas-fir (dense)
	Honeylocust
	Larch, western
	Oak, swamp chestnut
	Pine, eastern white
	Pine, longleaf
	Pine, slash
	Tamarack
Heartwood of little or no durability	Alders (all)
	Ashes (all)
	Aspens (all)
	Basswood
	Beech, American
	Birches (all)
	Buckeye
	Butternut
	Cottonwoods (all)
	Elms (all)
	Firs (true)
	Hackberry
	Hemlocks (all)
	Hickories (all)
	Magnolias (all)
	Maples (all)
	Oaks, red (all)
	Pines (other than eastern white, longleaf, and slash)
	Poplars (all)
	Spruces (all)
	Sweetgum
	Willows (all)
	Yellow-poplar

AMERICAN STANDARD DRESSED LUMBER SIZES

WIDTH. Though the standard calls for all of the widths listed below, lumberyards commonly stock only those in nominal 2″ sizes.

LENGTH. Dimension, Scaffold Plank, Decking, Timbers, and Selects and Commons S-Dry are manufactured in lengths of 6′ and over in increments of 1′. Flooring, Factory and Shop Lumber, several Sidings, and Ceiling stock come 4′ and longer in increments of 1′. Finish and Boards S-Dry are made 3′ and longer. In practice, general construction lumber is stocked by lumberyards in length multiples of 2′.

ROUGH. The dimensions of rough lumber are inexact but exceed the dimensions of finished lumber of the corresponding nominal size by the amount necessary to permit surfacing of one or two sides and/or one or two edges.

NOTE: Because of many factors, such as shrinkage, the "standard" dimensions listed do not always hold true, and variances of several 64ths of an inch may occur, especially in general construction lumber. If uniform size is required, careful piece-by-piece selection will be necessary.

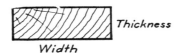

Thickness

Width

	Thickness			Width	
Nominal (Boards)	Dry	Green	Nominal (Boards)	Dry	Green
1	$3/4$	$25/32$	2	$1\frac{1}{2}$	$1\frac{9}{16}$
$1\frac{1}{4}$	1	$1\frac{1}{32}$	3	$2\frac{1}{2}$	$2\frac{9}{16}$
$1\frac{1}{2}$	$1\frac{1}{4}$	$1\frac{9}{32}$	4	$3\frac{1}{2}$	$3\frac{9}{16}$
			5	$4\frac{1}{2}$	$4\frac{5}{8}$
			6	$5\frac{1}{2}$	$5\frac{5}{8}$
			7	$6\frac{1}{2}$	$6\frac{5}{8}$
			8	$7\frac{1}{4}$	$7\frac{1}{2}$
			9	$8\frac{1}{4}$	$8\frac{1}{2}$
			10	$9\frac{1}{4}$	$9\frac{1}{2}$
			11	$10\frac{1}{4}$	$10\frac{1}{2}$
			12	$11\frac{1}{4}$	$11\frac{1}{2}$
			14	$13\frac{1}{4}$	$13\frac{1}{2}$
			16	$15\frac{1}{4}$	$15\frac{1}{2}$
(Dimension)			(Dimension)		
2	$1\frac{1}{2}$	$1\frac{9}{16}$	2	$1\frac{1}{2}$	$1\frac{9}{16}$
$2\frac{1}{2}$	2	$2\frac{1}{16}$	3	$2\frac{1}{2}$	$2\frac{9}{16}$
3	$2\frac{1}{2}$	$2\frac{9}{16}$	4	$3\frac{1}{2}$	$3\frac{9}{16}$
$3\frac{1}{2}$	3	$3\frac{1}{16}$	5	$4\frac{1}{2}$	$4\frac{5}{8}$
4	$3\frac{1}{2}$	$3\frac{9}{16}$	6	$5\frac{1}{2}$	$5\frac{5}{8}$
$4\frac{1}{2}$	4	$4\frac{1}{16}$	8	$7\frac{1}{4}$	$7\frac{1}{2}$
			10	$9\frac{1}{4}$	$9\frac{1}{2}$
			12	$11\frac{1}{4}$	$11\frac{1}{2}$
			14	$13\frac{1}{4}$	$13\frac{1}{2}$
			16	$15\frac{1}{4}$	$15\frac{1}{2}$
(Timber)			(Timber)		
5 and greater		$\frac{1}{2}$ less than nominal	5 and greater		$\frac{1}{2}$ less than nominal

SOFTWOOD LUMBER CLASSIFICATIONS

YARD LUMBER. This comprises lumber of all sizes and patterns under 5″ thick that is intended for ordinary construction and general building purposes. The grading of yard lumber is based on the intended use of the particular grade and is applied to each piece with reference to its size, length, and (often) appearance when graded at the mill—without considering further manufacture or later grading.

1. *Finish*—Yard lumber less than 3″ thick and 12″ or less in width.
2. *Boards*—Yard lumber less than 2″ thick and 2″ or more in width.
3. *Strips*—Yard lumber less than 2″ thick and less than 8″ wide.
4. *Dimension*—Yard lumber at least 2″ thick but less than 5″ thick, and 2″ or more in width.
   a. *Planks*—2″ to 4″ thick and 8″ or more in width.
   b. *Scantling*—2″ to less than 5″ thick and less than 8″ wide.
   c. *Heavy Joists*—4″ thick and 8″ or more in width.

STRUCTURAL LUMBER. This comprises lumber that is at least 2″ thick and at least 2″ wide (except in the case of *Joist and Plank*) intended for use where working stresses are required. The grading of structural lumber is based on the strength of the piece when used in its entirety. The *Heavy Joists* category above may be included in this classification, depending upon the use of the material.

1. *Joist and Plank*—Lumber from 2″ to 4″ thick and at least 4″ wide.
2. *Timbers*—Lumber 5″ or more in its smallest nominal dimension, classified as posts, sills, girders, beams, stringers, purlins, etc.

FACTORY AND SHOP LUMBER. This comprises lumber intended to be cut up for use in further manufacture. It is graded on the basis of the useful percentage of the whole area of the lumber required to produce a limited number of cuttings either of a specified or of a given minimum size and quantity.

1. *Factory Plank*—Lumber 1″ to 4″ thick and at least 5″ wide, intended for doors, sashes, etc.
2. *General*—Lumber for various general cut-up uses.

Note: Softwood lumber classification, quality standards, and grading are based upon the American Softwood Lumber Standard and the National Grading Rule for Dimension Lumber. However, more detailed and specific grading rules are formulated by various regional manufacturers' associations, mostly in conjunction with the American Lumber Standards Committee. Principal grading agencies are the West Coast Inspection Bureau, the Western Wood Products Association, the Southern Pine Inspection Bureau, and the Redwood Inspection Service. The considerable variation in classifications, nomenclature, and grading must be considered when particular lumber specifications must be met, and applicable grading authorities should be consulted as necessary.

SOFTWOOD LUMBER CLASSIFICATIONS

QUALITY CLASSIFICATION OF YARD LUMBER

SELECT
Suitable for natural finishes
> Grade A—Practically clear.
> Grade B—Of high quality—generally clear.

Suitable for paint finishes
> Grade C—Adapted to high-quality paint finishes.
> Grade D—Intermediate between higher finishing grades and common grades, and partaking somewhat of the nature of both.

COMMON
Suitable for use without waste
> No. 1 (Construction)—Sound and tight-knotted—may be considered watertight.
> No. 2 (Standard)—Less restricted in quality than No. 1, but of the same general character.

Permitting some waste
> No. 3 (Utility)—Prevailing grade characteristics larger than in No. 2.
> No. 4 (Economy)—Low quality.
> No. 5—Lowest recognized grade, but must be usable.

COMMON LUMBER ABBREVIATIONS

AD Air-dried	MBF 1,000 board feet
ALS American Lumber Standards	MC Moisture content
B&B B and Better	N Nosed
B&Btr B and Better	P Planed
BD Board	QTD Quartered
BD FT Board foot	RDM Random
BDL Bundle	RL Random lengths
BM Board measure	RW Random widths
CLR Clear	SAP Sapwood
COM Common	SEL Select
DF Douglas fir	STRUCT Structural
DIM Dimension	S&E Side and edge
E Edge	S1E Surfaced one edge
EG Edge grain	S2E Surfaced two edges
FAS Firsts and Seconds	S1S Surfaced one side
FG Flat grain	S2S Surfaced two sides
FT Foot	S4S Surfaced four sides
GR Green	S1S1E Surfaced one side, one edge
HDWD Hardwood	S1S2E Surfaced one side, two edges
HRTWD Heartwood	S2S&SM Surfaced two sides and
JTD Jointed	standard matched
KD Kiln-dried	T&G Tongue and groove
LBR Lumber	VG Vertical grain
LF Linear (or lineal) foot	WT Weight
M Thousand	WTH Width

LUMBER
NOMENCLATURE

So far as is practicable, it is important that different species of woods bear distinctive common names. This constitutes good trade practice and helps to protect the consumer from purchasing under the guise of misleading names species that might be inferior or inappropriate for the job at hand. These names should be uniformly used, and concerted efforts should be made to prevent adding to the present confusion by using improper names or by circulating further misleading or ill-chosen names.

The following list offers a means to acquaint lumber users with the standard names employed by the Forest Service and other recognized authorities for lumber and for the trees from which it is manufactured. The list will also alleviate some of the confusion among lumber consumers resulting from use, both in the trade and in casual conversation, of numerous regional and colloquial misnomers.

Correct Name of Lumber and Botanical Name of Tree	Other Names Loosely or Erroneously Employed
ALASKA-CEDAR *(Chamaecyparis nootkatensis)*	Alaska Yellow Cedar Nootka Cypress
BALDCYPRESS *(Taxodium distichum)*	Red Cypress Yellow Cypress White Cypress Black Cypress Louisiana Red Cypress Gulf Red Cypress Tidewater Red Cypress Gulf Coast Red Cypress Gulf Cypress Cypress
CEDAR, Atlantic White *(Chamaecyparis thyoides)*	White Cedar Juniper Swamp Cedar Boat Cedar
CEDAR, Northern White *(Thuja occidentalis)*	Cedar Arborvitae Michigan White Cedar New Brunswick Cedar
DOUGLAS-FIR *(Pseudotsuga menziesii* and var. *glauca)*	Douglas Yellow Fir Oregon Fir Fir Red Fir Pacific Coast Douglas-Fir Montana Fir National Yellow Fir Yellow Fir Oregon Pine Golden Rod Douglas-Fir Yellow Douglas-Fir "Santian" Quality Fir
FIR, Balsam *(Abies balsamea)*	Eastern Fir Balsam
FIR, California Red *(Abies magnifica)*	Golden Fir White Fir Red Fir Silvertip

27

LUMBER
NOMENCLATURE

Correct Name of Lumber and Botanical Name of Tree	*Other Names Loosely or Erroneously Employed*

FIR, Fraser . . . . Eastern Fir
   (Abies fraseri)   Balsam

FIR, Grand . . . . White Fir
   (Abies grandis)   Lowland White Fir
                            Lowland Fir

FIR, Noble . . . . Larch
   (Abies procera)   White Fir
                            Red Fir

FIR, Pacific Silver . . . . Larch
   (Abies amabilis)   White Fir
                            Amabilis Fir
                            Cascades Fir

FIR, White . . . . Balsam Fir
   (Abies concolor)   Silver Fir
                            Concolor Fir

HEMLOCK, Eastern . . . . Canadian Hemlock
   (Tsuga canadensis)   Hemlock Spruce
                            West Virginia Hemlock
                            Hemlock
                            Wisconsin White Hemlock
                            Pennsylvania Hemlock
                            Pennsylvania White Hemlock

HEMLOCK, Western . . . . West Coast Hemlock
   (Tsuga heterophylla)   Pacific Hemlock
                            Hemlock Spruce
                            Western Hemlock Spruce
                            Western Hemlock Fir
                            Pacific Coast Hemlock
                            Pacific (Western) Hemlock
                            Hemlock
                            Prince Albert Fir
                            Gray Fir
                            Silver Fir
                            Alaska Pine

INCENSE-CEDAR . . . .
   (Libocedrus decurrens)

LARCH, Western . . . . Larch
   (Larix occidentalis)   Montana Larch
                            Hackmatack
                            Western Tamarack

OAK, Red, comprises these (and other) species:

  Black Oak . . . . Tanbark Oak
     (Quercus velutina)

  Cherrybark Oak . . . . Swamp Red Oak
     (Quercus falcata var. *pagodaefolia)* Spanish Oak

LUMBER
NOMENCLATURE

Correct Name of Lumber and Botanical Name of Tree	*Other Names Loosely or Erroneously Employed*

OAK, Red, comprises these (and other) species: (continued)

Laurel Oak . . . .
(Quercus laurifolia)

Northern Red Oak . . . . West Virginia Soft Red Oak
(Quercus rubra)

Nuttall Oak . . . .
(Quercus nuttallii)

Pin Oak . . . . Swamp Oak
(Quercus palustris) Jack Oak

Scarlet Oak . . . .
(Quercus coccinea)

Shumard Oak . . . .
(Quercus shumardii)

Southern Red Oak . . . . Spanish Oak
(Quercus falcata)

Water Oak . . . . Pin Oak
(Quercus nigra) Shingle Oak
Spanish Oak

Willow Oak . . . . Pin Oak
(Quercus Phellos) Shingle Oak
Sand Oak

OAK, White, comprises these (and other) species:

Bur Oak . . . . Overcup Oak
(Quercus macrocarpa) Mossycup Oak

Chestnut Oak . . . . Tanbark Oak
(Quercus prinus) Basket Oak

Chinkapin Oak . . . . Pin Oak
(Quercus muehlenbergii) Chestnut Oak
Rock Oak

Live Oak . . . .
(Quercus virginiana)

Overcup Oak . . . . Swamp Post Oak
(Quercus lyrata) Mossycup Oak

Post Oak . . . . Blackjack Oak
(Quercus stellata)

Swamp Chestnut Oak . . . . Cow Oak
(Quercus michauxii) Basket Oak

Swamp White Oak . . . . Swamp Oak
(Quercus bicolor)

White Oak . . . . West Virginia Soft White Oak
(Quercus alba) Forked Leaf White Oak
Overcup Oak
Mossycup Oak

LUMBER
NOMENCLATURE

Correct Name of Lumber and Botanical Name of Tree	*Other Names Loosely or Erroneously Employed*
PINE, Eastern White *(Pinus strobus)*	Northern Pine Canadian White Pine Soft White Pine Weymouth Pine Soft Pine White Pine Wisconsin White Pine Soft Cork White Pine Minnesota White Pine
PINE, Jack *(Pinus banksiana)*	Scrub Pine Gray Pine Black Pine
PINE, Lodgepole *(Pinus contorta)*	Tamarack Knotty Pine Black Pine Spruce Pine Jack Pine
PINE, Ponderosa *(Pinus ponderosa)*	Western Soft Pine Western Pine Pondosa Pine California White Pine Bull Pine Blackjack Pine
PINE, Red *(Pinus resinosa)*	Norway Pine Pitch Pine Hard Pine

PINE, Southern, comprises these species:

Loblolly Pine *(Pinus taeda)*	North Carolina Pine Virginia Pine Arkansas Soft Pine Southern Pine Southern Yellow Pine Yellow Pine Old-field Pine
Longleaf Pine *(Pinus palustris)*	Florida Longleaf Yellow Pine Georgia Yellow Pine Pitch Pine Hard Pine Yellow Pine
Pitch Pine *(Pinus rigida)*	Southern Pine Hard Pine Shortleaf Pine Mountain Pine
Pond Pine *(Pinus serotina)*	Southern Pine
Shortleaf Pine *(Pinus echinata)*	Arkansas Soft Pine North Carolina Pine Pitch Pine Shortleaved Yellow Pine

LUMBER
NOMENCLATURE

Correct Name of Lumber and Botanical Name of Tree	Other Names Loosely or Erroneously Employed

PINE, Southern, comprises these species: (continued)

Slash Pine . . . .
(Pinus elliottii)

Virginia Pine . . . . Southern Pine
(Pinus virginiana)   Jersey Pine
                                Scrub Pine

PINE, Spruce . . . . Bottom White Pine
(Pinus glabra)   Poor Pine
                                Cedar Pine
                                Walter Pine

PINE, Sugar . . . . California Sugar Pine
(Pinus lambertiana)   Big Pine
                                Genuine White Pine

PINE, Western White . . . . Idaho White Pine
(Pinus monticola)   White Pine
                                Mountain White Pine

PORT-ORFORD-CEDAR . . . . White Cedar
(Chamaecyparis lawsoniana)   Lawson Cypress
                                Oregon Cedar

REDCEDAR, Eastern . . . . Juniper
(Juniperus virginiana)

REDCEDAR, Western . . . . Shinglewood
(Thuja plicata)   Canoe Cedar
                                Pacific Redcedar
                                Giant Arborvitae

REDWOOD . . . . Sequoia
(Sequoia sempivirens)   Coast Redwood
                                California Redwood

SPRUCE, Eastern, comprises the following species:

Black Spruce . . . .
(Picea mariana)

Red Spruce . . . . Adirondack Spruce
(Picea rubens)   Canadian Spruce

White Spruce . . . . Adirondack Spruce
(Picea glauca)   Canadian Spruce

SPRUCE, Englemann . . . . Balsam
(Picea engelmannii)   Mountain Spruce
                                Arizona Spruce
                                White Spruce
                                Silver Spruce

SPRUCE, Sitka . . . . Yellow Spruce
(Picea sitchensis)   Silver Spruce
                                Tideland Spruce
                                West Coast Spruce
                                Western Spruce

TAMARACK . . . . Larch
(Larix laricina)   Hackmatack
                                Eastern Larch

TYPES OF
NAILS

Common Nails
Sizes from 2d to 60d

Oval Head Spike, Chisel Point
Lengths to 16", various gauge

Flat Head Spike, Diamond Point
Lengths to 16", various gauge

Casing Nails
Sizes from 2d to 40d

Finishing Nails
Sizes from 2d to 20d

Siding Nails
Sizes from 5d to 10d

Fence Nails
Sizes from 5d to 20d

Oval Head Hinge Nails
Sizes from 4d to 20d

Flat Head Hinge Nails
Sizes from 4d to 20d

Boat Nails
Sizes from 4d to 20d

Sinkers
Sizes from 2d to 60d

Lath Nails
Size · 1 1/8"

Blued Lath Nails
Sizes 2d to 3d

Blued Plaster Board Nails
Sizes · 1" to 1 3/4"

Barbed Roofing Nails
Sizes · 3/4" to 2"

Flat Head Barbed Car Nails
Sizes from 4d to 60d

Oval Head Barbed Car Nails
Sizes 4d to 60d

Barbed Box Nails
Sizes 2d to 40d

Smooth Box Nails
Sizes 2d to 40d

Common Brads
Sizes from 2d to 60d

Flooring Brads
Sizes from 6d to 20d

Clinch Nails
Sizes from 2d to 20d

Smooth Foundry Nails
Sizes from 3/4" to 3" plus

Flooring Nails
Sizes from 6d to 20d

Duplex Head Nails
Sizes from 6d to 30d

Sheet Roofing Fasteners
Sizes from 6" to 15 1/2"

Wood Shingle Nails
Size · 1 3/4"

Dowel Pin
Sizes · 5/8" to 2"

Leak-Proof Roofing Nails
Sizes · 1 1/2" to 2"

Parquet Floor Nails
Sizes · 1 1/8" to 1 1/4"

NAILS AND NAILING
REQUIREMENTS

Use	Size	Nails	Kind	Length
Shiplap (or square-edged such as used for platforms, floors, or sheathing)	1 x 4 1 x 6 1 x 8 1 x 10 1 x 12	2 2 2 3 3	8d common or box	2½"
	2 x 4 2 x 6 2 x 8 2 x 10 2 x 12	2 2 2 3 3	20d common or box	4"
	3 x 4 3 x 6 3 x 8 3 x 10 3 x 12	2 2 2 3 3	60d common	6"
Base, Chair Rails	1¹/₁₆	2	6d finish	2"
Casing, per opening			6d & 8d casing or finish	2" 2½"
Ceiling	¾ x 4 ½–⅝	1 1	8d finish 6d finish	2½" 2"
Finish	²⁵/₃₂ 1¹/₁₆	2 2	8d finish 10d finish	2½" 3"
Flooring	1 x 3 1 x 4 1 x 5	1	8d floor	2½"
Framing	2 x 4 to 2 x 16 3 x 4 to 3 x 16		10d to 20d common or box 60d common	3" 4" 6"
Drop Siding	1 x 4 1 x 6 1 x 8	2	8d casing or finish	2½"
Bevel Siding	½ x 4 ½ x 6 ½ x 8	1	6d siding or box	2"
Lath	48"	16"o/c	3d lath	1¼"
Wood Shingles	16"–18" 24"	2"	3d shingle 4d shingle	1¼" 1½"
Wood Shakes, Reroofing	16"–18" 24"	2"	5d shingle 6d shingle	1¾" 2"

33

SUITABILITY OF WOODS
FOR TRIM

EXTERIOR HOUSE TRIM

Usual requirements: Medium decay resistance, good painting and weathering characteristics, easy working qualities, maximum freedom from warp.

Highly suitable: Cedars, baldcypress, redwood—adapted to blinds, rails, and balcony and porch trim and decking, where decay hazard and weathering are high (heartwood only). Eastern white pine, sugar pine, western white pine, yellow-poplar—adapted to ordinary trim where decay hazard is moderate or low (heartwood only).

Special architectural treatments: White oaks, red oaks, birches, other hardwoods—used with natural finish (heartwood only).

Good suitability: Hemlocks, ponderosa pine, spruces, white fir—when drainage is good. Douglas-fir, western larch, southern pines—special priming treatment is advisable to improve paint-holding qualities.

Grades used: A, B, or B and Better finish is used in the best construction, C and D finish in more economical construction. No. 1 or No. 2 boards used where appearance is not important.

INTERIOR HOUSE TRIM WITH NATURAL FINISH

Usual requirements: Pleasing figure, hardness, freedom from warp and shrinkage.

Highly suitable: Ashes, birches, cherries, oaks, quartered sycamore, walnuts, certain imported hardwoods.

Special architectural treatment: Pecky cypresses, etched or special-grain cypresses, Douglas-fir, western larch, southern pines, curly or bird's-eye maples. Other woods that are used but lack the hardness of the preceding group are knotty cedars, ponderosa pine, spruces, sugar pine, white pines, aspens, and lodgepole pine.

Good suitability: Baldcypress, western hemlock, Douglas-fir, western larch, southern pines, redwood, beeches, maples, sweetgum.

Grades used: High-class hardwood interior trim is usually of the highest Firsts and Seconds (FAS) grade, with Select being of slightly lower quality. The softwood grade A or grade B and Better is commonly used in high-class construction. In the more economical types of construction, C grade is serviceable. D grade requires special selection or some cutting to obtain clear material. Special grades of knotty pines, pecky baldcypress, sound wormy oaks, and others are available to meet special architectural requirements in some types of high-class construction.

INTERIOR HOUSE TRIM WITH PAINT FINISH

Usual requirements: Fine and uniform texture, hardness, absence of discoloring pitch, freedom from warp and shrinkage.

Highly suitable: Birches, cherries, walnuts, yellow-poplar. The following woods may be used where liability to marring is negligible and special priming is used —eastern white pine, ponderosa pine, sugar pine, western white pine.

Good suitability: Hemlocks, redwood, spruces, white fir, basswood, beech, sweetgum, maples, tupelo. Where requirements for smoothness of finish are not exacting, the following woods may be used satisfactorily—baldcypress, Douglas-fir, western larch, southern pines, ashes, oaks.

Grades used: C is the lowest softwood grade commonly used for high-class paint and enamel finish. D can be used but requires some selection or cutting. No. 1 is used for ordinary or rough paint finishes. In cheaper or more economical homes, No. 2 may be used for ordinary or rough paint finishes. Smooth paint finishes are difficult to obtain and maintain over knots in No. 1, No. 2, and No. 3 grades. The Firsts and Seconds (FAS) grade in the hardwoods is used for exacting requirements of high-class paint and enamel finish in high-cost homes. The Select grade is also used, but requires some selection, cutting, and piecing. No. 1 Common hardwoods are used for interior trim in the lower-cost home, but in this class of home softwoods are generally used for interior trim that is to be painted.

GRADES OF SOUTHERN PINE TO SPECIFY

These easily worked soft pines of high strength and wear resistance, produced chiefly from stands of loblolly, slash, shortleaf, and longleaf found through a range from Virginia to Texas, are collectively called Southern Pine, and are identified by trade and SPIB (Southern Pine Inspection Bureau) grade marks stamped on the material.

Use	Southern Pine Grade Recommended*
Sills†—on foundation	Utility
Sills†—on piers	No. 2
Plates, braces, blocking, etc.	Utility
Rafters, joists, headers, etc.	No. 3
Studding	Stud
Framing boards	No. 2
Roof decking	No. 2 KD
Roof sheathing	No. 3
Subflooring	No. 3
Wall sheathing	No. 3
Stair carriages	No. 1
Stair treads	No. 1 Dense
Beams, posts, columns	No. 2, No. 2 SR
Fence posts, boards	No. 3
Fence pickets	No. 2
Balcony and deck posts	No. 1 Dense, No. 1 SR
Balcony and deck stringers, beams	No. 1, No. 1 SR
Steps	No. 2
Railings and posts	No. 1
Decking	No. 2
Porch flooring	No. 2
Porch ceiling	No. 2 Ceiling
Finish flooring, exposed	C Flooring
Finish flooring, covered	No. 2 Flooring
Interior finish and trim	C
Shelving	No. 1
Paneling, rustic	No. 2 KD
Paneling, appearance	C
Exterior siding, rustic	No. 2
Exterior siding, appearance	C & Btr
Exterior trim	No. 1
Exterior moldings	C Moldings
Window and door parts	C

*Minimum grade recommended. Higher grades used as required.
†Sills must be pressure-treated with preservatives.

STANDARD
WOOD MOLDINGS

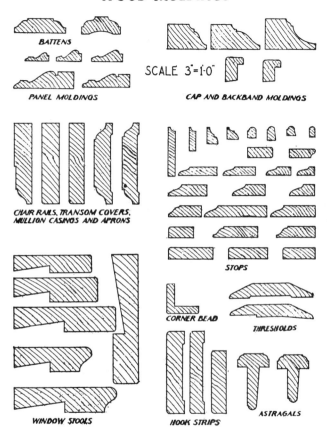

BATTENS

PANEL MOLDINGS

SCALE 3"=1'-0"

CAP AND BACKBAND MOLDINGS

CHAIR RAILS, TRANSOM COVERS, MULLION CASINGS AND APRONS

STOPS

CORNER BEAD

THRESHOLDS

WINDOW STOOLS

HOOK STRIPS

ASTRAGALS

LATTICE

The wood molding patterns shown on this and the following data sheet are typical of those available. Not all of the patterns shown are available at any one time, nor are all of the patterns that are currently available necessarily depicted. This is because patterns are always in a state of slow change as lumber sizes and design preferences change. The pattern numbering system also changes periodically. Until recently the WP/Series system was in use; it has now been supplanted by the WM/Series system. In addition, exact profiles and dimensions vary from time to time. Molding users should be guided by patterns actually available locally or should contact the Western Wood Moulding and Millwork Producers, Portland, OR, for current specifics.

STANDARD
WOOD MOLDINGS

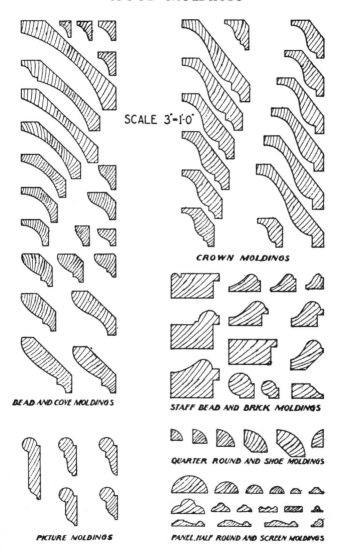

SCALE 3"=1'-0"

CROWN MOLDINGS

BEAD AND COVE MOLDINGS

STAFF BEAD AND BRICK MOLDINGS

QUARTER ROUND AND SHOE MOLDINGS

PICTURE MOLDINGS

PANEL, HALF ROUND AND SCREEN MOLDINGS

37

GENERAL INFORMATION ABOUT SOUTHERN PINE

WHAT IS SOUTHERN PINE?   Commercial southern pine is a general name for a number of closely related species consisting chiefly of:

Shortleaf	*Pinus echinata*
Longleaf	*Pinus palustris*
Loblolly	*Pinus taeda*
Slash	*Pinus elliottii*

Included in the group as minor species are:

Pond	*Pinus serotina*
Pitch	*Pinus rigida*
Virginia	*Pinus virginiana*

These pines are produced from stands ranging through the southern states from Virginia to Texas and are identified by grade and trade marks stamped on the material (see below).

WHERE TO BUY.   Southern pine can be obtained from responsible lumber dealers throughout the South, East, and Midwest. Full information can be obtained from the Southern Forest Products Association.

PHYSICAL CHARACTERISTICS.   Southern pine generally has a light, soft, lustrous texture and a fine grain with a distinctive and eye-appealing pattern. The wood does not mar or scar easily, is resistant to splintering, and holds up extremely well under hard usage and service; it is long wearing. It has particularly high nail-holding power, and the various southern pine species have among the highest strength ratings of all softwoods. In addition, southern pine is well-seasoned and highly retentive of wood preservatives.

WORKABILITY.   Southern pine possesses great toughness of fiber but works readily, holds nails and bolts securely, does not split readily when nailed, and takes finishes extremely well.

CHARACTER OF FINISH STOCK.   Top grade finish stock is made from the clear, lustrous, fine-figured sapwood of southern pine. Its resistance to marring and scratching, together with general superior workability, excellent figuring, and an ability to accept stains and natural applied finishes exceptionally well, allows this wood to attain a maximum of value and beauty when employed as interior trim and paneling.

MOISTURE CONTENT.   Southern pines, in all grades, are required to be seasoned to a moisture content that provides proper conditioning for the intended use of the material specified. This is a maximum of 19% for lumber 2″ or less in thickness, down to as little as 15% in some grades, and 23% for timbers 5″ and thicker. Requirements call for a maximum of 12% in some KD (kiln-dried) grades.

GRADE MARKING.   Each grade of lumber in any species is more than the individual manufacturer's idea of the grade. It represents the consensus of the industry, in the form (in this instance) of the American Softwood Lumber Standard (PS 20). Southern pines are graded under the rules of the Southern Pine Inspection Bureau, in conformity with this national standard. All grades are stamped with the grade mark symbols with which the building profession is familiar.

STRUCTURAL STRENGTH.   Southern pine possesses adequate strength for all stresses and loads to be expected in the construction of residences and of moderate-size commercial buildings and apartments. Tables of design values are too extensive to include here, but may be obtained from either the Southern Pine Inspection Bureau or the Southern Forest Products Association.

DROP SIDING AND CENTER-MATCHED SHEATHING

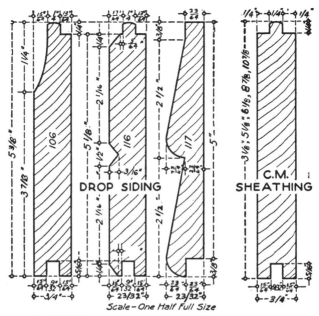

Scale — One Half Full Size

DROP SIDING. Southern pine drop siding is available in the above and several other patterns. It may be used either with or without sheathing (as economy may require). The use of drop siding as a decorative material is an interesting field for exploration. A few suggestions appear in the small drawing below.

CENTER-MATCHED SHEATHING. Center-matched sheathing surfaced 2 sides is available in No. 1, No. 2, No. 3, and No. 4 grades. The standard moisture content limit is 19%, except for kiln-dried (KD) stock, which is 15%. It is furnished in bundles when end-matched (No. 1, No. 2, and No. 3 grades only), and lengths are assorted with a minimum 15", unless otherwise specified. When plain-end, it is furnished in unbundled random lengths of 4' or more. Particular care should be taken when using end-matched for subflooring or sheathing to specify that each board be placed so that it is nailed to at least two framing members. The end-matching allows the use of this material at considerable savings in manufacture and labor during laying.

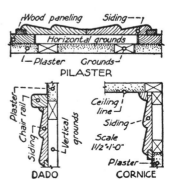

39

HARDWOOD AND DECORATIVE PLYWOODS

Hardwood plywoods are made for both utilitarian and decorative purposes; certain decorative softwoods are also included in this category when used as a face veneer. About fifty species of woods (mostly domestic) are most commonly used as face veneers, but approximately one hundred others (largely imported exotics) are also employed on a limited basis. In addition, numerous other species may be obtainable upon request. These woods can be obtained in a great variety of figures, grains, and colors, including reds, purples, greens, blacks, browns, and yellows. Grains, figuring, and patterns may be assembled in various ways on each sheet, further varying the decorative possibilities, as shown below.

TYPES. Hardwood and decorative plywoods are made in four types: Technical, Type I, Type II, and Type III. Technical is the strongest and is of exterior grade; strength and moisture resistance diminish from Type I through Type III.

CONSTRUCTIONS. There are seven different hardwood and decorative plywood constructions, the differences depending upon the type of core; all have an odd number of plies:

> 1. Hardwood veneer core, 3 plies and up.
> 2. Softwood veneer core, 3 plies and up.
> 3. Hardwood lumber core, 3, 5, and 7 plies.
> 4. Softwood lumber core, 3, 5, and 7 plies.
> 5. Particle board core, 3 and 5 plies.
> 6. Hardboard core, 3 plies.
> 7. Special core, 3 plies and up.

SIZES. The most common sheet size is 48″ x 96″, but other stock sizes are 48″ x 84″ and 48″ x 120″. However, many sizes are available on special order, including panels as long as 40′. The most common thicknesses are ¼″ and ¾″, but the possible range is from ⅛″ to over 2″.

GRADES. Full descriptions of the grades and specifications of hardwood veneers are long and complex; softwood veneer grades vary as well but are based upon different parameters. In terms of diminishing quality, they are: Premium Grade (A); Good Grade (1); Sound Grade (2); Utility Grade (3), and Backing Grade (4). There is also Specialty Grade (SP), where the characteristics of the veneers are agreed upon by buyer and seller; wall paneling face grades are often in this category. Core materials are graded separately under yet another system.

40

HARDWOOD AND DECORATIVE PLYWOODS

BLOCK MATCHED

REVERSED DIAMOND

"V" MATCHED

DIAMOND MATCHED

BURL

CENTER MATCHED

4-WAY BUTT

HALF ROUND CUT

COLORS OF
DECORATIVE PLYWOODS

BROWN—*Colors vary from light yellowish-brown to almost black.*

Madrone Burl	Red Gum	Teak
Maidhu Burl	Quartered Red Gum	Redwood Burl
Locust	Carpathian Elm Burl	Amargosa
Circassian Walnut	Koko	Carretta
French Walnut	Brown Ash	Acacia
Claro Walnut	Tamo	Myrtle Burl
American Black	Bellinga	Indian Laurel
Walnut	Marnut	Pollard Oak
Turkish Walnut	Gonzalo Alves	Spicewood
Black Sea Walnut	Cocobola	Acle
English Walnut	Bosse	Chaplash
Olivewood	Thuya Burl	Collmar
Iroko	Orientalwood	Dao
Crotch Walnut	English Brown Oak	Ipil
Burl Walnut	Butternut	Peartree
Butt Walnut	Levoa	Yuba

RED—*Colors vary from a pale pinkish-red to almost vermillion.*

Macacahula	Citron	Curly Birch
Bloodwood	Rosadura	Andaman Padouk
Australian Maple	Western Laurel	Brazilian Rosewood
African Padouk	Tiama	Bubinga
Blackwood	Red Oak	Macca
East Indian Rosewood	Honduras Mahogany	Lauan
Apple	Cuban Mahogany	French Satine
Canaletta	Mexican Mahogany	Rosewood Crotch
Bataan	African Mahogany	Faux Rose
Spanish Cedar	Maple Burl	Congo Rosewood
Aromatic Red Cedar	African Cherry	Sabicu
Birch	Black Cherry	Fiddleback Mahogany
Curarie	Lacewood	Letterwood

YELLOW—*Most of the woods listed under this color are on the dividing line between white, yellow and grey.*

East Indian Satinwood	Ayous	Canary
Amarillo	Eucalyptus	Prima Vera
Aspen	German Oak	Palo Blanco
Aspen Crotch	Russian Oak	Elm
Maple Burl	French Oak	Avodire
Blistered Maple	American White Oak	English Ash
Curly Maple	Sycamore	Poplar
Birdseye Maple	Koa	Yellow Pine
Plain Maple	Tasmanian Oak	Italian Olive
Birch	Yellow Padouk	Douglas Fir
Harewood, Natural	San Domingo Satinwood	

WHITE—*The woods in this group are as nearly white as it is possible to get.*

White Birch	Basswood	Camphorwood, has fine
Boxwood	Tupelo	pink stripe
Holly	Lemonwood	White Pine

STRIPED—*Woods carrying a prominent stripe of different color.*

Zebrawood,	Tulipwood,
Cream and Brown	Cream and Pink
(Brown Stripe)	(Pink Stripe)

BLACK—*Following are nearly clear black.*

Macassar Ebony, has	Ebonized Maple,	Baboon Ebony,
light brown striped	Jet Black	Jet Black
in black		

PURPLE

Purpleheart
East Indian Rosewood

GREEN

Greenheart

GREY

English Harewood
Empirewood
English Ash Burl

SOFTWOOD PLYWOOD

Softwood plywoods are manufactured in tremendous quantities and used throughout the country for a multitude of purposes. A large proportion is produced in conformity with the national standards set forth in *Product Standard PS 1 For Construction and Industrial Plywood,* as promulgated by the National Bureau of Standards. Some 75% of the softwood plywoods produced in the United States are trade and grade marked under the aegis of the American Plywood Association (APA, formerly the DFPA, or Douglas Fir Plywood Association). This organization represents the bulk of the U.S. softwood plywood manufacturers, and the APA markings on their products is an assurance of quality. Of the remaining 25% of the softwood plywoods produced, the greater part carries trade and grade marks that are also acceptable but less widely recognized.

TYPES OF SOFTWOOD PLYWOOD. There are two principal types of softwood plywood; interior and exterior. The interior type is subdivided into three categories. The first is interior with *interior* glue, wherein the plies are bonded with a relatively non-moisture-resistant glue; this plywood is intended only for protected interior applications. The second is interior with *intermediate* glue; this plywood is intended for protected interior use where protection may be delayed during construction, allowing the temporary presence of moisture, or where leakage or high humidity may be constant or recurring factors. The third is interior with *exterior* (waterproof) glue; this plywood is intended for protected interior use where long construction delays or constant and relatively severe moisture conditions are present. Exterior type plywood is produced solely with exterior (waterproof) glue and is intended for permanent exterior exposure applications. It is weatherproof and will withstand repeated wettings and dryings.

WOOD SPECIES USED. The most familiar species of wood used in softwood plywoods are the varieties of Douglas-fir, and they are indeed very widely used. However, there are more than seventy other species employed as well, many of them hardwoods. These woods are classified in five groups (see below). Those listed in Group 1 are the strongest and stiffest; those in Group 5, the weakest. Face and back plies can be made from any of the species in any of the groups, whereupon the group number becomes the classification number of the completed plywood sheet. Inner plies may be made of any of the species listed in the first four groups, regardless of the species used for the face and back plies. Sheets classed as Group 5 can be made up of any of the listed species. In addition, inner plies can be made from any species of softwood or hardwood that has a published average specific gravity of 0.41, based on green volume and oven-dry weight.

GROUP 1	GROUP 2	
Apitong	Baldcypress	Maple, Black
Beech, American	Douglas-Fir 2*	Menkulang
Birch, Sweet	Fir, California Red	Meranti, Red
Birch, Yellow	Fir, Grand	Mersawa
Douglas-Fir 1†	Fir, Noble	Pine, Pond
Kapur	Fir, Pacific Silver	Pine, Red
Keruing	Fir, White	Pine, Virginia
Larch, Western	Hemlock, Western	Pine, Western White
Maple, Sugar	Lauan, Almon	Port-Orford-Cedar
Pine, Caribbean	Lauan, Bagtikan	Spruce, Red
Pine, Ocote	Lauan, Mayapis	Spruce, Sitka
Pine, Loblolly	Lauan, Red	Sweetgum
Pine, Longleaf	Lauan, Tangile	Tamarack
Pine, Shortleaf	Lauan, White	Yellow-Poplar
Pine, Slash		
Tanoak		

GROUP 3	GROUP 4	GROUP 5
Alaska-cedar	Aspen, Bigtooth	Basswood
Alder, Red	Aspen, Quaking	Fir, Balsam
Birch, Paper	Cativo	Poplar, Balsam
Fir, Subalpine	Cottonwood, Eastern	
Hemlock, Eastern	Cottonwood, Black	
Maple, Bigleaf	Incense-cedar	
Pine, Jack	Pine, Eastern White	
Pine, Lodgepole	Pine, Sugar	
Pine, Ponderosa	Western Redcedar	
Pine, Spruce		
Redwood		
Spruce, Black		
Spruce, Engelmann		
Spruce, White		

*Washington, California, Oregon, Idaho.
†Nevada, Utah, Colorado, New Mexico, Arizona.

SOFTWOOD PLYWOOD

CONSTRUCTIONS. Softwood plywood panel constructions involve both plies and layers. Layers are always in odd numbers, with each layer placed at right angles to the next. A single layer may consist of one ply, or one thickness of veneer, or of two plies, or two thicknesses of veneer, laid with their grain parallel to make up one layer. The minimum number of both plies and layers ranges from three to seven, depending upon the thickness of the finished panel, but as many of each as are required for particular purposes can be built up.

SIZES. The most common softwood panel size is 48″ x 96″. However, 48″ x 108″ and 48″ x 120″ are also available. In addition, both 36″ and 60″ are considered standard panel widths, and lengths are available up to 144″ in 12″ increments. The most common panel thicknesses are ¼″, ⁵⁄₁₆″, ⅜″, ½″, ⅝″, and ¾″. However, ⅞″, 1⅛″, ¹⁹⁄₃₂″ and ²³⁄₃₂″ are available in certain types of panels. Other panel dimensions are available on special order; these vary depending upon the type.

VENEER GRADES. Unlike hardwood plywoods, which often employ a core type of construction, softwood plywood panels are entirely composed of veneer plies. All veneers used conform to a veneer grading system, as follows:

Grade N. This veneer is intended for natural finish. It is clear, free of defects, stain, splits, pitch pockets, knots, or knotholes. No more than two pieces can be used to make up a 48″-wide panel, and no more than three in larger panels, and the pieces must be well matched for grain and color. Tiny amounts of synthetic fillers may be present. Repairs are limited to six, and they must be small and well matched in. This is the top grade, presenting an excellent surface, and generally used only as a face veneer on interior-type plywood. Inner plies are Grade C or D, and the panels are special-order items.

Grade A. This grade presents a surface suitable for painting or similar finishing. It must be firm and smoothly cut, and free of open defects, knots, knotholes, pitch pockets, or open splits. Synthetic fillers can be used in small cracks, checks, and openings. Small pitch streaks are allowable, as is sapwood and discoloration. Repairs are limited to eighteen per panel, and must not exceed certain size requirements. This grade is used for appearance applications, and is available in the forms of both face and back veneers (mostly the former) in exterior and interior types of panels. The inner ply grade is D for interior, C for exterior.

Grade B. This grade must be solid and largely free from open defects and broken grain. It presents a reasonable appearance, but is primarily intended for utilitarian purposes or uses where appearance does not matter. Minor sanding or patching defects are allowable, and the surface may be slightly rough. Repairs of certain kinds and sizes may be made; they can be numerous and obvious. Inner plies are D for interior, C for exterior.

Grade C. This grade must not have defects that will impair either strength or general serviceability. Knots may be as large as 1½″ but must be tight. Open defects are restricted in size but are allowable, as are certain splits and voids. Repairs may be made, and discoloration may be present. Grade C is primarily used in both interior and exterior types intended for general construction purposes where appearance does not matter. Inner plies are grades D for interior, C for exterior.

Grade C-Plugged. This grade is of somewhat lesser quality than Grade C, and is used primarily for utility purposes and for inner plies. Open defects and repairs are relatively substantial, but strength and serviceability are unimpaired.

Grade D. This lowest grade is permitted to have any number of plugs, patches, open defects, shims, wormholes, and the like, so long as neither strength nor serviceability is seriously impaired. The primary use is for inner plies and for back plies, on interior-type panels only.

PANEL GRADE DESIGNATIONS. Softwood plywood panels are given certain overall grade designations that are dependent upon the grades of the face and back veneers, with the face grade being listed first. Thus, a panel with a Grade A face veneer and a Grade B back veneer is designated A-B. Other nomenclature is sometimes included in the designation to indicate a special kind of panel, such as Structural I C-D. Most panel grades are factory-sanded on both sides, a few are touch-sanded, and several utilitarian grades are left unsanded.

MAJOR GROUPINGS. The American Plywood Association has divided softwood plywood products into two major groupings.

1. *Appearance Grades.* This group of plywood panels, available in both interior and exterior types, is employed where appearance is of either primary or secondary importance. There is no diminution of strength, rigidity, etc.; appearance is an added factor.

2. *Engineered Grades.* This group of plywoods is used primarily for general construction purposes where appearance is of no consequence. Several specialized types are included. Both interior and exterior types are available.

A complete listing of the various softwood plywood APA grade designations for both Appearance and Engineered Grades will be found on the pages immediately following.

SOFTWOOD PLYWOOD
SELECTION

APA APPEARANCE GRADES

INTERIOR TYPE

N-N INT-APA. For top-quality cabinetwork, furniture-making, natural finishes. Special order. Common thickness—¾".

NGA INT-APA. Same as above.

N-B INT-APA. Same as above.

N-D INT-APA. For natural finish paneling (cabinets, walls, etc.). Special order. Common thickness—¼".

A-A INT-APA. For cabinetry, furniture, partitions, etc., where appearance of both faces is important. Suitable for natural, stained, painted finish. Common thicknesses—¼", ⅜", ½", ⅝", ¾".

A-B INT-APA. Same as above, appearance of one face less important.

A-D INT-APA. For racks, shelves, cabinetry, etc., where back is hidden or appearance does not matter. Thicknesses as for A-A.

B-B INT-APA. For utility purposes requiring two smooth sides. Thicknesses as for A-A.

B-D INT-APA. For utility purposes requiring one smooth side, for bins, shelves, racks, commercial/industrial use. Thicknesses as for A-A.

DECORATIVE PANELS INT-APA. Various decorative face-ply treatments (sawn, brushed, grooved, etc.) for paneling, cabinetry, displays, accents, etc. Common thicknesses—⁵⁄₁₆", ⅜", ½", ⅝".

PLYRON INT-APA. Hardboard face both sides, for flooring, shelves, doors, counters, worktops. Common thicknesses—½", ⅝", ¾".

EXTERIOR TYPE

A-A EXT-APA. For use where appearance of both sides is important, as in boats, signs, cabinets, etc. Common thicknesses—¼", ⅜", ½", ⅝", ¾".

A-B EXT-APA. Same as above, where back face is less important.

A-C EXT-APA. Same as above, where back face is unimportant.

B-B EXT-INT. Two solid faces, for utility purposes. Thicknesses as for A-A.

B-C EXT-APA. Utility, for commercial/industrial, agricultural purposes where appearance is unimportant. Thicknesses as for A-A.

HDO EXT-APA. High Density Overlay of smooth resin-fiber on one or both faces. For cabinetry, signs, fences, tanks, concrete forms, worktops. Common thicknesses—⅜", ½", ⅝", ¾".

MDO EXT-APA. Medium Density Overlay of smooth resin-fiber on one or both faces. Excellent for siding, general outdoor applications, paint base, signs, etc. Thicknesses as for HDO.

303 SIDING EXT-APA. Various face ply treatments (rough, grooved, channeled, etc.) For siding, fences, etc. Common thicknesses—⅜", ½", ⅝".

T 1-11 EXT-APA. 303 with ¼" deep grooves 4" or 8" o/c, or other spacing on special order; shiplap edges; textured, unsanded, or MDO face treatments available. For siding. Common thickness—⅝".

PLYRON EXT-APA. Hardboard two faces. Common thicknesses—½", ⅝", ¾".

MARINE EXT-APA. For boat hulls and similar marine applications. Thicknesses as for A-A. HDO and MDO faces available.

SOFTWOOD PLYWOOD
SELECTION

APA ENGINEERED GRADES

INTERIOR TYPE

C-D INT-APA. For wall and roof sheathing, subflooring, commercial/industrial purposes. Available with intermediate and exterior glue. Common thicknesses— $5/16''$, $3/8''$, $1/2''$, $5/8''$, $3/4''$.

STRUCTURAL I C-D INT-APA. Unsanded; exterior glue only. For engineered high-strength applications (containers, stressed-skin panels, box beams, etc.). Thicknesses as for C-D.

STRUCTURAL II C-D INT-APA. Same as above, different specifications.

UNDERLAYMENT INT-APA. For floor or subfloor underlayment. Intermediate and exterior glues available. Available in Structural I and Structural II. Common thicknesses— $1/4''$, $3/8''$, $1/2''$, $5/8''$, $3/4''$.

C-D PLUGGED INT-APA. For utility use in built-ins, wall or ceiling backing, cable reels, etc. Available with intermediate or exterior glue, and in Structural I and Structural II. Thicknesses as for C-D.

2-4-1. Use specifically as combination subfloor-underlayment. Available sanded or touch-sanded, and with exterior glue. Available in exterior type on special order, for decks, etc. Thickness— $1\frac{1}{8}''$.

EXTERIOR TYPE

C-C EXT-APA. For subflooring, roof decking, crating, utility building siding, etc. Unsanded; waterproof bond. Common thicknesses— $5/16''$, $3/8''$, $1/2''$, $5/8''$, $3/4''$.

STRUCTURAL I C-C EXT-APA. For engineered exterior high-strength applications. Common thicknesses as for C-C.

STRUCTURAL II C-C EXT-APA. Same as above, except different strength specifications. Common thicknesses as for C-C.

UNDERLAYMENT C-C PLUGGED EXT-APA. For underlayment or combination subfloor-underlayment in severe moisture conditions such as decks or balcony floors, or for commercial/industrial applications such as bins, cargo containers, pallets, and truck floors. Touch sanded; available tongue-and-groove. Common thicknesses— $1/4''$, $3/8''$, $1/2''$, $5/8''$, $3/4''$.

C-C PLUGGED EXT-APA. Same as above, except less satisfactory for underlayment.

B-B PLYFORM CLASS I EXT-APA. Use for reusable concrete forms. Sanded two sides; mill oiled. Available in HDO. Common thicknesses— $5/8''$, $3/4''$.

B-B PLYFORM CLASS II EXT-APA. Same as above, slightly different characteristics.

NOTE: The standard panel size for all of the above grades is 4' x 8'. However, other sizes are available on special order.

PLASTER TO
RECEIVE PAINT

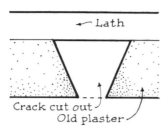

WASHING PLASTER WALLS.   New paint will not adhere well to greasy walls. Wash first with a vigorous alkali (e.g., ammonia solution or trisodium phosphate) and rinse with clean water to remove gloss as well as dirt from old paint or enamel. Remaining gloss should be sanded off so that the new paint will have good adhesion.

REPAIRING PLASTER CRACKS.   Repair all cracks and damaged plaster before repainting. Undercut the plaster along the cracks with a utility knife to form a dovetail joint that will retain the patching material. The areas of new patching material must be allowed to dry thoroughly and then be primed. When patching, match the texture of old sand or plastic-designed surfaces; the patched sections should be patted with a rough tool, piece of old carpet, scrub brush, or sponge. Hair cracks can be filled with patching plaster or, better, with vinyl paste spackling compound.

VERY BAD WALLS.   In old buildings, the plaster sometimes becomes so badly cracked, or piled up with repeated applications of paint or wallpaper, that a satisfactory redecorating job is impossible. In such cases it is more economical and satisfactory to cover the entire wall and/or ceiling with plywood, plasterboard, or other wallboard. Sometimes the insulating value of the wall (or ceiling) can be increased at the same time, by the choice of wallboard material or by furring out and retrofitting an insulating material.

PAINT-VARNISH REMOVER.   Numerous commercial removers are readily available; choice should be based on the type of finish to be removed, as well as the nature of the finished surfaces. Caution should be observed, as many removers are toxic, flammable, messy to work with, incompatible with the underlying material, or simply badly suited for some applications. One safe, easy, and inexpensive method that does work well if the paint or varnish is not too thick is to apply a solution of trisodium phosphate in a ratio of 1 pound per gallon of hot water. Brush or mop the solution on liberally, and after about a half hour remove the softened finish with a dull scraper or putty knife, taking care not to gouge the surface. Repeat the process as necessary. As soon as possible after stripping, wash the area with clean water and wipe dry to prevent excessive soaking into the plaster (or raising the grain in wood). Be sure to wear rubber or plastic gloves, and cover everything in the vicinity that might get spattered.

WOOD FLOOR
FINISHES

HOT LINSEED OIL. Years ago, floors were commonly finished with hot linseed oil. Each application was buffed by hand. When the surface finally became saturated with oil, it was waxed and maintained by further waxing at suitable intervals. Unbodied drying oils penetrate relatively deeply into wood, requiring a good many applications and making the process rather laborious. (An unbodied oil is one that has not been treated or heated to increase the viscosity substantially. Raw, refined, and boiled linseed oil, raw and refined soybean oil, tung oil, and perilla oil are all unbodied oils.) Hot linseed oil finish is durable, does not show scratches, can be readily patched at places of maximum wear, dries hard enough to be free from tackiness, and makes a floor easily kept clean by dry-mopping. In time, the finish darkens, deepening the original color.

As time passes, however, the finish becomes tacky with wax buildup and darkens to a color that is almost, if not completely, black. For this reason, linseed oil finish fell into disrepute and was replaced by other finishes. However, the old oil finishes have staged something of a comeback over recent years, and remain perfectly satisfactory in certain applications.

SHELLAC. Shellac too was once a favorite floor finish, but because of its several drawbacks and the advent of modern synthetic finishes it has fallen into disfavor. Despite this, it remains a useful finish in some applications. Its chief advantage is that it dries very rapidly. A floor may be finished or refinished and put back into service in just a few hours; a first coat can be recoated in 3 hours and put into service 24 hours later. It is also relatively inexpensive and easy to apply. Shellac forms a coating of substantial thickness over the surface of the wood, in contrast to finishes that penetrate into the surface of the wood. A shellac finish has the following characteristics: a highly lustrous appearance; extreme slipperiness unless the wax coating is kept very thin; a finish that turns white from water; and worn areas that can rarely be patched without showing edges.

PENETRATING SEALERS. Various types of sealers, which may be thinned varnishes or various combinations of oils, are available. They penetrate more deeply into varnishes or other surface-film types of finishes, saturating the surface layer of the wood. Two applications on new wood are usually all that is required. The sealer may be clear, or combined with a stain. Sealers are often waxed, but they also provide a fine base for varnish. In general, they are characterized by: minimum slipperiness when waxed; less luster than varnish or shellac; minimun maintenance requirements; and simple and (afterward) unnoticeable rewaxing or resealing and rewaxing of high traffic areas.

STAIN. Wood floors can be stained readily, but this does not constitute a complete finish. A wax top-coating alone is unsatisfactory; a sealer or varnish must be applied for protection. The stain must be oil-base or with a base of some other non-grain-raising type. Since penetration is slight, wear can produce areas of mismatched color that are very difficult to disguise. Thus, the protective top-coat of other finish must be carefully maintained.

WOOD FLOOR
FINISHES

VARNISH.   These coatings, even the quick-drying varieties, require fairly long intervals between applications and a long curing time before becoming serviceable. Varnish finishes have different bases. Most have a base of alkyd resin, clear and nonyellowing. Phenolic resin-base coatings (spar varnishes) are tougher and more moisture resistant; however, they tend to yellow and/or darken with age. Polyurethane varnishes are clear, thin, tough, and highly wear-resistant, making them ideal as floor finishes. There are many different formulations, each requiring somewhat different application techniques, so the manufacturer's instructions should always be closely followed. Both gloss and semigloss types are available. Varnish floor finishes are: relatively durable; lustrous to highly glossy; relatively unslippery when waxed; and easy to maintain.

SHELLAC-VARNISH.   In this type of finish, the shellac is used as a sealer on the bare wood, then covered by one or more coats of varnish. As a sealer, shellac is best when thinned to a substantial extent. Orange shellac should be used, since it is more moisture-resistant than the white (bleached) variety. Shellac also makes a good spot-sealer over knots in resinous woods, helping to prevent pitch and resins from bleeding through the top-coats. However, other sealers are tougher and more effective in many applications. Shellac cannot be used as a sealer under polyurethane varnishes; they are not compatible.

LACQUER.   Though lacquer is seldom used as a floor finish, it can be attractive and long-wearing if properly applied. A special primer must be used on bare wood. The lacquer can be applied by brush or roller, but is best sprayed on in several thin coats—the more, the better. The final misting coat should be reduced to 3 parts lacquer and 1 part thinner. The finish can be rubbed, using various techniques, to virtually any desired degree of luster.

GYM-TYPE FINISHES.   These finishes are special varnishes that have been developed and formulated for gymnasium floors and similar applications where a high degree of wear and abuse is a constant factor. Shellac is often used as a sealer, followed by two coats of the finish. These are perhaps the toughest (and most expensive) of all floor finishes and must be applied exactly as the manufacturer's instructions indicate.

PAINTS.   Although floors are not painted as much nowadays as they once were, paint remains a viable alternative. Decks and porches are often so treated, and Colonial house restorations, especially those with stenciled floors, demand it. Bare wood should first be treated with a recommended primer, which is followed usually by two applications of paint. Either gloss or semigloss paint may be used, but the latter is apt to be somewhat tougher and less prone to chipping. Special floor paint formulations should be used whenever possible; there are also different products for interior and exterior use. The surface must be completely clean and dry, and free of any oils or greases. Maintenance is simple and no waxing is required. However, worn areas are difficult to patch, and usually recoating is necessary.

BASIC SPECIFICATIONS
FOR PAINTING

Complete descriptions of paint and applied-finish products, together with detailed specifications for their use and application, can be obtained directly from manufacturers or through their dealers.

1. GENERAL CONDITIONS. The general conditions bound herewith are a part of this Section. The subcontractor for work in this Section is to read them and be bound thereby.

2. WORK INCLUDED. This Section includes all labor and materials necessary to complete the painting and finishing of the building.

3. WORK NOT INCLUDED. This Section does not include shop coats.

4. MATERIALS. Use materials manufactured by [insert name of manufacturer]. Deliver materials to the work in the original sealed containers. Do all required mixing on the premises. Do not reduce or change materials in any way except as and when specified.

5. SAMPLES. Prepare required samples well in advance of the work so as to cause no delay, to meet the approval of the architect as to color, and match the approved sample accurately in the finished work.

6. PROTECTION OF PROPERTY. Protect adjacent work and materials from damage.

7. PREPARATION OF SURFACES. The subcontractor for work in this Section is wholly responsible for the finish of his work. Do not commence any part of the work until the surface is in proper condition. If required, apply an approved and appropriate sealer to all knots or sappy spots, and do not apply any finish until after the manufacturer's recommended drying time has passed. Fill all nail holes, cracks, and blemishes with a recommended material, before or after priming/painting, as required. Filler shall match the shade of the final coat. Clean greasy or oily surfaces with turpentine or mineral spirits before applying any materials. Remove rust and scale by scraping, wire brushing, or sandblasting.

8. WORKMANSHIP. No exterior painting shall be done in rainy, damp, or frosty weather. Manufacturer's minimum/maximum temperatures for application shall be observed. No interior painting or finishing shall be done until the building has been thoroughly dried out, either naturally or by means of artificial heat. Allow all finishes to dry for the recommended time period between coats. Lightly sand or steel-wool surfaces and dust between coats, if and as required. Where paste wood fillers are used, all excess shall be carefully cleaned from the surfaces by wiping across the grain.

9. REMOVAL. All inflammable waste materials shall be removed from the premises at the end of each working day, and all other flammables carefully and properly stored. When the work is completed, remove all surplus materials and equipment. Clean all misplaced paint, varnish, and so forth, to leave the premises in perfect condition. This subcontract will not be deemed fulfilled until final approval of the architect.

U.S.S. GAUGE FOR IRON AND STEEL SHEETS

Number of Gauge	Approximate Thickness In Inch Fractions	Approximate Thickness In Inch Decimals	Weight per Square Foot
0000000	1/2	0.5	20.00 lbs.
000000	15/32	0.46875	18.75 "
00000	7/16	0.4375	17.50 "
0000	13/32	0.40625	16.25 "
000	3/8	0.375	15.00 "
00	11/32	0.34375	13.75 "
0	5/16	0.3125	12.50 "
1	9/32	0.28125	11.25 "
2	17/64	0.265625	10.63 "
3	1/4	0.25	10.00 "
4	15/64	0.234375	9.38 "
5	7/32	0.21875	8.75 "
6	13/64	0.203125	8.13 "
7	3/16	0.1875	7.50 "
8	11/64	0.171875	6.88 "
9	5/32	0.15625	6.25 "
10	9/64	0.140625	5.63 "
11	1/8	0.125	5.00 "
12	7/64	0.109375	4.38 "
13	3/32	0.09375	3.75 "
14	5/64	0.078125	3.13 "
15	9/128	0.0703125	2.81 "
16	1/16	0.0625	2.50 "
17	9/160	0.05625	2.25 "
18	1/20	0.05	2.00 "
19	7/160	0.04375	1.75 "
20	3/80	0.0375	1.50 "
21	11/320	0.034375	1.38 "
22	1/32	0.03125	1.25 "
23	9/320	0.028125	1.13 "
24	1/40	0.025	1.00 "
25	7/320	0.021875	14 oz.
26	3/160	0.01875	12 "
27	11/640	0.0171875	11 "
28	1/64	0.015625	10 "
29	9/640	0.0140625	9 "
30	1/80	0.0125	8 "
31	7/640	0.0109375	7 "
32	13/1280	0.01015625	6½ oz.
33	3/320	0.009375	6 "
34	11/1280	0.00859375	5½ "
35	5/640	0.0078125	5 "
36	9/1280	0.00703125	4½ "
37	17/2560	0.00664062	4¼ "
38	1/160	0.00625	4 "

SHEET GLASS SIZES, WEIGHTS, DESCRIPTION

DESCRIPTION.   Sheet glass is drawn vertically and held absolutely flat from molten state to finished sheet. During the drawing process no rolls or foreign substances of any kind touch the surface of the glass until it has cooled sufficiently to be beyond injury. It has a brilliant fire-polished surface that is reflective, unmarred, and resistant to scratching, but the faces are not perfectly parallel at all points and there is invariably some degree of distortion. However, these flat-glass products are made of pure and carefully selected ingredients that result in excellent transparency. This results in unobstructed visibility through it and in the transmission of the true colors of all objects seen through it. The transparency is permanent, and the glass retains its clarity indefinitely.

Sheet glass is separated into two categories; window glass, and heavy sheet glass. Both are graded at the factory by experts, in conformity to certain U.S. Government standards.

AA QUALITY.   This is the best quality of sheet glass obtainable and is specially selected for use in the highest-grade work. It is also the highest priced and is used primarily for residential and commercial glazing.

A QUALITY.   This quality level permits no imperfections that will appreciably interfere with straight vision. This is the standard quality grade for commercial purposes and is used primarily in residential and commercial glazing applications.

B QUALITY.   This quality level admits of the same kinds of defects as the A quality, but allows them to be larger, heavier, and more numerous. The primary use for B quality glass is in economical residential and commercial glazing.

Classification	Quality	Thickness	Oz. per Sq. Ft.	Max. Size*	% Light Xmission
Window: Single Strength	AA, A, B	$3/32''$	19.5	28 sq. ft.	91
Window: Double Strength	AA, A, B	$1/8''$	26.0	35 sq. ft.	91
Heavy Sheet	AA, A, B	$3/16''$	40.0	84″ x 120″	90
Heavy Sheet	AA, A, B	$7/32''$	45.6	84″ x 120″	89

*Maximum sheet sizes depend upon several factors and vary among different manufacturers.

GLASS QUALITY
STANDARDS

GENERAL PRINCIPLES. All flat glass contains some imperfections. The principles employed in grading are intended to exclude defects that would be objectionable in a given grade. This is difficult to accomplish because there are no sharp lines of demarcation between grades; even experienced inspectors will differ in judgment as a quality of the glass approaches the limits of the grades. Small lights must be quite free from imperfections as compared with larger ones. The center of any sheet should be clear, whereas the edges may contain more pronounced defects. For some materials, such as coated glass or tempered safety glass, requirements over and above the quality of the base glass itself must be met, such as for coating uniformity and quality or for strength.

STANDARDS. The principal standard for base glass is Federal Specification DD-G-451D, *Glass, Float or Plate, Sheet, Figured (Flat, for Glazing, Mirrors and Other Uses).* This standard sets forth the permissible number and size of such imperfections as scratches, bubbles, and inclusions, as well as the general overall dimensional and optical quality levels of the glass. It covers both clear and tinted glass, as well as patterned glass and glass that will undergo further treatment, such as coating or silvering. Wired glass must also conform to this specification.

Heat-strengthened and fully tempered glass is manufactured to conform to Federal Specification DD-G-1403B, *Glass, Plate (Float), Sheet, Figured, and Spandrel (Heat-Strengthened and Fully-Tempered).* In addition to regulating optical quality, this standard specifies the limits of surface and edge compression, the maximum degree of bow and kink, and similar factors. These types of glass must also conform to, and are used as required under, the Federal Safety Standard for Architectural Glazing Materials (Standard 16 CFR 1201).

Wired glass, laminated glass, and fully tempered glass (as well as rigid plastics) used in safety glazing applications must also conform to the requirements laid down under ANSI Z97.1, *Performance Specifications and Methods of Test for Safety Glazing Material Used in Buildings.* The thrust of this specification is to ensure proper failsafe properties in the material, to avoid personal injury and further life-safety.

Insulating glass formed into sealed units is governed by ASTM E-6 P3, along with attendant specifications for testing. This standard is concerned with the durability and longevity of the seal, freedom of the interior surfaces from fogging, etc.

QUALITY CERTIFICATION. The quality of safety glazing materials is certified by the Safety Glazing Certification Council (SGCC), which evaluates, randomly samples, and tests such materials on a continuous basis. Testing is done under the requirements of ANSI Z34.1, *American National Standard Practice for Certification Procedures,* and ensures compliance with ANSI Z97.1 (see above). Products meeting these standards carry an SGCC certification label; such certification should be specified.

The quality of insulating glass units is certified by the Insulating Glass Certification Council (IGCC) in accordance with ANSI Z34.1. Such products carry the IGCC label, and should be specified.

It should be noted that not all manufacturers of these glass products participate in certification programs, nor do all of them have the right to use IGCC or SGCC labels.

REGULAR FLOAT
AND PLATE GLASS

MANUFACTURE OF PLATE GLASS. Plate glass is transparent, flat, relatively thin glass having plane polished surfaces. It is made by casting large sheets over a smooth mold. The sheets are then ground and polished mechanically to true flat surfaces having great brilliance and high reflectivity. Because the two surfaces of the glass form true and parallel planes, polished plate glass affords virtually perfect undistorted vision or reflection from any angle.

MANUFACTURE OF FLOAT GLASS. Float glass is manufactured by a relatively new process, originally developed in England, whereby the glass is cast on a bed of molten metal. The resulting glass sheets are of a quality level almost undistinguishable from that of plate glass, even though the process allows the elimination of the grinding and polishing procedures and a consequently lower production cost. Float and plate glass are considered interchangeable for nearly all applications, and for anyone but a glass expert there is no detectable difference between the two types. Some manufacturers are now producing float glass in thicknesses that are direct substitutes for the sheet glass classifications.

QUALITY LEVELS. Both float and plate glass are available in three quality levels:

Silvering Quality. This is specially and carefully selected glass that meets the highest standards and is exceptionally free from defects of any kind. It is intended for use in applications where perfection is a necessity, such as in top-grade mirrors and optical products.

Mirror-glazing Quality. This is a superior grade of glass with only a few very minor defects. It is intended for use where the requirements are demanding, but somewhat below those for Silvering Quality. As the term implies, the principal use is for mirrors and similar applications.

Glazing Quality. Though it must meet certain minimum standards that ensure an excellent product, Glazing Quality glass is the standard "run-of-the-mill" grade. It is normally specified for ordinary commercial/industrial and residential glazing applications.

Classification	Thickness	Oz. per Sq. Ft.	Max. Size*	% Light Xmission
	$^3/_{32}$"	19.7	28 sq. ft.	90
	$^1/_8$"	26.0	35 sq. ft.	90
Float or Plate: Clear	$^5/_{32}$"	32.8	92 sq. ft.	89
	$^3/_{16}$"	39.2	92 sq. ft.	89
	$^1/_4$"	52.3	128" x 204"	88

*Maximum sheet sizes depend upon several factors and vary among different manufacturers.

HEAVY FLOAT
OR PLATE GLASS

DESCRIPTIONS OF HEAVY FLOAT OR PLATE GLASS.   Glass in thicknesses of ⁵⁄₁₆″ to ⅞″ is termed *Heavy Float* or *Heavy Plate* glass. Since the strength of the glass increases in direct proportion to the square of the thickness, this material is adaptable to many uses where substantial strength is required. Both heavy plate and heavy float glass have true plane surfaces with great brilliance and high reflectivity, and afford clear, perfect vision.

USES FOR HEAVY FLOAT OR PLATE GLASS.   Both types of glass are widely used for commercial and residential glazing, bookshelves, decorative panels, partitions, valances, lighting fixtures, showcase tops, aquaria, tabletops, modern furniture, displays, and in a host of other applications.

QUALITY LEVELS.   Heavy float or plate glass is manufactured only in one grade, *Glazing Quality*, which is also sometimes called *Select Quality*.

Classification	Thickness	Lb. per Sq. Ft.	Max. Size*	% Light Xmission
Heavy Float or Plate: Clear	⁵⁄₁₆″	4.08	124″ x 200″	87
	⅜″	4.90	124″ x 200″	86
	½″	6.54	120″ x 200″	84
	⅝″	8.17	120″ x 200″	82
	¾″	9.81	115″ x 200″	81
	⅞″	11.45	115″ x 200″	79

*Maximum sheet sizes depend upon several factors and vary among different manufacturers.

REFLECTIVE COATED GLASS

Reflective coated glass, often simply referred to as coated glass, is plate or float glass that has been given a special coating of a transparent metal or metal oxide film by means of a sophisticated vacuum-deposition process. The thickness of the film is carefully controlled during processing to ensure a uniform coating, and the opacity depends upon the type and thickness of film used. Numerous metals are employed for widely varying effects and performances. Harder films like chromium or cobalt oxide are durable enough to be satisfactorily applied to single glazing. Other, more delicate films such as gold, nickel, copper, and aluminum are generally relegated to an inside surface of double- or triple-glazing or of laminated glass.

The primary purpose of reflective coated glass lies in the reduction of solar heat and glare. Visibility from the inside out is excellent, while visible light transmittance through the glass from the outside ranges from as little as 8% to 50%, and daylight reflectance ranges from approximately 6% to 45%. The exact values depend upon the film itself, upon whether or not two different coated lights or coated glass and tinted glass are used together, and, if so, upon the characteristics of each. Secondarily, coated glass reduces the transmittance of ultraviolet rays into the building, thus minimizing fading of drapes, furnishings, paintwork, and the like, and reducing sunrot of fabrics. Coated glass also acts much as a one-way mirror does; whenever there is reasonable contrast between interior and exterior light, visibility from the more highly illuminated side through the glass to the lesser-illuminated side is obscured because of the high reflectivity of the coating. Thus, this type of glazing also affords a certain degree of privacy. Because of its striking appearance, coated glass can also be used to provide interesting architectural and decorative emphasis.

The base glass employed is float or plate, most commonly in the 1/4" thickness; 1/8" glass is also available. Typically, the maximum sheet size available in ordinary annealed glass is 117" x 140", except for 1/8" stock, which is generally about 66" x 114". Coated glass is readily available in 1/4" fully tempered sheets of a maximum size of approximately 72" x 110", as well as in certain tempered geometric shapes on special order. Coated glass may also be used in combination with tinted glass, or the tinted glass may itself be coated with a reflective metallic film. Insulating glass and laminated glass may be obtained with reflective coatings. Reflective films are not only applicable to vision glass, they are also widely used on spandrel or architectural glass for all manner of applications. Thus, a tremendous range of performance characteristics exists, from which a coated glass, alone or in combination with other types of glass, may be selected to suit the needs of virtually any application.

A wide range of colors and tones is offered, differing somewhat from manufacturer to manufacturer. The general color families are copper, bronze, gold, silver, blue, blue-green, and neutral; others may be introduced from time to time.

Maintenance of coated glass where the reflective film is not exposed is the same as for any other type of flat glass. However, if the film is exposed, cleaning must be done with care, using ordinary window cleaners or mild detergents and water, applied with clean cloths and dried with soft cloths or a quality squeegee. The greatest danger lies in scratches from particles of grit or from contact with cleaning tools.

TINTED GLASS

DESCRIPTION.   This glass is made by a special process that involves the inclusion of certain admixtures to the glass batch. The result is a glass with the capacity for significantly lowering the transmission of solar heat through the glass and into the building. At the same time, the transmission of visible light is reduced to a degree that depends upon the particular tint and thickness of glass. Light transmission may be as high as 78% or as low as 19%, while solar transmission can range from 65% to as little as 22%. The reduction of solar transmission can result in substantially lower air conditioning costs in a building whose windows, skylights, etc., are glazed with tinted glass. The quantity and quality of light transmission can also be controlled to a degree by proper glass selection, in that solar glare can be reduced considerably while as large a percentage of natural daylight component as is desired can be admitted. Careful selection of glazing and shading devices in combination can lead to improved visual performance and a reduction in costs of electric lighting. In addition, tinted glass offers certain aesthetic qualities that can be incorporated into both interior and exterior design and decor details of a building. The colors most commonly available are green or blue-green, gray, and bronze. These tints are available in sheet, float, and plate glass.

USES OF TINTED GLASS.   The low heat transmission of this glass, as well as its ability to reduce glare, suits it for a wide variety of applications. It can be employed to great advantage in southern and western exposures (and eastern, if desirable) of all types of buildings—schools, residences, factories, hotels, office buildings. It is particularly useful where the total glass expanses are large, particularly in situations where personnel are at work in close proximity to the glazing. It is also useful in applications where the admittance of excessive amounts of solar transmission might have a deleterious effect upon items directly behind the glazing, such as merchandise on display in a storefront.

Since several tint and thickness combinations of tinted glass are available, light and solar transmission characteristics can be tailored to a high degree to suit particular applications. Glazing selection can be further modified by using different tints, or tints and clear glass, together in double- or triple-glazing combinations. In such cases, however, considerable care must be taken to avoid edge damage and provide for thermal movement by way of careful and correct installation. Otherwise, thermal stresses will cause breakage. For the same reason, the glass must be protected from high temperature differentials, especially between centers and edges of the sheets, that might be caused by too-close heat outlets, drapes or venetian blinds positioned less than 2" from the glass surface and/or improperly ventilated, rapid warming by heating systems from low-temperature starts.

PHYSICS OF LIGHT.   The wavelength of visible light ranges from 350 to 750 millimicrons (0.35 to 0.75 microns; 1 micron equals one millionth of a meter), accounting for 46% of the total energy emitted from the sun.

TINTED GLASS

Another 49% of that radiation is in the infrared or heat range (750 to about 2,800 millimicrons). Below the visible light range (under 350 millimicrons) lies the ultraviolet band. Tinted glass is able to transmit high proportions of visible light while absorbing large proportions of the infrared radiation.

COLORS OF TINTED GLASS. The colors of tinted glass are hues of green or blue-green, gray, and bronze, which vary in depth or density with the thickness of the glass. The colors appear rather intense and "solid" when viewed from the exterior, especially from some distance, but they do not noticeably impinge upon the consciousness of a viewer looking out from the inside of the glazing, particularly after a short period of adjustment. There seems to be little or no distraction due to the coloring, except in the case of very thick or multiple-tint glazings designed to limit light and solar transmission severely (and consequently to limit vision considerably as well). Tinted glazings of the thinner variety allow incoming light to make colors appear natural, giving a quality of natural light that corresponds roughly to the artificial light from so-called daylight bulbs. It is neither glaring nor depressing, but instead rather soothing.

QUALITY. Tinted glass is produced in one quality level only—*Glazing* or *Select* (the two are synonymous). The grading requirements for this quality are the same as those for grading regular sheet, float, or plate glass.

Classification	Thickness	Lb. per Sq. Ft.	Maximum Size*	% Light Xmission	% Solar Xmission
	1/8″	1.64	35 sq. ft.	62	63
	3/16″	2.45	92 sq. ft.	51	53
Tinted Glass: Gray	1/4″	3.27	128″ x 204″	42	44
	3/8″	4.90	124″ x 200″	28	31
	1/2″	6.54	120″ x 200″	19	22
	1/8″	1.64	35 sq. ft.	68	65
	3/16″	2.45	92 sq. ft.	58	55
Tinted Glass: Bronze	1/4″	3.27	128″ x 204″	50	46
	3/8″	4.90	124″ x 200″	37	33
	1/2″	6.54	120″ x 200″	28	24
Tinted Glass: Green	3/16″	2.45	92 sq. ft.	78	55
	1/4″	3.27	128″ x 204″	74	48

*Maximum sheet sizes depend upon several factors and vary among different manufacturers.

STRENGTHENED GLASS

Plain float or plate glass as it comes directly from the basic manufacturing process is insufficiently strong to be fully useful. Several different processes are used to correct this situation, imparting sufficient strength to the glass to render it satisfactory for either ordinary or extraordinary service.

ANNEALED GLASS. After flat glass is formed, it goes through a process called annealing in order to become suitable for ordinary service as a glazing material. Unless this is done, the newly formed glass will cool at a nonuniform rate and is likely to fracture under normal temperature stresses encountered in ordinary glazing applications.

After the glass is formed, it moves through a long oven called a lehr, in a continuous strip. As the glass moves along, the temperature is first raised substantially to relieve stresses in the material, then lowered gradually and carefully under precise control. The glass cools slowly at a uniform rate over all parts until it eventually reaches room temperature, free from all internal strains. Cutting and processing is then easily accomplished. Standard flat glass is processed in this manner.

HEAT-STRENGTHENED GLASS. The strength of flat glass can be approximately doubled by subjecting it to a treatment known as heat-strengthening. This involves heating the glass to a temperature just below that at which it will soften, and then chilling it rapidly by subjecting it to washes of cold air on both sides. This quickly produces shrinking and cooling of the surfaces, followed more slowly by shrinking and cooling of the interior portions of the material. Since there is a differential in shrinking and cooling rates between the surfaces and the core of the glass, a "skin effect" is created whereby the surfaces are in compression and the core is in tension, and the two states are in balance. Thus, a partially tempered glass is created.

FULLY TEMPERED GLASS. The process used to create fully tempered glass is an extension of the heat-strengthening process. Fully tempered glass is commonly referred to as safety glass. Two methods are used; one places the glass sheets in a horizontal position, while in the other tongs suspend the sheets vertically. In either case there is a certain small amount of bow and warp and a consequent unavoidable degree of distortion. However, the amount is small and regulated by standards. Color, clarity, light transmission characteristics, and chemical composition remain unchanged. Fully tempered glass is four to five times as strong as ordinary annealed flat glass of the same thickness; it is highly resistant to impact and shock. If broken, it does not shatter, but instead

STRENGTHENED GLASS

crumbles into chunks somewhat like large pieces of rock salt; as a result, unlike the highly dangerous ordinary glass shards, broken tempered glass poses little hazard. Because of its nature, tempered glass cannot be cut or processed after tempering, since it would simply disintegrate. All processing, of whatever sort, must be done prior to tempering.

Standard sizes of clear tempered glass may vary somewhat, but typical dimensions are shown in the table. Larger sizes are available on special order. Some tinted glass types are also available in certain sizes of tempered glass, as are double insulating glass panels and standard sizes in either single or double glazing for residential sliding-glass or patio doors. Special shapes and sizes can be custom-tempered to suit particular applications, within limitations that vary with the item involved.

CHEMICALLY STRENGTHENED GLASS. Glass can also be strengthened to some degree by certain chemical treatments. One common process is to immerse the glass in a bath of molten potassium salt. This results in the small sodium ions on the surface of the glass being replaced by large potassium ions from the salt. This in turn leads to surface compression stresses and inner tension somewhat similar to, but less effective than, those produced by tempering. While the finished product is stronger than annealed glass, it is much weaker than tempered glass and is adversely affected by surface scratches or nicks, due to the thinness of the compression layer.

Classification	Thickness	Lb. per Sq. Ft.	Maximum Size*	% Light Xmission
	1/8"	1.64	42" x 76"	90
	5/32"	2.05	48" x 96"	89
	3/16"	2.45	60" x 96"	89
	1/4"	3.27	85" x 110"	88
	5/16"	4.08	90" x 110"	87
Clear Glass: Tempered	3/8"	4.90	96" x 120"	86
	1/2"	6.54	96" x 132"	84
	5/8"	8.17	96" x 132"	82
	3/4"	9.81	96" x 132"	81
	7/8"	11.45	74" x 110"	79

*Maximum sheet sizes depend upon several factors and vary among different manufacturers.

COMPOSITE
GLASS

TYPES.   There are two main types of composite glass: laminated and insulating. Insulating glass is composed of a factory-made assembly of two or three lights of glass with one or two sealed, dry-air spaces between them. They may be metal-edged or glass-edged. Laminated glass products may be subdivided into four main groups: general purpose, burglar resistant, bullet resistant, and acoustical. It should be noted, however, that there are many special uses for laminated glass not evident from these group titles. The principal feature of laminated glass is that it is highly resistant (though to different degrees, depending upon its construction) to impact and breakage; if broken, it will not shatter but will remain mostly within the opening. Perhaps the best-known and most common laminated glass product is the automotive safety-glass window. All laminated glass types can be made up from sheet, plate, or float glass, and may be clear, wired, heat-strengthened, fully tempered, tinted, or coated.

GENERAL-PURPOSE LAMINATED GLASS.   Ordinary laminated glass is composed of two layers of glass separated by a layer of clear plastic, such as polyvinyl butyral. The glass surfaces to which the plastic is bonded must not be patterned or imprinted—or anything else but fresh, clean glass. The plastic interlayer is generally 0.015" or 0.030" thick. The sheets may be made up of two lights of $3/32$" sheet glass (SS/SS), two lights of $7/64$" sheet glass (Lam/Lam), two lights of $1/8$" sheet glass (DS/DS), two lights of $1/8$" plate or float glass, or two lights of heavy plate or float glass. Thus, nominal thicknesses of the sheets range from approximately $13/64$" to 1". The framing system must sometimes be made up to suit, but some standard components are also available. Maximum sheet size varies with the specific product, and special sizes may be made to order.

BURGLAR-RESISTANT GLASS.   This is actually a variation on the heavier types of general-purpose laminated glass and can be made up in several different ways. One arrangement uses two lights of $1/8$" plate or float glass with an extra-thick interlayer of plastic (about 0.090"). The purpose is to make the glass nearly impossible to break by manual force or to cut through easily, as a safeguard against robbery or looting by smashing store or other windows. It is obviously equally useful wherever relatively inexpensive protection against casual but fairly severe manual impact is required.

BULLET-RESISTANT GLASS.   This type of laminated glass is composed of at least four lights of glass, of whatever kind or combination of kinds required, bonded together with three or more interlayers of plastic. The glass is generally $1/4$" plate or float, and the interlayering is usually 0.015" thick. Overall thicknesses of 1" to 2" are commonly available, and units of up to 7" or more in thickness can be obtained. Such glass panels (depending upon thickness) are proof against small-bore, medium- and high-power, and magnum firearms of all types. Manufacturers conduct tests, using various firearms, of the strength of bullet-resistant glass, and the results are available from them. Because of the nature of the construction of bullet-resistant glass, it can be made to virtually any specifications and degree of impact- and/or bullet-resistance. Though primarily used for bank teller windows and similar security applications, bullet-resistant glass can be used for other purposes as well, whenever the economics of the choice are justifiable.

ACOUSTICAL GLASS.   This type of laminated glass is available with various Sound Transmission Class (STC) ratings for different applications. The principal uses are for radio and television broadcasting studios, recording studios, office partitions, and the like, where clear vision is necessary along with a degree of soundproofing impossible to attain with ordinary glass lights. The lights generally consist of two or more layers of plate or float glass, interlayered with plastic 0.045" thick. Like other types of laminated glass, it can be (and usually is) tailor-made for different specific job applications.

REQUIREMENTS FOR
BENT GLASS

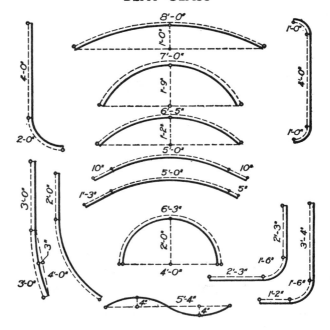

Almost any type or thickness of glass can be bent, including clear plate or float, tinted, obscure or patterned, wired, and Pyrex. Thermal insulating units can be made up, and safety glass can be fabricated into curved shapes. Certain kinds of reflective-coated glass can also be bent, but many cannot; applications of this sort require careful consideration. The process involves forming a sheet of manufactured glass to a mold or pattern by heating the glass until it softens and conforms properly, aided by gravity and the use of special equipment as required. The bent piece is then carefully reannealed and heat strengthened or fully tempered.

No piece of bent glass is free of minor defects. There is inevitably a departure from exact straightness, and typically the bend is made to a tolerance of plus or minus ⅛″ from the specified radius line. There is some distortion in the finished glass, and since the process of bending is a mechanical one, some minor tool abrasion is likely to be evident. Bent glass is not available in standard or stock sizes or forms but is custom-made to specification.

A template or pattern of the desired curve should always be submitted, even when regular curves are ordered. Glass should not be bent to a curve exceeding a half circle nor in obtuse angles approaching right angles, for such extreme curves greatly increase the risk of breakage and of injury to the glass surfaces. Where strongly curved or reverse-curved patterns are to be followed, in many cases a section of straight glass should be placed between the curved sections. Curved sections containing openings cannot be bent without risk of breakage. Requests for such pieces, as well as those of a particularly intricate nature, may be accepted for bending only with the understanding that the customer assumes the cost of pieces that may be broken during the bending process.

Some typical bends are shown in the drawing above, but do not indicate the immense range of shapes that can be obtained, including segments of ellipses, parabolas, and compound curves.

TYPES OF
MIRRORS

GLASS MIRRORS. Mirrors can be made of several different metals and plastics, but those normally used to produce true, undistorted reflections are made of glass. Though any glass can be used, most commercial mirrors are made from silvering or mirror grades of plate or float glass. Only small, inexpensive household mirrors are made of window glass. Most mirror stock is manufactured in ¼" thickness, especially in the larger sizes, but ⅛" and ³⁄₁₆" thicknesses are also in common use.

ORDINARY TYPE. Mirrors are made by passing the glass under a spray of silver nitrate and tin chloride to deposit the reflective surface. In ordinary types of mirrors, the silvering is then protected with a coating of mirror-back paint, varnish, or shellac. Though adequate for some applications, this type of silvering is easily damaged and subject to deterioration. The possible maximum size of these mirrors is governed by the size of the glass sheets available.

COPPER-BACK. Copper-back mirrors were developed to meet the demand for mirrors that would have high resistance against deterioration; they are in widespread use today. A heavy layer of silvering is first deposited on the glass, followed by an electroplated deposit of copper. If necessary, additional protective coatings or backings can be applied. For highest quality and longest service, particularly in demanding applications, copper-back mirrors should be specified. Where high strength is also a requirement, heat-strengthened or laminated glass may be obtained. Copper-back mirrors are made for all manner of applications, both structural and decorative, and can be custom-made to suit any design. The maximum size is limited to the glass sheet size available, usually about 84" x 120".

DURABILITY. Copper-back mirrors are very durable. Manufacturers offer various kinds of guarantees against defective workmanship and/or silver spoilage from ordinary climatic or atmospheric conditions, with free replacement in provable cases of product defectiveness. Under normal circumstances, this type of mirror can be expected to provide satisfactory service for many years. However, copper-back mirrors cannot be guaranteed when installed in conditions of excessive moisture, weather exposure, or other harsh environments beyond their design capabilities, or when mishandled or improperly installed.

BEVELING. The standard widths of bevels are ½" and 1¼", though other widths may be available or requested in custom designs. Polished edges, where the edges of the mirror stock are ground and polished to a slightly rounded contour, are also common and may be had alone or in conjunction with bevels.

WHEEL-CUTTING. Plate glass mirrors may have V-cut lines of any suitable design cut into the surface by a wheel. The cut surfaces may be either polished or unpolished, as desired.

DECORATIVE MIRRORS. Plate or float glass mirrors may be obtained in various colors, such as pink, gray, gunmetal, bronze, amber, smoke, and gold. They may also be gold-veined, marbleized, or otherwise treated for decorative effect. Designs of nearly any sort may be painted upon them, and etching and frosting are further possibilities. Many standard shapes, such as square or rectangular tiles, are available, and special shapes may be manufactured to order.

INSTALLATION
OF MIRRORS

MASTIC SET MIRRORS. At right is shown a method of setting mirrors with special mirror-setting mastic, useful when moldings or frames are objectionable because of the design. Ordinary mastic must not be used as it is likely to contain ingredients injurious to the mirror backing, and will also void any guarantee. Copper-back mirrors for mastic setting must be specified, and the manufacturer's instructions and recommendations for installation must be followed.

INSTALLATION OF MASTIC MIRRORS. When mirrors are to be installed against finished plaster, the plaster must be sound, firm, and thoroughly dry. Mirrors may be set over masonry as long as it is extremely smooth, with no projections that will penetrate the space allowed for mastic and potentially scratch the back of the mirror. The mastic must be applied in accordance with the manufacturer's instructions and product recommendations.

SUPPORT AND LEVELING OF MIRRORS. A bond coat of a composition that will ensure close affinity with the mastic must be applied to the wall. The mastic acts both as an adhesive and as a leveling medium, but the weight of the mirror must be supported by some structural method—such as by resting it on a chair rail, on a molding, or on metal inserts. The plan drawing at right shows grounds used for bearing to ensure the plumbness of adjacent mirrors, so that reflections of the room will be true. Single mirrors require no bearing grounds.

MIRRORS SET WITH WOOD MOLDINGS. Where the backs of the mirrors are likely to sweat, ventilation must be provided, as shown in the drawing below. All mirrors set in wood moldings are supported on the bottom edge by two tiny white pine blocks placed in the rabbet. Using felt in the setting of mirrors is never necessary; in fact, because of its ability to hold moisture, felt may do harm. The backs of all setting moldings and the rabbet should be *stained* black, not painted.

ROSETTES. Clean 3"-square white pine grounds, surfaced all sides, should be screwed to the wall with countersunk flat head screws for setting mirrors with rosettes. Drilled holes in mirrors should be located at least 2" from the edges. Grounds should be kept at least ¼" back of the mirror edges.

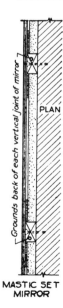

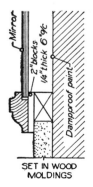

SET IN WOOD
MOLDINGS

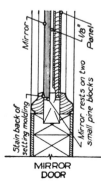

MIRROR
DOOR

MASTIC SET
MIRROR

DESCRIPTION OF
SEALED MULTIPLE GLAZING

Sealed multiple glazing is composed of two or more lights of glass separated from each other by ¼", ³⁄₁₆", or ½" of dehydrated air space and hermetically sealed at the edges; these panes provide high resistance to heat loss, reduce the radiant cooling effect, reduce street noises, cut down on condensation, and have many other advantages. Both metal-edged and glass-edged types are available.

Various combinations and types of glass may be used on special order. The glass sheet must have at least one smooth surface. Obtainable units can be composed of the following kinds of glass (among others), but it is wise to consult the manufacturer before deciding upon any given design:

Clear plate glass, two or three lights
Clear float glass, two or three lights
Clear and tinted glass, two or three lights
Clear and coated glass, two or three lights
Clear and diffuse or patterned glass
Tinted and coated glass
Tempered glass

Overall thermal conductivity or thermal resistance to heat loss will vary with the width of the air spaces between the pieces of glass that make up the unit. The standard widths for sealed multiple glazing air space are ¼", ³⁄₁₆", and ½". The chart below shows the value of U under conditions of an inside temperature of 70°F with natural convective air movement, and an outside temperature of 10°F with a 15 mph wind velocity. A convenient rule of thumb is that all standard double-pane insulating glass has a U of 0.60.

Air spaces 0.8" thick and greater show very little change in the curve (a fact determined by reputable testing laboratories). Panes with two air spaces and three sheets of glass have a $U = 0.47$.

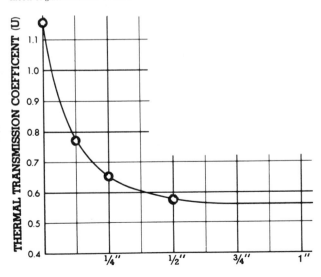

DEWPOINTS OF SEALED
MULTIPLE GLAZING

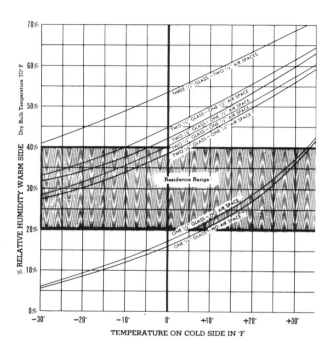

The chart above shows the conditions under which condensation will occur during cold weather on the inside of windows with single panes and with sealed multiple panes. The chart is calculated for free air movement of normal convection currents on the warm side. Condensation will occur at slightly higher temperatures than shown if air movement is restricted by curtains, shades, or other means.

The inside surface temperature of openings glazed with sealed multiple panes greatly reduces the amount of heat that must be supplied near such areas. This permits design flexibility, promotes physical comfort, allows visibility because of the absence of frost or mist, and eliminates the semiannual storm sash problem.

STRENGTH OF GLASS
IN WIND

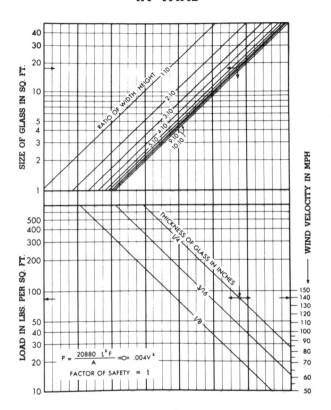

As an example of the use of this chart, suppose we wish to design an 18-square-foot glass panel with a width-to-height ratio of 7:10. Read across the upper part of the chart from the 18-square-foot size of glass to the diagonal line 7:10 for the ratio of width to height. At this intersection read down to the lower diagonal lines for the thickness of glass. The intersection with the diagonal line shows that a ¼″ thickness will stand 84 pounds per square foot on the left hand scale. Reading to the right from this intersection we find this is equivalent to 140 mph wind velocity.

SPANDREL GLASS

Spandrel glass is a special type of glass intended for use in covering the spandrel areas of buildings in place of or in conjunction with brick, stone, aluminum, precast concrete, or other materials commonly used to face the exterior walls of buildings. It is also used for cladding any opaque exterior wall area and is especially appropriate for re-covering old buildings in upgrading and modernization programs. Spandrel glass has numerous advantages, such as good energy efficiency, relative ease of installation, nonsusceptibility to corrosion or deterioration, ease and low cost of maintenance, excellent potential for architectural emphasis and impact, aesthetic appeal, and good uniformity of appearance.

Spandrel glass is float glass that has undergone a heat-strengthening process or is fully tempered. Two varieties are commonly available: glass coated with a metallic film and glass coated with a colored ceramic frit on the backside. Colors of metallic film-coated spandrel glass vary from one manufacturer to another, but generally they are in the gold, silver, copper, bronze, and blue or blue-green color groups. Frit-coated spandrel glass colors are generally more subdued—browns, bronzes, grays, and black—in the standard hues, but brighter colors are sometimes available, and special colors may be ordered. Spandrel glass is normally furnished in $\frac{1}{4}$" thickness; typically the maximum cut size is approximately 80" x 100", though larger sizes may be obtained (often on special order).

Spandrel glass may be used along with vision glass in such a way that the entire face of a building will present an acceptably uniform appearance. The spandrel glass is mounted with an air space behind it, allowing substantial introduction of diffused daylight. The light levels and the appearances of both spandrel and vision glass areas are thus quite similar. But as the contrast between interior and exterior lighting increases, the vision areas become visible.

In cases where the building substrate is visible or partially visible under normal exterior lighting conditions (a condition called read-through), a backup coating called an opacifier may be applied to the backside of the glass. This situation occurs with coated spandrel glass rather than frit-coated glass, which is largely opaque to begin with. However, when frit-coated spandrel glass is used for interior partitions, transoms, or similar purposes, a mottled effect and/or small clear-glass "pinholes" may be visible. In such cases, a backup material should be applied to enhance uniformity of appearance.

A totally uniform appearance such as would be expected of a silvered mirror or a painted surface is impossible to attain. The overall appearance of spandrel glass depends upon the uniformity of its reflectance, which cannot be made perfect. Disparities in exterior reflectance between vision and spandrel glass, for example, are inevitable. In addition, the degree of ripple, bow, and similar distortion of the reflected images in the spandrel glass is generally more obvious and pronounced than in vision glass, because of the effects of the heat treatment given to the spandrel glass. Thus, while numerous measures can be taken to assure a reasonably uniform appearance, perfection is not possible.

Properly installed, spandrel materials can be used to achieve excellent U-values and a good degree of thermal efficiency. For example, the combination of spandrel glass, a $\frac{1}{2}$" air space, and a backing of 2" rigid fiberglass insulation yields a U-value of 0.09 Btu/hr/ft²/°F. Another possibility is thermal insulation directly applied to the backside of the spandrel glass. In this instance, however, delamination and loss of efficiency of the insulation are a possibility because of moisture penetration and subsequent condensation. Other uses exist for high U-value spandrel areas; manufacturers should be consulted.

Both metallic film-coated and frit-coated spandrel glass may be obtained with an applied scrim backing. The scrim is a glass fiber-reinforced fabric adhered to the backside of the spandrel glass sheets. Its purpose is to provide an extra measure of safety in the event of glass breakage—the scrim tends to hold the glass fragments together and keep them from flying about (it also serves to increase the opacity of the glass). Such products should meet the requirements of Section 08810 of the Federal Construction Guide Specification for Safety Consideration. Architects should recognize the possibility that scrim backing might keep the entire sheet of spandrel glass intact, creating a decidedly dangerous situation if the sheet were to break free from its framing and scale away.

Various "standard" and proprietary glazing products and systems may be obtained for installation of spandrel glass; manufacturers should be consulted for complete details and specifications. In order to avoid breakage of spandrel glass sheets, the installation must be well engineered and competently done, using the proper materials. The causes of spandrel glass breakage include: too great wind loads; thermal stress caused by too rapid heat buildup and/or improper ventilation; an improperly designed glazing system, which can cause pressures and stresses on the glass; damage during handling and installation of the panels, resulting in either immediate or delayed breakage; improper protection of the glass edges; imperfections in the glass; and outside forces such as falling or flying objects. Though some breakage in a large installation is inevitable, it can be greatly minimized by calling for correct specifications and proper and thorough engineering, and by following the manufacturer's guidelines for installation.

DIMENSIONS AND SHAPES
OF GLASS BLOCK

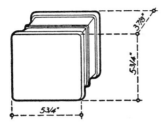

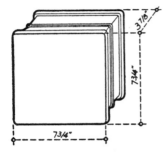

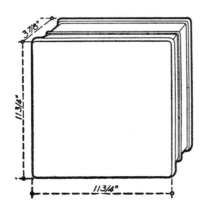

Not all patterns of glass block are available in all sizes. Though the three sizes represented here are probably the ones most commonly used, others are available, including 4½" x 4½", 7½" x 7½", and 9½" x 9½" in square blocks, and 3¾" x 7¾", 5¾" x 7¾", and 4½" x 9½" in rectangular blocks. The manufacturer of the block to be used should be consulted on types, patterns, and installation accessories (anchors, reinforcing, oakum, and so on) to find the material that is currently available, since changes do occur from time to time.

DESCRIPTION AND PROPERTIES
OF GLASS BLOCK

TYPES AND PATTERNS.   Several types of glass block are available: light-diffusing, light-directing, light-attenuating, vision, solid glass, decorative, and solar reflective. Various patterns are obtainable in one or more of the types, including decorative imprints, rounded flutes or ripples, crosshatching, pebbling, reflective finish, and multiple triangles. One type has white opal glass fused to the inside sidewalls; another has black glass similarly placed; a third contains fibrous glass diffusion inserts; still others are tinted.

CRUSHING STRENGTH.   Glass block panels should never be used to carry loads other than their own superimposed weight within the limits of allowable panel sizes. Glass blocks have unusual strength in compression (typically 400-600 pounds per square inch for standard blocks), but such factors as nonuniform distribution of load prevent their being used as a load-bearing material.

BOND TO MORTAR.   The edges of glass blocks are made in such a way as to provide a strong mechanical bond between the cement mortar and the blocks.

HEAT INSULATION.   One of the advantages of glass block construction over single-glazed windows is the former's superior heat insulation because of the dead air space within the blocks. Depending upon size and construction of the type of block, U-values range from about 0.64 to 0.44 (R-1.56 to R-2.27). For comparison, $\frac{1}{8}$" sheet glass is rated at U-1.04 (R-0.96) and 3" solid glass block at U-0.87 (R-1.15).

SURFACE CONDENSATION.   Tests show that moisture will not condense on the warm side of glass block panels in normal use, even under conditions of extreme exposure. In those special industries or cases where inside temperatures and humidities are higher than normal, humidities considerably greater than those possible with single-glazed sash can exist before condensation will form.

WIND RESISTANCE.   From tests on many glass block panels, it has been found that any panel (within the limits of areas recommended) will withstand a safe load of 20 lbs/ft^2 with a factor of safety of at least 2.7.

SOUND INSULATION.   Glass block panels typically have a sound transmission loss value of 38 to 45 decibels and will improve the acoustics of rooms where they replace single-glazed sash because of their insulating properties against *transmitted* sound.

LIGHT TRANSMISSION.   With the exception of certain decorative and solar-control types, glass blocks are made of clear, colorless glass, admitting light of full daylight tone. With proper pattern selection, the light and decorative effect can be controlled within a wide range.

SOLAR HEAT GAIN.   The use of glass block for light-transmitting areas results in a marked reduction in the total solar heat gain as compared with conventional windows. This factor is of considerable advantage in air-conditioned buildings. However, it does not eliminate the need for adequate ventilation or shading in rooms that are not air-conditioned.

The maximum hourly rate of relative heat gain through glass block depends entirely upon the specific type of block selected. Typically, the design values range from approximately 42 to 70 Btu/hr/ft^2 for solar reflective types, and on the order of 215 Btu/hr/ft^2 for clear glass vision types, in nominal 8" x 8" block size. However, much depends upon test conditions: specific data and all parameters should be obtained from manufacturers before computations are made.

DESCRIPTION AND PROPERTIES
OF GLASS BLOCK

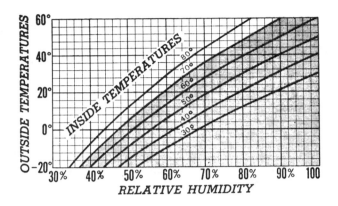

INTERIOR SURFACE CONDENSATION. For the average installation, the accompanying chart gives outside temperatures required to produce surface condensation on the inside surface of a glass block panel. The values shown are appropriate for standard nominal 8″ x 8″ blocks. They must be adjusted somewhat for different sizes; for a given set of inside conditions, the outside temperature that will cause condensation is lower for larger blocks, higher for smaller ones.

WIND RESISTANCE. Any panel with the area limits recommended by the manufacturers will withstand a safe load of 20 pounds per square foot with a factor of safety of at least 2.7. Tests at Purdue University on a panel 7′-3″ wide by 8′-8″ high showed that the panel is entirely elastic under repeated loadings with a pressure of 40 pounds per square foot, corresponding to a wind velocity of about 100 mph.

BUILDING CODES. Most building authorities' requirements for strength, wind resistance, fire and hose stream resistance, and other properties are fully met if the manufacturer's recommendations are followed. In some locales, however, smaller panel sizes and/or additional reinforcement may be required. Consult local authorities during the planning stage.

SOUND REDUCTION. Glass block panels are considerably more effective than most types of fenestration in reducing transmitted sounds, especially in comparison with most normal (not "sound wall") types of partition construction.

INSTALLATION OF
GLASS BLOCK

EXTERIOR INSTALLATION.   Glass blocks have unusual strength in compression but such factors as nonuniform distribution of load forbid their use as a load-bearing building material. The basic principles of installation are to provide (1) complete freedom of movement of the panel within the enframing construction, and (2) proper anchorage of the panel at head, sill, and jambs.

Basically the installation procedure is as follows:

1. The sill is coated with asphalt emulsion as a bond breaker.

2. Resilient expansion strips are placed around the perimeter of the panel opening except at the sills.

3. Blocks are laid with full mortar beds, using a preferred mix of 1 part portland cement, ¼ part lime, and 3 parts sand, or a prepared masonry mortar mix of low volume change.

4. Joint reinforcement is placed at 24″ intervals on the horizontal joints, in most cases, and in joints above and below all openings in the panel.

5. If glass block panels are not set in chases, crimped wall anchors are built from the enframing construction into the horizontal mortar joints of the glass block panels; crimping permits movement in the plane of the wall, but supports the panel against wind pressure.

6. Mortar joints are tooled.

7. Oakum is packed at jamb and head, if recessed, and at internal supports, for cushioning. The perimeter of the panel is caulked on the exterior and interior. Caulk must be a nonstaining, waterproof mastic.

LIGHT TRANSMISSION.   Percentage figures for light transmission do not convey an accurate impression of the physical light-transmitting capabilities of glass block. For example, light-directing units produce the least illumination on overcast days and the highest percentage of any block pattern on sunny days in direct sunlight. However, such figures are useful for comparative purposes and certain calculations. These values are generally accepted as accurate light transmission data under standard testing procedures: solid glass block, approximately 80%; view types of glass block, approximately 75% to 85%; diffusion types of blocks, approximately 28% to 40%; solar reflective types of glass block, approximately 5% to 20%. Values for specific models of blocks can be obtained from the manufacturer.

LIGHT-DIRECTING
GLASS BLOCK

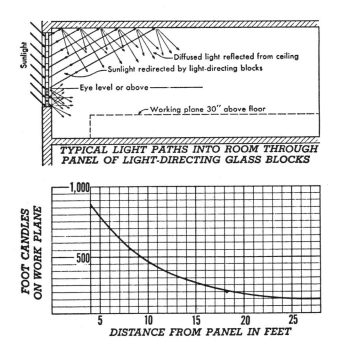

Sunlight

Diffused light reflected from ceiling

Sunlight redirected by light-directing blocks

Eye level or above

Working plane 30" above floor

TYPICAL LIGHT PATHS INTO ROOM THROUGH PANEL OF LIGHT-DIRECTING GLASS BLOCKS

FOOT CANDLES ON WORK PLANE

1,000

500

DISTANCE FROM PANEL IN FEET

5   10   15   20   25

Laboratory tests show unshaded single glazed steel sash to transmit 4900 fc at 4'-0" on the working plane. Intensity falls sharply to 900 foot candles at 8'-0". Intensity falls again to 500 foot candles at 13'-0". Intensity at 28'-0" is about 240 foot candles. All tests were made on a sunny day with an outdoor intensity 7200 fc.

DESCRIPTION. Light-directing glass blocks depend upon the optical refraction of light produced by horizontal prisms pressed into the interior faces. This refraction redirects the incident light upward where the ceiling reflection (especially if painted in a light color) helps to cast the illumination farther into the remote areas of the room. A uniform curve of illumination is obtained, and at the same time no excessive brightness-contrast is created, if the blocks are installed above eye-level.

SIZES AND SHAPES. Prismatic glass block can be obtained in several square and rectangular standard sizes; it is installed in the same manner as diffusion or reflective block.

GLASS BLOCK PANELS

SIZE AND SHAPE. Glass block panels may be square or rectangular, straight or curved. Minimum radii for curved panels are listed on page 76; there is no maximum radius. Panels may constitute an entire full-height wall, or the upper or lower portions in any arrangement, and may surround openings such as for doors, clear-glassed areas, or "pass-throughs." The maximum height for exterior panels is typically 20', the maximum width 25', and the maximum area 250 sq. ft. for panels securely braced and held by steel head and jamb, or 144 sq. ft. for other constructions. However, if the panel is held only by panel anchors, the maximums are 10' height, 10' width, and 100 sq. ft. area. For interior panels, the maximum height is 25' but width and area are variable. Panels constructed of blocks of less than standard thickness must be commensurately reduced in size. In any case, local building codes should be checked for possible limitations on panel sizes. Panels may be joined.

AIR LEAKAGE. Properly installed, a panel constructed of any type of glass block should permit no appreciable amount of air leakage, and calculations for infiltration may ordinarily be neglected. Entrance of dust and dirt is eliminated.

SECURITY. Standard glass block panels are remarkably strong and durable and cannot easily be broken through. If an outer surface is broken, normally the inner surface will remain intact; a new block can be installed readily. Solid glass blocks are virtually indestructible, and will even withstand rifle fire at close range. All glass block panels are fire resistant, and even view types render vision from the outside inward inaccurate and obscured.

MAINTENANCE. Glass block panels require virtually no maintenance beyond periodically hosing off the outside and cleaning the inside surfaces with a damp cloth or window cleaner.

LIGHT CONTROL. Glass block panels may be used for light control. Prismatic blocks will direct light from the outside into a room, diffusion blocks will reduce outside light and control glare, and solar reflective blocks will control light, heat, and glare coming from the outside. Used inside, view or diffusion blocks can be used to create softer, more even lighting in interior spaces without fostering a closed-in, boxy effect. Different types of blocks may be used in concert in the same panel or panel series to create various effects.

Since diffused light incident on a north exposed wall is of lower intensity than direct sunlight and is not refracted regularly, prismatic blocks cannot be used successfully on the north side of buildings. Since sunlight does not reach a north wall, solar reflective blocks are similarly ineffective in that location. Neither type is used in interior partition panels, for the same reason, unless chosen for decorative impact.

On east, south, and west sides of the building, and the light-directing blocks are used above eye level. Below eye level, a diffusing block or one of similar appearance is used. Where partial vision is required, a view block is installed from eye level downward to any desired height from the floor. Decorative blocks may be installed at any location appropriate to the overall design effect and to the block type and pattern.

GLASS BLOCK
PANELS

	Outside	East	West	South			
Sun Time	Air Temp. F°			North Latitude Degrees			
		30° to 45°		30°	35°	40°	45°
7:00	74	61.0	—	−4.5	−2.0	−0.5	1.0
8:00	76	77.5	—	0.0	2.0	4.0	5.0
9:00	79	73.5	5.0	5.0	7.0	10.0	12.0
10:00	83	57.5	6.5	11.0	15.0	18.0	20.8
11:00	87	45.0	7.5	16.5	22.0	25.5	32.0
12:00	90	36.5	10.5	21.5	28.0	33.8	40.8
1:00	93	30.0	22.0	25.0	31.8	38.5	46.0
2:00	94	24.0	35.0	26.0	32.0	39.0	47.0
3:00	95	19.5	55.0	24.0	29.8	36.5	45.0
4:00	95	15.5	77.0	20.0	25.5	31.5	40.5
5:00	93	12.5	85.5	15.0	20.0	25.2	33.5
6:00	91	10.5	55.0	9.5	13.5	18.0	25.5
7:00	89	8.0	18.5	3.5	7.0	11.0	18.0

HEAT GAIN THROUGH GLASS BLOCKS FOR AUGUST 1st
Btu per Sq. Ft. per Hour
(Solar Radiation Plus Normal Transmission — Inside Temperature 78° F.)

SOLAR HEAT GAIN. The above table is representative of solar heat-gain values for standard glass block. The approximate percentage transmission of solar heat according to an ASHRAE test conducted on August 25th is 27.1% for south exposure. This compares with single glazing in steel sash for south exposure of 56.5%. Thus, in that case the glass block admitted about half of the solar heat of single glazing.

The thermal performance of glass block depends upon several factors, including block size, opacity of the glass, U-value, shading coefficient, and coatings (if any). Valid comparisons can be made only if the different types of blocks being tested are subjected to the same test conditions and parameters. However, it is safe to say that the relative heat gain of view-type block is approximately one third less than that of $\frac{1}{8}''$ flat glass, while that of solar reflective block ranges from about one quarter to about one third the relative heat gain of $\frac{1}{8}''$ flat glass, depending upon the coatings applied to the block. Specific values may be obtained from manufacturers.

Three methods may be used to control solar heat input further: shading, ventilation, and air conditioning. Curved panels should be placed with care, since the center of curvature may create a point of solar heat focus.

When cooling loads are to be calculated, a data base of values that is as complete and accurate as possible should be employed if the results are to be satisfactory. This involves careful assessment of the environmental conditions at the building site, as well as the use of proper block values obtained from the manufacturer, in support of the particular block panel designs being considered.

THERMAL CONDUCTIVITY. Tests by both Purdue University and Pittsburgh Testing Laboratory on panels of 8″ x 8″ glass blocks give the overall transmission coefficients U in Btu/ft²/hour/°F approximately as follows:

Still air 0.42
15 mph wind 0.49

However, the U-values of glass blocks differ with the type and size of block. As noted previously, the range can be from 0.87 (R-1.15) or more to as little as 0.44 (R-2.27), or perhaps a bit less, not including the surface film factors. The manufacturer should be consulted for specific values for blocks being considered.

WEATHERING. Tests in which large panels of glass blocks have been subjected to a 15 mph wind-driven water spray for 8 hours, followed by 16 hours of spray without wind, have shown that the construction is capable of withstanding severe conditions of storm exposure without water penetration. Large test panels exposed to repeated cycles of alternate water spray and freezing (at temperatures down to −30° F) have withstood this treatment without evidence of leakage, cracking, or other structural deterioration.

CURVED WALLS OF GLASS BLOCK

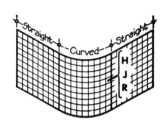

Blocks Needed:
90° Curved Panel

o.s. radius	number
6″ Square Blocks	
52½″	13
56¼″	14
56¾″	14
60″	15
61″	15
63¾″	16
65″	16
67½″	17
69″	17
71¼″	18
73″	18
8″ Square Block	
69″	13
74″	14
74¾″	14
79″	15
80″	15
84″	16
85¼″	16

Combinations of flat and curved panels forming integral glass block areas can be installed in the manner described for the respective limitations shown on detail sheets. However, curved areas should be separated from flat areas by means of intermediate expansion joints and supports, as indicated on the small diagram above. For intermediate expansion joints and supports, see detail sheets.

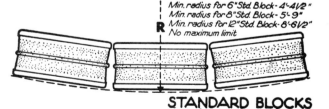

Min. radius for 6″ Std. Block- 4′-4½″
Min. radius for 8″ Std. Block- 5′-9″
Min. radius for 12″ Std. Block- 8′-6½″
No maximum limit

STANDARD BLOCKS

Standard or thin blocks (either square or rectangular) of any size can be used to construct curved panels. The diagram shows the limits resulting from a recommended minimum thickness of ⅛″ and a maximum thickness of ⅝″. The table lists typical radii arrangements and the number of blocks required for one course, using the recommended joint-thickness limitations. Notice that half of each listed radius for the 6″ blocks can be attained by using 3″ x 6″ blocks stacked vertically, and half of each listed radius for the 8″ blocks can be attained by using 4″ x 8″ or 4″ x 12″ blocks stacked vertically.

KEY TO REQUIREMENTS
FOR EXTERIOR PANELS

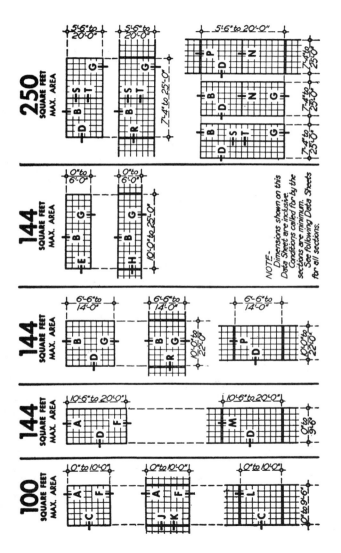

ANCHORS, TIES, CUSHIONS

EXPANSION STRIPS

"Fiberglas" 2'-1" — ½" — 4⅛"
Cork 3'-0" — ¼", ½" — 3'-4"

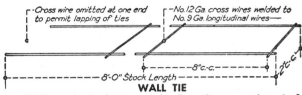

Cross wire omitted at one end to permit lapping of ties

No.12 Ga. cross wires welded to No. 9 Ga. longitudinal wires

8"c-c.

8'-0" Stock Length

WALL TIE

Wall Ties are made of galvanized wire. For continuous use lap ends of wall ties 6"min. Wall ties must not bridge expansion joints and shall run from end to end of panels.

No.20 Gauge galvanized Wall Anchor 1-3/4" wide x 2'-0" long

Wall anchors must be crimped within expansion joint. When this type of lateral support is employed anchors should occur in same joints as Wall Ties

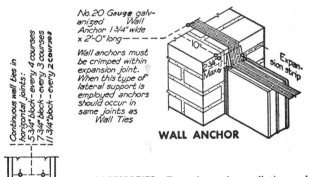

WALL ANCHOR

Continuous wall ties in horizontal joints:
5-3/4" block - every 4 courses
7-3/4" block - every 3 courses
11-3/4" block - every 2 courses

¼" Joint

WALL TIE DATA

ACCESSORIES. Expansion strips, wall ties, and wall anchores have been especially developed for the erection of glass block panels.

EXPANSION STRIPS. *Glass Blocks* are non-load bearing. Adequate provision must be made for support of construction above glass block panels. Expansion is provided for by the use of expansion strips at the jambs and head. These expansion strips are premolded in sizes shown above.

WALL TIES. The lower half of mortar bed is first placed without furrowing. Then the wall tie is placed and covered with the upper half of the mortar bed without furrowing.

WALL ANCHORS. In new structures wall anchors should be built into the wall 10". In existing construction wall anchors may be attached with 4" expansion bolts.

EXTERIOR PANELS
HEADS A, B

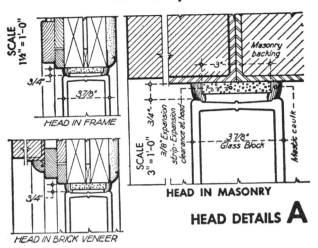

SCALE 1½"=1'-0"

3/4"

HEAD IN FRAME

3/4"

HEAD IN BRICK VENEER

SCALE 3"=1'-0"

Masonry backing

3"

3/4"

3/8" Expansion strip-Expansion clearance at head

3 7/8" Glass Block

Mastic caulk

HEAD IN MASONRY

HEAD DETAILS A

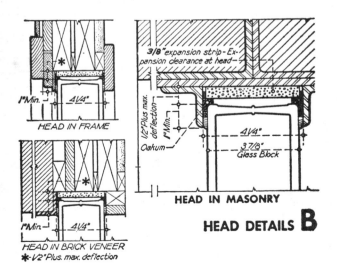

1"Min.

4¼"

HEAD IN FRAME

1"Min.

4¼"

HEAD IN BRICK VENEER
*-½"Plus. max. deflection

3/8" expansion strip - Expansion clearance at head

½"Plus max. deflection

Oakum

1" Min.

4¼"

3 7/8" Glass Block

HEAD IN MASONRY

HEAD DETAILS B

79

EXTERIOR PANELS
JAMBS C, D, E

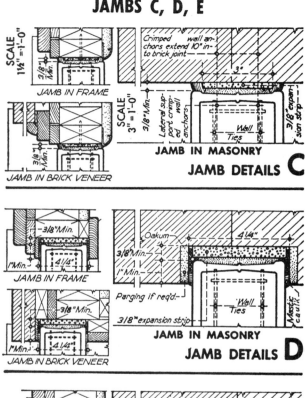

SCALE 1½" = 1'-0"

3/8" Min.

JAMB IN FRAME

JAMB IN BRICK VENEER

SCALE 3" = 1'-0"

3/8" Min.

Lateral support crimped wall anchors

Crimped wall anchors extend 10" into brick joint

3"

Wall Ties

3/8" expansion strip

JAMB IN MASONRY

JAMB DETAILS C

−3/8" Min.

4¼"

1" Min.

JAMB IN FRAME

−3/8" Min.

4¼"

1" Min.

JAMB IN BRICK VENEER

Oakum

4¼"

3/8" Min.

1" Min.

Parging if req'd.

3/8" expansion strip

Wall Ties

Mastic caulk

JAMB IN MASONRY

JAMB DETAILS D

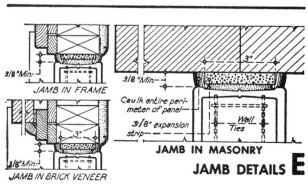

3/8" Min.

JAMB IN FRAME

3"

3/8" Min.

JAMB IN BRICK VENEER

3/8" Min.

Caulk entire perimeter of panel

3/8" expansion strip

3"

Wall Ties

JAMB IN MASONRY

JAMB DETAILS E

80

EXTERIOR PANELS
SILLS F, G

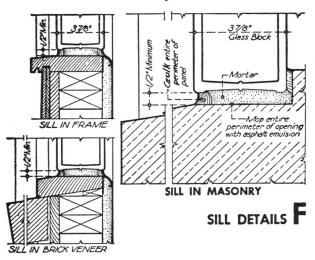

SILL IN FRAME

1/2" Min.

3 7/8"

1/2" Minimum

Caulk entire
perimeter of
panel

3 7/8"
Glass Block

Mortar

Mop entire
perimeter of opening
with asphalt emulsion

SILL IN BRICK VENEER

SILL IN MASONRY

SILL DETAILS F

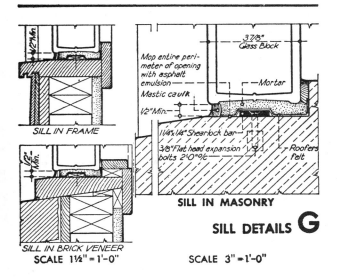

SILL IN FRAME

Mop entire peri-
meter of opening
with asphalt
emulsion

Mastic caulk

1/2" Min.

1 1/4"x 1/4" Shearlock bar

3/8" Flat head expansion
bolts 2'-0" o/c

3 7/8"
Glass Block

Mortar

Roofers
felt

SILL IN BRICK VENEER
SCALE 1½" = 1'-0"

SILL IN MASONRY

SILL DETAILS G

SCALE 3" = 1'-0"

81

EXTERIOR PANELS
INTERMEDIATES H, J, K

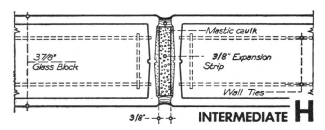

Intermediate H:
- Mastic caulk
- 3 7/8" Glass Block
- 3/8" Expansion Strip
- Wall Ties
- 3/8"

INTERMEDIATE H

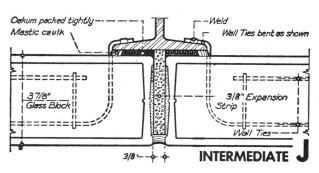

Intermediate J:
- Oakum packed tightly
- Mastic caulk
- Weld
- Wall Ties bent as shown
- 3 7/8" Glass Block
- 3/8" Expansion Strip
- Wall Ties
- 3/8"

INTERMEDIATE J

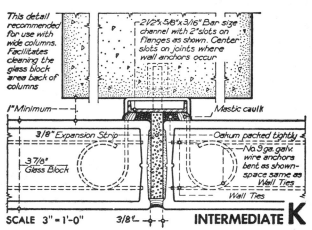

This detail recommended for use with wide columns. Facilitates cleaning the glass block area back of columns

2 1/2"x 5/8"x 3/16" Bar size channel with 2" slots on flanges as shown. Center slots on joints where wall anchors occur

1" Minimum

Mastic caulk

3/8" Expansion Strip

3 7/8" Glass Block

Oakum packed tightly

No. 9 ga. galv. wire anchors bent as shown space same as Wall Ties

Wall Ties

SCALE 3"=1'-0"   3/8"

INTERMEDIATE K

EXTERIOR PANELS
INTERMEDIATES L, N, M, P

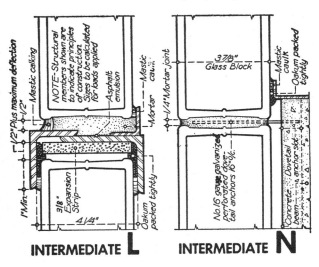

SCALE 3" = 1'-0"

INTERMEDIATE L

NOTE—Structural members shown are to indicate principles of construction. Sizes to be calculated for loads applied

Mastic calking

Mastic calk

Asphalt emulsion

Mortar

1/2" Plus maximum deflection

1/2"

1" Min.

3/8" Expansion Strip

4 1/4"

Oakum packed tightly

INTERMEDIATE N

3 7/8" Glass Block

1/4" Mortar joint

Mastic caulk

Oakum packed tightly

No. 16 gauge galvanized perforated dove-tail anchors 16% c.

Concrete beam

Dovetail anchor slot

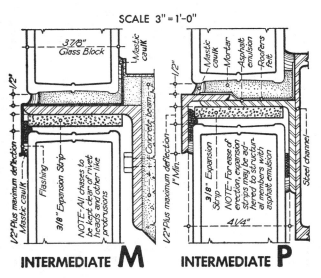

INTERMEDIATE M

3 7/8" Glass Block

Mastic caulk

1/2"

1/2" Plus maximum deflection

Mastic caulk

Flashing

3/8" Expansion Strip

NOTE—All chases to be kept clear of rivet heads and other like protrusions

Concrete beam

INTERMEDIATE P

Mastic caulk

Mortar

Asphalt emulsion

Roofers felt

1/2"

1/2" Plus maximum deflection

1" Min.

3/8" Expansion Strip

NOTE—For ease of erection, expansion strips may be ad-hered to structur-al members with asphalt emulsion

4 1/4"

Steel channel

EXTERIOR PANELS
INTERMEDIATES R, S, T

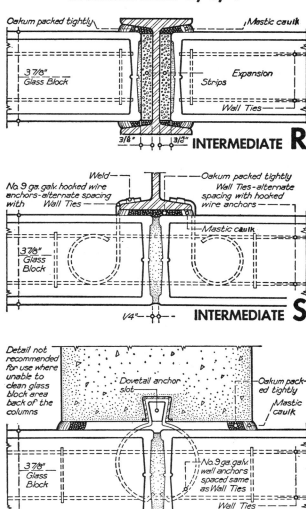

Oakum packed tightly — Mastic caulk

3 7/8" Glass Block — Expansion Strips — Wall Ties

3/8" · 3/4" — **INTERMEDIATE R**

Weld — Oakum packed tightly
No. 9 ga. galv. hooked wire anchors-alternate spacing with Wall Ties — Wall Ties-alternate spacing with hooked wire anchors — Mastic caulk

3 7/8" Glass Block

1/4" — **INTERMEDIATE S**

Detail not recommended for use where unable to clean glass block area back of the columns — Dovetail anchor slot — Oakum packed tightly — Mastic caulk

3 7/8" Glass Block — No. 9 ga. galv. wall anchors spaced same as Wall Ties — Wall Ties

SCALE 3" = 1'-0"   1/4" — **INTERMEDIATE T**

INTERIOR PANELS
100 AND 144 SQ. FT. MAX.

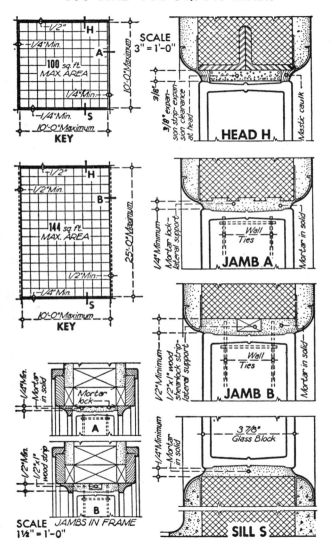

SCALE 3" = 1'-0"

KEY

100 sq. ft. MAX. AREA

1/2" H
1/4" Min. A
1/4" Min.
1/4" Min. S
10'-0" Maximum
10'-0" Maximum

KEY

144 sq. ft. MAX. AREA

1/2" H
1/2" Min. B
1/2" Min.
1/4" Min. S
10'-0" Maximum
25'-0" Maximum

1/4" Min.
Mortar in solid
Mortar lock
A
1/2"x1" wood strip

1/2" Min.
1/2"x1" wood strip
B

SCALE JAMBS IN FRAME
1½" = 1'-0"

3/8" expansion strip-expansion clearance at head
Mastic caulk

HEAD H

1/4" Minimum
Mortar lock-lateral support
Wall Ties
Mortar in solid

JAMB A

1/2" Minimum
1/2"x1" wood sheanlock strip-lateral support
Wall Ties
Mortar in solid

JAMB B

1/4" Minimum
Mortar in solid
3 7/8" Glass Block

SILL S

85

INTERIOR PANELS
250 SQ. FT. MAX.

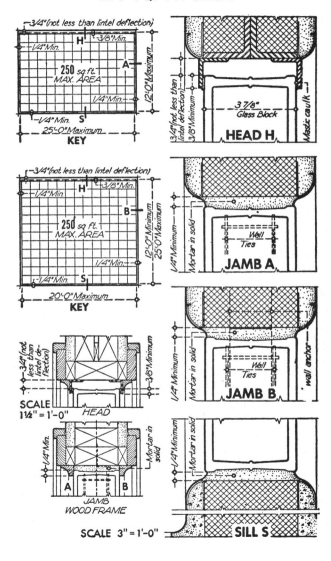

KEY

-3/4"(not less than lintel deflection)

-3/8"Min.

H

1/4"Min.

250 sq ft. MAX. AREA

A

12'-0" Maximum

1/4"Min.

-1/4" Min. S

25'-0" Maximum

HEAD H

3/4"(not less than lintel deflection)

3/8" Minimum

3 7/8" Glass Block

Mastic caulk

KEY

-3/4"(not less than lintel deflection)

1/4"Min.

3/8"Min.

H

B

250 sq ft. MAX. AREA

12'-0" Minimum / 25'-0" Maximum

1/4"Min.

1/4"Min. S

20'-0" Maximum

JAMB A

1/4" Minimum

Mortar in solid

Wall Ties

JAMB B

1/4" Minimum

Mortar in solid

Wall Ties

Wall anchor

HEAD

3/4"(not less than lintel deflection)

3/8" Minimum

SCALE 1½" = 1'-0"

JAMB WOOD FRAME

1/4" Min.

Mortar in solid

A   B

SCALE 3" = 1'-0"

SILL S

1/4" Minimum

Mortar in solid

INTERIOR PANELS
250 SQ. FT. MAX.

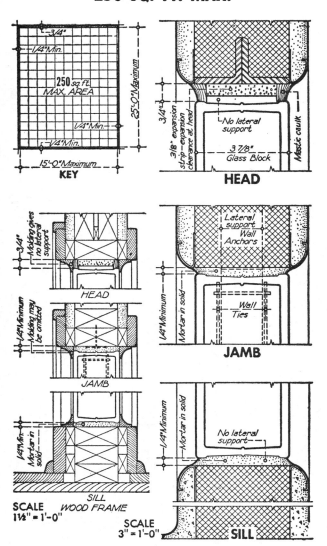

KEY

250 sq ft. MAX. AREA

3/4"

1/4" Min.

25'-0" Maximum

15'-0" Maximum

1/4" Min.

1/4" Min.

HEAD

3/4"

3/8" expansion strip-expansion clearance at head

No lateral support

3 7/8" Glass Block

Mastic caulk

HEAD

3/4"

Molding gives no lateral support

JAMB

1/4" Minimum

Molding may be omitted

SILL

1/4" Min. Mortar in solid

SCALE 1½" = 1'-0"

WOOD FRAME

JAMB

1/4" Minimum

Mortar in solid

Lateral support Wall Anchors

Wall Ties

SILL

1/4" Minimum

Mortar in solid

No lateral support

SCALE 3" = 1'-0"

87

METAL FRAMES ADJACENT TO GLASS BLOCK PANELS

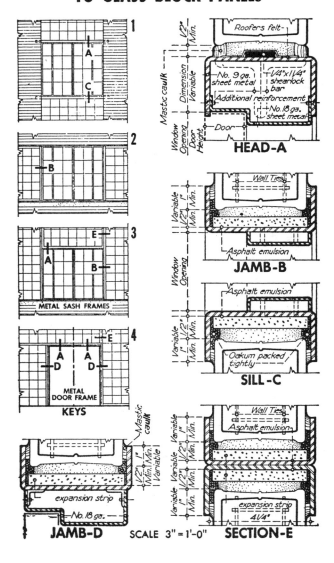

KEYS

1

2

3 — METAL SASH FRAMES

4 — METAL DOOR FRAME

HEAD-A

Roofers felt

Mastic caulk

Dimension Variable

No. 9 ga. sheet metal

1/4"x1/4" shearlock bar

Additional reinforcement

No. 18 ga. sheet metal

Window Opening

Door Height

Door

JAMB-B

Wall Ties

Variable 1/2" Min. 1" Min.

Asphalt emulsion

Window Opening

SILL-C

Asphalt emulsion

Variable 1/2" Min. 1" Min.

Oakum packed tightly

JAMB-D

Mastic caulk

1/2" Min. 1" Min. Variable

expansion strip

No. 18 ga.

SECTION-E

Wall Ties

Asphalt emulsion

Variable 1/2" Min. 1" Min. Variable

Variable 1" Min.

expansion strip

4 1/4"

SCALE 3" = 1'-0"

88

WOOD FRAMES IN
GLASS BLOCK PANELS

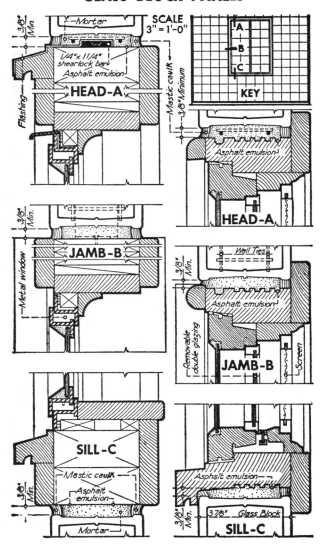

SCALE 3" = 1'-0"

HEAD-A

Mortar

3/8" Min.

1/4" x 1 1/4" shearlock bar
Asphalt emulsion

Flashing

Mastic caulk

3/8" Minimum

KEY

HEAD-A

Asphalt emulsion

JAMB-B

3/8" Min.

Metal window

JAMB-B

Wall Ties

3/8" Min.

Asphalt emulsion

Removable double glazing

Screen

SILL-C

Mastic caulk

Asphalt emulsion

3/8" Min.

Mortar

SILL-C

Asphalt emulsion

3/8" Min.

3 7/8" Glass Block

89

WOOD FRAMES IN
GLASS BLOCK PANELS

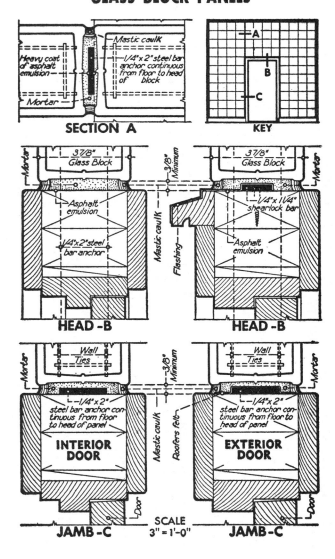

SECTION A

Mastic caulk

Heavy coat of asphalt emulsion

1/4"x 2" steel bar anchor continuous from floor to head of block

Mortar

KEY

A

B

C

HEAD -B

3 7/8" Glass Block

Mortar

Asphalt emulsion

1/4"x 2" steel bar anchor

3/8" Minimum

Mastic caulk

Flashing

HEAD -B

3 7/8" Glass Block

Mortar

1/4"x 1 1/4" shearlock bar

Asphalt emulsion

JAMB -C

Wall Ties

Mortar

1/4"x 2" steel bar anchor continuous from floor to head of panel

INTERIOR DOOR

Door

3/8" Minimum

Mastic caulk

Roofers felt

JAMB -C

Wall Ties

Mortar

1/4"x 2" steel bar anchor continuous from floor to head of panel

EXTERIOR DOOR

Door

SCALE
3" = 1'-0"

STEEL FRAMES IN
INTERIOR PANELS

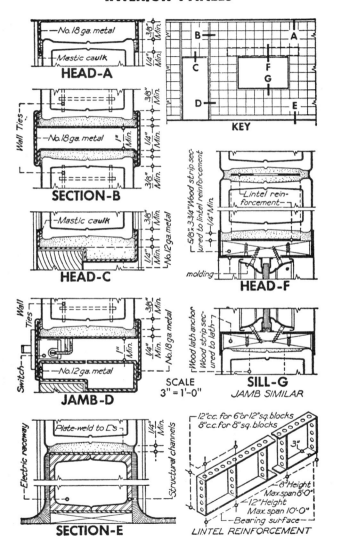

No.18 ga. metal · 3/8" Min. · 1/4" Min.
Mastic caulk
HEAD-A

Wall Ties · No.18 ga. metal · 3/8" Min. · 1/4" Min. · 1" Min. · 3/8" Min.
SECTION-B

Mastic caulk · 3/8" Min. · 1/4" Min. · No.12 ga. metal
HEAD-C

B · A
C · F · G
D · E
KEY

5/8"x 3-3/4" Wood strip secured to lintel reinforcement · Lintel reinforcement · 1/4" Min. · molding
HEAD-F

Wall Ties · Switch · 3/8" Min. · 1/4" Min. · 1" Min. · No.18 ga. metal · No.12 ga. metal
JAMB-D

SCALE
3" = 1'-0"

Wood lath anchor · Wood strip secured to lath
SILL-G
JAMB SIMILAR

Electric raceway · Plate-weld to ∟'s · 1/4" Min. · Structural channels
SECTION-E

12"c.c. for 6"or 12"sq. blocks · 8"c.c. for 8"sq. blocks · 3" · 8"Height Max.span 8'-0" · 12"Height Max.span 10'-0" · Bearing surface
LINTEL REINFORCEMENT

PATTERNED AND
WIRED GLASS

PATTERNED GLASS.   Patterned glass, also called figured glass, is sheet glass that has a geometric or linear pattern embossed into it during the rolling process. The base material is the same as in the case of plate or float glass, and so is the manufacturing method, but the finished product is neither ground nor polished. The primary purpose of patterned glass is to allow a good degree of light transmission while at the same time obscuring vision and affording privacy. A secondary purpose is to achieve a decorative effect.

Patterned glass is available in numerous textures and patterns, depending upon the manufacturer. Patterns may be grouped roughly as follows: ribbed, stippled, fluted, hammered, granular, striped, and floral. The pattern may be rolled into one or both sides of the glass. Etching or sandblasting may be employed to provide further obscurity. However, patterned glass is itself weaker than plate or float glass of the same thickness, and etching or sandblasting weakens the glass even more. This must be taken into consideration in specifying size and thickness of lights.

Patterned glass generally transmits about 80% to 90% of the light incident upon it. Thicknesses of $\frac{1}{8}''$ and $\frac{7}{32}''$ are most common, but others are also manufactured. The weight of $\frac{1}{8}''$ patterned glass ranges from about 1.60 to about 2.10 pounds per square foot, while the weight of $\frac{7}{32}''$ may vary from about 2.40 to about 3.00 pounds per square foot. Maximum size is approximately 60" x 132". Some patterns are available in heat-strengthened or fully tempered glass.

WIRED GLASS.   Wired glass is designed primarily as a safety glass. It is plate glass into which a wire mesh has been rolled during the manufacturing process. Wired glass may be obtained in clear form, allowing full vision (and full visual exposure of the wire mesh) or in some of the styles of patterned glass (in a few of which the wire mesh is barely visible). Generally the mesh pattern is square or diamond; occasionally parallel stranding is employed. The strand spacing of the mesh patterns varies with the specific product.

Wired glass has good resistance to low-impact blows and is also used as a fire-resistant material if it conforms to certain standards promulgated by Underwriters' Laboratories, Inc. It transmits approximately 80% to 85% of the light incident upon it. Nearly all wired glass is made in $\frac{1}{4}''$ thickness, weighing about 3.5 pounds per square foot, and the maximum sheet size is approximately 60" x 144".

SELECTION.   There are nine factors involved in the selection of the proper patterned or wired glass for a specific purpose. These factors are:

1. Obscurity/privacy
2. Glare reduction
3. Light transmission
4. Solar heat reduction
5. Fire protection
6. Accidental breakage protection
7. Safety glazing
8. Security
9. Appearance

OBSCURING GLASS.   For windows, partitions, and doors that must transmit daylight illumination, but where obscurity is required, the rolled, patterned sheet glasses offer a wide variety of textures and figures. The degree of obscurity or privacy required will govern the relative opacity of the glass selected; there is a reasonable range from which to choose. The size of the pane and possible vibration will govern the thickness of glass to be used. If protection against fire and/or breakage is also desirable, a patterned wire glass may be selected. If solar heat radiation is to be minimized, a patterned glass with appropriate properties may be selected, or a patterned glass may be used in combination with a solar reflective coated plate or float glass. Where glare is a problem, a patterned glass with a glare reducing surface may be employed.

SAWTOOTH SKYLIGHTS OR CLERESTORIES.   Unless designed especially for the collection of solar radiation in a solar heating design, sawtooth skylights or clerestories generally face north. Since it is impossible to improve the thoroughly diffused light from the north, a durable glass with relatively smooth surfaces that is easy to keep clean, such as a hammered type, is recommended. Sawtooth skylights or clerestories facing in other directions and not intended as part of a solar design can be fitted with patterned glass of relatively greater obscurity, increasing as required as the facing direction approaches south, where the amount of direct incident light is greatest. Where accidental breakage of the lights is of concern, a wired type of patterned glass can be installed.

PATTERNED AND WIRED GLASS

MONITOR SKYLIGHTS OR CLERESTORIES. These skylights are usually designed to run east and west so that one side receives light from the south and the other side receives north light. As with sawtooth skylights, a hammered type of patterned glass is excellent for the north side of a monitor skylight. If it is necessary to increase the distribution of the light, a type of patterned glass that has been especially designed to build up the light intensity on each side of a light source should be selected. With such a glass, it is possible to increase the illumination at a point 50 feet from the light source by as much as 100%, compared to rough glass. On the south side of any skylight, glare and solar heat transmission may be critical factors. In such cases, glazing may be done with a glare-reducing or a solar-radiation-reducing type of patterned glass. Or, patterned glass may be used in combination with a metallic film-coated plate or float glass to achieve even better results.

REGULAR SKYLIGHTS. Ordinary skylights of either the flat or the bubble or dome type, when let into flat roofs, may receive light from all directions. When they are let into pitched roofs, the compass orientation of the roofline determines the directions from which light is received, but the directional spread of light is generally greater than in the case of sawtooth or monitor skylights. Flat skylights may be directly glazed with wired and/or patterned glass, while clear-glazed bubble or dome skylights must be underglazed from the inside. The selection of glass will depend upon the size of the skylight, the type of light it receives, and the desirability of spreading the light over a wide area by installing a light-distributing type of patterned glass. Solar heat transmission may be reduced, or heat and glare may be reduced, by using an appropriate type of patterned glass, either alone or in combination with coated plate or float glass. Added strength and protection against breakage by accident or vandalism may be gained by using a wired patterned glass.

SIDEWALL SASH. Many buildings are relatively narrow units, not requiring overhead or skylight illumination. In buildings of this type it may be desirable to cut down the illumination in the space within a few feet of the windows and build it up at points farther away, while at the same time diffusing the light uniformly to reduce shadows and contrasts. Certain types of patterned glass are especially designed for windows that have east, west, and south exposures where reducing solar heat transmission and glare as much as possible is not essential. Solar heat and glare problems would indicate the use of a patterned glass type (alone or in combination with plate or float glass) that is designed to combat these problems.

FIRE AND BREAKAGE. The objects of wire glass are to afford constant fire protection at minimum cost and to provide a certain minimum level of protection against breakage. Whether to specify wired glass instead of other, more fire-resistant or stronger materials such as laminated glass or plastics must be determined from the dictates of the specific application balanced against cost-effectiveness. Windows, doors, transoms, skylights, and all places where fire or breakage is a special consideration are likely candidates for wired glass. Wired glass may be fractured by severe heat or sudden shock. The wire mesh holds the shattered pieces in place, preventing serious injury or loss of life. It prevents draft and holds fire within the bounds of its origin. It also provides added strength against willful or accidental breakage, as well as a certain measure of added security against casual breaking-and-entering and similar incidents. The regular *Mississippi* wire glass has borne the approval of Underwriters' Laboratories, Inc., since 1906. Even for locations requiring the very finest appearance, types of wired glass may be found suitable.

MISCELLANEOUS APPLICATIONS. Patterned glass is excellent for use in interior partitions where it is desirable to maintain an even distribution of light between rooms. It also may be employed to provide privacy between work areas, as in a series of office cubicles contained in a single room. It is also sometimes used simply for its decorative impact. Some types make fine diffusers beneath recessed lighting fixtures or luminous ceiling installations. In fully tempered form, patterned glass is widely installed in bathtub and shower doors and/or enclosures.

WEIGHTS OF
MATERIALS

Substance	Weight, Pounds per Cubic Foot	Substance	Weight, Pounds per Cubic Foot	Substance	Weight, Pounds per Cubic Foot
Ashlar Masonry		***Minerals—Continued***		***Various Solids***	
Granite, syenite, gneiss	165	Pumice, natural	40	Cereal, oats, bulk	32
Limestone, marble	160	Quarts, flint	165	Cereal, barley, bulk	39
Sandstone, bluestone	140	Sandstone, bluestone	147	Cereal, corn, rye, bulk	48
		Shale, slate	175	Cereal, wheat, bulk	48
Mortar Rubble		Soapstone, talc	169	Hay and Straw, bales	20
Masonry				Cotton, Flax, Hemp	93
Granite, syenite, gneiss	155	***Stone, Quarried, Piled***		Fats	58
Limestone, marble	150	Basalt, granite, gneiss	96	Flour, loose	28
Sandstone, bluestone	130	Limestone, marble,		Flour, pressed	47
		quarts	95	Glass, common	156
Dry Rubble Masonry		Sandstone	82	Glass, plate or crown	161
Granite, syenite, gneiss	130	Shale	92	Glass, crystal	184
Limestone, marble	125	Greenstone, hornblende	107	Leather	59
Sandstone, bluestone	110			Paper	58
		Bituminous Substances		Potatoes, piled	42
Brick Masonry		Asphaltum	81	Rubber, caoutchouc	59
Pressed brick	140	Coal, anthracite	97	Rubber, goods	94
Common brick	120	Coal, bituminous	84	Salt, granulated, piled	48
Soft brick	100	Coal, lignite	78	Saltpeter	67
		Coal, peat, turf, dry	47	Starch	96
Concrete Masonry		Coal, charcoal, pine	23	Sulphur	125
Cement, stone, sand	144	Coal, charcoal, oak	33	Wool	82
Cement, slag, etc	130	Coal, coke	75		
Cement, cinder, etc	100	Graphite	131	***Timber, U. S. Seasoned***	
		Paraffine	56	Ash, white-red	40
Various Building		Petroleum	54	Cedar, white-red	22
Materials		Petroleum refined	50	Chestnut	41
Ashes, cinders	40-45	Petroleum benzine	46	Cypress	30
Cement, P'rtl'd, loose	90	Petroleum gasoline	42	Elm, White	45
Cement, Portland, set	183	Pitch	69	Fir, Douglas spruce	32
Lime, gypsum, loose	53-64	Tar, bituminous	75	Fir, eastern	25
Mortar, set	103			Hemlock	29
Slags, bank slag	67-72	***Coal and Coke, Piled***		Hickory	49
Slags, bank screenings	98-117	Coal, anthracite	47-58	Locust	46
Slags, machine slag	96	Coal, bituminous, lig-		Maple, hard	43
Slags, slag sand	49-55	nite	50-54	Maple, white	33
		Coal, peat, turf	20-26	Oak, chestnut	54
Earth, Etc., Excavated		Coal, charcoal	10-14	Oak, live	59
Clay, dry	63	Coke	23-32	Oak, red, black	41
Clay, damp, plastic	110			Oak, white	46
Clay and gravel, dry	100	***Metals, Alloys, Ores***		Pine, Oregon	32
Earth, dry, loose	76	Aluminum, cast-ham-		Pine, red	30
Earth, dry, packed	95	mered	165	Pine, white	26
Earth, moist, loose	78	Aluminum, bronze	481	Pine, yellow, long-leaf	44
Earth, moist, packed	96	Brass, cast-rolled	534	Pine, yellow, short-leaf	38
Earth, mud, flowing	108	Bronze, 7.9 to 14% Sn.	509	Poplar	30
Earth, mud, packed	115	Copper, cast-rolled	556	Redwood, California	26
Riprap, limestone	80-115	Copper, ore pyrites	262	Spruce, white, black	27
Riprap, sandstone	90	Gold, cast-hammered	1205	Walnut, black	38
Riprap, shale	105	Iron, cast, pig	450	Walnut, white	26
Sand, gravel, dry, loose	90-105	Iron, wrought	485	Moisture Contents:	
Sand, gravel, dry, p'k'd	100-120	Iron, steel	490	Seasoned timber 15	
Sand, gravel, dry, wet	118-120	Iron, spiegel-eisen	468	to 20%	
		Iron, ferro-silicon	437	Green timber up	
Minerals		Iron, ore, hematite	325	to 50%	
Asbestos	153	Iron, ore limonite	237		
Barytes	281	Iron, ore magnetite	315	***Various Liquids***	
Basalt	184	Iron, slag	172	Alcohol, 100%	49
Bauxite	159	Lead	710	Acids, muriatic, 40%	75
Borax	109	Lead, ore, galena	465	Acids, nitric, 91%	94
Chalk	137	Manganese	475	Acids, sulphuric, 87%	112
Clay, marl	137	Manganese ore, pyro-		Lye, Soda, 66%	106
Dolomite	181	lusite	259	Oils, vegetable	58
Feldspar, orthoclase	159	Mercury	849	Oils, mineral, lubricants	57
Gneiss, serpentine	159	Nickel	565	Water, 4°C, max.	
Granite, syenite	175	Nickel monel metal	556	density	62.428
Greenstone, trap	187	Platinum, cast-ham-		Water, 100°C	59.830
Gypsum, alabaster	159	mered	1330	Water, ice	56
Hornblende	187	Silver, cast-hammered	656	Water, snow, fresh	
Limestone, marble	165	Tin, cast-hammered	459	fallen	8
Magnesite	187	Tin, ore, cassiterite	418	Water, sea water	64
Phosphate rock, apatite	200	Zinc, cast-rolled	440		
Porphyry	172	Zinc, ore, blende	253	***Gases, Air = 1***	
				Air, 0°C, 760 mm	.0807

DEAD LOADS OF
BUILDING MATERIALS

	(pounds per sq. ft.)
Roofs	
5-ply felt and gravel	6
4-ply felt and gravel	5
Roll roofing, light	0.50
Asphalt shingles	2.50
Wood shingles	2.50
¾″ softwood sheathing	2.50
1½″ softwood decking	5
¼″ slate	10
Clay Spanish tiles	10
Sheet metal	1–4

	(pounds per sq. ft.)
Ceilings	
Lath and ¾″ plaster	8
Suspended metal lath and plaster	10
Painted sheet metal	3–4
½″ gypsum board	2
Gypsum lath and plaster	5.50

	(pounds per sq. ft.)
Floors	
¾″ hardwood finish flooring	2.8
¾″ softwood finish flooring	2
⁵⁄₁₆″ hardwood floor tile	1
⅝″ plywood subflooring	1.75
Cement or terrazzo finish per 1″ thickness	12
4″ solid concrete slab	48
Softwood joists, 2 x 12 per linear foot	3.75
Softwood joists, 2 x 8 per linear foot	2.50

	(pounds per cu. ft.)
Walls and Partitions	
Stone or gravel concrete	144
Cinder concrete, well tamped	108
Cinder concrete, not tamped	60
Ordinary brickwork	120
Hollow clay tile masonry	48
Sandstone rubble	132
Sandstone ashlar	144
Limestone rubble	144
Limestone ashlar	156
Granite rubble	156
Granite ashlar	168

	(pounds per sq. ft.)
½″ gypsum board	2
½″ sand plaster	4.4
½″ cement plaster	4.8
¾″ lath and plaster	8
2 x 4 stud wall, 16″ o/c, with plates, 8′ high	1.5
2 x 6 stud wall, 24″ o/c, with plates, 8′ high	2

MINIMUM ALLOWABLE
LIVE LOADS

(pounds per sq. ft.)

Floor Loads

Dwelling attic floors for light storage only	20
Dwelling floors, solid or ribbed monolithic slab	30
Dwelling floors used for living purposes	40
Hospital rooms and wards	40
Hotel guest rooms	40
Lodging and tenement houses	40
Assembly rooms with fixed seats	50
Office floors should be designed to support either a concentrated load of 2,000 pounds located on any area of 2½ square feet, or a uniform load of	50
Light manufacturing	75
Retail salesrooms, light merchandise	75
Garages, passenger cars only	50
Printing plants	100
Spaces where crowds may collect	100
Fire escapes and exit passageways	100
Aisles, corridors, lobbies	100
Public spaces in hotels and public buildings	100
Assembly halls without fixed seats	100
Banquet rooms	100
Grandstands (linear foot of seating)	120
Theater stages	125
Stairways	100
Gymnasiums	100
Wholesale stores, light merchandise	100
Garages, all types of vehicles	100
Light storage purposes	125
Heavy storage purposes	250
Sidewalks	250
Library stacks	125
Library reading room	60
Classrooms	40
Residential balconies	60

Roof Loads

Rise 4″ or less per foot, pounds per square foot of horizontal projection	20
Rise 4″ to 12″ per foot, pounds per square foot of horizontal projection	16
Rise over 12″ per foot	12

These live loads are typical of minimums given in the Uniform Building Code and similar model codes for general construction. In the absence of local laws or building codes, the above loads may be used as a basis for the engineering design. Otherwise, check local requirements.

A live load is defined as the weight resulting from furniture, persons, or other movable and varying loads that are not a permanent part of the structure. Note that wind, snow, earthquake, impact, dead, and other loads are not considered a part of the live load of a structure.

HOW TO FIGURE
REACTIONS

Usually the first step in the design of a beam is to find the value of the reactions (the upward forces exerted by the supports), by applying the principle that *the sum of the clockwise moments equals the sum of the counter-clockwise moments, at any point.*

By "moment" is meant the product of a force and its perpendicular distance from the given point. In *Figure a* the moment is 150 × 6 = 900 ft. lbs. In this instance it is counter-clockwise as it tends to produce rotation opposite to the hands of a clock.

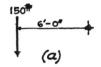

To find the reactions of a simple beam (beam with two supports, one at each end) loaded as in *Figure b,* take moments at

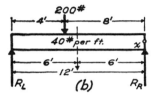

point x. This eliminates the unknown R_R from our computation, as it passes through point x and hence its perpendicular distance is zero. The clockwise moment in the example is R_L × 12. The moment of the concentrated load is 200 × 8 and is counter-clockwise. The uniform load of 40 lbs. per ft. acts as if its whole weight were exerted at

its center of gravity as shown by the dotted arrow, and its moment is (40 × 12) × 6 and is counter-clockwise. Set the clockwise moments opposite the counter-clockwise moments and solve:—

$$R_L \times 12 = 200 \times 8 + (40 \times 12) \times 6$$
$$R_L = (1600 + 2880)/12 = 373 \text{ lbs.}$$

Since in any beam, the sum of the upward forces exerted by the reactions must equal the sum of the downward forces exerted by the loads, we can find the value of R_R as follows:—

$$R_R = 200 + (40 \times 12) - 373 = 307 \text{ lbs.}$$

This procedure is typical for any simple beam, however loaded. It should be evident from *Figure c* that if the loads are symmetrical the reactions will be equal, each having a value equivalent to half the total load, making moment calculations needless.

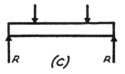

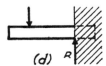

In cantilever beams *Figure d,* the one reaction at the wall is equal to the sum of all the loads, without the necessity of moment calculations. In continuous beams (beams with 3 or more supports), the computations to find the reactions are generally too involved for the architect to attempt.

VERTICAL SHEAR

Relatively short and heavily loaded beams sometimes fail by shearing. Vertical Shear is the measure of the tendency of the two parts of a beam on opposite sides of a given section to slide in opposite directions as in *Fig. a*, and is equal to the reaction on the left of the section minus the loads on the left of the section.

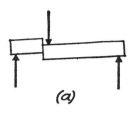

(a)

$$V = +R - \Sigma W$$

Applying this definition to a simple beam loaded as in *Fig. b* we can find the value of the vertical shear at any section. For example, the shear at a section 5' from the left support will be equal to—

$$V_{5'} = +530 - (50 \times 5) - 100 = +180 \text{ lbs.}$$

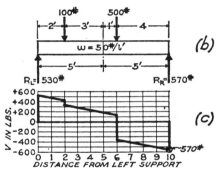

(b)

(c)

It is often necessary in unsymmetrically loaded beams to draw a shear diagram to show how the shear (V) varies. This is not only so that the beam may be designed to resist failure from shear, but because we will later have to determine from it where the maximum bending moment (another cause of failure) occurs. The diagram is constructed by calculating the shear at several sections along the beam and plotting the results as in *Fig. c*. V is positive when the reaction is greater, negative when the loads are greater. The maximum shear in the example occurs an infinitesimal distance to the left of R_R and is equal to 570 lbs.

The maximum shear in a cantilever beam, *Fig. d*, is equal to the reaction, which is also equal to the sum of the loads. The maximum shear for continuous beams involves computations too complicated for the average architect.

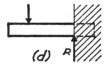

(d)

BENDING
MOMENTS

To determine the size a beam must be to resist failure by "transverse rupture" (see *Figure a*) under given loads, we must investigate the Bending Moment, which is the measure of the tendency of external

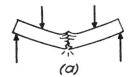

(a)

loads to cause such failure. The bending moment at any section of a simple beam is equal to the moment of the reaction to the left of the section, minus the moments of the loads to the left of the section. This definition may be expressed:

$$M_X = M_R - \Sigma M_W$$

The word "moment" alone will be used in referring both to bending moment and to moment of a force, the kind of moment meant being evident from the context of the discussion.

Applying the definition to a simple beam loaded as in *Figure b*, we can find the numerical value of the moment at any section. For example the moment at the section 5' from the left support will be: $M_5 = (530 \times 5) - (50 \times 5 \times 2.5) - (100 \times 3) = 2,650 - 625 - 300 = 1,725$ ft. lbs.

Since the usual calculations for design are made in inch pounds, we have: 1,725 ft. lbs. × 12 = 20,700 in. lbs.

The Moment Diagram is made by calculating the moment at intervals along the beam and plotting the results as in *Figure c*. The position of the worst moment condition can be determined from the shear diagram without drawing a moment diagram, for *the*

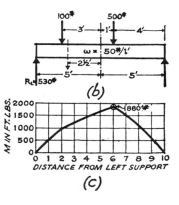

(b)

(c)

maximum moment occurs where the shear is zero. From an examination of the shear diagram (page 98), we could have determined that the maximum moment would occur 6' from the left support. One calculation would then have shown its value to be 1,880 ft. lbs. The beam must be designed so that it is strong enough to resist this maximum value.

The maximum moment for a cantilever beam occurs at the wall. It is calculated in exactly the same manner as for simple beams. Always view the drawing with the reaction at the right to eliminate it from the calculations so that its value need not be known.

The computation of maximum moments for continuous beams is somewhat complicated. For such beams having approximately equal spans L, and uniform loads per foot of beam w, the usual building code requirement of a maximum moment of $12wL^2/10$ inch lbs. for end spans, and wL^2 inch lbs. for interior spans, will generally be on the side of safety.

CHANGING CONCENTRATED
INTO UNIFORM LOADS*

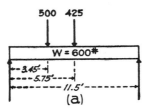

(a)

Handbooks which give sizes of beams to resist bending and deflection are based on *uniformly distributed* loads. To use such data for a beam with *concentrated* loads it is necessary to find the uniformly distributed load that produces an equivalent bending effect.

For example, in *Figure a*, the location of the 500 lbs. concentrated load with respect to the nearer support is 3.45/11.5 = .3 of the total span. Consulting the graph below it is found that the factor for a load in this location is about 1.70. This means that a uniformly distributed load of 1.70 × 500 lbs. or 850 lbs. produces approximately the same bending moment in the beam as the 500 lbs. concentrated load produces at the .3 position.

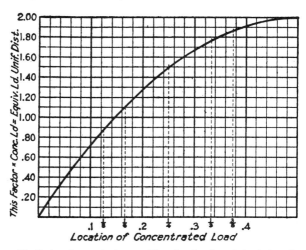

Location of Concentrated Load

Similarly the equivalent uniformly distributed load for the 425 lbs. concentrated load at the center of the span is found to be 2.00 × 425 lbs. which equals 850 lbs.

The *total* equivalent uniformly distributed load will be the sum of the separate uniform loads which have been found and are given, or 850 + 850 + 600 = 2300 lbs. (= 200 lbs./1' of span). The beam can be designed as if it were loaded as in *Figure b*. Note, however, that the reactions in the two cases are *not* the same.

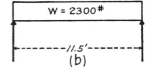

(b)

For simply supported beams.

100

STRESSES FOR USUAL
LOADING CONDITIONS

UNIFORMLY LOADED SIMPLE BEAM.

Maximum Shear at ends $= W/2$ lbs.
Maximum Moment at center $= Wl/8$ in. lbs.
Maximum Deflection $= 5Wl^3/384EI$ inches.

SIMPLE BEAM WITH CONCENTRATED LOAD AT CENTER.

Maximum Shear $= P/2$ lbs.
Maximum Moment at center $= Pl/4$ in. lbs.
Maximum Deflection $= Pl^3/48EI$ inches.

SIMPLE BEAM WITH CONCENTRATED LOAD AT ANY POINT.

Maximum Shear $= Pb/l$ lbs.
Maximum Moment at P $= Pab/l$ in. lbs.
Maximum Deflection at D $= Pabc(l+a)/9EIl$.
Note b is greater than a.

UNIFORMLY LOADED CANTILEVER BEAM.

Maximum Shear at wall $= Wx/l$ lbs.
Maximum Moment at wall $= Wl/2$ in. lbs.
Maximum Deflection at end $= Wl^3/8EI$ in. lbs.

CANTILEVER BEAM WITH CONCENTRATED LOAD AT END.

Maximum Shear $= P$ lbs.
Maximum Moment at wall $= Pl$ in. lbs.
Maximum Deflection at end $= Pl^3/3EI$.

$P =$ Concentrated Load in lbs.
$W =$ Total Load in lbs. uniformly distributed.
$l =$ Span of beam in inches.
$I =$ Moment of Inertia in inches^4. (For rectangular beam $= bd^3/12$. For steel beams consult handbook.)
$E =$ Modulus of Elasticity. (Steel $= 29,000,000$ — other materials vary.)

STRESSES FOR NO. 1
DIMENSION STOCK

The following stresses may be used in the design of wood-framed buildings, for joists and rafters, where No. 1 or equivalent grade (under visual grading rules) is to be used. This is a grade often used for buildings, but if wood having higher stress capabilities is desired or required, it is advisable to opt for a higher grade (such as Select Structural) or to choose a suitable material graded under the structural grading rules.

The figures given are for members in repetitive use rather than single or engineered use; they represent the allowable unit stresses in pounds per square inch and are for members 2″ to 4″ thick and 5″ wide (or wider), only. They are typical of the values required by model building codes. Complete tables of values for other grades, species, and lumber sizes can be found in model building codes, various reference books, official grading agency publications, and in the *Supplement* to the *National Design Specifications for Wood Construction* published by the National Forest Products Association.

Species	F_b	F_v	E	$F_c \perp$
Cedar, Northern White	1,000	65	800,000	205
Cedars, Western	1,300	75	1,100,000	265
Douglar-Fir (north)	1,750	95	1,800,000	385
Douglas-Fir (south)	1,650	90	1,400,000	335
Fir, Alpine	1,150	70	1,300,000	195
Fir, Balsam	1,150	60	1,200,000	170
Hemlock, Eastern	1,500	85	1,300,000	365
Hemlock, Western	1,550	90	1,600,000	280
Larch, Western	1,750	95	1,800,000	385
Pine, Eastern White	1,150	65	1,200,000	220
Pine, Lodgepole	1,300	70	1,300,000	250
Pine, Ponderosa	1,200	70	1,200,000	235
Pine, Red	1,150	70	1,300,000	280
Pines, Southern	1,700	95	1,700,000	405
Pine, Sugar	1,200	70	1,200,000	235
Pine, Western White	1,100	70	1,400,000	190
Redwood	1,700	100	1,400,000	425
Spruces, Eastern	1,250	65	1,400,000	255
Spruce, Engelmann	1,150	70	1,300,000	195
Spruce, Sitka	1,300	75	1,500,000	280
Spruce, Sitka (coast)	1,250	65	1,700,000	290
Tamarack	1,500	85	1,300,000	365

WOOD BEAMS AND JOISTS
HORIZONTAL SHEAR

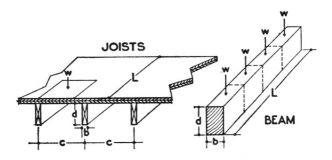

JOIST FORMULAS

(1) $F_v bd = 0.0625cwL$

(2) $c = \dfrac{F_v bd}{0.0625wL}$

(3) $d = \dfrac{0.0625cwL}{F_v b}$

(4) $w = \dfrac{F_v bd}{0.0625cL}$

BEAM FORMULAS

(1) $F_v bd = 0.75wL$

(2) $d = \dfrac{0.75wL}{F_v b}$

(3) $w = \dfrac{F_v bd}{0.75L}$

c = Spacing of joists in inches on center.
F_v = Allowable horizontal shear.
b = Actual breadth of joists or beam unless code allows use of nominal breadth.
d = Actual depth of joists or beam unless code allows use of nominal depth.
w = For *joists*, total dead and live load in lbs. per square foot, uniformly distributed. For *beams*, total dead and live load in lbs. per linear foot of beam, uniformly distributed. To get the safe superimposed load, subtract the weight of the joists or beam and other dead loads.
L = Span of joists or beam in feet.

EXAMPLE OF JOIST DESIGN

Given: w = 105 lbs. per sq. ft., L = 8.0′, western larch (F_v = 95).
To find: Spacing required for 2 x 8s determined by horizontal shear.
Solution: Substituting proper values in Joist Formula (2) above:

$$c = \frac{95 \times 1.5 \times 7.25}{0.0625 \times 105 \times 8} = 19.68'' \text{ o/c, actual}$$

Note: Cases of joist failure from horizontal shear are extremely rare and shear calculations are not necessary in ordinary work.

EXAMPLE OF BEAM DESIGN

Given: w = 500 lbs. per linear foot, L = 11.0′, nominal b = 6″, Sitka spruce (F_v = 70).
To find: Depth of beam required as determined by horizontal shear.
Solution: Substituting proper values in Beam Formula (2) above,

$$d = \frac{0.75 \times 500 \times 11}{70 \times 5.5} = 10.7'', \text{ actual}$$

WOOD BEAMS AND JOISTS
BENDING

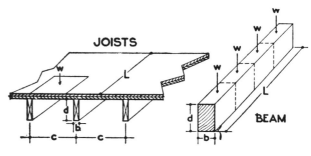

JOIST FORMULAS

(1) $F_b bd^2 = 0.75 cwL^2$

(2) $c = \dfrac{F_b bd^2}{0.75wL^2}$

(3) $d = \sqrt{\dfrac{0.75cwL^2}{F_b b}}$

(4) $w = \dfrac{F_b bd^2}{0.75cL^2}$

BEAM FORMULAS

(1) $F_b bd^2 = 9wL^2$

(2) $d = \sqrt{\dfrac{9wL^2}{F_b b}}$

(3) $w = \dfrac{F_b bd^2}{9L^2}$

c = Spacing of joists in inches on center.
F_b = Allowable extreme fiber in bending.
b = Actual breadth of joists or beam unless code allows use of nominal breadth.
d = Actual depth of joists or beam unless code allows use of nominal depth.
w = For *joists*, total dead and live loads in lbs. per square foot, uniformly distributed. For *beams*, total dead and live load in lbs. per linear foot of beam, uniformly distributed. To get the safe superimposed load, subtract the weight of the joists or beam and other dead loads.
L = Span of joists or beam in feet.

EXAMPLE OF JOIST DESIGN

Given: w = 105 lbs. per sq. ft., L = 8.0', western larch (F_b = 1,750).
To find: Spacing required for 2 x 8s as determined by bending.
Solution: Substituting proper values in Joist Formula (2) above,

$$d = \frac{1,750 \times 1.5 \times 7.25 \times 7.25}{0.75 \times 105 \times 8.0 \times 8.0} = 27.4'' \text{ o/c, actual}$$

EXAMPLE OF BEAM DESIGN

Given: w = 500 lbs. per linear foot, L = 11.0', nominal b = 6", Sitka spruce (F_b = 1,000).
To find: Depth of beam required as determined by bending.
Solution: Substituting proper values in Beam Formula (2) above,

$$d = \sqrt{\frac{9 \times 500 \times 11.0 \times 11.0}{1,000 \times 5.5}} = 9.95'', \text{ actual}$$

Note: The F_b for No. 1 grade Sitka spruce beams in single use is typically listed as 1,000 (different than for repetitive use as joists).

WOOD BEAMS AND JOISTS
DEFLECTION

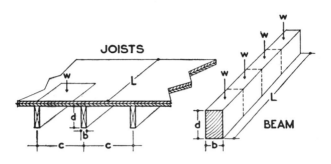

JOIST FORMULAS

(1) $Ebd^3 = 675cwL^3$

(2) $c = \dfrac{Ebd^3}{675wL^3}$

(3) $d = \sqrt[3]{\dfrac{675cwL^3}{Eb}}$

(4) $w = \dfrac{Ebd^3}{675cL^3}$

BEAM FORMULAS

(1) $Ebd^3 = 8100wL^3$

(2) $d = \sqrt[3]{\dfrac{8100wL^3}{Eb}}$

(3) $w = \dfrac{Ebd^3}{8100L^3}$

c = Spacing of joists in inches on center.
E = Allowable modulus of elasticity.
b = Actual breadth of joists or beam.
d = Actual depth of joists or beam to limit deflection to $\frac{1}{360}$ of the span.
w = For *joists*, total dead and live load in lbs. per square foot, uniformly distributed. For *beams*, total dead and live load in lbs. per linear foot of beam, uniformly distributed. To get the allowable superimposed load, subtract the weight of the joists or beam and other dead loads.
L = Span of joists or beam in feet.

EXAMPLE OF JOIST DESIGN

Given: w = 105 lbs. per sq. ft., L = 8.0', western larch (E = 1,800,000).
To find: Spacing required for 2 x 8s to limit deflection to $\frac{1}{360}$ of the span.
Solution: Substituting proper values in Joist Formula (2) above,

$$c = \frac{1,800,000 \times 1.5 \times 7.25 \times 7.25 \times 7.25}{675 \times 105 \times 8.0 \times 8.0 \times 8.0} = 28.4'' \text{ actual}$$

EXAMPLE OF BEAM DESIGN

Given: w = 500 lbs. per linear foot, L = 11.0', nominal b = 6", Sitka spruce (E = 1,300,000).
To find: Depth of beam required to limit deflection to span/360.
Solution: Substituting proper values in Beam Formula (2) above,

$$d = \sqrt[3]{\frac{8,100 \times 500 \times 11.0 \times 11.0 \times 11.0}{1,300,000 \times 5.5}} = 9.1'' \text{ actual}$$

Note: The E for No. 1 grade Sitka spruce beams in single use is typically listed as 1,300,000 (different than for repetitive use as joists).

HOW TO DESIGN
WOOD RAFTERS

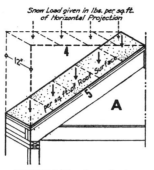

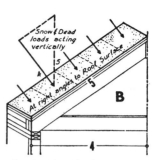

SNOW LOAD. Snow loads are usually given in codes in pounds per square foot of horizontal projection. To translate this figure into pounds per square foot of roof surface, consider the strip of roof 1'-0" wide, as shown in *Figure A*. If the horizontal projection were 4'-0" and the snow load 15 pounds, the total load on the section would be $4 \times 15 = 60$. If the length of the roof along the slope is 5'-0", the load per square foot of roof surface becomes $60 \div 5 = 12$ lbs., acting vertically.

WEIGHT OF CONSTRUCTION. Weights of roof construction materials are given in lbs. per square foot of roof surface. Suppose in this case that the weight of construction is 13 lbs. Adding this value to the snow load, we have a total of 25 lbs. per square foot of roof surface, acting vertically. We must find the component of this vertical load which acts at *right angles* to the roof surface. The component is found by similar triangles, as shown in *Figure B*, as follows:

$$\frac{25 \times 4}{5} = 20 \text{ lbs. per sq. ft. acting at right angles to roof surface.}$$

WIND LOAD. Wind loads are ordinarily given in pounds per square foot acting at right angles to the roof surface. If the wind load is 10 lbs., it can be added to the component of the snow and dead loads, making a total of 30 lbs. per square foot.

DESIGN OF RAFTERS. The span of rafters is taken to be their unsupported length measured on the slope. Plate, ridge, and collar beams (if at every rafter) are regarded as supports. The design of rafters becomes an identical calculation to that of joists, using the greatest unsupported length as the span, and the load in pounds per square foot, acting at right angles to the roof surface, as explained in the foregoing.

Snow load changed to lbs. per sq. ft. of roof surface.

Weight of materials added to snow load.

Normal component of combined loads calculated.

Wind load added to get total normal load.

ANGLE LINTELS
IN BRICKWORK

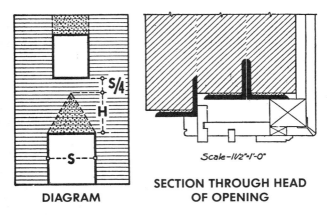

DIAGRAM	SECTION THROUGH HEAD OF OPENING

Scale–1½"–1'–0"

CORBELING ACTION OF BRICK MASONRY. When brick masonry is laid on a lintel over an opening, the brickwork bond will have enough strength to create a self-supporting corbeled arch. From the experience of wreckers and the results of fire, many examples can be adduced to show that only a small triangular area of the wall over an opening is actually dependent upon the lintel for support. The size of this triangle and the stresses acting in it are not susceptible of exact analysis; engineers variously assume the height of the triangle as 0.50, 0.67, or 0.865 times the span. Since header or soldier courses do not bond to create corbeling effect, the height of the triangle should be taken from the top of such courses and not from the top of the opening.

WHEN CORBELING ACTION MAY BE ASSUMED. There must be a sufficient amount of brickwork over the triangle to permit the arch effect to act. One writer has given a minimum for this distance as ¼ of the opening span.

WHEN CORBELING ACTION MAY NOT BE ASSUMED. If the triangle over the opening does not have a sufficient amount of brickwork over it (as in the upper window in the illustration), then the lintel must carry the entire load within the dotted lines.

USE OF STEEL ANGLE LINTELS. Calculations can be made on the assumption of a wall thickness of 4"—two of the selected lintels would be used for an 8" wall; three for a 12" wall, as shown in the illustration.

TABLE OF LINTEL SIZES. The following table assumes that the triangular space above the opening is equilateral, having a height of 0.865 times the span; the weight of brickwork, 120 lbs. per cu. ft.; there is sufficient brickwork above the triangle to insure that it will act as a corbeled arch to span the opening.

Span in Feet	Total Triangular Load	Equivalent Uniformly Distributed Load	Angle to Use for Each 4" Thickness of Brickwork
7'-0"	848#	1130#	3½ x 3½ x ¼
8'-0"	1106#	1475#	3½ x 3½ x 5⁄16
10'-0"	1730#	2307#	3½ x 5 x 5⁄16

107

AREAS OF
REINFORCING RODS

Spacing in Inches	Area of Steel, Sq. In. per Foot Width of Slab				
	¼″ (#2)	⅜″ (#3)	½″ (#4)	⅝″ (#5)	¾″ (#6)
3	0.20	0.44	0.78	1.23	1.77
3½	0.17	0.38	0.67	1.05	1.51
4	0.15	0.33	0.59	0.92	1.33
4½	0.13	0.29	0.52	0.82	1.18
5	0.12	0.26	0.47	0.74	1.06
5½	0.11	0.24	0.43	0.67	0.96
6	0.10	0.22	0.39	0.61	0.88
6½	0.09	0.20	0.36	0.57	0.82
7	0.08	0.19	0.34	0.53	0.76
7½	0.08	0.18	0.31	0.49	0.71
8	0.07	0.17	0.29	0.46	0.66
8½	0.07	0.16	0.28	0.43	0.62
9	0.07	0.15	0.26	0.41	0.59
9½	0.06	0.14	0.25	0.39	0.56
10	0.06	0.13	0.24	0.37	0.53
10½	0.06	0.13	0.22	0.35	0.51
11	0.05	0.12	0.21	0.33	0.48
11½	0.05	0.11	0.20	0.32	0.46
12	0.05	0.11	0.20	0.31	0.44

Spacing in Inches	Area of Steel, Sq. In. per Foot Width of Slab				
	⅞″ (#7)	1″ (#8)	1⅛″ (#9)	1¼″ (#10)	1⅜″ (#11)
3	2.40	3.14	3.98	4.91	5.94
3½	2.06	2.69	3.41	4.21	5.09
4	1.80	2.36	2.98	3.68	4.46
4½	1.60	2.09	2.65	3.27	3.96
5	1.44	1.88	2.39	2.95	3.56
5½	1.31	1.71	2.17	2.68	3.24
6	1.20	1.57	1.99	2.45	2.97
6½	1.11	1.45	1.84	2.27	2.74
7	1.03	1.35	1.70	2.10	2.55
7½	0.96	1.26	1.59	1.96	2.38
8	0.90	1.18	1.49	1.84	2.23
8½	0.85	1.11	1.40	1.73	2.10
9	0.80	1.05	1.33	1.64	1.98
9½	0.76	0.99	1.26	1.55	1.88
10	0.72	0.94	1.19	1.47	1.78
10½	0.69	0.90	1.14	1.40	1.70
11	0.66	0.86	1.08	1.34	1.62
11½	0.63	0.82	1.04	1.28	1.55
12	0.60	0.79	0.99	1.23	1.49

Calculations are based upon nominal rod size. The old fractional rod sizes have been replaced by rod numbers, which represent the number of ⅛″ increments per equivalent nominal rod diameter. Sizes 9, 10, and 11 are round and replace the old square rod sizes 1″, 1⅛″, and 1¼″.

CONCRETE DWELLING CONSTRUCTION

Construction practice over the years has developed four general types of concrete dwelling construction, along with a few variants and combinative styles. Though above-ground concrete dwellings are only infrequently constructed nowadays, interest in concrete earth-sheltered houses, both partially and wholly covered, is enjoying rapid growth. As with other types of dwellings, concrete houses of whatever type must conform to local building codes wherever such codes are in force. The four general types of concrete dwellings are as follows:

1. CONCRETE STRUCTURAL FRAME WITH CURTAIN WALLS.

This type has a structural frame of reinforced concrete columns, beams, and girders and floor slabs cast in place; thin enclosure walls are plastered and backplastered or shot with a concrete gun on wire mesh or metal lath attached to columns and beams.

Dwellings constructed with monolithic reinforced concrete frames cast in metal lath or other forms and with either enclosing walls carried by such frames or reinforced concrete bearing walls must be designed in accordance with standard methods of reinforced concrete design to carry safely the dead weight of the structure and the live loads that may be imposed. Enclosure or panel walls must be of sufficient strength and rigidity to resist lateral forces and transmit them to the framework.

The adequacy of a structural concrete frame proposed for dwellings is susceptible of analysis according to principles of reinforced concrete design and should conform to building code requirements for same. Inasmuch as the structural frame carries all the loads, the enclosure walls need have only such strength as is necessary to transmit wind loads to the structural frame. This has been successfully accomplished by a thickness of $1\frac{1}{2}''$ of cement mortar plastered and backplastered on metal lath that is attached to the structural frame of the building. The interior portion of exterior walls is formed by plastering on metal lath to a thickness of $\frac{7}{8}''$ to $1''$. An air space is thus provided for insulation. The total thickness of such exterior walls is governed by the width required for window and door frames; it is usually not less than $6''$.

Instead of constructing the enclosure or curtain walls by plastering on metal lath, the builder may use a concrete gun or other mechanical means of applying concrete or mortar.

In view of the relatively light types of reinforcement customary for concrete dwelling construction, it is wise to make the concrete covering over such reinforcement of sufficient thickness to provide full protection at all points against corrosion. Metal lath or other lightweight metal reinforcing fabric should be thoroughly galvanized or painted.

CONCRETE DWELLING CONSTRUCTION

2. MASONRY.   This form consists of blocks of brick or concrete laid into walls with mortar joints, with mortarless interlocking units and with grouted cores, or with stacked standard units parged with fiberglass-reinforced bonding mix. Building codes may specify wall thickness and may not cover interlocking or bonded systems.

3. MONOLITHIC CONCRETE WALLS.   The vertical loads on bearing walls not more than three stories high are comparatively small. The stability of the completed structure as a whole should be considered in any analysis of wall thickness requirements for dwellings.

Experience in the construction of houses having plain concrete bearing walls has shown that a thickness of 6″ is sufficient for abovegrade construction. Reinforcement of not less than 0.2%, computed on a vertical height of 12″, should be placed over all wall openings and at corners of the structure to prevent cracks. Foundations and earth-sheltered structures must be engineered to suit prevailing conditions. Building code requirements must be satisfied for monolithic concrete construction.

Several systems of construction that produce double concrete walls have been successfully used. Such systems produce two walls, each 4″ thick, with an air space or insulating material between the two thicknesses. (Wall and air space thickness dimensions are variable.) Wall openings and corners should be reinforced in the same manner as is used for solid single walls. The inner and outer parts of such walls must be securely braced and tied together with noncorrodible ties or other means that bring them into common action. Positive means must be provided to transmit floor and roof loads to walls.

4. PRECAST CONSTRUCTION.   In precast construction, precast units or parts different from ordinary concrete block or similar modules are employed. These structural parts range from the special forms of small units, which serve merely as enclosure walls between members of a load-carrying framework, to large slabs forming an entire wall of the building (as in the tilt-up construction system) or the members for an entire house that are preengineered, precast, and transported to the building site.

Precast concrete units for construction of dwellings must be of sufficient strength—and where necessary must be reinforced and/or tensioned—to carry safely the loads imposed. Connections between the several parts of such structures must be sufficiently strong and rigid to resist the vertical and horizontal forces that might be imposed during and after assembly.

The strength of large precast concrete units can be computed and verified by tests. The structural adequacy of a system employing units of sufficient strength will depend largely on the details of the connections, the support afforded by adjacent units, and the stability of the structure as a whole. Systems that employ relatively small units should be judged on the basis of the structural adequacy of the framework carrying the units. If the units themselves are reinforced concrete structural members, they are susceptible of theoretical analysis, and a decision as to structural adequacy will therefore be based on engineering design.

BREAST AND
RETAINING WALLS

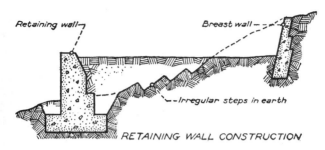

RETAINING WALL CONSTRUCTION

BREAST WALLS. These are erected only to prevent weathering or disruption of earth or other material which is in its undisturbed natural position and which is sufficiently cohesive and stable to support itself unless disturbed. Obviously, breast walls cannot be used to support earth whose angle is greater than the natural angle of repose. The following table gives these values.

Kind of Earth	Angle of Repose	Weight in lbs. per cu. ft.
Sand, clean, dry	33° 41′	90
Sand & Clay	36° 53′	95
Clay, dry	36° 53′	100
Clay, plastic	26° 34′	100
Gravel, clean	36° 53′	100
Gravel & Clay, dry	36° 53′	100
Gravel, Sand & Clay, dry	36° 53′	100
Soil	36° 53′	100
Soft Rotten Rock	36° 53″	110
Hard Rotten Rock	45° 00′	100
Bituminous Cinders	45° 00′	45
Anthracite Ashes	45° 00′	30

Where the ground to be supported is firm and the strata are horizontal, breast walls are usually built more to protect than to sustain the earth. A trifling force skilfully applied to unbroken ground will keep in place a mass of material which, if once allowed to move, would crush a heavy wall. The strength of a breast wall must be increased when the strata to be supported incline down toward the wall.

RETAINING WALLS. These are constructed so that rotation or overturning due to the pressure of material behind the wall will be prevented. Where the ground freezes to an appreciable depth, the back of the wall should be sloped from below the frost line toward its top surface. This slope should be quite smooth to lessen the hold of the frost and prevent displacement. If the original ground is made irregular with steps and the earth well rammed in layers, the pressure will be less than where the earth is placed in layers sloping toward the earth.

WATERPROOFING. The action of acids or alkalis in the ground water is destructive to concrete and in such locations a standard waterproofing material should be applied. If finished brick parapets occur on top of retaining walls, a dampproof or waterproof course should be laid under them.

GRAVITY TYPE
RETAINING WALL

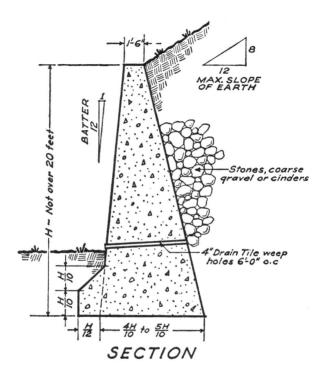

SECTION

The Gravity retaining wall is perhaps the most common type and requires no complicated reinforcing. It depends upon its own shape and weight to resist earth pressure. It is the simplest to construct and for walls under 20 feet in height, it is often the most economical. Excavate to below frost line and to firm enough soil to withstand the pressure at the toe of the wall due to the tendency to overturn.

Since retaining walls do not withstand any pressure during construction the forms can be stripped as soon as the concrete has set enough to sustain its own weight. This allows the most economical and satisfactory finishing, which is accomplished by simply rubbing with a wooden float dipped in water and sand. In this way the form marks are rubbed off and a smooth surface obtained.

SUGGESTED VOLUMETRIC MIXES FOR CONCRETE

KIND OF CONCRETE WORK *	MIX BY VOLUME JOB DAMP MATERIALS			Workability or Consistency	Water Added at Mixer Per Bag, Gallons	A One Bag Batch Makes This Volume of Concrete Cu. Ft.	MATERIALS FOR ONE CUBIC YARD OF CONCRETE			
	Cement Bags	Sand Cu. Ft.	Stone, Gravel Cu. Ft				Cement Bags	Sand Cu. Ft.	Stone, Gravel Cu. Ft.	Water Added at Mixer Gallons
Footings Heavy Foundations	1	3.75	5	stiff	6.4	6.2	4.3	16.3	21.7	27.6
Watertight Concrete for Cellar Walls and Walls Above Ground	1	2.5	3.5	medium	4.9	4.5	6.0	15.0	21.0	29.5
Driveways Floors Walks } One course	1	2.5	3	stiff	4.4	4.1	6.5	16.3	19.5	28.7
Driveways Floors Walks } Two course	1	Top 2	0	stiff	3.6	2.14	12.6	25.2		45.3
	1	Base 2.5	4	stiff	4.9	4.8	5.7	14.2	22.8	27.8
Pavements	1	2.2	3.5	stiff	4.3	4.2	6.4	14.1	22.4	27.5
Watertight Concrete for Tanks, Cisterns and Precast Units (piles, posts, thin reinforced slabs, etc.)	1	2	3	medium	4.1	3.8	7.1	14.2	21.3	29.3
	1	2	3	wet	4.9	3.9	6.9	13.8	20.7	33.7
Heavy Duty Floors	1	1.25	2	stiff	3.4	2.8	9.8	12.3	19.6	33.9
Mortar for Laying Concrete Building Units	1	6 plaster sand	1 sack 50 lbs. Hydrated Lime	medium	12.5	5.5	4.9	29.4	4.9 sacks of lime	61.2

Many specifications require proportioning and measuring by weight with accurate control of a specified maximum allowable water content.

FOUNDATION DESIGN CHART

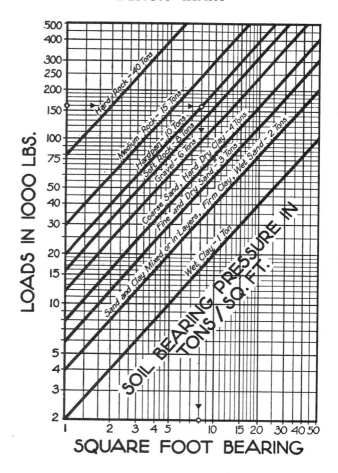

EXAMPLE. For a load of 160,000 lbs. on hardpan, the chart shows that a footing of 8 sq. ft. would be required. The values given for various soils are averages and may not agree with your local code. Be sure to check local requirements before using this chart. Values falling between the diagonal lines can be readily interpolated.

UNDERPINNING
ABUTTING FOUNDATIONS

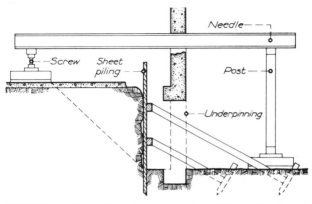

NEEDLING — Where old walls are in weak condition and/or the soil is not stable, the underpinning is accomplished with the aid of "needling." Shoring (*see below*) is usually necessary.

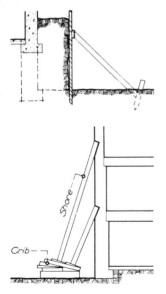

SECTIONING — If the old walls are sound and the soil stable, a short excavation is made and a 6 ft. length of new wall is built under the old wall. When this new section will bear weight, another section is added, and so continued until the old wall has a continuous foundation under it.

SHORING — Sockets are cut in the old wall and *shores*, also called *spur braces*, are inserted. These rest on a crib of timbering. Shores prevent slipping, bulging, and reduce the load to be supported while underpinning is placed.

115

PREVENTING CAVE-INS
IN EXCAVATIONS

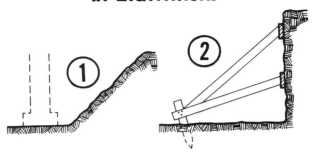

1. SLOPED BANK — In this type the earth takes its natural angle of repose where the soil lacks the stability to stand vertically when cut. Such excavation is undesirable and is frequently forbidden in specifications because the undisturbed earth remaining creates a bowl for the collection of water (both before and after the backfilling is done) and an undesirably large amount of soil removal and backfill is required.

2. BRACED BANK — If the soil has some stability but will not stand unaided, very simple bracing may be sufficient.

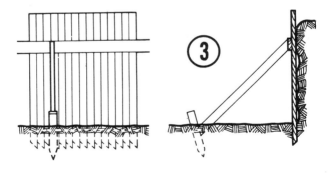

3. SHEET PILING — In very fluid soils sheet piling is driven and braced, and may be used as the outside form for poured foundations. Wood, steel, or concrete sheet piling are available.

116

PREVENTING CAVE-INS
IN EXCAVATIONS

4. VERTICAL BANK — In moist, clayey soil the earth may stand vertically without support. This makes it the most economical type where it is feasible. If no space is needed outside the foundation wall for inspection, waterproofing, piping, or other work, such earth may serve as an outside form for concrete as shown in the upper illustration. This procedure is not recommended, however, because in the pouring operation, earth particles may be too easily knocked off into the concrete. The absorption of the soil may draw water from the concrete and weaken it.

If poured concrete foundations are to be used, it is better to employ the method shown in the lower illustration, using an outside form.

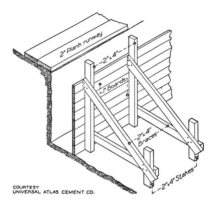

COURTESY
UNIVERSAL ATLAS CEMENT CO.

5. TRENCH WALL — A trench is excavated and the encircling foundation wall is built. Then the general excavations are carried on inside the walls, which may require bracing.

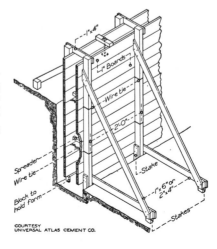

COURTESY
UNIVERSAL ATLAS CEMENT CO.

INCH AND FOOT
EQUIVALENTS

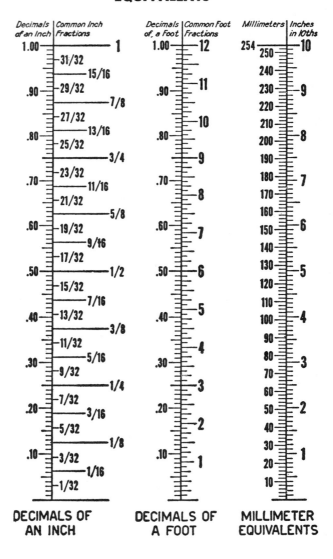

Decimals of an Inch	Common Inch Fractions
1.00	1
	31/32
	15/16
.90	29/32
	7/8
	27/32
	13/16
.80	25/32
	3/4
	23/32
.70	11/16
	21/32
	5/8
.60	19/32
	9/16
	17/32
.50	1/2
	15/32
	7/16
.40	13/32
	3/8
	11/32
.30	5/16
	9/32
	1/4
	7/32
.20	3/16
	5/32
	1/8
.10	3/32
	1/16
	1/32

DECIMALS OF AN INCH

Decimals of a Foot	Common Foot Fractions
1.00	12
	11
.90	10
.80	9
.70	8
.60	7
.50	6
.40	5
.30	4
	3
.20	2
.10	1

DECIMALS OF A FOOT

Millimeters	Inches in 10ths
254	10
250	
240	
230	9
220	
210	
200	8
190	
180	7
170	
160	
150	6
140	
130	5
120	
110	
100	4
90	
80	3
70	
60	
50	2
40	
30	
20	1
10	

MILLIMETER EQUIVALENTS

118

DECIMALS
OF A FOOT

0″	.0000	1″	.0833	2″	.166667	3″	.2500
1/16	.0052	1 1/16	.0885	2 1/16	.171875	3 1/16	.2552
1/8	.0104	1 1/8	.09375	2 1/8	.1771	3 1/8	.2604
3/16	.015625	1 3/16	.0990	2 3/16	.1823	3 3/16	.265625
1/4	.0208	1 1/4	.1042	2 1/4	.1875	3 1/4	.2708
5/16	.0260	1 5/16	.109375	2 5/16	.1927	3 5/16	.2760
3/8	.03125	1 3/8	.1146	2 3/8	.1979	3 3/8	.28125
7/16	.0365	1 7/16	.1198	2 7/16	.203125	3 7/16	.2865
1/2	.0417	1 1/2	.1250	2 1/2	.2083	3 1/2	.2917
9/16	.046875	1 9/16	.1302	2 9/16	.2135	3 9/16	.296875
5/8	.0521	1 5/8	.1354	2 5/8	.21875	3 5/8	.3021
11/16	.0573	1 11/16	.140625	2 11/16	.2240	3 11/16	.3073
3/4	.0625	1 3/4	.1458	2 3/4	.2292	3 3/4	.3125
13/16	.0677	1 13/16	.1510	2 13/16	.234375	3 13/16	.3177
7/8	.0729	1 7/8	.15625	2 7/8	.2396	3 7/8	.3229
15/16	.078125	1 15/16	.1615	2 15/16	.2448	3 15/16	.328125

4″	.3333	5″	.416667	6″	.5000	7″	.5833
4 1/16	.3385	5 1/16	.421875	6 1/16	.5052	7 1/16	.5885
4 1/8	.34375	5 1/8	.4271	6 1/8	.5104	7 1/8	.59375
4 3/16	.3490	5 3/16	.4323	6 3/16	.515625	7 3/16	.5990
4 1/4	.3542	5 1/4	.4375	6 1/4	.5208	7 1/4	.6042
4 5/16	.359375	5 5/16	.4427	6 5/16	.5260	7 5/16	.6093
4 3/8	.3646	5 3/8	.4479	6 3/8	.53125	7 3/8	.6146
4 7/16	.3698	5 7/16	.453125	6 7/16	.5365	7 7/16	.6198
4 1/2	.3750	5 1/2	.4583	6 1/2	.5417	7 1/2	.6250
4 9/16	.3802	5 9/16	.4635	6 9/16	.546875	7 9/16	.6302
4 5/8	.3854	5 5/8	.46875	6 5/8	.5521	7 5/8	.6354
4 11/16	.390625	5 11/16	.4740	6 11/16	.5573	7 11/16	.640625
4 3/4	.3958	5 3/4	.4792	6 3/4	.5625	7 3/4	.6458
4 13/16	.4010	5 13/16	.484375	6 13/16	.5677	7 13/16	.6510
4 7/8	.40625	5 7/8	.4896	6 7/8	.5729	7 7/8	.65625
4 15/16	.4115	5 15/16	.4948	6 15/16	.578125	7 15/16	.6615

8″	.666667	9″	.7500	10″	.8333	11″	.916667
8 1/16	.671875	9 1/16	.7552	10 1/16	.8385	11 1/16	.921875
8 1/8	.6771	9 1/8	.7604	10 1/8	.84375	11 1/8	.9271
8 3/16	.6823	9 3/16	.765625	10 3/16	.8490	11 3/16	.9323
8 1/4	.6875	9 1/4	.7708	10 1/4	.8542	11 1/4	.9375
8 5/16	.6927	9 5/16	.7760	10 5/16	.859375	11 5/16	.9427
8 3/8	.6979	9 3/8	.78125	10 3/8	.8646	11 3/8	.9479
8 7/16	.703125	9 7/16	.7865	10 7/16	.8698	11 7/16	.953125
8 1/2	.7083	9 1/2	.7917	10 1/2	.8750	11 1/2	.9583
8 9/16	.7135	9 9/16	.796875	10 9/16	.8802	11 9/16	.9635
8 5/8	.71875	9 5/8	.8021	10 5/8	.8854	11 5/8	.96875
8 11/16	.7240	9 11/16	.8073	10 11/16	.890625	11 11/16	.9740
8 3/4	.7292	9 3/4	.8125	10 3/4	.8958	11 3/4	.9792
8 13/16	.734375	9 13/16	.8177	10 13/16	.9010	11 13/16	.984375
8 7/8	.7396	9 7/8	.8229	10 7/8	.90625	11 7/8	.9896
8 15/16	.7448	9 15/16	.828125	10 15/16	.9115	11 15/16	.9948

DECIMALS
OF AN INCH

Fraction	64ths	Decimal	Fraction	64ths	Decimal
			½	32	.500
—	1	.015625	—	33	.515625
⅟₃₂	2	.03125	17⁄₃₂	34	.53125
—	3	.046875	—	35	.546875
⅟₁₆	4	.0625	9⁄₁₆	36	.5625
—	5	.078125	—	37	.578125
3⁄₃₂	6	.09375	19⁄₃₂	38	.59375
—	7	.109375	—	39	.609375
⅛	8	.125	⅝	40	.625
—	9	.140625	—	41	.640625
5⁄₃₂	10	.15625	21⁄₃₂	42	.65625
—	11	.171875	—	43	.671875
3⁄₁₆	12	.1875	11⁄₁₆	44	.6875
—	13	.203125	—	45	.703125
7⁄₃₂	14	.21875	23⁄₃₂	46	.71875
—	15	.234375	—	47	.734375
¼	16	.250	¾	48	.750
—	17	.265625	—	49	.765625
9⁄₃₂	18	.28125	25⁄₃₂	50	.78125
—	19	.296875	—	51	.796875
5⁄₁₆	20	.3125	13⁄₁₆	52	.8125
—	21	.328125	—	53	.828125
11⁄₃₂	22	.34375	27⁄₃₂	54	.84375
—	23	.359375	—	55	.859375
⅜	24	.375	⅞	56	.875
—	25	.390625	—	57	.890625
13⁄₃₂	26	.40625	29⁄₃₂	58	.90625
—	27	.421875	—	59	.921875
7⁄₁₆	28	.4375	15⁄₁₆	60	.9375
—	29	.453125	—	61	.953125
15⁄₃₂	30	.46875	31⁄₃₂	62	.96875
—	31	.484375	—	63	.984375

CONVERSION FACTORS

One board foot	=144	cubic inches
One centimeter	=0.3937	inches
One centimeter	=0.01	meters
One centimeter	=10	millimeters
One cubic centimeter	=3.531×10^{-5}	cubic feet
One cubic centimeter	=0.06102	cubic inches
One cubic foot	=28317	cubic cms.
One cubic foot	=1728	cubic inches
One cubic foot	=7.481	gallons
One cubic foot	=28.32	liters
One cubic inch	=16.39	cubic cms.
One degree (angle)	=0.01745	radians
One foot per second	=0.6818	miles per hour
One gallon	=231	cubic inches
One gallon	=3.785	liters
One gram	=2.205×10^{-3}	pounds
One gram per cu. cm.	=62.43	pounds per cubic foot
One horsepower	=550	foot-pounds per second
One horsepower	=0.7457	kilowatts
One inch	=2.540	centimeters
One kilogram	=1000	grams
One kilogram	=2.205	pounds
One kilogram per sq. mm.	=1422	pounds per square inch
One mile	=5280	feet
One pound	=453.6	grams
One pound per sq. in.	=0.068	atmospheres
One pound per sq. in.	=2.307	feet of water
One pound per sq. in.	=2.036	inches of mercury
One pound per sq. in.	=7.031×10^{-4}	kilograms per sq. mm.
One radian	=57.30	degrees
One square inch	=6.452	square cms.
One ton (long)	=2240	pounds
One ton (long) per sq. in.	=1.575	kilograms per sq. mm.

MULTIPLICATION AND DIVISION

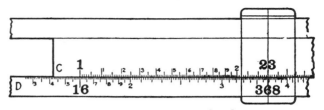

1600 × .23 = 368

$$\frac{36.8}{23} = 1.6$$

MULTIPLICATION

In this problem the slide projects to the right, and the position of the decimal point in the result is found by taking one less than the sum of the whole digits in the two factors. Thus 1600 has 4 whole digits, .23 has 0 whole digits, so $(4+0)-1=3$, and there are 3 whole digits in the result.

DIVISION

Division is exactly the opposite of multiplication. The position of the decimal point in the result, when the slide projects to the right, is found by subtracting the number of whole digits in the divisor from the number of whole digits in the dividend and then adding one. Thus 36.8 has 2 whole digits, 23 has 2 whole digits, so $(2-2)+1=1$.

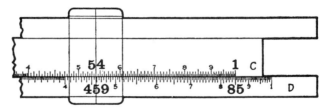

850 × .054 = 45.9

$$\frac{45\,900}{5.4} = 8\,500$$

MULTIPLICATION

When the slide projects to the left as in this example, the decimal point is found by adding the whole digits in both factors. Thus 850 has 3 whole digits, .054 has −1 whole digits, so the result will have $3+(-1)=2$ whole digits.

DIVISION

When the slide projects to the left in division, the number of whole digits in the result will be found to equal the number of whole digits in the dividend less the number of whole digits in the divisor. Thus 45900 has 5 whole digits, 5.4 has 1 whole digit, so the result will have $5-1=4$ whole digits in the result.

NOTE

321.9876 has 3 whole digits	.3219 has 0 whole digits
32.1987 has 2 whole digits	.0321 has −1 whole digits
3.219̊8 has 1 whole digit	.0032 has −2 whole digits

CIRCULAR ARCS, CHORDS AND SEGMENTS

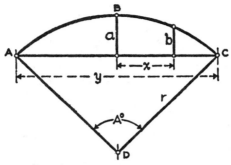

Area of circular sector ABCD = arc L $\frac{r}{2}$

Length of arc ABC $= \frac{\pi r A^0}{180} = 0.017453 A^0 r$

$r = \frac{y^2}{8a} + \frac{a}{2}$

$x = \sqrt{r^2 - (r + b - a)^2}$

$b = \sqrt{r^2 - x^2} - (r - a)$

$a = r - \sqrt{r^2 - \frac{y^2}{4}} = \frac{y}{2} \tan \frac{A}{4} = r + b - \sqrt{r^2 - x^2}$

$y = 2 \sqrt{2ar - a^2} = 2r \sin \frac{A}{2}$

$A^0 = \frac{57.29578 \text{ arc}}{r}$

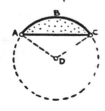

Area of the segment ABC = area of the sector ABCD minus the area of the triangle ACD.

Area of the segment ABC = area of the sector ABCDA plus the area of the triangle ADC.

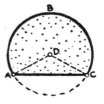

123

THE CIRCLE AND
THE SQUARE

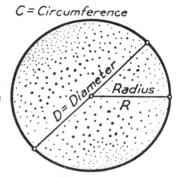

$C = Circumference$

$\pi = 3.14159265+$
$C = \pi D = 2 \pi R$
$C = 3.5446 \sqrt{area}$
$D = 0.3183 C = 2 R$
$D = 1.1283 \sqrt{area}$
Area $= \pi R^2 = 0.785398 D^2$
Area $= 0.07958 C^2 = \dfrac{\pi D^2}{4}$

Area of square $= 1.2732 \times$ **area of circle**

CIRCUMSCRIBED SQ.

Area of square $= 0.6366 \times$ **area of circle**
$S = 0.7071 D$
$D = 1.4142 S$

INSCRIBED SQUARE

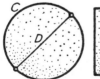

EQUAL AREAS

$S = 0.8862 D$
$S = 0.2821 C$
$D = 1.1284 S$

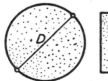

EQUAL PERIPHERIES

$S = 0.7854 D$
$D = 1.2732 S$

124

AREAS OF
PLANE FIGURES

Area of triangle $= \dfrac{a\,b}{2}$

Area of parallelogram $=$ a b

Area of trapezoid $= a\left(\dfrac{b+c}{2}\right)$

Area of trapezium $=$ Divide into two triangles and find the area of each separately.

Area of regular polygon having n sides $= r\left(\dfrac{n\,s}{2}\right)$

Area $= .2146\ r^2$

Area $= 0.7854\ (d_2^2 - d_1^2)$

Area of ellipse $= 0.7854\ d_1d_2$

Area of parabola $= \dfrac{2\ a\ b}{3}$

SOLID, DRY, AND LIQUID MEASURE

CUBIC OR SOLID MEASURE

United States and British.

1 cubic inch = .0005787 cubic foot = .000021433 cubic yard.

1 cubic foot = 1728 cubic inches = .03703704 cubic yard.

1 cubic yard = 27 cubic feet = 46656 cubic inches.

1 cord of wood = 128 cubic feet = 4 feet by 4 feet by 8 feet.

1 perch of masonry = 24.75 cubic feet = 16.5 feet by 1.5 feet by 1 foot. It is usually taken as 25 cubic feet.

DRY MEASURE

United States only.

Pints	Quarts	Gallons	Pecks	Bushels	Cubic Inches
1	.50	.125	.0625	.015625	33.6003125
2	1.	.25	.125	.03125	67.200625
8	4.	1.	.05	.125	268.8025
16	8.	2.	1.	.25	537.605
64	32.	8.	4.	1.	2150.42

1 heaped bushel = 1.25 struck bushel, and the cone must be not less than 6 inches high.

LIQUID MEASURE

United States only.

Gills	Pints	Quarts	Gallons	Barrels	Cubic Inches
1	.25	.125	.03125	.000992	7.21875
4	1.	.5	.125	.003968	28.875
8	2.	1.	.25	.007937	57.75
32	8.	4.	1.	.031746	231.
1008	252.	126.	31.5	1.	7276.5

The British imperial gallon = 277.410 cubic inches or 10 pounds avoirdupois of pure water at 62° F. and barometer at 30 inches.

The British imperial gallon = 1.20091 United States gallons.

1 fluid dram = 60 minims = .125 fluid ounce = .0078125 pint.

1 fluid ounce = 480 minims = 8 drams = .0625 pint.

LAND, LINEAR, AND MISC. MEASURE

LINEAR MEASURE
United States and British.

Inches	Feet	Yards	Rods	Furlongs	Miles
1	.08333	.02778	.0050505	.00012626	.00001578
12	1.	.33333	.0606061	.00151515	.00018939
36	3.	1.	.1818182	.00454545	.00056818
198	16.5	5.5	1.	.025	.003125
7920	660.	220.	40.	1.	.125
63360	5280.	1760.	320.	8.	1.

ROPE AND CABLE MEASURE

1 inch = .111111 span = .013889 fathom = .0001157 cable's length.
1 span = 9 inches = .125 fathom = .00104167 cable's length.
1 fathom = 6 feet = 8 spans = 72 inches = .008333 cable's length.
1 cable's length = 120 fathoms = 720 feet = 960 spans = 8640 inches.

NAUTICAL MEASURE

1 nautical mile is an international unit, used officially in the United States since July 1, 1959, equal to the length of one minute of arc of a great circle of the earth, or 6,076.115 feet = 1,852 meters = 1.1508 statute miles.

1 league is now generally accepted as 3 statute miles = 15,840 feet, but also as 3 nautical miles = 18,228.345 feet. Depending upon the historical time period and country, 1 league might equal a distance of from 2.4 to 4.6 statute miles.

1 fathom = 6 feet; 1,000 fathoms = 1 nautical mile (approximately).

GUNTER'S CHAIN

1 link = 7.92 inches = .01 chain = .000125 mile.
1 chain = 100 links = 66 feet = 4 rods = .0125 mile.
1 mile = 80 chains = 8,000 links.

SQUARE OR LAND MEASURE
United States and British.

Square Inches	Square Feet	Square Yards	Square Rods	Acres	Square Miles
1	.006944	.0007716	....	....	....
144	1.	.111111	....	....	....
1296	9.0	1.	.03306	.0002066	....
39204	272.25	30.25	1.	.00625	.0000097
6272640	43560.	4840.	160.	1.	.0015625
	27878400.	3097600.	102400.	640.	1.

1 square rood = 40 square rods.

1 acre = 4 square roods.

1 square acre = 208.71 feet square.

U. S. AND BRITISH WEIGHTS

AVOIRDUPOIS WEIGHT.

United States and British.

Grains*	Drams	Ounces	Pounds	Hundred-weight	Gross Tons
1.	.03657	.002286	.000143	.00000128	.000000064
27.34375	1.	.0625	.003906	.00003488	.000001744
437.5	16.	1.	.0625	.00055804	.00002790
7000.	256.	16.	1.	.0089286	.0004464
784000.	28672	1792.	112.	1.	.05
15680000.	573440.	35840.	2240.	20.	1.

1 pound avoirdupois = 1.215278 pounds troy.
1 net ton = 2000 pounds = .892857 gross ton.

TROY WEIGHT.

United States and British.

Grains*†	Pennyweight	Ounces†	Pounds†
1	.041667	.0020833	.0001736
24	1.	.05	.0041667
480	20.	1.	.0833333
5760	240.	12.	1.

1 pound troy = .822857 pound avoirdupois.
175 ounces troy = 192 ounces avoirdupois.

APOTHECARIES' WEIGHT.

United States and British.

Grains*†	Scruples	Drams	Ounces†	Pounds†
1	.05	.016667	.0020833	.000173611
20	1.	.333333	.0416667	.0034722
60	3.	1.	.125	.0104167
480	24.	8.	1.	.0833333
5760	288.	96.	12.	1.

†The pound, ounce and grain are the same as in troy weight.
*The avoirdupois grain = troy grain = apothecaries' grain.

METRIC WEIGHTS
AND MEASURES

LENGTH, CAPACITY, AND WEIGHT

Length	Kilometer	Hecto-meter	Decameter	Meter	Decimeter	Centimeter	Millimeter
Capacity	Kiloliter or Stere	Hectoliter or Decistere	Decaliter or Centistere	Liter or Millistere	Deciliter	Centiliter	Milliliter
Weight	Kilogram	Hecto-gram	Decagram	Gram	Decigram	Centigram	Milligram
	1	10	100	1,000	10,000	100,000	1,000,000
		1	10	100	1,000	10,000	100,000
			1	10	100	1,000	10,000
				1	10	100	1,000
				.1	1	10	100
				.01	.1	1	10
				.001	.01	.1	1

1 myriameter = 10 kilometers = 10,000 meters.
1 tonne = 1000 kilograms = 100 quintals = 10 myriagrams.
1 gram = weight of 1 cubic centimeter of distilled water at its maximum density at sea level in latitude of Paris and barometer at 760 millimeters.
1 liter = 1 cubic decimeter.

SQUARE OR SURFACE MEASURE

Square Kilometer	Square Hectometer or Hectare	Square Decameter or Are	Square Meter or Centiare	Square Decimeter	Square Centimeter	Square Millimeter
1	100	10,000	1,000,000			
	1	100	10,000	1,000,000		
	.01	1	100	10,000	1,000,000	
	.0001	.01	1	100	10,000	1,000,000
	.000001	.0001	.01	1	100	10,000
		.000001	.0001	.01	1	100
			.000001	.0001	.01	1

1 square myriameter = 100 square kilometers = 10,000,000 square meters.

CUBIC MEASURE

Cubic Decameter	Cubic Meter	Cubic Decimeter	Cubic Centimeter	Cubic Millimeter
1	1,000	1,000,000	1,000,000,000	
.001	1	1,000	1,000,000	1,000,000,000
.000001	.001	1	1,000	1,000,000
.000000001	.000001	.001	1	1,000
	.000000001	.000001	.001	1

1 cubic meter = 1 kiloliter = 1 stere.

EQUIVALENTS
OF MEASUREMENT

ap = apothecary; av = avoirdupois; Br = British; US = United States.

ACRE equals:
a square 208.71 feet on a side
43,560 square feet
4,840 square yards
1/640th square mile
0.404687 hectare
4,046.87 square meters

BARREL (US) equals:
196 pounds average, flour, customary value
105 dry quarts, dried fruits and vegetables

BARREL (liquid, US) equals:
No legal value
31½ and 31 gallons (US), customary value to some extent
42 gallons (US), customary value for petroleum
36 gallons (US), customary value for ale and beer

BOARD FOOT equals:
1 square foot × 1 inch (may be nominal or actual)

BUSHEL (Br) equals:
4 pecks (Br)
8 gallons (Br)
32 quarts (Br)
64 pints (Br)
2,219.28 cu. inches
1.28431 cu. feet
1.03202 bushels (US)
36.3677048 liters or cu. decimeters

BUSHELS (US) equals:
4 pecks (US)
32 quarts (dry; US)
64 pints (dry; US)
2,150.420 cu. inches
1.24446 cu. feet
35.23928 liters or cu. decimeters
0.3523928 hectoliter
0.968972 bushels (Br)
7.75178 gallons (Br)

CABLE (cable length, Br) equals:
0.1 knot or nautical mile (Br)
608 feet (sometimes taken as 608.6 feet)

CABLE (cable length, US) equals:
720 feet
120 fathoms (US)
219.457 meters

CENTIMETER equals:
0.01 meter
0.0328083 foot
0.393700 inch
393.700 mils

CENTIMETER³ (cu. cm.) or milliliter, equals:
0.001 liter or cu. decimeter
0.0616234 cu. inch

CHAIN, engineer's, equals:
100 links
100 feet
30.480 meters

EQUIVALENTS
OF MEASUREMENT

ap = apothecary; av = avoirdupois; Br = British; US = United States.

CHAIN, Gunter's or surveyor's, equals:
   100 links
   66 feet
   4 rods, perches or poles
   0.1 furlong
   1/80 statute mile (US)
   20.117 meters

CHAIN, metric, equals:
   20 meters
   100 links
   65.61667 feet

CIRCULAR MIL, CIRCULAR INCH, CIRCULAR CENTIMETER, ETC.
   See mil, inch, centimeter, etc.

CORD (of wood) equals:
   4 feet x 4 x 8 feet
   128 cu. feet
   8 cord feet
   3.62458 cu. meters

DRAM (av) equals:
   1/16 ounce (av)
   27.34375 grains
   0.455729 dram (ap)
   1.77185 grams

DRAM (ap) equals:
   1/8 ounce (troy or ap)
   3 scruples
   60 grains
   2.19429 drams (av)
   3.887934 grams

FATHOM (US) equals:
   6 feet
   1.8288 meters

FOOT (US) equals:
   12,000 mils
   12 inches
   1/3 yard
   1/5280 or 0.000189394 statute mile (US)
   30.4801 centimeters
   1.0000029 feet (Br)

FOOT² (sq. ft.) (US) equals:
   144 sq. inches
   1/9 or 0.111111 sq. yard
   183.346 cir. inches
   1.27324 cir. feet
   929.034 sq. centimeters
   1.0000057 sq. feet (Br)

FOOT³ (cu. ft.) equals:
   1.728 cu. inches
   0.0370370 cu. yard
   28.3170 liters or cu. decimeters
   7.48052 gallons (US)
   0.803564 bushel (US)

GALLON (liquid; US) equals:
   231 cu. inches
   0.133681 cu. foot
   3.78543 liters or cu. decimeters
   3,785.43 cu. centimeters
   32 gills (US)
   8 pints (liquid; US)
   4 quarts (liquid; US)
   0.8327024 gallon (Br)

EQUIVALENTS
OF MEASUREMENT

ap = apothecary; av = avoirdupois; Br = British; US = United States.

GILL (liquid; US) equals:
1/4 pint (liquid; US)
1/32 gallon (US)

GRAIN (same in av, troy and ap weights) equals:
1/7000 pound (av)
1/5760 pound (troy or ap)
0.00228571 ounce (av)
0.0647989 gram

GRAM equals:
0.001 kilogram
15.43235639 grains
0.564383 drams (av)
0.0352740 ounce (av)
0.00220462 pound (av)
0.771618 scruple
0.257206 drams (ap)
0.0321507 ounce (troy or ap)

HOGSHEAD (liquid; US) equals:
63 gallons (US)
2 barrels of 31 1/2 gallons (US)
238.48 liters

HUNDREDWEIGHT, short, equals:
100 pounds (av)
1/20 or 0.05 short or net ton
45.35924 kilograms

HUNDREDWEIGHT, long, equals:
112 pounds (av)
1/20 or 0.05 long or gross ton
50.8024 kilograms

INCH equals:
1,000 mils
1/12 foot
1/36 yard
2.540005 centimeters

INCH² (sq. in.) equals:
1/144 or 0.00694444 sq. foot
1,000,000 sq. mils
1,273,240 cir. mils
1.27324 circular inches
6.45163 sq. centimeters
8.21447 cir. centimeters

INCH³ (cu. in.) equals:
1/1728 or 0.000578704 cu. foot
16.38716 cu. centimeters or milliliters
0.01638716 liter or cu. decimeter

KILOGRAM OR KILO equals:
1,000 grams
0.001 metric ton
15,432.35639 grains
35.2740 ounces (av)
2.20462 pounds (av)
0.0220462 hundredweight (short)
0.0196841 hundredweight (long)
0.00110231 short or net ton
0.000984206 long or gross ton
32.1507 ounces (troy or ap)

EQUIVALENTS
OF MEASUREMENT

ap = apothecary; av = avoirdupois; Br = British; US = United States.

KILOMETER equals:
1,000 meters
3,280.83 feet
1,093.61 yards
0.621370 statute miles (US)
0.539957 knot or nautical mile (US)

KNOT (US) equals:
1 nautical mile (US) per hour, or 1 nautical mile

LEAGUE (US) equals:
15,840 feet
5,280 yards
3 statute miles (US)
4.82805 kilometers
Sometimes taken as 3 knots or nautical miles (US)

LINK equals:
0.01 of measuring chain (In the engineer's chain, each link is
12 inches long; in the Gunter's or surveyor's chain, each link is
7.92 inches long; in the metric chain, each link is 20 centimeters
long.)

LITER equals:
1 cu. decimeter
10 deciliters
1,000 cu. centimeters
0.01 hectoliter
0.001 cu. meter
61.0234 cu. inches
0.0353145 cu. foot
2.11336 pints (liquid; US)
1.05668 quarts (liquid; US)
0.264170 gallon (US)
1.81616 pints (dry; US)
0.908078 quart (dry; US)
0.113510 peck (US)
0.0283774 bushel (US)
1.75980 pints (Br)
0.879902 quart (Br)
0.219975 gallon (Br)
0.109988 peck (Br)
0.0274969 bushel (Br)

METER (international) equals:
0.001 kilometer
0.01 hectometer
0.1 dekameter
10 decimeters
100 centimeters
1,000 millimeters
1,000,000 micrometers
39.370113 inches (Br)
39.37 inches exact legal value (US)
3.28083 feet (US)
1.09361 yards (US)
0.000621370 statute mile (US)

MIL, circular, equals:
0.000001 circular inch
0.785398 sq. mil
0.000000785398 sq. inch
0.000645163 cir. millimeter
.000506709 sq. millimeter

EQUIVALENTS
OF MEASUREMENT

ap = apothecary; *av* = avoirdupois; *Br* = British; *US* = United States.

MILE, statute or land (US) equals:
5,280 feet
1.60935 kilometers

MILE² (sq. mile) equals:
640 acres
3,097,600 sq. yards
2.59000 sq. kilometers

MILLIMETER equals:
0.001 meter
39.370 mils
0.039370 inch

OUNCE (ap) same as troy ounce, equals:
480 grains
24 scruples
8 drams (ap)
1/12 or 0.0833333 pound (troy or ap)
31.1035 grams

OUNCE (av) equals:
16 drams (av)
1/16 or 0.062500 pound (av)
437.500 grains
28.3495 grams
0.911458 ounce (troy or ap)

OUNCE (troy, gold and silver) same as ap ounce, equals:
480 grains
20 pennyweights
1/12 or 0.0833333 pound (troy or ap)
0.6085714 pound (av)
31.1035 grams
1.09714 ounces (av)

OUNCE, fluid (ap US) equals:
480 minims (ap US)
8 fluid drams (ap US)
1/16 pint (ap US)
1.80469 cu. inches
0.0295737 liter

PECK (US) equals
8 quarts (dry; US)
0.25 bushel (US)
8.80982 liters

PENNYWEIGHT (troy) equals:
24 grains
1.55517 grams
1/20 ounce (troy or ap)

PERCH, linear, see rod

PERCH, of masonry, equals:
16 1/2 feet x 1 1/2 feet x 1 foot
24 3/4 cu. feet (generally taken as 25 cu. feet, sometimes 22 cu. feet)
0.70085 cu. meter

EQUIVALENTS
OF MEASUREMENT

ap = apothecary; av = avoirdupois; Br = British; US = United States.

PINT (dry; US) equals:
0.5 quart (dry; US)
0.550614 liter

PINT (liquid; US) equals:
0.125 gallon (US)
0.473179 liter

PIPE or butt (liquid; US) equals:
126 gallons (US)
2 hogsheads (US)
476.96 liters

POLE, see rod

POUND (av) equals:
7,000 grains
16 ounces (av)
256 drams (av)
14.5833 ounces (troy or ap)
453.5924277 grams
0.4535924277 kilogram
7000/5760 or 0.21528 pounds (troy or ap)

POUND (troy or ap) equals:
5,760 grains 12 ounces (troy or ap)
0.373242 kilogram
5760/7000 or 1.21528 pound (av)

QUART (dry; US) equals:
2 pints (dry; US)
1/8 or 0.125 peck (US)
1/32 or 0.031250 bushel (US)
67.200625 cu. inches
0.0388893 cu. foot
1.10123 liters
1,101.23 cu. centimeters
11.0123 deciliters
1.16365 quarts (liquid; US)
0.968972 quart (Br)
0.242243 gallon (Br)

QUART (liquid; US) equals:
0.25 gallon (US)
0.946359 liter

ROD or perch or pole, equals:
16 1/2 feet
5 1/2 yards
1/40 furlong
1/320 statute mile (US)
5.0292 meters

ROOD equals:
1/4 acre
40 sq. rods, poles or perches
1,210 sq. yards
1,011.72 sq. meters

SECTION (of land) equals:
1 mile square
640 acres

EQUIVALENTS
OF MEASUREMENT

ap = apothecary; av = avoirdupois; Br = British; US = United States.

SQUARE (building) equals:

100 sq. feet

TON (gross) displacement of water, equals:

35.8813 cu. feet
1.01605 cu. meters

TON register (shipping for whole vessels) equals:

100 cu. feet
2.8317 cu. meters

TON, long or gross, equals:

2,240 pounds (av)
1.12 short or net tons
1,016.05 kilograms
1.01605 metric tons

TON, short or net, equals:

2,000 pounds (av)
20 hundredweights (short)
907.185 kilograms
0.907185 metric ton
17.8571 hundredweights (long)
0.892857 long or gross ton

TON, metric, (tonne, tonneau, millier or bar) equals:

2,204.62 pounds (av)
1.10231 short or net tons
0.984206 long or gross ton
1,000 kilograms

YARD (US) equals:

36 inches
3 feet
1.0000029 yards (Br)
0.914402 meter

YARD² (sq. yd.) (US) equals:

1,296 sq. inches
9 sq. feet
1/4840 or 0.000206612 acre
0.836131 sq. meter
1.0000057 sq. yards (Br)

YARD³ (cu. yd.) equals:

27 cu. feet
46,656 cu. inches
0.764559 cu. meter

SOLUTION OF
RIGHT TRIANGLES

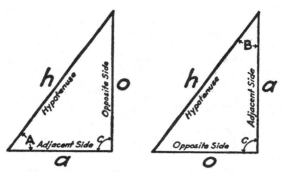

In any right triangle, if the side o = 4 and the hypotenuse h = 5, the ratio of o/h will be 4/5, which equals 0.80, and is called the *sine* of the given angle. Regardless of the size of the triangle the sine will always be the same if the angle is the same, and vice versa. For instance, if the sides were o = 8 and h = 10, or if o = 1.12 and h = 1.40, the sine in either case would be 0.80 and the angle would be 53°-07'-48".

This principle holds true for all of the six ratios, or *functions*, that can be made from the three sides. These functions with the usual abbreviations are as follows:

$$\frac{o}{h} = \text{Sine (sin)} \qquad \frac{h}{o} = \text{Cosecant (csc)}$$

$$\frac{a}{h} = \text{Cosine (cos)} \qquad \frac{h}{a} = \text{Secant (sec)}$$

$$\frac{o}{a} = \text{Tangent (tan)} \qquad \frac{a}{o} = \text{Cotangent (cot)}$$

Following *Data Sheets* give the numerical values of the various functions for different angles between 0° and 90° with which any unknown part of a right triangle can be found if two other parts are known. Knowing two of the three sides one of the functions is calculated and the corresponding angle can be found from the tables, or from geometry the third side can be computed. Knowing an angle, the proper function can be found from the tables to use as a multiplier of the known side to find the unknown side.

$$o = h \sin \qquad\qquad a = h \cos$$
$$ = a \tan \qquad\qquad = o \cot$$
$$ = \sqrt{(h+a)(h-a)} \qquad = \sqrt{(h+o)(h-o)}$$
$$h = a \sec \qquad\qquad C = 90°$$
$$ = o \csc \qquad\qquad = A + B$$
$$ = \sqrt{o^2 + a^2}$$

SINES 0° TO 45°
COSINES 45° TO 90°

COSINE		60'	50'	40'	30'	20'	10'	0'
	SINE	0'	10'	20'	30'	40'	50'	60'
89°	0°	.00000	.00291	.00582	.00873	.01164	.01454	.01745
88	1	.01745	.02036	.02327	.02618	.02908	.03199	.03490
87	2	.03490	.03781	.04071	.04362	.04653	.04943	.05234
86	3	.05234	.05524	.05814	.06105	.06395	.06685	.06976
85	4	.06976	.07266	.07556	.07846	.08136	.08426	.08716
84	5	.08716	.09005	.09295	.09585	.09874	.10164	.10453
83	6	.10453	.10742	.11031	.11320	.11609	.11898	.12187
82	7	.12187	.12476	.12764	.13053	.13341	.13629	.13917
81	8	.13917	.14205	.14493	.14781	.15069	.15356	.15643
80	9	.15643	.15931	.16218	.16505	.16792	.17078	.17365
79	10	.17365	.17651	.17937	.18224	.18509	.18795	.19081
78	11	.19081	.19366	.19652	.19937	.20222	.20507	.20791
77	12	.20791	.21076	.21360	.21644	.21928	.22212	.22495
76	13	.22495	.22778	.23062	.23345	.23627	.23910	.24192
75	14	.24192	.24474	.24756	.25038	.25320	.25601	.25882
74	15	.25882	.26163	.26443	.26724	.27004	.27284	.27564
73	16	.27564	.27843	.28123	.28402	.28680	.28959	.29237
72	17	.29237	.29515	.29793	.30071	.30348	.30625	.30902
71	18	.30902	.31178	.31454	.31730	.32006	.32282	.32557
70	19	.32557	.32832	.33106	.33381	.33655	.33929	.34202
69	20	.34202	.34475	.34748	.35021	.35293	.35565	.35837
68	21	.35837	.36108	.36379	.36650	.36921	.37191	.37461
67	22	.37461	.37730	.37999	.38268	.38537	.38805	.39073
66	23	.39073	.39341	.39608	.39875	.40142	.40408	.40674
65	24	.40674	.40939	.41204	.41469	.41734	.41998	.42262
64	25	.42262	.42525	.42788	.43051	.43313	.43575	.43837
63	26	.43837	.44098	.44359	.44620	.44880	.45140	.45399
62	27	.45399	.45658	.45917	.46175	.46433	.46690	.46947
61	28	.46947	.47204	.47460	.47716	.47971	.48226	.48481
60	29	.48481	.48735	.48989	.49242	.49495	.49748	.50000
59	30	.50000	.50252	.50503	.50754	.51004	.51254	.51504
58	31	.51504	.51753	.52002	.52250	.52498	.52745	.52992
57	32	.52992	.53238	.53484	.53730	.53975	.54220	.54464
56	33	.54464	.54708	.54951	.55194	.55436	.55678	.55919
55	34	.55919	.56160	.56401	.56641	.56880	.57119	.57358
54	35	.57358	.57596	.57833	.58070	.58307	.58543	.58779
53	36	.58779	.59014	.59248	.59482	.59716	.59949	.60182
52	37	.60182	.60414	.60645	.60876	.61107	.61337	.61566
51	38	.61566	.61795	.62024	.62251	.62479	.62706	.62932
50	39	.62932	.63158	.63383	.63608	.63832	.64056	.64279
49	40	.64279	.64501	.64723	.64945	.65166	.65386	.65606
48	41	.65606	.65825	.66044	.66262	.66480	.66697	.66913
47	42	.66913	.67129	.67344	.67559	.67773	.67987	.68200
46	43	.68200	.68412	.68624	.68835	.69046	.69256	.69466
45	44	.69466	.69675	.69883	.70091	.70298	.70505	.70711

SINES 45° TO 90°
COSINES 0° TO 45°

COSINE		60'	50'	40'	30'	20'	10'	0'
	SINE	0'	10'	20'	30'	40'	50'	60'
44°	45°	.70711	.70916	.71121	.71325	.71529	.71732	.71934
43	46	.71934	.72136	.72337	.72537	.72737	.72937	.73135
42	47	.73135	.73333	.73531	.73728	.73924	.74120	.74314
41	48	.74314	.74509	.74703	.74896	.75088	.75280	.75471
40	49	.75471	.75661	.75851	.76041	.76229	.76417	.76604
39	50	.76604	.76791	.76977	.77162	.77347	.77531	.77715
38	51	.77715	.77897	.78079	.78261	.78442	.78622	.78801
37	52	.78801	.78980	.79158	.79335	.79512	.79688	.79864
36	53	.79864	.80038	.80212	.80386	.80558	.80730	.80902
35	54	.80902	.81072	.81242	.81412	.81580	.81748	.81915
34	55	.81915	.82082	.82248	.82413	.82577	.82741	.82904
33	56	.82904	.83066	.83228	.83389	.83549	.83708	.83867
32	57	.83867	.84025	.84182	.84339	.84495	.84650	.84805
31	58	.84805	.84959	.85112	.85264	.85416	.85567	.85717
30	59	.85717	.85866	.86015	.86163	.86310	.86457	.86603
29	60	.86603	.86748	.86892	.87036	.87178	.87321	.87462
28	61	.87462	.87603	.87743	.87882	.88020	.88158	.88295
27	62	.88295	.88431	.88566	.88701	.88835	.88968	.89101
26	63	.89101	.89232	.89363	.89493	.89623	.89752	.89879
25	64	.89879	.90007	.90133	.90259	.90383	.90507	.90631
24	65	.90631	.90753	.90875	.90996	.91116	.91236	.91355
23	66	.91355	.91472	.91590	.91706	.91822	.91936	.92050
22	67	.92050	.92164	.92276	.92388	.92499	.92609	.92718
21	68	.92718	.92827	.92935	.93042	.93148	.93253	.93358
20	69	.93358	.93462	.93565	.93667	.93769	.93869	.93969
19	70	.93969	.94068	.94167	.94264	.94361	.94457	.94552
18	71	.94552	.94646	.94740	.94832	.94924	.95015	.95106
17	72	.95106	.95195	.95284	.95372	.95459	.95545	.95630
16	73	.95630	.95715	.95799	.95882	.95964	.96046	.96126
15	74	.96126	.96206	.96285	.96363	.96440	.96517	.96593
14	75	.96593	.96667	.96742	.96815	.96887	.96959	.97030
13	76	.97030	.97100	.97169	.97237	.97304	.97371	.97437
12	77	.97437	.97502	.97566	.97630	.97692	.97754	.97815
11	78	.97815	.97875	.97934	.97992	.98050	.98107	.98163
10	79	.98163	.98218	.98272	.98325	.98378	.98430	.98481
9	80	.98481	.98531	.98580	.98629	.98676	.98723	.98769
8	81	.98769	.98814	.98858	.98902	.98944	.98986	.99027
7	82	.99027	.99067	.99106	.99144	.99182	.99219	.99255
6	83	.99255	.99290	.99324	.99357	.99390	.99421	.99452
5	84	.99452	.99482	.99511	.99540	.99567	.99594	.99619
4	85	.99619	.99644	.99668	.99692	.99714	.99736	.99756
3	86	.99756	.99776	.99795	.99813	.99831	.99847	.99863
2	87	.99863	.99878	.99892	.99905	.99917	.99929	.99939
1	88	.99939	.99949	.99958	.99966	.99973	.99979	.99985
0	89	.99985	.99989	.99993	.99996	.99998	1.0000	1.0000

TANGENTS 0° TO 45°
COTANGENTS 45° TO 90°

COTAN		60′	50′	40′	30′	20′	10′	0′
	TAN	0′	10′	20′	30′	40′	50′	60′
89°	0°	.00000	.00291	.00582	.00873	.01164	.01455	.01746
88	1	.01746	.02036	.02328	.02619	.02910	.03201	.03492
87	2	.03492	.03783	.04075	.04366	.04658	.04949	.05241
86	3	.05241	.05533	.05824	.06116	.06408	.06700	.06993
85	4	.06993	.07285	.07578	.07870	.08163	.08456	.08749
84	5	.08749	.09042	.09335	.09629	.09923	.10216	.10510
83	6	.10510	.10805	.11099	.11394	.11688	.11983	.12278
82	7	.12278	.12574	.12869	.13165	.13461	.13758	.14054
81	8	.14054	.14351	.14648	.14945	.15243	.15540	.15838
80	9	.15838	.16137	.16435	.16734	.17033	.17333	.17633
79	10	.17633	.17933	.18233	.18534	.18835	.19136	.19438
78	11	.19438	.19740	.20042	.20345	.20648	.20952	.21256
77	12	.21256	.21560	.21864	.22169	.22475	.22781	.23087
76	13	.23087	.23393	.23700	.24008	.24316	.24624	.24933
75	14	.24933	.25242	.25552	.25862	.26172	.26483	.26795
74	15	.26795	.27107	.27419	.27732	.28046	.28360	.28675
73	16	.28675	.28990	.29305	.29621	.29938	.30255	.30573
72	17	.30573	.30891	.31210	.31530	.31850	.32171	.32492
71	18	.32492	.32814	.33136	.33460	.33783	.34108	.34433
70	19	.34433	.34758	.35085	.35412	.35740	.36068	.36397
69	20	.36397	.36727	.37057	.37388	.37720	.38053	.38386
68	21	.38386	.38721	.39055	.39391	.39727	.40065	.40403
67	22	.40403	.40741	.41081	.41421	.41763	.42105	.42447
66	23	.42447	.42791	.43136	.43481	.43828	.44175	.44523
65	24	.44523	.44872	.45222	.45573	.45924	.46277	.46631
64	25	.46631	.46985	.47341	.47698	.48055	.48414	.48773
63	26	.48773	.49134	.49495	.49858	.50222	.50587	.50953
62	27	.50953	.51320	.51688	.52057	.52427	.52798	.53171
61	28	.53171	.53545	.53920	.54296	.54674	.55051	.55431
60	29	.55431	.55812	.56194	.56577	.56962	.57348	.57735
59	30	.57735	.58124	.58513	.58905	.59297	.59691	.60086
58	31	.60086	.60483	.60881	.61280	.61681	.62083	.62487
57	32	.62487	.62892	.63299	.63707	.64117	.64528	.64941
56	33	.64941	.65355	.65771	.66189	.66608	.67028	.67451
55	34	.67451	.67875	.68301	.68728	.69157	.69588	.70021
54	35	.70021	.70455	.70891	.71329	.71769	.72211	.72654
53	36	.72654	.73100	.73547	.73996	.74447	.74900	.75355
52	37	.75355	.75812	.76272	.76733	.77196	.77661	.78129
51	38	.78129	.78598	.79070	.79544	.80020	.80498	.80978
50	39	.80978	.81461	.81946	.82434	.82923	.83415	.83910
49	40	.83910	.84407	.84906	.85408	.85912	.86419	.86929
48	41	.86929	.87441	.87955	.88473	.88992	.89515	.90040
47	42	.90040	.90569	.91099	.91633	.92170	.92709	.93252
46	43	.93252	.93797	.94345	.94896	.95451	.96008	.96569
45	4	.96569	.97133	.97700	.98270	.98843	.99420	1.0000

TANGENTS 45° TO 90°
COTANGENTS 0° TO 45°

COTAN		60'	50'	40'	30'	20'	10'	0'
	TAN	0'	10'	20'	30'	40'	50'	60'
44°	45°	1.0000	1.0058	1.0117	1.0176	1.0236	1.0295	1.0355
43	46	1.0355	1.0416	1.0477	1.0538	1.0599	1.0661	1.0724
42	47	1.0724	1.0786	1.0850	1.0913	1.0977	1.1041	1.1106
41	48	1.1106	1.1171	1.1237	1.1303	1.1369	1.1436	1.1504
40	49	1.1504	1.1572	1.1640	1.1709	1.1778	1.1847	1.1918
39	50	1.1918	1.1988	1.2059	1.2131	1.2203	1.2276	1.2349
38	51	1.2349	1.2423	1.2497	1.2572	1.2647	1.2723	1.2799
37	52	1.2799	1.2876	1.2954	1.3032	1.3111	1.3190	1.3270
36	53	1.3270	1.3351	1.3432	1.3514	1.3597	1.3680	1.3764
35	54	1.3764	1.3848	1.3934	1.4020	1.4106	1.4193	1.4282
34	55	1.4282	1.4370	1.4460	1.4550	1.4641	1.4733	1.4826
33	56	1.4826	1.4919	1.5013	1.5108	1.5204	1.5301	1.5399
32	57	1.5399	1.5497	1.5597	1.5697	1.5798	1.5900	1.6003
31	58	1.6003	1.6107	1.6213	1.6319	1.6426	1.6534	1.6643
30	59	1.6643	1.6753	1.6864	1.6977	1.7090	1.7205	1.7321
29	60	1.7321	1.7438	1.7556	1.7675	1.7796	1.7917	1.8041
28	61	1.8041	1.8165	1.8291	1.8418	1.8546	1.8676	1.8807
27	62	1.8807	1.8940	1.9074	1.9210	1.9347	1.9486	1.9626
26	63	1.9626	1.9768	1.9912	2.0057	2.0204	2.0353	2.0503
25	64	2.0503	2.0655	2.0809	2.0965	2.1123	2.1283	2.1445
24	65	2.1445	2.1609	2.1775	2.1943	2.2113	2.2286	2.2460
23	66	2.2460	2.2637	2.2817	2.2998	2.3183	2.3369	2.3559
22	67	2.3559	2.3750	2.3945	2.4142	2.4342	2.4545	2.4751
21	68	2.4751	2.4960	2.5172	2.5387	2.5605	2.5826	2.6051
20	69	2.6051	2.6279	2.6511	2.6746	2.6985	2.7228	2.7475
19	70	2.7475	2.7725	2.7980	2.8239	2.8502	2.8770	2.9042
18	71	2.9042	2.9319	2.9600	2.9887	3.0178	3.0475	3.0777
17	72	3.0777	3.1084	3.1397	3.1716	3.2041	3.2371	3.2709
16	73	3.2709	3.3052	3.3402	3.3759	3.4124	3.4495	3.4874
15	74	3.4874	3.5261	3.5656	3.6059	3.6471	3.6891	3.7321
14	75	3.7321	3.7760	3.8208	3.8667	3.9136	3.9617	4.0108
13	76	4.0108	4.0611	4.1126	4.1653	4.2193	4.2747	4.3315
12	77	4.3315	4.3897	4.4494	4.5107	4.5736	4.6383	4.7046
11	78	4.7046	4.7729	4.8430	4.9152	4.9894	5.0658	5.1446
10	79	5.1446	5.2257	5.3093	5.3955	5.4845	5.5764	5.6713
9	80	5.6713	5.7694	5.8708	5.9758	6.0844	6.1970	6.3138
8	81	6.3138	6.4348	6.5606	6.6912	6.8269	6.9682	7.1154
7	82	7.1154	7.2687	7.4287	7.5958	7.7704	7.9530	8.1444
6	83	8.1444	8.3450	8.5556	8.7769	9.0098	9.2553	9.5144
5	84	9.5144	9.7882	10.078	10.385	10.712	11.059	11.430
4	85	11.430	11.826	12.251	12.706	13.197	13.727	14.301
3	86	14.301	14.924	15.605	16.350	17.169	18.075	19.081
2	87	19.081	20.206	21.470	22.904	24.542	26.432	28.636
1	88	28.636	31.242	34.368	38.188	42.964	49.104	57.290
0	90	57.290	68.750	85.940	114.59	171.89	343.77	infin.

SECANTS 0° TO 45°
COSECANTS 45° TO 90°

COSEC	SEC	60′	50′	40′	30′	20′	10′	0′
		0′	10′	20′	30′	40′	50′	60′
89°	0°	1.0000	1.0000	1.0000	1.0000	1.0001	1.0001	1.0002
88	1	1.0002	1.0002	1.0003	1.0003	1.0004	1.0005	1.0006
87	2	1.0006	1.0007	1.0008	1.0010	1.0011	1.0012	1.0014
86	3	1.0014	1.0015	1.0017	1.0019	1.0021	1.0022	1.0024
85	4	1.0024	1.0027	1.0029	1.0031	1.0033	1.0036	1.0038
84	5	1.0038	1.0041	1.0044	1.0046	1.0049	1.0052	1.0055
83	6	1.0055	1.0058	1.0061	1.0065	1.0068	1.0072	1.0075
82	7	1.0075	1.0079	1.0083	1.0086	1.0090	1.0094	1.0098
81	8	1.0098	1.0102	1.0107	1.0111	1.0116	1.0120	1.0125
80	9	1.0125	1.0129	1.0134	1.0139	1.0144	1.0149	1.0154
79	10	1.0154	1.0160	1.0165	1.0170	1.0176	1.0182	1.0187
78	11	1.0187	1.0193	1.0199	1.0205	1.0211	1.0217	1.0223
77	12	1.0223	1.0230	1.0236	1.0243	1.0249	1.0256	1.0263
76	13	1.0263	1.0270	1.0277	1.0284	1.0291	1.0299	1.0306
75	14	1.0306	1.0314	1.0321	1.0329	1.0337	1.0345	1.0353
74	15	1.0353	1.0361	1.0369	1.0377	1.0386	1.0394	1.0403
73	16	1.0403	1.0412	1.0421	1.0430	1.0439	1.0448	1.0457
72	17	1.0457	1.0466	1.0476	1.0485	1.0495	1.0505	1.0515
71	18	1.0515	1.0525	1.0535	1.0545	1.0555	1.0566	1.0576
70	19	1.0576	1.0587	1.0598	1.0609	1.0620	1.0631	1.0642
69	20	1.0642	1.0653	1.0665	1.0676	1.0688	1.0700	1.0712
68	21	1.0712	1.0724	1.0736	1.0748	1.0760	1.0773	1.0785
67	22	1.0785	1.0798	1.0811	1.0824	1.0837	1.0850	1.0864
66	23	1.0864	1.0877	1.0891	1.0904	1.0918	1.0932	1.0946
65	24	1.0946	1.0961	1.0975	1.0990	1.1004	1.1019	1.1034
64	25	1.1034	1.1049	1.1064	1.1079	1.1095	1.1110	1.1126
63	26	1.1126	1.1142	1.1158	1.1174	1.1190	1.1207	1.1223
62	27	1.1223	1.1240	1.1257	1.1274	1.1291	1.1308	1.1326
61	28	1.1326	1.1343	1.1361	1.1379	1.1397	1.1415	1.1434
60	29	1.1434	1.1452	1.1471	1.1490	1.1509	1.1528	1.1547
59	30	1.1547	1.1567	1.1586	1.1606	1.1626	1.1646	1.1666
58	31	1.1666	1.1687	1.1708	1.1728	1.1749	1.1770	1.1792
57	32	1.1792	1.1813	1.1835	1.1857	1.1879	1.1901	1.1924
56	33	1.1924	1.1946	1.1969	1.1992	1.2015	1.2039	1.2062
55	34	1.2062	1.2086	1.2110	1.2134	1.2158	1.2183	1.2208
54	35	1.2208	1.2233	1.2258	1.2283	1.2309	1.2335	1.2361
53	36	1.2361	1.2387	1.2413	1.2440	1.2467	1.2494	1.2521
52	37	1.2521	1.2549	1.2577	1.2605	1.2633	1.2662	1.2690
51	38	1.2690	1.2719	1.2748	1.2778	1.2808	1.2837	1.2868
50	39	1.2868	1.2898	1.2929	1.2960	1.2991	1.3022	1.3054
49	40	1.3054	1.3086	1.3118	1.3151	1.3184	1.3217	1.3250
48	41	1.3250	1.3284	1.3318	1.3352	1.3386	1.3421	1.3456
47	42	1.3456	1.3492	1.3527	1.3563	1.3600	1.3636	1.3673
46	43	1.3673	1.3711	1.3748	1.3786	1.3824	1.3863	1.3902
45	44	1.3902	1.3941	1.3980	1.4020	1.4061	1.4101	1.4142

SECANTS 45° TO 90°
COSECANTS 0° TO 45°

COSEC		60′	50′	40′	30′	20′	10′	0′
	SEC	0′	10′	20′	30′	40′	50′	60′
44°	45°	1.4142	1.4184	1.4225	1.4267	1.4310	1.4352	1.4396
43	46	1.4396	1.4439	1.4483	1.4527	1.4572	1.4617	1.4663
42	47	1.4663	1.4709	1.4755	1.4802	1.4849	1.4897	1.4945
41	48	1.4945	1.4993	1.5042	1.5092	1.5142	1.5192	1.5243
40	49	1.5243	1.5294	1.5346	1.5398	1.5450	1.5504	1.5557
39	50	1.5557	1.5611	1.5666	1.5721	1.5777	1.5833	1.5890
38	51	1.5890	1.5948	1.6005	1.6064	1.6123	1.6183	1.6243
37	52	1.6243	1.6304	1.6365	1.6427	1.6489	1.6553	1.6616
36	53	1.6616	1.6681	1.6746	1.6812	1.6878	1.6945	1.7013
35	54	1.7013	1.7082	1.7151	1.7221	1.7291	1.7362	1.7435
34	55	1.7435	1.7507	1.7581	1.7655	1.7730	1.7806	1.7883
33	56	1.7883	1.7960	1.8039	1.8118	1.8198	1.8279	1.8361
32	57	1.8361	1.8444	1.8527	1.8612	1.8697	1.8783	1.8871
31	58	1.8871	1.8959	1.9049	1.9139	1.9230	1.9323	1.9416
30	59	1.9416	1.9511	1.9606	1.9703	1.9801	1.9900	2.0000
29	60	2.0000	2.0101	2.0204	2.0308	2.0413	2.0519	2.0627
28	61	2.0627	2.0736	2.0846	2.0957	2.1070	2.1185	2.1301
27	62	2.1301	2.1418	2.1537	2.1657	2.1779	2.1902	2.2027
26	63	2.2027	2.2154	2.2282	2.2412	2.2543	2.2677	2.2812
25	64	2.2812	2.2949	2.3088	2.3228	2.3371	2.3515	2.3662
24	65	2.3662	2.3811	2.3961	2.4114	2.4269	2.4426	2.4586
23	66	2.4586	2.4748	2.4912	2.5078	2.5247	2.5419	2.5593
22	67	2.5593	2.5770	2.5949	2.6131	2.6316	2.6504	2.6695
21	68	2.6695	2.6888	2.7085	2.7285	2.7488	2.7695	2.7904
20	69	2.7904	2.8118	2.8334	2.8555	2.8779	2.9006	2.9238
19	70	2.9238	2.9474	2.9714	2.9957	3.0206	3.0458	3.0716
18	71	3.0716	3.0977	3.1244	3.1516	3.1792	3.2074	3.2361
17	72	3.2361	3.2653	3.2951	3.3255	3.3565	3.3881	3.4203
16	73	3.4203	3.4532	3.4867	3.5209	3.5559	3.5915	3.6280
15	74	3.6280	3.6652	3.7032	3.7420	3.7817	3.8222	3.8637
14	75	3.8637	3.9061	3.9495	3.9939	4.0394	4.0859	4.1336
13	76	4.1336	4.1824	4.2324	4.2837	4.3362	4.3901	4.4454
12	77	4.4454	4.5022	4.5604	4.6202	4.6817	4.7448	4.8097
11	78	4.8097	4.8765	4.9452	5.0159	5.0886	5.1636	5.2408
10	79	5.2408	5.3205	5.4026	5.4874	5.5749	5.6653	5.7588
9	80	5.7588	5.8554	5.9554	6.0589	6.1661	6.2772	6.3925
8	81	6.3925	6.5121	6.6363	6.7655	6.8998	7.0396	7.1853
7	82	7.1853	7.3372	7.4957	7.6613	7.8344	8.0157	8.2055
6	83	8.2055	8.4047	8.6138	8.8337	9.0652	9.3092	9.5668
5	84	9.5668	9.8391	10.128	10.433	10.759	11.105	11.474
4	85	11.474	11.868	12.291	12.746	13.235	13.763	14.336
3	86	14.336	14.958	15.637	16.380	17.198	18.103	19.107
2	87	19.107	20.230	21.494	22.926	24.562	26.451	28.654
1	88	28.654	31.258	34.382	38.202	42.976	49.114	57.299
0	89	57.299	68.757	85.946	114.59	171.89	343.78	inf.

HOW TO USE
LOGARITHMS

EXPONENT OF NUMBERS. An exponent, or power, or index, is a small number written slightly above and to the right of a number to indicate how many times the number is to be taken as a factor in the product. Suppose we assume that a = 2, then;

$$a^2 = aa = 2 \times 2 = 4$$
$$a^3 = aaa = 2 \times 2 \times 2 = 8$$

If we wish to multiply like numbers, we add the exponents, thus;

$$a^2 \times a^3 = aa \times aaa = a^{2+3} = a^5 = (2 \times 2) \times (2 \times 2 \times 2) = 32$$

If we wish to divide like numbers, we subtract exponents, thus;

$$\frac{a^3}{a^2} = \frac{aaa}{aa} = a^{3-2} = a^1 = \frac{2 \times 2 \times 2}{2 \times 2} = \frac{8}{4} = 2$$

LOGARITHMS ARE EXPONENTS OF 10. Any number can be expressed as a power of 10. This power, or exponent, consists of two parts known as the *characteristic* and the *mantissa*.

THE CHARACTERISTIC. The characteristic of a number that has one or more digits to the left of the decimal point will be one less than the number of digits, and will be positive.

The characteristic of a decimal fraction will be a number representing the position of the first significant figure to the right of the decimal point, and will be negative.

The exponents of 10 in the following table are the characteristics for the numbers given in the left column;

$$100,000.000 = 10^5$$
$$10,000.000 = 10^4$$
$$1,000.000 = 10^3$$
$$100.000 = 10^2$$
$$10.000 = 10^1$$
$$1.000 = 10^0$$
$$0.100 = 10^{-1} \text{ (also written } \overline{10}^1, \text{ or } 10^{9-10})$$
$$0.010 = 10^{-2} \text{ (also written } \overline{10}^2, \text{ or } 10^{8-10})$$
$$0.001 = 10^{-3} \text{ (also written } \overline{10}^3, \text{ or } 10^{7-10})$$

THE MANTISSA. Now, suppose we want to express some number such as 246 in terms of 10 raised to a power. The number 246 falls between 100 and 1,000, or between 10^2 and 10^3. In other words, the exponent will be 2 plus some fraction. Such a fraction is called a mantissa. So-called Tables of Logarithms are not tables of *Logarithms* at all—they are really Tables of Mantissas or that fractional part of the exponent of 10 for any number that is not an exact multiple of 10. Thus we find that the *mantissa* of the number 246 is 39094, which means that the *logarithm* is 2.39094. This is written;

$$\log 246 = 2.39094$$

The mantissa is always positive, depends only upon the sequence of the digits in the number without regard to the position of any decimal point. The mantissa for 2,460,000 is the same as for 246 or 0.0000246 or 2.46. But the characteristics are different;

$$\log 2,460,000 = 6.39094$$
$$\log 246 = 2.39094$$
$$\log 0.0000246 = 5.39094$$
$$\log 2.46 = 0.39094$$

Thus, in looking up the mantissa of a number in the tables, remember it is the same for 24 as for 2400. This eliminates the need for values below 100 in a table of 100 to 1.000. In a table from 1,000 to 10,000 values below 1,000 are not needed.

HOW TO USE
LOGARITHMS

HOW TO USE THE TABLES. In the tables that follow, notice that under the column headed "O" there are some mantissas with 5 digits, while those under the remaining columns have but 3 digits. The first two digits of the 5-digit mantissas apply to all values reading across the page until reaching a series with asterisks, or until the next 5-digit mantissa is reached.

To find the mantissa of the number 1234: Read across from 123 to the column headed 4. You will find the value *132. In the next line under the "O" column you will find a 5-digit value whose 1st two figures are 09. Therefore the mantissa of the number 1234 is 09132.

USE OF THE "D" COLUMN. The mantissas in the tables are given for numbers with 4 significant figures. The 5th significant figure can be found by using the column of differences which gives the average numerical difference in each line between the values given for the 4-figure numbers.

To find the mantissa of the number 24643: This will fall between the mantissas for the numbers 2464 and 2465, which are 39164 and 39182. (*See Diagram*). The difference is 18. The 5th figure we want in the number is 3. so .3 x 18 = 5.4. 39164 + 5 = 39169, which is the mantissa of 24643.

To find the number whose mantissa is 39169: This is 5 more than the nearest lower mantissa 39164, for the number 2464. (*See Diagram*). Since the difference between this and the next higher mantissa is 18, take 5/18 x 10 = 2.78 which we can call 3. Therefore the number sought is 24643.

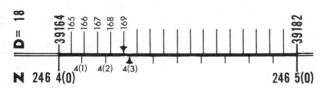

EXAMPLE OF MULTIPLICATION. What is the product of 246 x 10.43?

$$\begin{aligned} \log 246 &= 2.39094 \\ \log 10.43 &= 1.01828 \\ \hline \text{total} &= 3.40922 \end{aligned}$$

Since the characteristic is 3, we know there will be 4 integers in the result. The table shows that 40922 is the mantissa for 25658, and the number sought is therefore 2565.8.

EXAMPLE OF DIVISION. What is 246 divided by 10.43?

$$\begin{aligned} \log 246 &= 2.39094 \\ \log 10.43 &= 1.01828 \\ \hline \text{subtract} &= 1.37266 \end{aligned}$$

Since the characteristic is 1, we know there will be 2 integers in the quotient. The table shows that 37266 is the mantissa for 23586 and the number sought is 23.586.

N	0	1	2	3	4	5	6	7	8	9	D
100	00000	043	087	130	173	217	260	303	346	389	43
101	432	475	518	561	604	647	689	732	775	817	43
102	860	903	945	988	*030	*072	*115	*157	*199	*242	42
103	01284	326	368	410	452	494	536	578	620	662	42
104	703	745	787	828	870	912	953	995	*036	*078	42
105	02119	160	202	243	284	325	366	407	449	490	41
106	531	572	612	653	694	735	776	816	857	898	41
107	938	979	*019	*060	*100	*141	*181	*222	*262	*302	40
108	03342	383	423	463	503	543	583	623	663	703	40
109	743	782	822	862	902	941	981	*021	*060	*100	40
110	04139	179	218	258	297	336	376	415	454	493	39
111	532	571	610	650	689	727	766	805	844	883	39
112	922	961	999	*038	*077	*115	*154	*192	*231	*269	39
113	05308	346	385	423	461	500	538	576	614	652	38
114	690	729	767	805	843	881	918	956	994	*032	38
115	06070	108	145	183	221	258	296	333	371	408	38
116	446	483	521	558	595	633	670	707	744	781	37
117	819	856	893	930	967	*004	*041	*078	*115	*151	37
118	07188	225	262	298	335	372	408	445	482	518	37
119	555	591	628	664	700	737	773	809	846	882	36
120	918	954	990	*027	*063	*099	*135	*171	*207	*243	36
121	08279	314	350	386	422	458	493	529	565	600	36
122	636	672	707	743	778	814	849	884	920	955	35
123	991	*026	*061	*096	*132	*167	*202	*237	*272	*307	35
124	09342	377	412	447	482	517	552	587	621	656	35
125	691	726	760	795	830	864	899	934	968	*003	35
126	10037	072	106	140	175	209	243	278	312	346	34
127	380	415	449	483	517	551	585	619	653	687	34
128	721	755	789	823	857	890	924	958	992	*025	34
129	11059	093	126	160	193	227	261	294	327	361	34
130	394	428	461	494	528	561	594	628	661	694	33
131	727	760	793	826	860	893	926	959	992	*024	33
132	12057	090	123	156	189	222	254	287	320	352	33
133	385	418	450	483	516	548	581	613	646	678	33
134	710	743	775	808	840	872	905	937	969	*001	32
135	13033	066	098	130	162	194	226	258	290	322	32
136	354	386	418	450	481	513	545	577	609	640	32
137	672	704	735	767	799	830	862	893	925	956	32
138	988	*019	*051	*082	*114	*145	*176	*208	*239	*270	31
139	14301	333	364	395	426	457	489	520	551	582	31
140	613	644	675	706	737	768	799	829	860	891	31
141	922	953	983	*014	*045	*076	*106	*137	*168	*198	31
142	15229	259	290	320	351	381	412	442	473	503	31
143	534	564	594	625	655	685	715	746	776	806	30
144	836	866	897	927	957	987	*017	*047	*077	*107	30
145	16137	167	197	227	256	286	316	346	376	406	30
146	435	465	495	524	554	584	613	643	673	702	30
147	732	761	791	820	850	879	909	938	967	997	29
148	17026	056	085	114	143	173	202	231	260	289	29
149	319	348	377	406	435	464	493	522	551	580	29
N	0	1	2	3	4	5	6	7	8	9	D

N	0	1	2	3	4	5	6	7	8	9	D
150	17609	638	667	696	725	754	782	811	840	869	29
151	898	926	955	984	*013	*041	*070	*099	*127	*156	29
152	18184	213	241	270	298	327	355	384	412	441	29
153	469	498	526	554	583	611	639	667	696	724	28
154	752	780	808	837	865	893	921	949	977	*005	28
155	19033	061	089	117	145	173	201	229	257	285	28
156	312	340	368	396	424	451	479	507	535	562	28
157	590	618	645	673	700	728	756	783	811	838	28
158	866	893	921	948	976	*003	*030	*058	*085	*112	27
159	20140	167	194	222	249	276	303	330	358	385	27
160	412	439	466	493	520	548	575	602	629	656	27
161	683	710	737	763	790	817	844	871	898	925	27
162	952	978	*005	*032	*059	*085	*112	*139	*165	*192	27
163	21219	245	272	299	325	352	378	405	431	458	27
164	484	511	537	564	590	617	643	669	696	722	26
165	748	775	801	827	854	880	906	932	958	985	26
166	22011	037	063	089	115	141	167	194	220	246	26
167	272	298	324	350	376	401	427	453	479	505	26
168	531	557	583	608	634	660	686	712	737	763	26
169	789	814	840	866	891	917	943	968	994	*019	26
170	23045	070	096	121	147	172	198	223	249	274	25
171	300	325	350	376	401	426	452	477	502	528	25
172	553	578	603	629	654	679	704	729	754	779	25
173	805	830	855	880	905	930	955	980	*005	*030	25
174	24055	080	105	130	155	180	204	229	254	279	25
175	304	329	353	378	403	428	452	477	502	527	25
176	551	576	601	625	650	674	699	724	748	773	25
177	797	822	846	871	895	920	944	969	993	*018	25
178	25042	066	091	115	139	164	188	212	237	261	24
179	285	310	334	358	382	406	431	455	479	503	24
180	527	551	575	600	624	648	672	696	720	744	24
181	768	792	816	840	864	888	912	935	959	983	24
182	26007	031	055	079	102	126	150	174	198	221	24
183	245	269	293	316	340	364	387	411	435	458	24
184	482	505	529	553	576	600	623	647	670	694	24
185	717	741	764	788	811	834	858	881	905	928	23
186	951	975	998	*021	*045	*068	*091	*114	*138	*161	23
187	27184	207	231	254	277	300	323	346	370	393	23
188	416	439	462	485	508	531	554	577	600	623	23
189	646	669	692	715	738	761	784	807	830	852	23
190	875	898	921	944	967	989	*012	*035	*058	*081	23
191	28103	126	149	171	194	217	240	262	285	307	23
192	330	353	375	398	421	443	466	488	511	533	23
193	556	578	601	623	646	668	691	713	735	758	22
194	780	803	825	847	870	892	914	937	959	981	22
195	29003	026	048	070	092	115	137	159	181	203	22
196	226	248	270	292	314	336	358	380	403	425	22
197	447	469	491	513	535	557	579	601	623	645	22
198	667	688	710	732	754	776	798	820	842	863	22
199	885	907	929	951	973	994	*016	*038	*060	*081	22
N	0	1	2	3	4	5	6	7	8	9	D

MANTISSA OF NUMBERS
2000 TO 2499

N	0	1	2	3	4	5	6	7	8	9	D
200	30103	125	146	168	190	211	233	255	276	298	22
201	320	341	363	384	406	428	449	471	492	514	22
202	535	557	578	600	621	643	664	685	707	728	21
203	750	771	792	814	835	856	878	899	920	942	21
204	963	984	*006	*027	*048	*069	*091	*112	*133	*154	21
205	31175	197	218	239	260	281	302	323	345	366	21
206	387	408	429	450	471	492	513	534	555	576	21
207	597	618	639	660	681	702	723	744	765	785	21
208	806	827	848	869	890	911	931	952	973	994	21
209	32015	035	056	077	098	118	139	160	181	201	21
210	222	243	263	284	305	325	346	366	387	408	21
211	428	449	469	490	510	531	552	572	593	613	20
212	634	654	675	695	715	736	756	777	797	818	20
213	838	858	879	899	919	940	960	980	*001	*021	20
214	33041	062	082	102	122	143	163	183	203	224	20
215	244	264	284	304	325	345	365	385	405	425	20
216	445	465	486	506	526	546	566	586	606	626	20
217	646	666	686	706	726	746	766	786	806	826	20
218	846	866	885	905	925	945	965	985	*005	*025	20
219	34044	064	084	104	124	143	163	183	203	223	20
220	242	262	282	301	321	341	361	380	400	420	20
221	439	459	479	498	518	537	557	577	596	616	20
222	635	655	674	694	713	733	753	772	792	811	19
223	830	850	869	889	908	928	947	967	986	*005	19
224	35025	044	064	083	102	122	141	160	180	199	19
225	218	238	257	276	295	315	334	353	372	392	19
226	411	430	449	468	488	507	526	545	564	583	19
227	603	622	641	660	679	698	717	736	755	774	19
228	793	813	832	851	870	889	908	927	946	965	19
229	984	*003	*021	*040	*059	*078	*097	*116	*135	*154	19
230	36:73	192	211	229	248	267	286	305	324	342	19
231	361	380	399	418	436	455	474	493	511	530	19
232	549	568	586	605	624	642	661	680	698	717	19
233	736	754	773	791	810	829	847	866	884	903	19
234	922	940	959	977	996	*014	*033	*051	*070	*088	18
235	37107	125	144	162	181	199	218	236	254	273	18
236	291	310	328	346	365	383	401	420	438	457	18
237	475	493	511	530	548	566	585	603	621	639	18
238	658	676	694	712	731	749	767	785	803	822	18
239	840	858	876	894	912	931	949	967	985	*003	18
240	38021	039	057	075	093	112	130	148	166	184	18
241	202	220	238	256	274	292	310	328	346	364	18
242	382	399	417	435	453	471	489	507	525	543	18
243	561	578	596	614	632	650	668	686	703	721	18
244	739	757	775	792	810	828	846	863	881	899	18
245	917	934	952	970	987	*005	*023	*041	*058	*076	18
246	39094	111	129	146	164	182	199	217	235	252	18
247	270	287	305	322	340	358	375	393	410	428	18
248	445	463	480	498	515	533	550	568	585	602	18
249	620	637	655	672	690	707	724	742	759	777	17
N	0	1	2	3	4	5	6	7	8	9	D

N	0	1	2	3	4	5	6	7	8	9	D
250	39794	811	829	846	863	881	898	915	933	950	17
251	967	985	*002	*019	*037	*054	*071	*088	*106	*123	17
252	40140	157	175	192	209	226	243	261	278	295	17
253	312	329	346	364	381	398	415	432	449	466	17
254	483	500	518	535	552	569	586	603	620	637	17
255	654	671	688	705	722	739	756	773	790	807	17
256	824	841	858	875	892	909	926	943	960	976	17
257	993	*010	*027	*044	*061	*078	*095	*111	*128	*145	17
258	41162	179	196	212	229	246	263	280	296	313	17
259	330	347	363	380	397	414	430	447	464	481	17
260	497	514	531	547	564	581	597	614	631	647	17
261	664	681	697	714	731	747	764	780	797	814	17
262	830	847	863	880	896	913	929	946	963	979	16
263	996	*012	*029	*045	*062	*078	*095	*111	*127	*144	16
264	42160	177	193	210	226	243	259	275	292	308	16
265	325	341	357	374	390	406	423	439	455	472	16
266	488	504	521	537	553	570	586	602	619	635	16
267	651	667	684	700	716	732	749	765	781	797	16
268	813	830	846	862	878	894	911	927	943	959	16
269	975	991	*008	*024	*040	*056	*072	*088	*104	*120	16
270	43136	152	169	185	201	217	233	249	265	281	16
271	297	313	329	345	361	377	393	409	425	441	16
272	457	473	489	505	521	537	553	569	584	600	16
273	616	632	648	664	680	696	712	727	743	759	16
274	775	791	807	823	838	854	870	886	902	917	16
275	933	949	965	981	996	*012	*028	*044	*059	*075	16
276	44091	107	122	138	154	170	185	201	217	232	16
277	248	264	279	295	311	326	342	358	373	389	16
278	404	420	436	451	467	483	498	514	529	545	16
279	560	576	592	607	623	638	654	669	685	700	16
280	716	731	747	762	778	793	809	824	840	855	15
281	871	886	902	917	932	948	963	979	994	*010	15
282	45025	040	056	071	086	102	117	133	148	163	15
283	179	194	209	225	240	255	271	286	301	317	15
284	332	347	362	378	393	408	423	439	454	469	15
285	484	500	515	530	545	561	576	591	606	621	15
286	637	652	667	682	697	712	728	743	758	773	15
287	788	803	818	834	849	864	879	894	909	924	15
288	939	954	969	984	*000	*015	*030	*045	*060	*075	15
289	46090	105	120	135	150	165	180	195	210	225	15
290	240	255	270	285	300	315	330	345	359	374	15
291	389	404	419	434	449	464	479	494	509	523	15
292	538	553	568	583	598	613	627	642	657	672	15
293	687	702	716	731	746	761	776	790	805	820	15
294	835	850	864	879	894	909	923	938	953	967	15
295	982	997	*012	*026	*041	*056	*070	*085	*100	*114	15
296	47129	144	159	173	188	202	217	232	246	261	15
297	276	290	305	319	334	349	363	378	392	407	15
298	422	436	451	465	480	494	509	524	538	553	15
299	567	582	596	611	625	640	654	669	683	698	15
N	0	1	2	3	4	5	6	7	8	9	D

N	0	1	2	3	4	5	6	7	8	9	D
300	47712	727	741	756	770	784	799	813	828	842	14
301	857	871	885	900	914	929	943	958	972	986	14
302	48001	015	029	044	058	073	087	101	116	130	14
303	144	159	173	187	202	216	230	244	259	273	14
304	287	302	316	330	344	359	373	387	401	416	14
305	430	444	458	473	487	501	515	530	544	558	14
306	572	586	601	615	629	643	657	671	686	700	14
307	714	728	742	756	770	785	799	813	827	841	14
308	855	869	883	897	911	926	940	954	968	982	14
309	996	*010	*024	*038	*052	*066	*080	*094	*108	*122	14
310	49136	150	164	178	192	206	220	234	248	262	14
311	276	290	304	318	332	346	360	374	388	402	14
312	415	429	443	457	471	485	499	513	527	541	14
313	554	568	582	596	610	624	638	651	665	679	14
314	693	707	721	734	748	76e	776	790	803	817	14
315	831	845	859	872	886	900	914	927	941	955	14
316	969	982	996	*010,	*024	*037	*051	*065	*079	*092	14
317	50106	120	133	147	161	174	188	202	215	229	14
318	243	256	270	284	297	311	325	338	352	365	14
319	379	393	406	420	433	447	461	474	488	501	14
320	515	529	542	556	569	583	596	610	623	637	14
321	651	664	678	691	705	718	732	745	759	772	14
322	786	799	813	826	840	853	866	880	893	907	13
323	920	934	947	961	974	987	*001	*014	*028	*041	13
324	51055	068	081	095	108	121	135	148	162	175	13
325	188	202	215	228	242	255	268	282	295	308	13
326	322	335	348	362	375	388	402	415	428	441	13
327	455	468	481	495	508	521	534	548	561	574	13
328	587	601	614	627	640	654	667	680	693	706	13
329	720	733	746	759	772	786	799	812	825	838	13
330	851	865	878	891	904	917	930	943	957	970	13
331	983	996	*009	*022	*035	*048	*061	*075	*088	*101	13
332	52114	127	140	153	166	179	192	205	218	231	13
333	244	257	270	284	297	310	323	336	349	362	13
334	375	388	401	414	427	440	453	466	479	492	13
335	504	517	530	543	556	569	582	595	608	621	13
336	634	647	660	673	686	699	711	724	737	750	13
337	763	776	789	802	815	827	840	853	866	879	13
338	892	905	917	930	943	956	969	982	994	*007	13
339	53020	033	046	058	071	084	097	110	122	135	13
340	148	161	173	186	199	212	224	237	250	263	13
341	275	288	301	314	326	339	352	364	377	390	13
342	403	415	428	441	453	466	479	491	504	517	13
343	529	542	555	567	580	593	605	618	631	643	13
344	656	668	681	694	706	719	732	744	757	769	13
345	782	794	807	820	832	845	857	870	882	895	13
346	908	920	933	945	958	970	983	995	*008	*020	13
347	54033	045	058	070	083	095	108	120	133	145	13
348	158	170	183	195	208	220	233	245	258	270	12
349	283	295	307	320	332	345	357	370	382	394	12
N	0	1	2	3	4	5	6	7	8	9	D

MANTISSA OF NUMBERS
3500 TO 3999

N	0	1	2	3	4	5	6	7	8	9	D
350	54407	419	432	444	456	469	481	494	506	518	12
351	531	543	555	568	580	593	605	617	630	642	12
352	654	667	679	691	704	716	728	741	753	765	12
353	777	790	802	814	827	839	851	864	876	888	12
354	900	913	925	937	949	962	974	986	998	*011	12
355	55023	035	047	060	072	084	096	108	121	133	12
356	145	157	169	182	194	206	218	230	242	255	12
357	267	279	291	303	315	328	340	352	364	376	12
358	388	400	413	425	437	449	461	473	485	497	12
359	509	522	534	546	558	570	582	594	606	618	12
360	630	642	654	666	678	691	703	715	727	739	12
361	751	763	775	787	799	811	823	835	847	859	12
362	871	883	895	907	919	931	943	955	967	979	12
363	991	*003	*015	*027	*038	*050	*062	*074	*086	*098	12
364	56110	122	134	146	158	170	182	194	205	217	12
365	229	241	253	265	277	289	301	312	324	336	12
366	348	360	372	384	396	407	419	431	443	455	12
367	467	478	490	502	514	526	538	549	561	573	12
368	585	597	608	620	632	644	656	667	679	691	12
369	703	714	726	738	750	761	773	785	797	808	12
370	820	832	844	855	867	879	891	902	914	926	12
371	937	949	961	972	984	996	*008	*019	*031	*043	12
372	57054	066	078	089	101	113	124	136	148	159	12
373	171	183	194	206	217	229	241	252	264	276	12
374	287	299	310	322	334	345	357	368	380	392	12
375	403	415	426	438	449	461	473	484	496	507	12
376	519	530	542	553	565	576	588	600	611	623	12
377	634	646	657	669	680	692	703	715	726	738	11
378	749	761	772	784	795	807	818	830	841	852	11
379	864	875	887	898	910	921	933	944	955	967	11
380	978	990	*001	*013	*024	*035	*047	*058	*070	*081	11
381	58092	104	115	127	138	149	161	172	184	195	11
382	206	218	229	240	252	263	274	286	297	309	11
383	320	331	343	354	365	377	388	399	410	422	11
384	433	444	456	467	478	490	501	512	524	535	11
385	546	557	569	580	591	602	614	625	636	647	11
386	659	670	681	692	704	715	726	737	749	760	11
387	771	782	794	805	816	827	838	850	861	872	11
388	883	894	906	917	928	939	950	961	973	984	11
389	995	*006	*017	*028	*040	*051	*062	*073	*084	*095	11
390	59106	118	129	140	151	162	173	184	195	207	11
391	218	229	240	251	262	273	284	295	306	318	11
392	329	340	351	362	373	384	395	406	417	428	11
393	439	450	461	472	483	494	506	517	528	539	11
394	550	561	572	583	594	605	616	627	638	649	11
395	660	671	682	693	704	715	726	737	748	759	11
396	770	780	791	802	813	824	835	846	857	868	11
397	879	890	901	912	923	934	945	956	966	977	11
398	988	999	*010	*021	*032	*043	*054	*065	*076	*086	11
399	60097	108	119	130	141	152	163	173	184	195	11
N	0	1	2	3	4	5	6	7	8	9	D

N	0	1	2	3	4	5	6	7	8	9	D
400	60206	217	228	239	249	260	271	282	293	304	11
401	314	325	336	347	358	369	379	390	401	412	11
402	423	433	444	455	466	477	487	498	509	520	11
403	531	541	552	563	574	584	595	606	617	627	11
404	638	649	660	670	681	692	703	713	724	735	11
405	746	756	767	778	788	799	810	821	831	842	11
406	853	863	874	885	895	906	917	927	938	949	11
407	959	970	981	991	*002	*013	*023	*034	*045	*055	11
408	61066	077	087	098	109	119	130	140	151	162	11
409	172	183	194	204	215	225	236	247	257	268	11
410	278	289	300	310	321	331	342	352	363	374	11
411	384	395	405	416	426	437	448	458	469	479	11
412	490	500	511	521	532	542	553	563	574	584	11
413	595	606	616	627	637	648	658	669	679	690	11
414	700	711	721	731	742	752	763	773	784	794	10
415	805	815	826	836	847	857	868	878	888	899	10
416	909	920	930	941	951	962	972	982	993	*003	10
417	62014	024	034	045	055	066	076	086	097	107	10
418	118	128	138	149	159	170	180	190	201	211	10
419	221	232	242	252	263	273	284	294	304	315	10
420	325	335	346	356	366	377	387	397	408	418	10
421	428	439	449	459	469	480	490	500	511	521	10
422	531	542	552	562	572	583	593	603	613	624	10
423	634	644	655	665	675	685	696	706	716	726	10
424	737	747	757	767	778	788	798	808	818	829	10
425	839	849	859	870	880	890	900	910	921	931	10
426	941	951	961	972	982	992	*002	*012	*022	*033	10
427	63043	053	063	073	083	094	104	114	124	134	10
428	144	155	165	175	185	195	205	215	225	236	10
429	246	256	266	276	286	296	306	317	327	337	10
430	347	357	367	377	387	397	407	417	428	438	10
431	448	458	468	478	488	498	508	518	528	538	10
432	548	558	568	579	589	599	609	619	629	639	10
433	649	659	669	679	689	699	709	719	729	739	10
434	749	759	769	779	789	799	809	819	829	839	10
435	849	859	869	879	889	899	909	919	929	939	10
436	949	959	969	979	988	998	*008	*018	*028	*038	10
437	64048	058	068	078	088	098	108	118	128	137	10
438	147	157	167	177	187	197	207	217	227	237	10
439	246	256	266	276	286	296	306	316	326	335	10
440	345	355	365	375	385	395	404	414	424	434	10
441	444	454	464	473	483	493	503	513	523	532	10
442	542	552	562	572	582	591	601	611	621	631	10
443	640	650	660	670	680	689	699	709	719	729	10
444	738	748	758	768	777	787	797	807	816	826	10
445	836	846	856	865	875	885	895	904	914	924	10
446	933	943	953	963	972	982	992	*002	*011	*021	10
447	65031	040	050	060	070	079	089	099	108	118	10
448	128	137	147	157	167	176	186	196	205	215	10
449	225	234	244	254	263	273	283	292	302	312	10
N	0	1	2	3	4	5	6	7	8	9	D

N	0	1	2	3	4	5	6	7	8	9	D
450	65321	331	341	350	360	369	379	389	398	408	10
451	418	427	437	447	456	466	475	485	495	504	10
452	514	523	533	543	552	562	571	581	591	600	10
453	610	619	629	639	648	658	667	677	686	696	10
454	706'	715	725	734	744	753	763	772	782	792	9
455	801	811	820	830	839	849	858	868	877	887	9
456	896	906	916	925	935	944	954	963	973	982	9
457	992	*001	*011	*020	*030	*039	*049	*058	*068	*077	9
458	66087	096	106	115	124	134	143	153	162	172	9
459	181	191	200	210	219	229	238	247	257	266	9
460	276	285	295	304	314	323	332	342	351	361	9
461	370	380	389	398	408	417	427	436	445	455	9
462	464	474	483	492	502	511	521	530	539	549	9
463	558	567	577	586	596	605	614	624	633	642	9
464	652	661	671	680	689	699	708	717	727	736	9
465	745	755	764	773	783	792	801	811	820	829	9
466	839	848	857	867	876	885	894	904	913	922	9
467	932	941	950	960	969	978	987	997	*006	*015	9
468	67025	034	043	052	062	071	080	089	099	108	9
469	117	127	136	145	154	164	173	182	191	201	9
470	210	219	228	237	247	256	265	274	284	293	9
471	302	311	321	330	339	348	357	367	376	385	9
472	394	403	413	422	431	440	449	459	468	477	9
473	486	495	504	514	523	532	541	550	560	569	9
474	578	587	596	605	614	624	633	642	651	660	9
475	669	679	688	697	706	715	724	733	742	752	9
476	761	770	779	788	797	806	815	825	834	843	9
477	852	861	870	879	888	897	906	916	925	934	9
478	943	952	961	970	979	988	997	*006	*015	*024	9
479	68034	043	052	061	070	079	088	097	106	115	9
480	124	133	142	151	160	169	178	187	196	205	9
481	215	224	233	242	251	260	269	278	287	296	9
482	305	314	323	332	341	350	359	368	377	386	9
483	395	404	413	422	431	440	449	458	467	476	9
484	485	494	502	511	520	529	538	547	556	565	9
485	574	583	592	601	610	619	628	637	646	655	9
486	664	673	681	690	699	708	717	726	735	744	9
487	753	762	771	780	789	797	806	815	824	833	9
488	842	851	860	869	878	886	895	904	913	922	9
489	931	940	949	958	966	975	984	993	*002	*011	9
490	69020	028	037	046	055	064	073	082	090	099	9
491	108	117	126	135	144	152	161	170	179	188	9
492	197	205	214	223	232	241.	249	258	267	276	9
493	285	294	302	311	320	329	338	346	355	364	9
494	373	381	390	399	408	417	425	434	443	452	9
495	461	469	478	487	496	504	513	522	531	539	9
496	548	557	566	574	583	592	601	609	618	627	9
497	636	644	653	662	671	679	688	697	705	714	9
498	723	732	740	749	758	767	775	784	793	801	9
499	810	819	827	836	845	854	862	871	880	888	9
N	0	1	2	3	4	5	6	7	8	9	D

N	0	1	2	3	4	5	6	7	8	9	D
500	69897	906	914	923	932	940	949	958	966	975	9
501	984	992	*001	*010	*018	*027	*036	*044	*053	*062	9
502	70070	079	088	096	105	114	122	131	140	148	9
503	157	165	174	183	191	200	209	217	226	234	9
504	243	252	260	269	278	286	295	303	312	321	9
505	329	338	346	355	364	372	381	389	398	406	9
506	415	424	432	441	449	458	467	475	484	492	9
507	501	509	518	526	535	544	552	561	569	578	9
508	586	595	603	612	621	629	638	646	655	663	9
509	672	680	689	697	706	714	723	731	740	749	9
510	757	766	774	783	791	800	808	817	825	834	9
511	842	851	859	868	876	885	893	902	910	919	9
512	927	935	944	952	961	969	978	986	995	*003	9
513	71012	020	029	037	046	054	063	071	079	088	8
514	096	105	113	122	130	139	147	155	164	172	8
515	181	189	198	206	214	223	231	240	248	257	8
516	265	273	282	290	299	307	315	324	332	341	8
517	349	357	366	374	383	391	399	408	416	425	8
518	433	441	450	458	466	475	483	492	500	508	8
519	517	525	533	542	550	559	567	575	584	592	8
520	600	609	617	625	634	642	650	659	667	675	8
521	684	692	700	709	717	725	734	742	750	759	8
522	767	775	784	792	800	809	817	825	834	842	8
523	850	858	867	875	883	892	900	908	917	925	8
524	933	941	950	958	966	975	983	991	999	*008	8
525	72016	024	032	041	049	057	066	074	082	090	8
526	099	107	115	123	132	140	148	156	165	173	8
527	181	189	198	206	214	222	230	239	247	255	8
528	263	272	280	288	296	304	313	321	329	337	8
529	346	354	362	370	378	387	395	403	411	419	8
530	428	436	444	452	460	469	477	485	493	501	8
531	509	518	526	534	542	550	558	567	575	583	8
532	591	599	607	616	624	632	640	648	656	665	8
533	673	681	689	697	705	713	722	730	738	746	8
534	754	762	770	779	787	795	803	811	819	827	8
535	835	843	852	860	868	876	884	892	900	908	8
536	916	925	933	941	949	957	965	973	981	989	8
537	997	*006	*014	*022	*030	*038	*046	*054	*062	*070	8
538	73078	086	094	102	111	119	127	135	143	151	8
539	159	167	175	183	191	199	207	215	223	231	8
540	239	247	255	263	272	280	288	296	304	312	8
541	320	328	336	344	352	360	368	376	384	392	8
542	400	408	416	424	432	440	448	456	464	472	8
543	480	488	496	504	512	520	528	536	544	552	8
544	560	568	576	584	592	600	608	616	624	632	8
545	640	648	656	664	672	679	687	695	703	711	8
546	719	727	735	743	751	759	767	775	783	791	8
547	799	807	815	823	830	838	846	854	862	870	8
548	878	886	894	902	910	918	926	933	941	949	8
549	957	965	973	981	989	997	*005	*013	*020	*028	8
N	0	1	2	3	4	5	6	7	8	9	D

N	0	1	2	3	4	5	6	7	8	9	D
550	74036	044	052	060	068	076	084	092	099	107	8
551	115	123	131	139	147	155	162	170	178	186	8
552	194	202	210	218	225	233	241	249	257	265	8
553	273	280	288	296	304	312	320	327	335	343	8
554	351	359	367	374	382	390	398	406	414	421	8
555	429	437	445	453	461	468	476	484	492	500	8
556	507	515	523	531	539	547	554	562	570	578	8
557	586	593	601	609	617	624	632	640	648	656	8
558	663	671	679	687	695	702	710	718	726	733	8
559	741	749	757	764	772	780	788	796	803	811	8
560	819	827	834	842	850	858	865	873	881	889	8
561	896	904	912	920	927	935	943	950	958	966	8
562	974	981	989	997	*005	*012	*020	*028	*035	*043	8
563	75051	059	066	074	082	089	097	105	113	120	8
564	128	136	143	151	159	166	174	182	189	197	8
565	205	213	220	228	236	243	251	259	266	274	8
566	282	289	297	305	312	320	328	335	343	351	8
567	358	366	374	381	389	397	404	412	420	427	8
568	435	442	450	458	465	473	481	488	496	504	8
569	511	519	526	534	542	549	557	565	572	580	8
570	587	595	603	610	618	626	633	641	648	656	8
571	664	671	679	686	694	702	709	717	724	732	8
572	740	747	755	762	770	778	785	793	800	808	8
573	815	823	831	838	846	853	861	868	876	884	8
574	891	899	906	914	921	929	937	944	952	959	8
575	967	974	982	989	997	*005	*012	*020	*027	*035	8
576	76042	050	057	065	072	080	087	095	103	110	8
577	118	125	133	140	148	155	163	170	178	185	8
578	193	200	208	215	223	230	238	245	253	260	8
579	268	275	283	290	298	305	313	320	328	335	8
580	343	350	358	365	373	380	388	395	403	410	8
581	418	425	433	440	448	455	462	470	477	485	7
582	492	500	507	515	522	530	537	545	552	559	7
583	567	574	582	589	597	604	612	619	626	634	7
584	641	649	656	664	671	678	686	693	701	708	7
585	716	723	730	738	745	753	760	768	775	782	7
586	790	797	805	812	819	827	834	842	849	856	7
587	864	871	879	886	893	901	908	916	923	930	7
588	938	945	953	960	967	975	982	989	997	*004	7
589	77012	019	026	034	041	048	056	063	070	078	7
590	085	093	100	107	115	122	129	137	144	151	7
591	159	166	173	181	188	195	203	210	217	225	7
592	232	240	247	254	262	269	276	283	291	298	7
593	305	313	320	327	335	342	349	357	364	371	7
594	379	386	393	401	408	415	422	430	437	444	7
595	452	459	466	474	481	488	495	503	510	517	7
596	525	532	539	546	554	561	568	576	583	590	7
597	597	605	612	619	627	634	641	648	656	663	7
598	670	677	685	692	699	706	714	721	728	735	7
599	743	750	757	764	772	779	786	793	801	808	7
N	0	1	2	3	4	5	6	7	8	9	D

MANTISSA OF NUMBERS
6000 TO 6499

N	0	1	2	3	4	5	6	7	8	9	D
600	77815	822	830	837	844	851	859	866	873	880	7
601	887	895	902	909	916	924	931	938	945	952	7
602	960	967	974	981	988	996	*003	*010	*017	*025	7
603	78032	039	046	053	061	068	075	082	089	097	7
604	104	111	118	125	132	140	147	154	161	168	7
605	176	183	190	197	204	211	219	226	233	240	7
606	247	254	262	269	276	283	290	297	305	312	7
607	319	326	333	340	347	355	362	369	376	383	7
608	390	398	405	412	419	426	433	440	447	455	7
609	462	469	476	483	490	497	504	512	519	526	7
610	533	540	547	554	561	569	576	583	590	597	7
611	604	611	618	625	633	640	647	654	661	668	7
612	675	682	689	696	704	711	718	725	732	739	7
613	746	753	760	767	774	781	789	796	803	810	7
614	817	824	831	838	845	852	859	866	873	880	7
615	888	895	902	909	916	923	930	937	944	951	7
616	958	965	972	979	986	993	*000	*007	*014	*021	7
617	79029	036	043	050	057	064	071	078	085	092	7
618	099	106	113	120	127	134	141	148	155	162	7
619	169	176	183	190	197	204	211	218	225	232	7
620	239	246	253	260	267	274	281	288	295	302	7
621	309	316	323	330	337	344	351	358	365	372	7
622	379	386	393	400	407	414	421	428	435	442	7
623	449	456	463	470	477	484	491	498	505	511	7
624	518	525	532	539	546	553	560	567	574	581	7
625	588	595	602	609	616	623	630	637	644	650	7
626	657	664	671	678	685	692	699	706	713	720	7
627	727	734	741	748	754	761	768	775	782	789	7
628	796	803	810	817	824	831	837	844	851	858	7
629	865	872	879	886	893	900	906	913	920	927	7
630	934	941	948	955	962	969	975	982	989	996	7
631	80003	010	017	024	030	037	044	051	058	065	7
632	072	079	085	092	099	106	113	120	127	134	7
633	140	147	154	161	168	175	182	188	195	202	7
634	209	216	223	229	236	243	250	257	264	271	7
635	277	284	291	298	305	312	318	325	332	339	7
636	346	353	359	366	373	380	387	393	400	407	7
637	414	421	428	434	441	448	455	462	468	475	7
638	482	489	496	502	509	516	523	530	536	543	7
639	550	557	564	570	577	584	591	598	604	611	7
640	618	625	632	638	645	652	659	665	672	679	7
641	686	693	699	706	713	720	726	733	740	747	7
642	754	760	767	774	781	787	794	801	808	814	7
643	821	828	835	841	848	855	862	868	875	882	7
644	889	895	902	909	916	922	929	936	943	949	7
645	956	963	969	976	983	990	996	*003	*010	*017	7
646	81023	030	037	043	050	057	064	070	077	084	7
647	090	097	104	111	117	124	131	137	144	151	7
648	158	164	171	178	184	191	198	204	211	218	7
649	224	231	238	245	251	258	265	271	278	285	7
N	0	1	2	3	4	5	6	7	8	9	D

N	0	1	2	3	4	5	6	7	8	9	D
650	81291	298	305	311	318	325	331	338	345	351	7
651	358	365	371	378	385	391	398	405	411	418	7
652	425	431	438	445	451	458	465	471	478	485	7
653	491	498	505	511	518	525	531	538	544	551	7
654	558	564	571	578	584	591	598	604	611	617	7
655	624	631	637	644	651	657	664	671	677	684	7
656	690	697	704	710	717	723	730	737	743	750	7
657	757	763	770	776	783	790	796	803	809	816	7
658	823	829	836	842	849	856	862	869	875	882	7
659	889	895	902	908	915	921	928	935	941	948	7
660	954	961	968	974	981	987	994	*000	*007	*014	7
661	82020	027	033	040	046	053	060	066	073	079	7
662	086	092	099	105	112	119	125	132	138	145	7
663	151	158	164	171	178	184	191	197	204	210	7
664	217	223	230	236	243	249	256	263	269	276	7
665	282	289	295	302	308	315	321	328	334	341	7
666	347	354	360	367	373	380	387	393	400	406	7
667	413	419	426	432	439	445	452	458	465	471	7
668	478	484	491	497	504	510	517	523	530	536	7
669	543	549	556	562	569	575	582	588	595	601	7
670	607	614	620	627	633	640	646	653	659	666	7
671	672	679	685	692	698	705	711	718	724	730	6
672	737	743	750	756	763	769	776	782	789	795	6
673	802	808	814	821	827	834	840	847	853	860	6
674	866	872	879	885	892	898	905	911	918	924	6
675	930	937	943	950	956	963	969	975	982	988	6
676	995	*001	*008	*014	*020	*027	*033	*040	*046	*052	6
677	83059	065	072	078	085	091	097	104	110	117	6
678	123	129	136	142	149	155	161	168	174	181	6
679	187	193	200	206	213	219	225	232	238	245	6
680	251	257	264	270	276	283	289	296	302	308	6
681	315	321	327	334	340	347	353	359	366	372	6
682	378	385	391	398	404	410	417	423	429	436	6
683	442	448	455	461	467	474	480	487	493	499	6
684	506	512	518	525	531	537	544	550	556	563	6
685	569	575	582	588	594	601	607	613	620	626	6
686	632	639	645	651	658	664	670	677	683	689	6
687	696	702	708	715	721	727	734	740	746	753	6
688	759	765	771	778	784	790	797	803	809	816	6
689	822	828	835	841	847	853	860	866	872	879	6
690	885	891	897	904	910	916	923	929	935	942	6
691	948	954	960	967	973	979	985	992	998	*004	6
692	84011	017	023	029	036	042	048	055	061	067	6
693	073	080	086	092	098	105	111	117	123	130	6
994	136	142	148	155	161	167	173	180	186	192	6
695	198	205	211	217	223	230	236	242	248	255	6
696	261	267	273	280	286	292	298	305	311	317	6
697	323	330	336	342	348	354	361	367	373	379	6
698	386	392	398	404	410	417	423	429	435	442	6
699	448	454	460	466	473	479	485	491	497	504	6
N	0	1	2	3	4	5	6	7	8	9	D

N	0	1	2	3	4	5	6	7	8	9	D
700	84510	516	522	528	535	541	547	553	559	566	6
701	572	578	584	590	597	603	609	615	621	628	6
702	634	640	646	652	658	665	671	677	683	689	6
703	696	702	708	714	720	726	733	739	745	751	6
704	757	763	770	776	782	788	794	800	807	813	6
705	819	825	831	837	844	850	856	862	868	874	6
706	880	887	893	899	905	911	917	924	930	936	6
707	942	948	954	960	967	973	979	985	991	997	6
708	85003	009	016	022	028	034	040	046	052	058	6
709	065	071	077	083	089	095	101	107	114	120	6
710	126	132	138	144	150	156	163	169	175	181	6
711	187	193	199	205	211	217	224	230	236	242	6
712	248	254	260	266	272	278	285	291	297	303	6
713	309	315	321	327	333	339	345	352	358	364	6
714	370	376	382	388	394	400	406	412	418	425	6
715	431	437	443	449	455	461	467	473	479	485	6
716	491	497	503	509	516	522	528	534	540	546	6
717	552	558	564	570	576	582	588	594	600	606	6
718	612	618	625	631	637	643	649	655	661	667	6
719	673	679	685	691	697	703	709	715	721	727	6
720	733	739	745	751	757	763	769	775	781	788	6
721	794	800	806	812	818	824	830	836	842	848	6
722	854	860	866	872	878	884	890	896	902	908	6
723	914	920	926	932	938	944	950	956	962	968	6
724	974	980	986	992	998	*004	*010	*016	*022	*028	6
725	86034	040	046	052	058	064	070	076	082	088	6
726	094	100	106	112	118	124	130	136	141	147	6
727	153	159	165	171	177	183	189	195	201	207	6
728	213	219	225	231	237	243	249	255	261	267	6
729	273	279	285	291	297	303	308	314	320	326	6
730	332	338	344	350	356	362	368	374	380	386	6
731	392	398	404	410	415	421	427	433	439	445	6
732	451	457	463	469	475	481	487	493	499	504	6
733	510	516	522	528	534	540	546	552	558	564	6
734	570	576	581	587	593	599	605	611	617	623	6
735	629	635	641	646	652	658	664	670	676	682	6
736	688	694	700	705	711	717	723	729	735	741	6
737	747	753	759	764	770	776	782	788	794	800	6
738	806	812	817	823	829	835	841	847	853	859	6
739	864	870	876	882	888	894	900	906	911	917	6
740	923	929	935	941	947	953	958	964	970	976	6
741	982	988	994	999	*005	*011	*017	*023	*029	*035	6
742	87040	046	052	058	064	070	075	081	087	093	6
743	099	105	111	116	122	128	134	140	146	151	6
744	157	163	169	175	181	186	192	198	204	210	6
745	216	221	227	233	239	245	251	256	262	268	6
746	274	280	286	291	297	303	309	315	320	326	6
747	332	338	344	349	355	361	367	373	379	384	6
748	390	396	402	408	413	419	425	431	437	442	6
749	448	454	460	466	471	477	483	489	495	500	6
N	0	1	2	3	4	5	6	7	8	9	D

N	0	1	2	3	4	5	6	7	8	9	D
750	87506	512	518	523	529	535	541	547	552	558	6
751	564	570	576	581	587	593	599	604	610	616	6
752	622	628	633	639	645	651	656	662	668	674	6
753	679	685	691	697	703	708	714	720	726	731	6
754	737	743	749	754	760	766	772	777	783	789	6
755	795	800	806	812	818	823	829	835	841	846	6
756	852	858	864	869	875	881	887	892	898	904	6
757	910	915	921	927	933	938	944	950	955	961	6
758	967	973	978	984	990	996	*001	*007	*013	*018	6
759	88024	030	036	041	047	053	058	064	070	076	6
760	081	087	093	098	104	110	116	121	127	133	6
761	138	144	150	156	161	167	173	178	184	190	6
762	195	201	207	213	218	224	230	235	241	247	6
763	252	258	264	270	275	281	287	292	298	304	6
764	309	315	321	326	332	338	343	349	355	360	6
765	366	372	377	383	389	395	400	406	412	417	6
766	423	429	434	440	446	451	457	463	468	474	6
767	480	485	491	497	502	508	513	519	525	530	6
768	536	542	547	553	559	564	570	576	581	587	6
769	593	598	604	610	615	621	627	632	638	643	6
770	649	655	660	666	672	677	683	689	694	700	6
771	705	711	717	722	728	734	739	745	750	756	6
772	762	767	773	779	784	790	795	801	807	812	6
773	818	824	829	835	840	846	852	857	863	868	6
774	874	880	885	891	897	902	908	913	919	925	6
775	930	936	941	947	953	958	964	969	975	981	6
776	986	992	997	*003	*009	*014	*020	*025	*031	*037	6
777	89042	048	053	059	064	070	076	081	087	092	6
778	098	104	109	115	120	126	131	137	143	148	6
779	154	159	165	170	176	182	187	193	198	204	6
780	209	215	221	226	232	237	243	248	254	260	6
781	265	271	276	282	287	293	298	304	310	315	6
782	321	326	332	337	343	348	354	360	365	371	6
783	376	382	387	393	398	404	409	415	421	426	6
784	432	437	443	448	454	459	465	470	476	481	6
785	487	492	498	504	509	515	520	526	531	537	6
786	542	548	553	559	564	570	575	581	586	592	6
787	597	603	609	614	620	625	631	636	642	647	6
788	653	658	664	669	675	680	686	691	697	702	6
789	708	713	719	724	730	735	741	746	752	757	6
790	763	768	774	779	785	790	796	801	807	812	5
791	818	823	829	834	840	845	851	856	862	867	5
792	873	878	883	889	894	900	905	911	916	922	5
793	927	933	938	944	949	955	960	966	971	977	5
794	982	988	993	998	*004	*009	*015	*020	*026	*031	5
795	90037	042	048	053	059	064	069	075	080	086	5
796	091	097	102	108	113	119	124	129	135	140	5
797	146	151	157	162	168	173	179	184	189	195	5
798	200	206	211	217	222	227	233	238	244	249	5
799	255	260	266	271	276	282	287	293	298	304	5
N	0	1	2	3	4	5	6	7	8	9	D

N	0	1	2	3	4	5	6	7	8	9	D
800	90309	314	320	325	331	336	342	347	352	358	5
801	363	369	374	380	385	390	396	401	407	412	5
802	417	423	428	434	439	445	450	455	461	466	5
803	472	477	482	488	493	499	504	509	515	520	5
804	526	531	536	542	547	553	558	563	569	574	5
805	580	585	590	596	601	607	612	617	623	628	5
806	634	639	644	650	655	660	666	671	677	682	5
807	687	693	698	703	709	714	720	725	730	736	5
808	741	747	752	757	763	768	773	779	784	789	5
809	795	800	806	811	816	822	827	832	838	843	5
810	849	854	859	865	870	875	881	886	891	897	5
811	902	907	913	918	924	929	934	940	945	950	5
812	956	961	966	972	977	982	988	993	998	*004	5
813	91009	014	020	025	030	036	041	046	052	057	5
814	062	068	073	078	084	089	094	100	105	110	5
815	116	121	126	132	137	142	148	153	158	164	5
816	169	174	180	185	190	196	201	206	212	217	5
817	222	228	233	238	243	249	254	259	265	270	5
818	275	281	286	291	297	302	307	312	318	323	5
819	328	334	339	344	350	355	360	365	371	376	5
820	381	387	392	397	403	408	413	418	424	429	5
821	434	440	445	450	455	461	466	471	477	482	5
822	487	492	498	503	508	514	519	524	529	535	5
823	540	545	551	556	561	566	572	577	582	587	5
824	593	598	603	609	614	619	624	630	635	640	5
825	645	651	656	661	666	672	677	682	687	693	5
826	698	703	709	714	719	724	730	735	740	745	5
827	751	756	761	766	772	777	782	787	793	798	5
828	803	808	814	819	824	829	834	840	845	850	5
829	855	861	866	871	876	882	887	892	897	903	5
830	908	913	918	924	929	934	939	944	950	955	5
831	960	965	971	976	981	986	991	997	*002	*007	5
832	92012	018	023	028	033	038	044	049	054	059	5
833	065	070	075	080	085	091	096	101	106	111	5
834	117	122	127	132	137	143	148	153	158	163	5
835	169	174	179	184	189	195	200	205	210	215	5
836	221	226	231	236	241	247	252	257	262	267	5
837	273	278	283	288	293	298	304	309	314	319	5
838	324	330	335	340	345	350	355	361	366	371	5
839	376	381	387	392	397	402	407	412	418	423	5
840	428	433	438	443	449	454	459	464	469	474	5
841	480	485	490	495	500	505	511	516	521	526	5
842	531	536	542	547	552	557	562	567	572	578	5
843	583	588	593	598	603	609	614	619	624	629	5
844	634	639	645	650	655	660	665	670	675	681	5
845	686	691	696	701	706	711	716	722	727	732	5
846	737	742	747	752	758	763	768	773	778	783	5
847	788	793	799	804	809	814	819	824	829	834	5
848	840	845	850	855	860	865	870	875	881	886	5
849	891	896	901	906	911	916	921	927	932	937	5
N	0	1	2	3	4	5	6	7	8	9	D

N	0	1	2	3	4	5	6	7	8	9	D
850	92942	947	952	957	962	967	973	978	983	988	5
851	993	998	*003	*008	*013	*018	*024	*029	*034	*039	5
852	93044	049	054	059	064	069	075	080	085	090	5
853	095	100	105	110	115	120	125	131	136	141	5
854	146	151	156	161	166	171	176	181	186	192	5
855	197	202	207	212	217	222	227	232	237	242	5
856	247	252	258	263	268	273	278	283	288	293	5
857	298	303	308	313	318	323	328	334	339	344	5
858	349	354	359	364	369	374	379	384	389	394	5
859	399	404	409	414	420	425	430	435	440	445	5
860	450	455	460	465	470	475	480	485	490	495	5
861	500	505	510	515	520	526	531	536	541	546	5
862	551	556	561	566	571	576	581	586	591	596	5
863	601	606	611	616	621	626	631	636	641	646	5
864	651	656	661	666	671	676	682	687	692	697	5
865	702	707	712	717	722	727	732	737	742	747	5
866	752	757	762	767	772	777	782	787	792	797	5
867	802	807	812	817	822	827	832	837	842	847	5
868	852	857	862	867	872	877	882	887	892	897	5
869	902	907	912	917	922	927	932	937	942	947	5
870	952	957	962	967	972	977	982	987	992	997	5
871	94002	007	012	017	022	027	032	037	042	047	5
872	052	057	062	067	072	077	082	086	091	096	5
873	101	106	111	116	121	126	131	136	141	146	5
874	151	156	161	166	171	176	181	186	191	196	5
875	201	206	211	216	221	226	231	236	240	245	5
876	250	255	260	265	270	275	280	285	290	295	5
877	300	305	310	315	320	325	330	335	340	345	5
878	349	354	359	364	369	374	379	384	389	394	5
879	399	404	409	414	419	424	429	433	438	443	5
880	448	453	458	463	468	473	478	483	488	493	5
881	498	503	507	512	517	522	527	532	537	542	5
882	547	552	557	562	567	571	576	581	586	591	5
883	596	601	606	611	616	621	626	630	635	640	5
884	645	650	655	660	665	670	675	680	685	689	5
885	694	699	704	709	714	719	724	729	734	738	5
886	743	748	753	758	763	768	773	778	783	787	5
887	792	797	802	807	812	817	822	827	832	836	5
888	841	846	851	856	861	866	871	876	880	885	5
889	890	895	900	905	910	915	919	924	929	934	5
890	939	944	949	954	959	963	968	973	978	983	5
891	988	993	998	*002	*007	*012	*017	*022	*027	*032	5
892	95036	041	046	051	056	061	066	071	075	080	5
893	085	090	095	100	105	109	114	119	124	129	5
894	134	139	143	148	153	158	163	168	173	177	5
895	182	187	192	197	202	207	211	216	221	226	5
896	231	236	240	245	250	255	260	265	270	274	5
897	279	284	289	294	299	303	308	313	318	323	5
898	328	332	337	342	347	352	357	361	366	371	5
899	376	381	386	390	395	400	405	410	415	419	5
N	0	1	2	3	4	5	6	7	8	9	D

N	0	1	2	3	4	5	6	7	8	9	D
900	95424	429	434	439	444	448	453	458	463	468	5
901	472	477	482	487	492	497	501	506	511	516	5
902	521	525	530	535	540	545	550	554	559	564	5
903	569	574	578	583	588	593	598	602	607	612	5
904	617	622	626	631	636	641	646	650	655	660	5
905	665	670	674	679	684	689	694	698	703	708	5
906	713	718	722	727	732	737	742	746	751	756	5
907	761	766	770	775	780	785	789	794	799	804	5
908	809	813	818	823	828	832	837	842	847	852	5
909	856	861	866	871	875	880	885	890	895	899	5
910	904	909	914	918	923	928	933	938	942	947	5
911	952	957	961	966	971	976	980	985	990	995	5
912	999	*004	*009	*014	*019	*023	*028	*033	*038	*042	5
913	96047	052	057	061	066	071	076	080	085	090	5
914	095	099	104	109	114	118	123	128	133	137	5
915	142	147	152	156	161	166	171	175	180	185	5
916	190	194	199	204	209	213	218	223	227	232	5
917	237	242	246	251	256	261	265	270	275	280	5
918	284	289	294	298	303	308	313	317	322	327	5
919	332	336	341	346	350	355	360	365	369	374	5
920	379	384	388	393	398	402	407	412	417	421	5
921	426	431	435	440	445	450	454	459	464	468	5
922	473	478	483	487	492	497	501	506	511	515	5
923	520	525	530	534	539	544	548	553	558	562	5
924	567	572	577	581	586	591	595	600	605	609	5
925	614	619	624	628	633	638	642	647	652	656	5
926	661	666	670	675	680	685	689	694	699	703	5
927	708	713	717	722	727	731	736	741	745	750	5
928	755	759	764	769	774	778	783	788	792	797	5
929	802	806	811	816	820	825	830	834	839	844	5
930	848	853	858	862	867	872	876	881	886	890	5
931	895	900	904	909	914	918	923	928	932	937	5
932	942	946	951	956	960	965	970	974	979	984	5
933	988	993	997	*002	*007	*011	*016	*021	*025	*030	5
934	97035	039	044	049	053	058	063	067	072	077	5
935	081	086	090	095	100	104	109	114	118	123	5
936	128	132	137	142	146	151	155	160	165	169	5
937	174	179	183	188	192	197	202	206	211	216	5
938	220	225	230	234	239	243	248	253	257	262	5
939	267	271	276	280	285	290	294	299	304	308	5
940	313	317	322	327	331	336	340	345	350	354	5
941	359	364	368	373	377	382	387	391	396	400	5
942	405	410	414	419	424	428	433	437	442	447	5
943	451	456	460	465	470	474	479	483	488	493	5
944	497	502	506	511	516	520	525	529	534	539	5
945	543	548	552	557	562	566	571	575	580	585	5
946	589	594	598	603	607	612	617	621	626	630	5
947	635	640	644	649	653	658	663	667	672	676	5
948	681	685	690	695	699	704	708	713	717	722	5
949	727	731	736	740	745	749	754	759	763	768	5
N	0	1	2	3	4	5	6	7	8	9	D

N	0	1	2	3	4	5	6	7	8	9	D
950	97772	777	782	786	791	795	800	804	809	813	5
951	818	823	827	832	836	841	845	850	855	859	5
952	864	868	873	877	882	886	891	896	900	905	5
953	909	914	918	923	928	932	937	941	946	950	5
954	955	959	964	968	973	978	982	987	991	996	5
955	98000	005	009	014	019	023	028	032	037	041	5
956	046	050	055	059	064	068	073	078	082	087	5
957	091	096	100	105	109	114	118	123	127	132	5
958	137	141	146	150	155	159	164	168	173	177	5
959	182	186	191	195	200	204	209	214	218	223	5
960	227	232	236	241	245	250	254	259	263	268	5
961	272	277	281	286	290	295	299	304	308	313	5
962	318	322	327	331	336	340	345	349	354	358	5
963	363	367	372	376	381	385	390	394	399	403	5
964	408	412	417	421	426	430	435	439	444	448	5
965	453	457	462	466	471	475	480	484	489	493	4
966	498	502	507	511	516	520	525	529	534	538	4
967	543	547	552	556	561	565	570	574	579	583	4
968	588	592	597	601	605	610	614	619	623	628	4
969	632	637	641	646	650	655	659	664	668	673	4
970	677	682	686	691	695	700	704	709	713	717	4
971	722	726	731	735	740	744	749	753	758	762	4
972	767	771	776	780	784	789	793	798	802	807	4
973	811	816	820	825	829	834	838	843	847	851	4
974	856	860	865	869	874	878	883	887	892	896	4
975	900	905	909	914	918	923	927	932	936	941	4
976	945	949	954	958	963	967	972	976	981	985	4
977	989	994	998	*003	*007	*012	*016	*021	*025	*029	4
978	99034	038	043	047	052	056	061	065	069	074	4
979	078	083	087	092	096	100	105	109	114	118	4
980	123	127	131	136	140	145	149	154	158	162	4
981	167	171	176	180	185	189	193	198	202	207	4
982	211	216	220	224	229	233	238	242	247	251	4
983	255	260	264	269	273	277	282	286	291	295	4
984	300	304	308	313	317	322	326	330	335	339	4
985	344	348	352	357	361	366	370	374	379	383	4
986	388	392	396	401	405	410	414	419	423	427	4
987	432	436	441	445	449	454	458	463	467	471	4
988	476	480	484	489	493	498	502	506	511	515	4
989	520	524	528	533	537	542	546	550	555	559	4
990	564	568	572	577	581	585	590	594	599	603	4
991	607	612	616	621	625	629	634	638	642	647	4
992	651	656	660	664	669	673	977	682	686	691	4
993	695	699	704	708	712	717	721	726	730	734	4
994	739	743	747	752	756	760	765	769	774	778	4
995	782	787	791	795	800	804	808	813	817	822	4
996	826	830	835	839	843	848	852	856	861	865	4
997	870	874	878	883	887	891	896	900	904	909	4
998	913	917	922	926	930	935	939	944	948	952	4
999	957	961	965	970	974	978	983	987	991	996	4
N	0	1	2	3	4	5	6	7	8	9	D

BADMINTON
COURTS

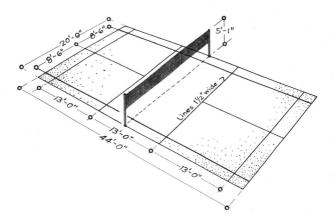

OUTDOOR COURTS. A level area of lawn 30′ x 50′ may be used without special preparation. Since only volley shots are played, there is no need for the perfect turf required for lawn tennis. Concrete, asphalt, and clay surfaces provide about an equally secure footing. Provide 1,500-watt lighting units on 25′ poles located 24″ from each net post for night play.

INDOOR COURTS. An unobstructed space of 4′ to 6′ should be allowed along both long sides of the court, and from 6′ to 10′ at the ends. An area of 30′ x 60′ provides ideal conditions for a doubles court. Lockers, seats for spectators, and other objects should not encroach on this unobstructed space.

The clear overhead space at the net line should be not less than 25′; 30′ is preferable. End walls 20′ high will be adequate. This allows gable roof construction if desired.

Light green is the most desirable color for walls and ceiling. Artificial lighting consists of three units 5′ o/c at each end of the net, 25′ high. A wood floor of fir T&G boards laid lengthwise is best. Skylights with diffusing and glare-reducing glass will furnish daylight illumination.

TENNIS COURTS

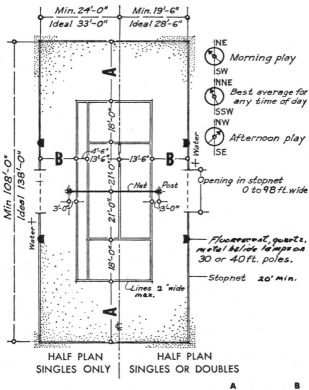

| Min. 24'-0" | Min. 19'-6" |
| Ideal 33'-0" | Ideal 28'-6" |

Min 108'-0"
Ideal 138'-0"

A

18'-0"

4'-6"
13'-6"   13'-6"

21'-0"

Net   Post

3'-0"
3'-0"

21'-0"

18'-0"

A
C

+ Water

NE
SW — Morning play

NNE
SSW — Best average for any time of day

NW
SE — Afternoon play

Opening in stopnet
0 to 98 ft. wide

Fluorescent, quartz, metal halide lamps on 30 or 40 ft. poles.

Stopnet 20' min.

Water +

Lines 2" wide max.

| HALF PLAN | HALF PLAN |
| SINGLES ONLY | SINGLES OR DOUBLES |

	A	B
Minimum recommended	15'-0"	8'-0"
Limited space	18'-0"	9'-0"
Standard	21'-0"	12'-0"
Courts in batteries	21'-0"	see note
Minimum for championship play	26'-0"	15'-0"
Ideal where space allows	30'-0"	15'-0"

Space courts in batteries 39' o/c for singles, and 48' o/c for doubles with no stopnet between them. Courts should be graded for drainage from net to stopnet, or from stopnet to net, or from long center axis laterally to stopnets, or from end stopnet to opposite end stopnet. Drainage lines should be sloped at a rate of $\frac{1}{10}''$ per foot for nonporous surfaces, and $\frac{1}{20}''$ to $\frac{1}{30}''$ for porous surfaces. Provide catch basins and drainage line to carry off surface water to dry well, sewer system, or drainage ditch. Subsurface drains may connect to the same sewer line.

TENNIS COURT
CONSTRUCTION

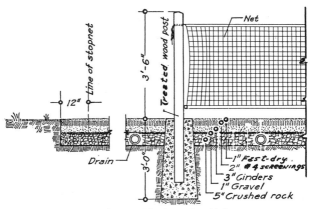

Net

Line of stopnet

Treated wood post

3'-6"

12"

Drain

3'-0"

1" Fast-dry
2" #4 screenings
3" Cinders
1" Gravel
5" Crushed rock

QUICK-DRYING CLAY SURFACE

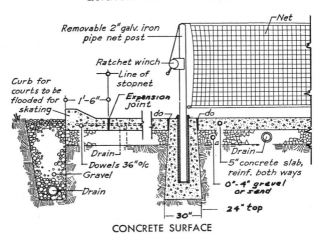

Removable 2" galv. iron
pipe net post

Net

Ratchet winch

Line of
stopnet

Curb for
courts to be
flooded for
skating

1'-6"

Expansion
joint

do

do

Drain

Drain

5" concrete slab,
reinf. both ways

Drain

Dowels 36" o/c

Gravel

0"- 4" gravel
or sand

Drain

24" top

30"

CONCRETE SURFACE

Clay, concrete, grass, dirt, gravel-and-pitch or so-called "black-top," asphalt, macadam, and various proprietary materials are all used for tennis court surfaces. Indoors, wood is most often used. Under impervious surfaces such as concrete, 4" agricultural tile lines should be laid 10'-0" o/c to prevent heaving from frost in winter.

SQUASH
COURTS

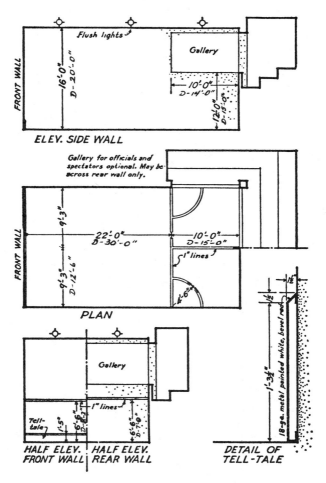

ELEV. SIDE WALL

Flush lights

Gallery

FRONT WALL

16'-0"
D-20'-0"

10'-0"
D-14'-0"

12'-0"
D-15'-0"

Gallery for officials and
spectators optional. May be
across rear wall only.

FRONT WALL

9'-3"

22'-0"
D-30'-0"

10'-0"
D-15'-0"

9'-3"
D-12'-6"

1" lines

4-6"

PLAN

Gallery

1" lines

Tell-
tale

1'-5"

6'-6"
D-7'-9"

6'-6"
D-7'-9"

HALF ELEV. | **HALF ELEV.**
FRONT WALL | **REAR WALL**

1½"

½"

1'-3½"

18-ga. metal painted white, bevel rod

DETAIL OF
TELL-TALE

Floor to be maple, machine-nailed, $^{25}/_{32}$" x 2¼" (real dimensions), running lengthwise, laid flat on cushioned sleeper, cushioned sleeper-plywood, resilient underlayment, channel, or padded plywood flooring system, or other approved surfacing/underlayment system. Walls to be panel system of fiberglass, glass, composite construction, or other appropriate material, made for the purpose, applied to frame or masonry support. Lines to be vivid red.

167

BASEBALL
DIAMOND

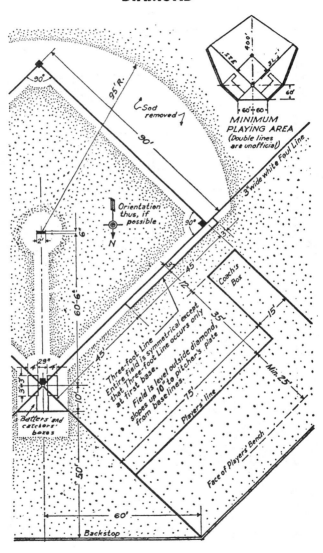

MINIMUM
PLAYING AREA
(Double lines
are unofficial)

Three-foot Line is symmetrical except that three-foot Line occurs only at first base.

Entire field is level outside diamond, slopes up 10' to pitcher's plate from base lines.

Sod removed

Orientation thus, if possible

3' wide white Foul Line

Coach's Box

Batters' and catchers' boxes

Players' line

Min 25'

Face of Players' Bench

Backstop

CROQUET
COURTS

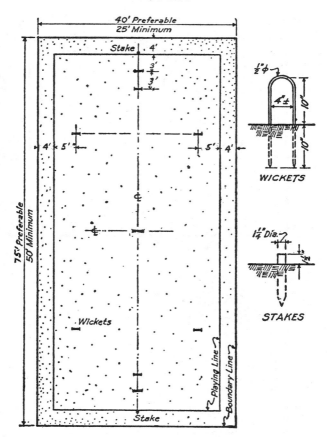

The court should consist of a level lawn closely cropped. The boundary line is a strong white cord stapled in place, or a chalk or lime line extending around the court. Some additional free space beyond the borders of the playing field is desirable. A total unobstructed area of approximately 50' x 85' provides an ideal condition. The wicket layout and the size of the playing field vary in different versions of the game. British croquet (not illustrated here) requires an entirely different court arrangement from that of lawn croquet. In any case, courts in batteries should have at least 10' or more between their boundary lines.

LACROSSE
FIELD

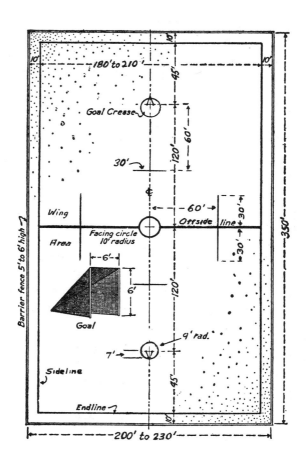

180' to 210'

10'

10'

10'

45'

Goal Crease

60'

120'

30'

60'

30'

Wing

Offside line

Facing circle
10' radius

Area

6'

6'

120'

Goal

9' rad.

7'

Sideline

45'

Endline

10'

Barrier fence 5' to 6' high

350'

200' to 230'

HORSESHOE COURTS
OFFICIAL STANDARD

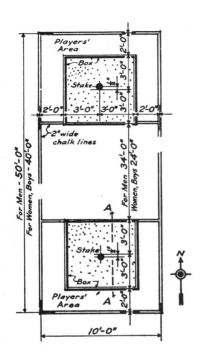

Multiple courts should be arranged in batteries of four, spaced 10'-0" o.c. 10'-0" or more should be allowed between the batteries.

Players' Area

Box

Stake

2" wide chalk lines

For Men – 50'-0"
For Women, Boys – 40'-0"

For Men 34'-0"
Women, Boys 24'-0"

Stake

Box

Players' Area

10'-0"

N

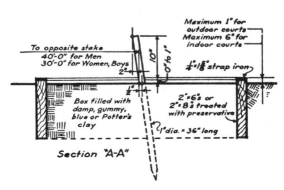

To opposite stake
40'-0" for Men
30'-0" for Women, Boys

Maximum 1" for outdoor courts
Maximum 6" for indoor courts

¼"×1⅜" strap iron

Box filled with damp, gummy, blue or Potter's clay

2"×6's or 2"×8's treated with preservative

1"dia. × 36" long

Section "A-A"

171

BILLIARD
ROOMS

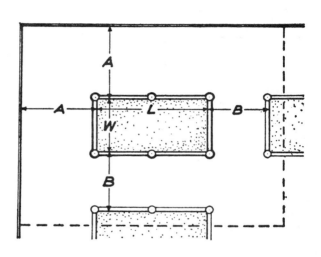

Size (W x L)	Where Used	Unobstructed Space (A)	Between Tables (B)
3 x 6	Jr. Pool	4'-6"	—
3½ x 7	Jr. Pool	5'-0"	—
4 x 8	Home; commercial standard in South America, Mexico, and Spain	5'-6"	4'-6"
4½ x 9	Popular U.S. commercial standard	5'-0"	5'-6"
5 x 10	U.S. professional standard	5'-0"	5'-6"
6 x 12	Commercial standard in Canada and England	5'-0"	5'-6"

STANDARD SWIMMING
POOL DIMENSIONS

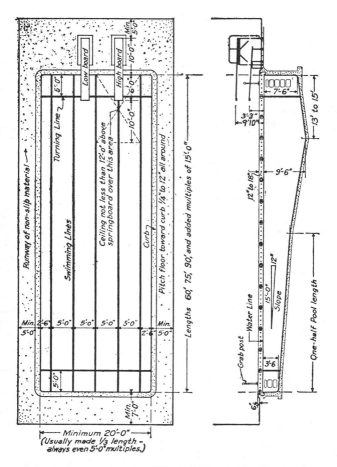

Swimming lanes must be at least 5'-0" wide. The pool should be at least four lanes wide and 60'-0" long; it can be extended in multiples of 5' or 15'. However, to meet general racing requirements, lanes must be 6'-0" wide (7'-0" for pools designed strictly for competition). A minimum length of 75'-0" is required for American competition, plus a small added amount for electronic timing panels. A length of 50 meters (164'-½") is required for pools designed for international competition. Minimum widths are 36'-0" (6' lanes) and 42'-0" (7' lanes).

SWIMMING POOL
SIZES AND CAPACITIES

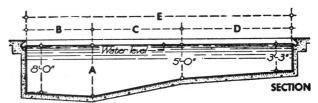

F · *Width of pool*
Q · *Approximate number of people at one time*
8Q ·*Maximum daily load*

The number of people admitted to the pool at one time is **given in** the table and is subject to variation, depending upon ages of the swimmers. For simultaneous use by small active boys and dignified older persons, the limit could be very much lower than that given—since a sense of overcrowding would result. If all the swimmers are of the same age, a larger number of users at one time would be tolerable.

The maximum daily load and the capacity of the pool in gallons are given to facilitate calculations involving water purifications, drainage and supply.

"STANDARD" SWIMMING POOL DIMENSIONS

A	B	C	D	E	F	Gallons	Q
9′	15′	20′	25′	60′	20′	55,000	32
9′	15′	20′	40′	75′	25′	80,000	42
9½′	18′	25′	47′	90′	30′	120,000	75
10′	18′	25′	62′	105′	35′	155,000	100
10′	20′	30′	70′	120′	40′	207,000	130
10′	20′	30′	85′	135′	45′	248,000	170
10′	20′	30′	100′	150′	50′	310,000	250
10′	20′	30′	130′	180′	60′	420,000	360
10′	20′	30′	160′	210′	70′	558,000	490

INDOOR
SWIMMING POOL

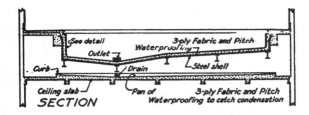

See detail
Outlet
Waterproofing
3-ply Fabric and Pitch
Curb
Drain
Steel shell
Ceiling slab
Pan of
3-ply Fabric and Pitch
Waterproofing to catch condensation

SECTION

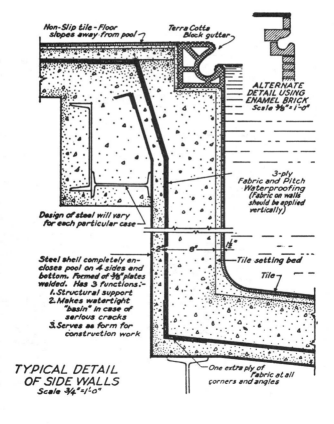

Non-Slip tile - Floor
slopes away from pool

Terra Cotta
Block gutter

ALTERNATE
DETAIL USING
ENAMEL BRICK
Scale 3/8"= 1'-0"

3-ply
Fabric and Pitch
Waterproofing
(Fabric on walls
should be applied
vertically)

Design of steel will vary
for each particular case

Tile setting bed

Tile

Steel shell completely en-
closes pool on 4 sides and
bottom. Formed of 3/8"plates
welded. Has 3 functions:-
1. Structural support
2. Makes watertight
"basin" in case of
serious cracks
3. Serves as form for
construction work

TYPICAL DETAIL
OF SIDE WALLS
Scale 3/4"= 1'-0"

One extra ply of
Fabric at all
corners and angles

175

SMALL OUTDOOR SWIMMING POOL

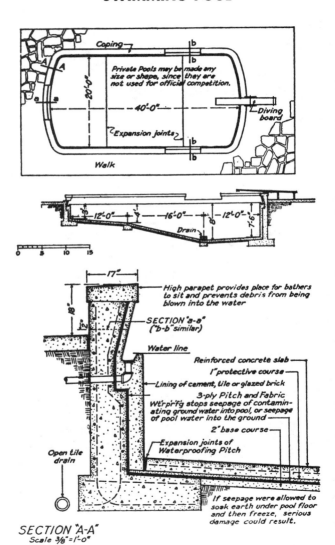

Coping

Private Pools may be made any size or shape, since they are not used for official competition.

20'-0"

40'-0"

Diving board

Expansion joints

b

b

Walk

12'-0" — 16'-0" — 12'-0"

Drain

8"

7'-6"

0   5   10   15

17"

High parapet provides place for bathers to sit and prevents debris from being blown into the water

18"

12"

SECTION "a-a"
("b-b" similar)

Water line

Reinforced concrete slab

1" protective course

Lining of cement, tile or glazed brick

3-ply Pitch and Fabric Wtr'prf'g stops seepage of contaminating ground water into pool, or seepage of pool water into the ground

2" base course

Expansion joints of Waterproofing Pitch

Open tile drain

If seepage were allowed to soak earth under pool floor and then freeze, serious damage could result.

SECTION "A-A"
Scale 3/8"=1'-0"

176

UNDERWATER LIGHTING
SWIMMING POOLS

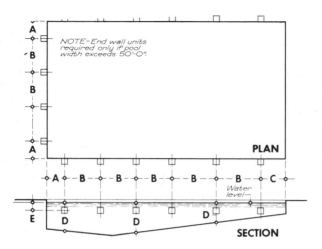

SIZE OF LAMP	A	B (Maximum)		C	E	
		Where D is More Than 5'-0"	Where D is Less Than 5'-0"		Min.	Max.
250-watt	4'-0"	8'-0"	10'-0"	5'-0"	1'-0"	1'-3"
400-watt						
500-watt	6'-0"	12'-0"	15'-0"	7'-6"	1'-6"	2'-0"
1000-watt						
1500-watt						

Total wattage should equal pool length multiplied by watts per square foot as recommended in the table below. To determine correct lamp size, divide the total wattage thus found by the number of units. Choose the nearest standard lamp size and respace the number of units required. The A and C dimensions given in the table above should be maintained, and the B dimensions should not be exceeded. All floodlights should be equipped with lenses that give a horizontal spread of light.

Notice that in most cases underwater lights are not recommended for ends of pools. If they are installed, they should be separately switched so that they may be turned off for racing.

All swimming pool lighting installations must comply with provisions of the National Electrical Code generally, and Article 680 in particular, as well as with local codes.

LOCATION OF POOL	RECOMMENDED WATTS PER SQUARE FOOT.	
	Good Practice	Minimum
OUTDOORS	3	1
INDOORS	5	2.5

UNDERWATER LIGHTING
SWIMMING POOLS

DRY NICHE METHOD.
Recommended for brick- or
tile-finished pools. Floodlights are mounted behind
watertight portholes and
are serviced from above
through manholes or a tunnel in the rear. A cast
bronze niche lining is cast
into the wall of the pool to
receive the porthole ring
and watertight door. Alu-
minum is lower priced but
can be used only in fresh,
chemical-free water.

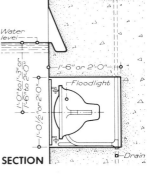

SECTION

WET NICHE METHOD.
Less expensive and simpler than
the dry niche method. The wet
niche is commonly used for outdoor concrete pools. Doors cover
the niche and are opened when
the lighting units need to be
raised to the surface for servicing.
Any construction that prevents
free circulation of water behind
the unit should be avoided.

Floodlights can be furnished
in cast aluminum casing and door
parts at a lower price than
bronze, but can be used only in
fresh, chemical-free water.

NOTE. The dimensions and
drawings shown are examples
only. Various specific lighting
arrangements and equipment
exist, and provisions of local
codes and the National Electrical Code, especially Article 680,
must be checked before final
plans are drawn.

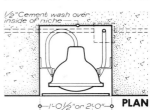

SECTION

PLAN

NOTE—Size of niche varies according
to the type of floodlight used.

PLAYGROUND POOLS

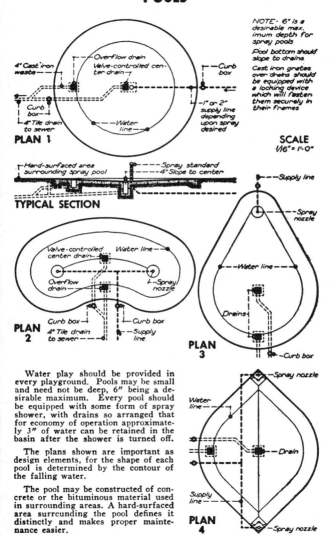

NOTE:- 6" is a desirable max-imum depth for spray pools

Pool bottom should slope to drains.

Cast iron grates over drains should be equipped with a locking device which will fasten them securely in their frames

PLAN 1

4" Cast iron waste

Overflow drain

Valve-controlled cen-ter drain

Curb box

Curb bar

4" Tile drain to sewer

Water line

1" or 2" supply line depending upon spray desired

SCALE
1/16" = 1'-0"

Hard-surfaced area surrounding spray pool

Spray standard

4" Slope to center

TYPICAL SECTION

Supply line

PLAN 2

Valve-controlled center drain

Water line

Overflow drain

Spray nozzle

Curb box

Curb box

4" Tile drain to sewer

Supply line

PLAN 3

Supply line

Spray nozzle

Water line

Drains

Curb bar

PLAN 4

Spray nozzle

Water line

Drain

Supply line

Spray nozzle

Water play should be provided in every playground. Pools may be small and need not be deep, 6" being a de-sirable maximum. Every pool should be equipped with some form of spray shower, with drains so arranged that for economy of operation approximate-ly 3" of water can be retained in the basin after the shower is turned off.

The plans shown are important as design elements, for the shape of each pool is determined by the contour of the falling water.

The pool may be constructed of con-crete or the bituminous material used in surrounding areas. A hard-surfaced area surrounding the pool defines it distinctly and makes proper mainte-nance easier.

179

BASKETBALL COURT

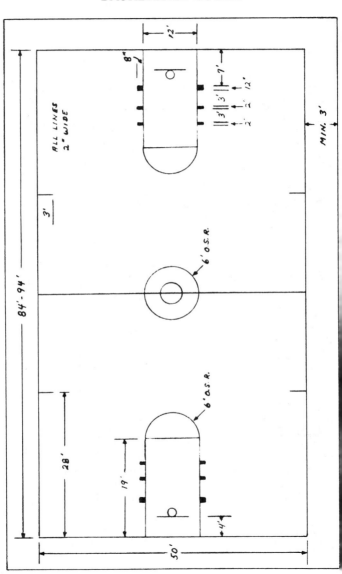

VOLLEYBALL COURT

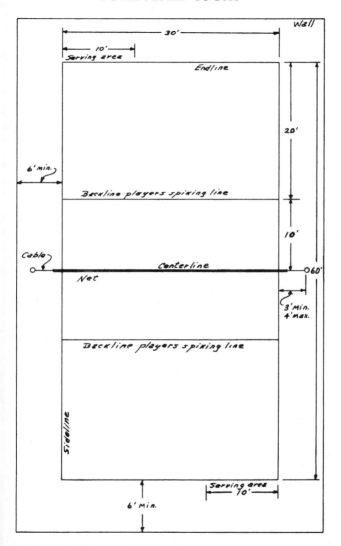

Net height 8'-0" for men, 7'-4¼" for women and secondary school players, 7'-0" or less for younger players. Minimum clear area above courts must be 26'-0". Special rules for coed and doubles play.

ICE HOCKEY
RINK

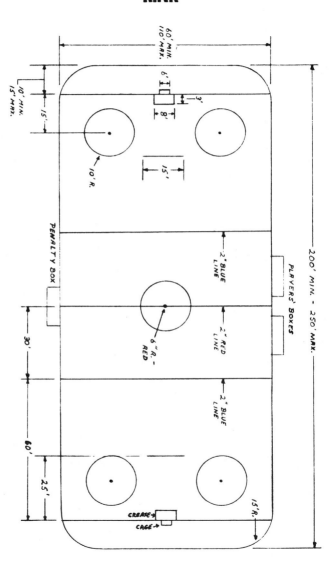

TABLE
TENNIS

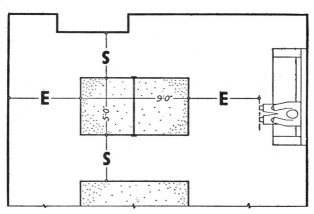

PLAN

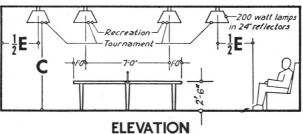

ELEVATION

	Unobstructed space		
	E	**S**	**C**
Advanced tournament play, late rounds	20'-0"	10'-0"	8'-6"
Advanced tournament play, early rounds	10'-0"	6'-0"	8'-6"
Beginners for tournament play	6'-0"	6'-0"	8'-6"
Advanced and average players for recreation	10'-0"	6'-0"	7'-6"
Dub players for recreation	5'-0"	3'-0"	6'-6"

BOWLING
ALLEYS

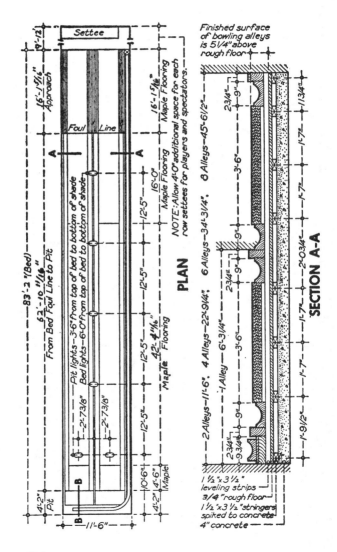

PLAN

Settee

9'-2"

16'-1 5/16" Approach

Foul Line

A — A

83'-2" (Bed)

62'-10 11/16" From Bed Foul Line to Pit

Pit lights — 3'-6" from top of bed to bottom of shade
Bed lights — 6'-0" from top of bed to bottom of shade

2'-7 3/8" 2'-7 3/8"

16'-1 5/16" Maple Flooring

16'-0" Maple Flooring

12'-5"

12'-5"

12'-5"

42'-4 11/16" Maple Flooring

12'-5"

10'-6" Maple

4'-6"

4'-2" Pit

B — B

11'-6"

NOTE: Allow 4'-0" additional space for each row settees for players and spectators.

SECTION A-A

Finished surface of bowling alleys is 5 1/4" above rough floor

2 3/4" 9"

2 3/4" 9"

2 3/4" 9"

2 3/4" 9"

9 3/4"

8 Alleys — 45'-6 1/2"

6 Alleys — 34'-3 3/4"

4 Alleys — 22'-9 1/4"

2 Alleys — 11'-6"

1 Alley — 6'-3 1/4"

3'-6"

3'-6"

3'-6"

1 1/4"

1'-7"

1'-7"

1'-7"

2'-0 3/4"

1'-9 1/2"

1 1/2"x3 1/2" leveling strips
3/4" rough floor
1 1/2"x3 1/2" stringers spiked to concrete
4" concrete

BOWLING
ALLEYS

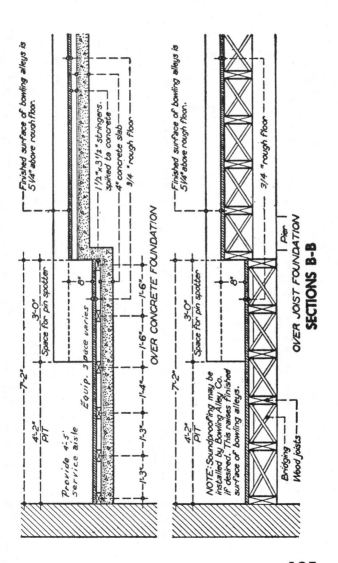

OVER CONCRETE FOUNDATION

- Finished surface of bowling alleys is 5¼" above rough floor.
- 1½"x3½" stringers, spiked to concrete
- 4" concrete slab
- ¾" rough floor
- 8"
- 1'-6"
- 1'-6"
- 1'-4"
- 1'-3"
- 1'-3"
- 7'-2"
- 3'-0" Space for pin spotter
- Equip. space varies
- 4'-2" PIT
- Provide 4'-5' service aisle

OVER JOIST FOUNDATION

- Finished surface of bowling alleys is 5¼" above rough floor.
- ¾" rough floor
- Pier
- 8"
- 7'-2"
- 3'-0" Space for pin spotter
- 4'-2" PIT
- Bridging
- Wood joists
- NOTE: Soundproofing may be installed by Bowling Alley Co. if desired. This raises finished surface of bowling alleys.

SECTIONS B-B

185

SHOOTING RANGE

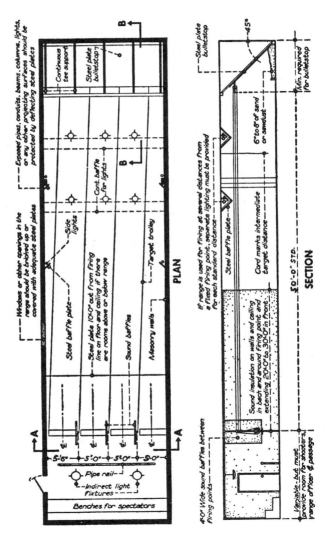

PLAN

- Exposed pipes, conduits, beams, columns, lights, or any other projecting surfaces should be protected by deflecting steel plates
- Continuous tee support
- Steel plate bulletstop
- Windows or other openings in the range should be bricked up or covered with adequate steel plates
- Cont. baffle for lights
- Side lights
- Steel baffle plate
- Steel plate 10'-0" out from firing line on floor and ceiling if there are rooms above or below range
- Sound baffles
- Target trolley
- Masonry walls
- 5'-6" — 5'-0" — 5'-0" — 5'-0"
- Pipe rail
- Indirect light fixtures
- Benches for spectators

SECTION

- Steel plate bulletstop
- 45°
- Steel plate bulletstop
- Min. required for bulletstop
- 6" to 8" of sand or sawdust
- If range is used for firing at several distances from a fixed firing point, separate lighting must be provided for each standard distance
- Steel baffle plate
- Cord marks intermediate target distance
- 50'-0" STD.
- Sound insulation on walls and ceiling in back and around firing point and extending 20'-0" to 30'-0" in front
- 4'-0" Wide sound baffles between firing points
- Variable - but must provide room for shooters, range of fixed & passage

186

SHOOTING RANGE

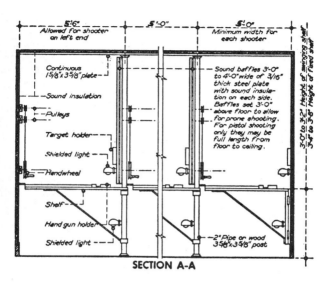

5'-6"
Allowed for shooter on left end

5'-0"

5'-0"
Minimum width for each shooter

3'-0 to 3'-6" Height of swinging steel
3'-4 to 3'-8" Height of fixed shelf

Continuous 1 5/8" x 3 5/8" plate

Sound insulation

Pulleys

Target holder

Shielded light

Handwheel

Shelf

Hand gun holder

Shielded light

Sound baffles 3'-0" to 4'-0" wide of 3/16" thick steel plate with sound insulation on each side. Baffles set 3'-0" above floor to allow for prone shooting. For pistol shooting only they may be full length from floor to ceiling.

2" Pipe or wood 3 5/8" x 3 5/8" post

SECTION A-A

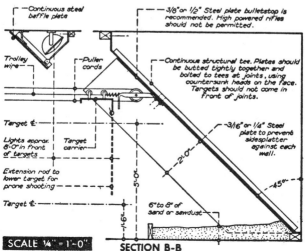

Continuous steel baffle plate

3/8" or 1/2" Steel plate bulletstop is recommended. High powered rifles should not be permitted.

Trolley wire

Puller cords

Continuous structural tee. Plates should be butted tightly together and bolted to tees at joints, using countersunk heads on the face. Targets should not come in front of joints.

Target ℄

Lights approx. 8'-0" in front of targets

Target carrier

3/16" or 1/4" Steel plate to prevent sidesplatter against each wall.

Extension rod to lower target for prone shooting

2'-0"

5'-0"

45°

Target ℄

6" to 8" of sand or sawdust

4'-6"

SCALE 1/4" = 1'-0"

SECTION B-B

SIZES OF PLAYGROUND APPARATUS

In the following table are given the dimensions and approximate use areas of several types of apparatuses frequently installed on children's playgrounds. Since the types of equipment made by various manufacturers differ somewhat, the dimensions and areas given are merely suggested. Furthermore, it is not likely that all of the apparatuses listed will be found on a single playground. It is desirable to provide safety zones around all apparatuses, especially movable ones.

Type of Apparatus	Length of Apparatus	Height of Apparatus	Space Required
Circular traveling rings..	10' dia.	12'	25' dia.
Gang slide	16'	8'	20'x45'
Giant stride		12'	32' dia
Horizontal bar	6'	8' upright	12'x20'
Horizontal ladder	16'	7'-6"	8'x24'
Merry-go-round	10' dia.		30' dia.
Sand box on table	6'x10' to 10'x20'		12'x16' to 16'x30'
Slide	16'	8'	12'x30'
Slide-spiral	35'	18'	25'x35'
Swings—set of 3	15' at top	12'	30'x35'
Swings—set of 6	30' at top	12'	30'x50'
Teeters—set of 4	12' to 15'	2'-6"	20'x20'
Traveling rings—set of 6.	40' at top	14'	20'x60'

Jungle gyms and other outdoor gymnasium outfits are manufactured in several sizes and combinations which occupy widely different areas. It is advisable to have all such equipment placed at least 15' from the nearest fence, building, or other apparatus.

The wading pool may be any desired size or shape although it is usually rectangular or circular. The circular pools generally have a diameter of from 40' to 75'.

The platform for dancing may be in any desired dimension. An average size would be 20' x 30' to 30' x 40'. According to a number of authorities, 40 to 50 square feet of space per child should be provided for apparatus play.

PLAYGROUND
GAME AREAS

The playing area dimensions given in this table are the accepted standards of averages at the time of compilation. However, the figures can and do change from time to time, though usually only slightly. Therefore, before final dimensions are laid down, the proper athletic organization or other authority should be consulted for up-to-date exact figures.

Name	Size of Marked Area (in Feet)	Size Field Required (in Feet)
Archery	90 to 300 long	50 (min. width) x 450 (max. length)
*Badminton	17 x 44 (single)	25 x 60
	20 x 44 (double)	30 x 60
*Baseball	90' diamond	300 x 300 min.
		350 x 350 (aver.)
*Basketball (men)	50 x 84 (min.)	56 x 90 (min.)
	50 x 94 (max.)	70 x 114 (max.)
Basketball (women)	42 x 74 (min.)	48 x 80 (min.)
	50 x 94 (max.)	70 x 114 (max.)
Boccie	18 x 62	30 x 80
Bowling Green	20 x 120 (1 alley)	20 x 120
	160 x 120 (8 alleys)	160 x 120
Cricket	Wickets 66 apart	420 x 420
*Croquet	40 x 75 (varies)	40 x 75
Deck Tennis	12 x 40 (single)	16 x 48
	18 x 40 (double)	22 x 48
Field Hockey	150 x 270 (min.)	200 x 350 (aver.)
	180 x 300 (max.)	
Football	160 x 360	180 x 420
Football (6-man)	120 x 300	140 x 360
*Handball	20 x 34	30 x 45
Horse Ring	120 x 240	120 x 240
*Horseshoes (men)	Stakes 40' apart	12 x 50
*Horseshoes (women)	Stakes 30' apart	12 x 40
*Lacrosse	230 x 380 (min.)	
Paddle Tennis	16 x 44 (single)	28 x 60
	20 x 44 (double)	32 x 60
Polo	600 x 900 (max.)	600 x 900
Quoits	Stakes 54' apart	25 x 80
Roque	30 x 60	30 x 60
Rugby	195 x 450 (min.)	
	225 x 480 (max.)	225 x 480
*Shuffleboard	6 x 52	10 x 57
Soccer (men)	195 x 330 (min.)	240 x 360 (aver.)
	225 x 360 (max.)	
Soccer (women)	120 x 240 (min.)	200 x 320 (aver.)
	180 x 300 (max.)	
Softball	60'diamond	250 x 250 (min.)
Speedball (men)	160 x 240 (min.)	180 x 300
	160 x 360 (max.)	180 x 420
Speedball (women)	180 x 300	200 x 340
*Squash	20 x 45	20 x 45
*Table Tennis	5 x 9	12 x 20
*Tennis (ideal)	27 x 78 (single)	57 x 138
	36 x 78 (double)	66 x 138
Touch Football	160 x 300	175 x 330
*Volleyball	30 x 60	42 x 72

*Detailed drawings appear elsewhere in this book.

SIZES OF
PLAYGROUND EQUIPMENT

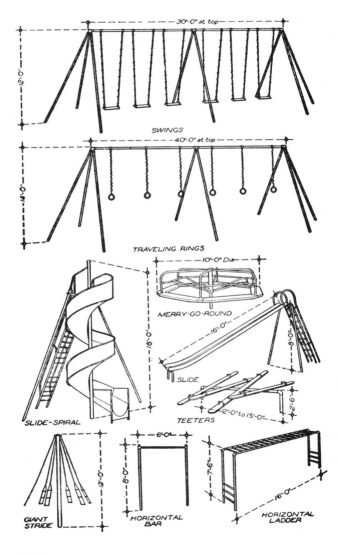

SWINGS

TRAVELING RINGS

MERRY-GO-ROUND

SLIDE

SLIDE-SPIRAL

TEETERS

GIANT STRIDE

HORIZONTAL BAR

HORIZONTAL LADDER

SHUFFLEBOARD COURTS

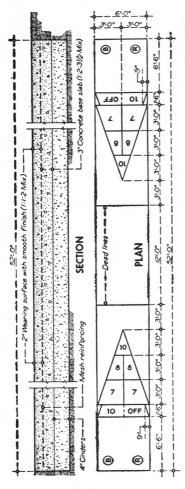

SECTION

PLAN

NOTE - All lines 3/4" wide
Dimensions center to
center of lines
No expansion joints

Ⓑ Black ⎫ When
Ⓡ Red ⎬ - playing
      ⎭ doubles

BASE COURSE. Well-drained level ground is the best site for a court. The site should be stripped of all sod down to uniformly firm ground and refilled with at least 4" of well tamped cinders. Where the sub-soil is firm sand, the cinder fill may be omitted. The base layer of concrete should be struck off to grade.

WEARING COURSE. Expanded metal or wire mesh weighing not less than 60 pounds per 100 square feet should be put down on the base course. The wearing or playing layer should be mixed and placed within 45 minutes after the base layer is struck off. It should be carefully brought to grade with a straightedge and wood floated. Grinding the surface with a machine shod with free, rapid cutting abrasive stones will give the smoothest and most satisfactory playing surface.

Curing is very important and the surface should be kept continually wet for 7 days. Since no expansion joints are used, careful curing will prevent surface checking or cracks. After curing the court should dry for 4 or 5 days before the playing lines are painted on.

MARKING. All lines are 3/4" wide and the dimensions on the drawing are from center to center of lines. A high quality paint made with an oil or varnish base is satisfactory for use on concrete surfaces. For new construction less than 6 months old, a zinc sulphate wash consisting of 3 pounds of crystals to 1 gallon of water should be applied to the concrete surface to be painted. Allow 48 hours for the zinc sulphate treatment to dry. Remove any crystals that appear on the surface before painting lines.

STADIUM
SEATING

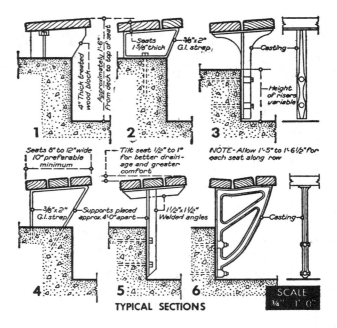

TYPICAL SECTIONS

TREADS. The dimensions of treads will have to be an economic compromise between the 2 conflicting factors: (1) Increasing the width of tread increases comfort by providing more leg room, but (2) reduces the sight line clearance. Treads vary from 2'-0" to 2'-6" with 2'-2" as an average.

RISERS. Increasing the riser height increases the total height of the structure and its cost. The number of rows of seats and the assumed sight line clearance produce dimensions of 6" to 1'-6" for risers.

WIDTH ALLOWED PER SPECTATOR. The complete disregard of spectator comfort is nowhere better shown than in the allotment of from 1'-5" to 1'-6½" in width to each spectator. Even the cheapest movie theater usually allows 19" and the better theaters have a substantial proportion of 20", 21" and 22" wide seat spacings.

SEATS AND SUPPORTS. Seats of 2 or 3 pieces are recommended as being less likely to warp than a single plank. Comfort and drainage are improved by tilting the seats slightly. Douglas fir, redwood, and Southern cypress which are free of pitch and kiln-dried or air-seasoned are most commonly used. Painting and preservative treatment increase the life of the wood. Supports attached to the risers facilitate cleaning, are easily placed. Supports are spaced 4'-0" o/c. Seats should be cut at expansion joints.

NBFU RULES ON
GASOLINE STORAGE

The position of outside aboveground gasoline storage tanks relative to property lines, public ways, and important buildings on the same site is regulated. In no case can a storage tank be placed closer than 5′ from a property line that is or can be built upon, or from the nearest side of a public way or the nearest important building. Further, certain greater minimum distances are imposed, depending upon the type and construction of the storage tank and its capacity; the latter ranges from 5′ to 175′ from property lines and from 5′ to 60′ from buildings and public ways, depending upon gallonage.

Underground gasoline storage tanks, either steel or nonmetallic, may be buried as shown in the drawings below. The distance from any part of the tank to the exterior surface of any basement or pit wall must be no less than 1 foot; the distance to any property line that can be built upon must be no less than 3′.

For further information, consult National Fire Protection Association publication NFPA 30, *Flammable and Combustible Liquids Code.*

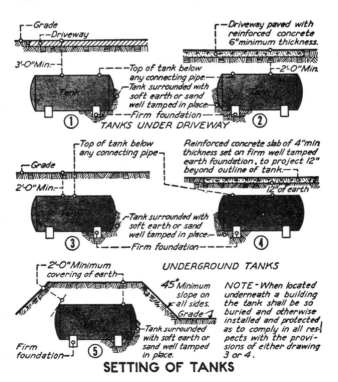

SETTING OF TANKS

FREIGHT TRAIN
CLEARANCE DIAGRAM

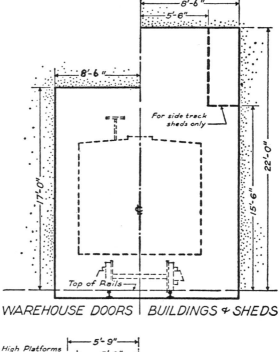

WAREHOUSE DOORS | BUILDINGS & SHEDS

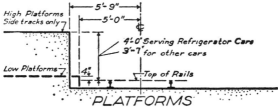

PLATFORMS

Clearances shown are minimum for straight track. Increase clearances on curved track for the overhang and tilting of a car 85′ long and 16′ high, with the superelevation of outer rail conforming to American Railway Engineering Association recommendations. The distance from the top of the rails to the top of the ties is 8″. Legal requirements should govern distances in excess of those shown. Current A.R.E.A. clearances and other recommendations should be checked before design is completed.

AIRPORT DESIGN
CHECKLIST

LANDING AREA
Size and Design
Grades
Drainage
Surfaces
Runways

APPROACHES
Zoned Area
Freedom from Obstructions

MARKING
Boundary Markers
Obstruction Marking
Identification Marker
Runway Marking

LIGHTING
Beacons
Boundary and Range Lights
Obstruction Lights
Floodlighting
Contact Lights
Instrument-landing Lighting
Course Light
Building Interior Lighting
Instrument Lighting
Emergency Power Supply
Remote Control
Miscellaneous Requirements

AIRCRAFT SERVICING FACILITIES
Fuel
Repairs
Storage

TRAFFIC CONTROL FACILITIES
Control Tower
Airport Traffic Control Room Equipment

BUILDINGS
Terminal
Passenger & Administration
Waiting Room
Rest Rooms
Dining Rooms
Ticket Office
Post Office
U. S. Weather Bureau
Airways Communication Office
Administration Offices
Employee Facilities
Hangars
Additional Buildings

OTHER FACILITIES
Aprons, Taxiways and Loading Areas
Road, Parking Lot, and Fence
Facilities for Visiting Public
Fire Protection Equipment
First-aid Facilities
Wreckage Equipment

JUDGING LAND FOR
PRIVATE AIR FIELDS

Some fields provide natural landing facilities for private planes. Some can be utilized with just a little preparation. Others need so much processing, that expense puts them out of consideration. However, modern bulldozers are equipped to level land and vegetation in a matter of hours. The table below rates land on its possible use as a private air field.

Factors of Land	GOOD Little cost to process	AVERAGE Medium cost to process	POOR Highest cost to process
LEVEL	Flat. Gentle rise. Long gentle swells	Slightly uneven, no more than 3' difference. Not over a 2% grade. (2' rise per 100'.)	Uneven, over 3' in difference. A grade in excess of 2%.
TYPE OF SOIL	Sandy or clay mixture. Rock-free.	Moderately rocky. Small rocks that can be scraped off easily.	Excessively rocky, with large, loose rocks. Rock ledge that needs grading.
VEGETA-TION	Tough, low grass.	Small shrubs, grass, and small trees.	Thick shrubs. Many large trees.
DRAINAGE	Natural drainage with gentle slope, rounded center rise, or flat top of plateau.	Absolutely flat, or low-lying, necessitating tiles to carry off water.	Hollow field, needing leveling, or tiles with possible pumping.
TYPE OF FINISH ON FIELD	Short, natural grass to help drainage and hold soil. A field previously used for agriculture needs less special drainage.	Seeded sod, attractive, needs occasional mowing. Clay needs no upkeep, but is dusty in Summer, muddy in wet seasons.	Concrete, good for heavy traffic, but expensive and unnecessary for private field.

AREA PER SEAT
FOR THEATERS

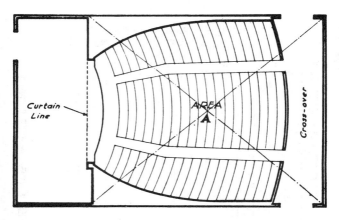

The following table gives the seating area of a number of auditoriums. It will be seen that the square feet to be allowed for each seat varies between fairly wide limits. The highest figure shown represents a 25% increase over the lowest. In making rough seating calculations it would be safer to allow 7 square feet per seat than the usual 6 square feet that is recommended by some authorities. Note that the seating area has been taken as the distance from the curtain line to the rear wall of the cross-over.

Name of Theater	Floor	Area "A"	No. of Seats	Sq. Ft. per Seat
Fred W. Wehrenburg Theater, St. Louis, Mo.	Main Floor	8,549	1,308	6.53
Ritz Theater, Baltimore, Md.	Main Floor	7,484	1,004	7.45
25th St. Theater, Newport News, Va.	Main Floor	3,278	549	5.97
Teatro de Comedia, Mexico	Mn. Fl. Balcony	3,662 2,134	490 349	7.47 6.11
Junior and Senior High School, Dobbs Ferry, N. Y.	Mn. Fl. Balcony	4,666 1,737	627 257	7.44 6.75
Ritz Theater, Columbus, Ohio	Main Floor	4,669	702	6.65
Virginia Polytechnic Institute, Blacksburg, Va.	Main Floor	21,098	3,003	7.02

LOCATION OF
FIRST ROW OF SEATS

S=SCREEN IMAGE HEIGHT
(= ⅟7 MAX. VIEWING DISTANCE)

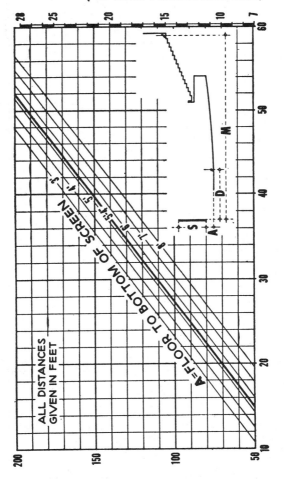

M=MAX. VIEWING DISTANCE

DETERMINING THE
MAIN FLOOR SLOPE

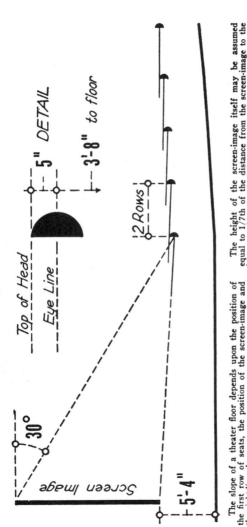

Top of Head

Eye Line

5" DETAIL

3'-8" to floor

2 Rows

30°

Screen Image

5'-4"

The slope of a theater floor depends upon the position of the first row of seats, the position of the screen-image and the sight-line clearance.

To preserve the illusion of reality, the screen-image must not be too high with respect to first row of spectators. The stage level in American theaters has been, almost without exception, taken as 3'-4" above the level of the first row of seats; the bottom of the screen-image being 2'-0" above the level of the stage. Therefore, a point 5'-4" above the floor is the focus of all eye lines for determining the main floor.

The height of the screen-image itself may be assumed equal to 1/7th of the distance from the screen-image to the last row of seats. A 30° angle with the horizontal from the top of the screen-image will intersect the horizontal eye line 3'-8" above the floor, and this will determine first row of seats.

The conditions for sight-line clearance should allow any seat occupant of anatomically average dimensions to see over the head of a spectator sitting in the second row ahead. A distance from eye to top of head of 5" is a safe assumption.

199

CONVENTIONAL VS. IDEAL
THEATER FLOOR SLOPE

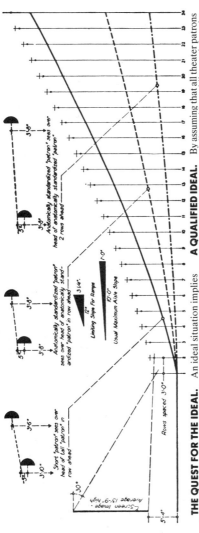

THE QUEST FOR THE IDEAL. An ideal situation implies that a short patron should be able to see over the head of a tall patron in the row immediately ahead. The chart above shows the floor slope needed to attain this condition, for a theater representing the average of United States movies houses.

DIFFICULTIES WITH THE IDEAL. The theoretically ideal floor slope is steeper than the usual building code limit for aisle slopes; it also exceeds the inclination beyond which the aisle is no longer a ramp. Given building code provisos regulating the number, height, and regularity of aisle steps, the theoretically ideal slope is effectively prohibited.

A QUALIFIED IDEAL. By assuming that all theater patrons are of anatomically average dimensions while retaining the premise that first-row sight-line clearance is desirable, we get the second curve shown above. This presents the same difficulties as the first case, although to a lesser degree.

USUAL COMPROMISE. The third curve shows the floor slope as originally recommended in "The Design of the Cinema," Part 2, which appeared in PENCIL POINTS in June 1938. This floor slope falls within the usual legally acceptable limits of 1:7–1:10 slopes and at the same time represents a practical precedent used in the design of movie theaters serving millions of patrons daily.

SIDE SEAT LIMITS FOR
SATISFACTORY VIEWING

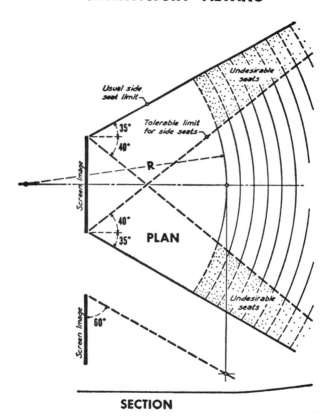

Usual side seat limit

Tolerable limit for side seats

Undesirable seats

35°
40°
R
Screen Image

40°
35°
PLAN

Undesirable seats

Screen Image
60°

SECTION

SIDE SEAT DISTORTION. Side seats from which the observer sees any part of the screen-image at an angle greater than 40° have been found, in a limited test, to destroy the illusion of reality. The usual side seat limit employed in motion-picture theater design has been a 35° line from the near edge of the screen, as shown. The hatched area indicates undesirable seats and this portion should be kept to an absolute minimum.

RADIUS OF SEATS. The seats in both the balcony and on the main floor follow a series of concentric circular segments so that the observers may sit approximately facing the action taking place on the stage or screen. The smallest usual radius for the chair size line of the first row is about 30'-0". So far as is known to this author, no logical rule exists for locating the center of the concentric circular segments.

THEATER CHAIRS FOR
MAIN FLOOR

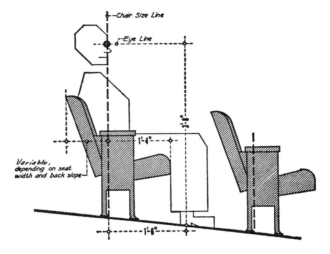

Theater chairs are arranged with legs to fit all floor slopes in ¼″ intervals from 0″ to 2¾″ per foot for conventional inclines, and from 0″ to minus 2½″ for reversed inclines. The legs are shortened or lengthened so that the theoretical eye-level will come 3′-8″ above the level of the heel, as shown in the drawing above.

In calculations of floor slope, the eye is assumed to be 3′-8″ above the floor on a vertical line through the eye, involving an apparent discrepancy. However, no important error results since it is only the equivalent of moving the entire floor (as designed) a distance of 1′-6″ nearer the screen. Lines in plan and section to indicate seat rows should represent "chair-size" lines rather than the backs of the seats because in the latter case nominal chair width coincides with actual width.

Some building laws specify 2′-6″ from back to back of seats as the minimum allowed. For extremely low-admission-price theaters equipped with veneer-wood-back seats, this distance probably represents an economic feasibility. However, for the average theater employing padded-back seats—and in the interest of comfort and maneuverability of the clientele—a back-to-back spacing of 2′-8″ is barely adequate; 2′-10″ represents a better normal condition, and 3′-0″ might be regarded as an attainable ideal.

The width of seats on the "chair-size line" varies from 18″ to 24″ in 1″ intervals. 1½″ is allowed for each end standard. Seats 18″ and 19″ wide are uncomfortable, and their use should be limited. Since row lengths vary in any given seating layout, the variation in seat widths allows for individualized adjustment. A width of from 20″ to 22″ is usual in all seating locations.

DETERMINING THE BALCONY SLOPE

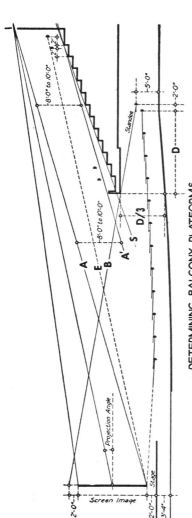

DETERMINING BALCONY PLATFORMS

The capacity of the balcony is most often between 33⅓ % to 50% of that of the main floor. The balcony section will be the result of a number of limiting conditions: *First,* the projection angle should not exceed 12° with the horizontal for ideal conditions. However, many theaters use a 20° angle and projectors allow up to about 30°. The balcony steps should fall below A'. *Second,* sight-line B of a standee should clear the bottom front edge of the balcony. *Third,* the balcony should not overhang the last row of main floor seats by more than 3 times the height, as indicated by D/3, for acoustical reasons. *Fourth,* the slope S should not necessitate aisle steps that are illegal or uncomfortable. *Fifth,* the sight-line E must clear the head of the occupant of the seat in the next row in front, to a focus at the bottom of the screen-image.

The common method of determining the balcony slope by projecting the line S to a point 7'-6" below stage level at the curtain line has resulted in highly unsatisfactory balcony vision, and the method should be avoided.

THEATER CHAIRS
FOR BALCONY

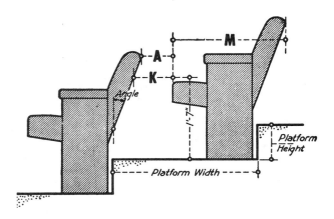

Steps upon which the balcony seats are placed are usually referred to as *platforms*. In the balcony aisles steps are introduced to make circulation possible. Building laws usually limit the height of a single step to 7½" or 8". With 2 such risers for each balcony seat platform, the maximum slope of the balcony would be between 15" and 16" in height for each 32" to 36" horizontally.

Older theaters for legitimate productions often have 3 steps for each seat platform, making a rise of 21" to 24" per platform. Such a pitch results in discomfort.

The knee room for balcony seats is measured on a line 1'-7" above the platform. Whereas raising the platform height does not affect the aisle width "A" for any given platform width, it does reduce the knee room "K". Thus it becomes important to have definite chair dimensions in mind before deciding on the platform width. The platform width is established so that "A" will be not less than 6½" nor "K" less than 8½".

The usual chair-back slope for the platform seats is 5¼" in the height of the seat-back, making an angle of 14°-8'. This requires that platforms from 2" to 11" high should be not less than 31" wide; platforms 11⅜" to 16" high should have 33" as a minimum platform width. The overall dimension "M" and the slope of the seat-back should be known before the balcony platforms are decided upon.

The first row in the balcony should have from 2" to 6" wider platform, so that toe room is provided and also so that people passing between the balcony rail and the seat occupants will not feel any danger of tripping and falling. If there is no rear cross-over, the platform width for the last row of seats against the rear wall will have to be from 6¼" to 9½" greater to accommodate the pitch of the back seat.

EXISTING PROPORTIONS
SCREEN-TO-SEATING

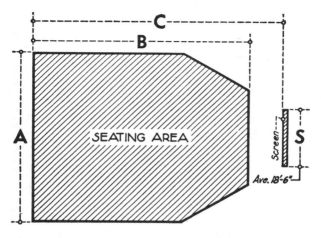

RATIO	MINIMUM	AVERAGE	MAXIMUM
B/A	1.52	1.98	2.35
A/S	2.50	3.00	3.50
C/S	4.65	5.20	5.85

EXISTING MOVIE THEATERS.   A survey conducted by the Society of Motion Picture Engineers, covering about 600 theaters throughout the United States, was undertaken to determine the existing conditions under which many millions of persons enjoy and pay for motion-picture entertainment. The value of the survey lies in the entirely safe assumption that characteristics of the 50% group that clusters about the gross average represent tolerable practice at that time. Care was taken that the theaters covered would represent a fair cross section of all the theaters operating in the country.

SURVEY RESULTS.   The results of the SMPE survey are shown above diagrammatically and represent the limits of the 50% group of theaters falling about the total group average. A disparity will be noted if the A or S values are calculated from the two ratios in which they both appear. In a statistical compilation of this type, such a disparity is natural. The shape of the seating area shown is for diagrammatic purposes only —it does not necessarily represent the forms encountered in the survey.

STAGE HEIGHT AND
GRID LOCATION

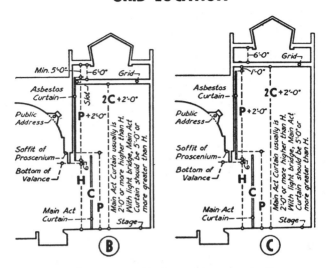

STAGE HEIGHT. The height of the stage is a matter of real importance and one that cannot be given too much consideration. The height of the gridiron above the stage floor depends upon the treatment of the proscenium arch.

SIMPLEST CASE. When the soffit of the proscenium construction is also the top of the clear stage opening, the bottom of the grid needs to be twice the proscenium height plus 3'-6".

SLOTTED GRID. Many architects, however, prefer to build the arch high—especially on a wide stage, to give it a more graceful effect. A wide valance is then hung in the archway to cut the proscenium opening down to a suitable height. This arrangement is a common one and produces a pleasing effect. If the valance is a fabric or other nonstructural material, the fireproof curtain must lap the actual proscenium soffit 2'-0" in the down position.

A savings can be effected, as shown in Figure B, by slotting the gridiron to allow passage of the fireproof curtain. Usually 3'-0" or 4'-0" in the height of the building can be saved, and by this method the weight of the fireproof curtain is carried by the proscenium wall instead of the gridiron. With deep stages this is particularly important.

GRID NOT SLOTTED. As shown in Figure C, the gridiron without slotting for the fireproof curtain necessitates added height in the stage construction.

PUBLIC ADDRESS SPEAKER. Notice particularly that, in the diagrams, the public address speakers are placed in front of the proscenium arch. Often they are put backstage, meaning that the performers using the microphone on the stage apron are in front of the loudspeakers. This arrangement creates a feedback of energy that completely destroys the intelligibility of the voice and creates an unpleasant effect. If the suggestion in the diagrams is followed, this difficulty is obviated.

SEATING CAPACITY
WITH PORTABLE CHAIRS

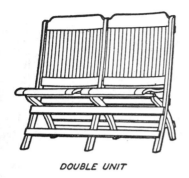

DOUBLE UNIT

Table A

ROWS	DEPTH
10	32'-6"
12	37'-6"
16	47'-6"
18	52'-6"
20	57'-6"
22	62'-6"
24	67'-6"
26	72'-6"
28	77'-6"
30	82'-6"

Figuring 2'-6" back to back; 3'-6" cross aisle at rear and 4'-0" cross aisle at front.

Table B

WIDTH	ARRANGEMENT	CHAIRS
18'-4"	Aisle 3'-0"	10
21'-4"	Aisle 3'-0"	12
24'-4"	Aisle 3'-0"	14
30'-4"	Aisle 3'-0" ... Aisle 3'-0"	16
33'-6"	Aisle 3'-0" ... Aisle 3'-0"	18
33'-4"	Aisle 3'-0" ... Aisle 3'-0" ... Aisle 3'-0"	16
36'-6"	Aisle 3'-0" ... Aisle 3'-0"	20
39'-7"	Aisle 3'-0" ... Aisle 3'-0"	22
39'-6"	Aisle 3'-0" ... Aisle 3'-0" ... Aisle 3'-0"	20
46'-0"	Aisle 3'-0" ... Aisle 3'-6" ... Aisle 3'-0"	24
56'-4"	Aisle 3'-6" ... Aisle 3'-6" ... Aisle 3'-6"	30
65'-4"	Aisle 3'-6" ... Aisle 3'-6" ... Aisle 3'-6"	36

The tables above will help you to determine the seating capacity of your auditorium. They show the most popular grouping and spacing arrangement of the various widths of auditoriums. Table *B* shows the total number of chairs which can be placed across the width of the room, the grouping arrangement, and the size and location of the aisles. Table *A* shows the number of rows in the depth of the room. By multiplying the number of chairs in the width of the room by the number of rows in the depth of the room, the total seating capacity is obtained. These tables makes no allowance for posts, obstructions, etc.

PORTABLE CHAIRS
CLEARING AND STORAGE

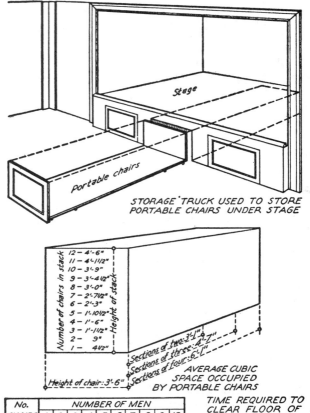

STORAGE TRUCK USED TO STORE
PORTABLE CHAIRS UNDER STAGE

Number of chairs in stack / Height of stack

12 — 4'-6"
11 — 4'-1½"
10 — 3'-9"
9 — 3'-4½"
8 — 3'-0"
7 — 2'-7½"
6 — 2'-3"
5 — 1'-10½"
4 — 1'-6"
3 — 1'-1½"
2 — 9"
1 — 4½"

Sections of two: 3'-1"
Sections of three: 4'-7"
Sections of four: 6'-1"
Height of chair: 3'-6"

AVERAGE CUBIC
SPACE OCCUPIED
BY PORTABLE CHAIRS

No.	NUMBER OF MEN									
CHAIRS	1	2	3	4	5	6	7	8	9	10
100	12	6	4							
200	24	12	8	6						
300	36	18	12	9	7					
400	48	24	16	12	10	8	7			
500	60	30	20	15	12	10	9			
600	72	36	24	18	14	12	10			
700	84	42	28	21	18	14	12	11		
800	96	48	32	24	20	16	14	12		
900	108	54	36	27	22	18	16	14	12	
1000	120	60	40	30	25	20	18	16	14	12

TIME REQUIRED TO
CLEAR FLOOR OF
PORTABLE CHAIRS.

This table, while purely theoretical, will prove of assistance in figuring the approximate time for clearing an auditorium of portable chairs. It is made up on the basis that one man can fold and move one section in 15 seconds or 8 chairs per minute.

DISPLAY FRAMES WITH
CONCEALED LIGHTING

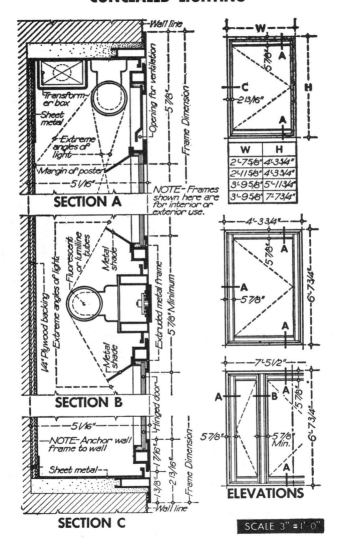

Wall line
Opening for ventilation
5 7/8"
Frame Dimension

Transformer box
Sheet metal
Extreme angles of light
Margin of poster
5 1/16"

SECTION A

NOTE—Frames shown here are for interior or exterior use.

W	H
2'-7 5/8"	4'-3 3/4"
2'-11 5/8"	4'-3 3/4"
3'-9 5/8"	5'-11 3/4"
3'-9 5/8"	7'-7 3/4"

W
5 7/8"
A
C
2 13/16"
H
A

1/4" Plywood backing
Extreme angles of light
Fluorescent or lumiline tubes
Metal shade
Extruded metal frame
5 7/8" Minimum
Metal shade

SECTION B

4'-3 3/4"
5 7/8"
A
5 7/8"
6'-7 3/4"
A

5 1/16"
NOTE—Anchor wall frame to wall
Hinged door
1-7/16"
Sheet metal
1-3/8"
2 13/16"
Frame Dimension
Wall line

SECTION C

7'-5 1/2"
A
B
A
5 7/8"
5 7/8"
5 7/8" Min.
6'-7 3/4"
A

ELEVATIONS

SCALE: 3" = 1'-0"

209

FLUSH AND SURFACE
DISPLAY FRAMES

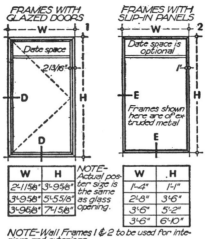

FRAMES WITH GLAZED DOORS

W	H
2'-11 5/8"	3'-9 5/8"
3'-9 5/8"	5'-5 5/8"
3'-9 5/8"	7'-1 5/8"

NOTE—Actual poster size is the same as glass opening.

NOTE—Wall Frames 1 & 2 to be used for interiors and exteriors.

FRAMES WITH SLIP-IN PANELS

Date space is optional

Frames shown here are of extruded metal

W	H
1'-4"	1'-1"
2'-8"	3'-6"
3'-6"	5'-2"
3'-6"	6'-10"

FRAMES FOR SLIP-IN CARDS

W	H
1'-3 1/8"	4 1/8"
1'-3 1/8"	5 1/8"
11 1/8"	9 1/8"
1'-3 1/8"	9 1/8"
1'-3 1/8"	1'-0 1/8"
1'-6 1/8"	1'-3 1/8"
1'-11 1/8"	1'-3 1/8"
3'-1 1/8"	1'-3 1/8"
2'-5 1/8"	1'-11 1/8"
2'-7 1/8"	3'-5 1/8"
3'-5 1/8"	5'-4 1/8"
3'-5 1/8"	6'-9 1/8"

NOTE—For interior use only.

ELEVATIONS

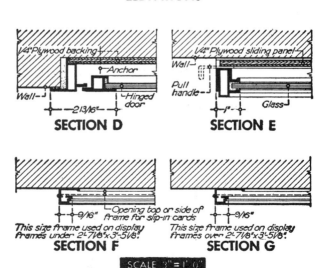

1/4" Plywood backing — Anchor — Wall — Hinged door — 2 13/16"

SECTION D

1/4" Plywood sliding panel — Wall — Pull handle — Glass — 1"

SECTION E

Opening top or side of frame for slip-in cards — 9/16"

This size frame used on display frames under 2'-7 1/8"x3'-5 1/8".

SECTION F

9/16"

This size frame used on display frames over 2'-7 1/8"x3'-5 1/8".

SECTION G

SCALE 3" = 1'-0"

210

LUNCH COUNTERS

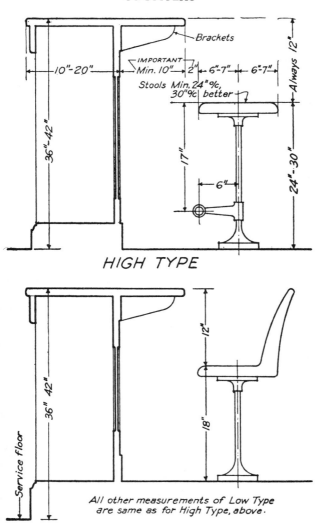

Brackets

IMPORTANT
Min. 10"

10"-20"

2" 6"-7" 6"-7"

Always 12"

Stools Min. 24" %,
30" % better

36"-42"

17"

6"

24"-30"

HIGH TYPE

36" 42"

12"

Service floor

18"

All other measurements of Low Type
are same as for High Type, above.

LOW TYPE

211

CAFÉ DOORS, WINDOWS, STORAGE

In the design of a café or similar eating establishment, the following points must be considered. In addition, the relevant state and local health requirements, which vary considerably from place to place, must be fully investigated and observed in order for the establishment to qualify for a Certificate of Occupancy and/or a License to Operate (and for the protection of customers).

STORAGE AND DISPLAY OF FOOD AND DRINK. All food and drink should be so stored and displayed as to be protected from dust, flies, vermin, handling, droplet infection, overhead leakage, and other contamination. All means necessary for the elimination of flies should be used.

Public-health reason. Food or drink not properly protected from contamination may become a public-health hazard.

Satisfactory compliance. Satisfactory compliance is implied when the following conditions prevail:

(1) There is no indication of the presence of rodents, roaches, ants, or other vermin. Food and drink are not stored in locations, such as below floor level, that might be subject to flooding or other causes of contamination.

(2) All unwrapped or unenclosed food and drink on display is properly refrigerated if necessary, and protected from public handling or other contamination, with approved hand openings, as required, on counter fronts, cases, and so forth.

(3) All supplementary and approved means necessary for the elimination of flies are employed.

(4) All enclosed spaces within double walls, between ceilings and floors, in fixtures and equipment that might harbor rodents have been eliminated by the removal of the sheathing that forms the enclosed space; or all exposed edges of such walls, floors, and sheathing have been protected against gnawing by rats by the installation of approved ratproof material, and all openings in walls, floors, and ceilings through which pipes, cables, and other conduits pass have been properly sealed with snug-fitting collars of metal or other approved ratproof material, securely fastened in place.

DOORS AND WINDOWS. When flies are prevalent, all openings into the outer air should be effectively screened, and entrance doors and/or screen doors should be self-closing, unless other effective means are provided to prevent entry by flies.

Public-health reason. Flies may contaminate food with disease organisms, thus nullifying the effectiveness of all other public health safeguards.

Satisfactory compliance. The following requirements are implied conditions of satisfactory compliance:

(1) All openings to the outer air are effectively screened with not less than 16-mesh wire cloth, and

(2) All doors are self-closing and screen doors to the outer air open outward; or

(3) Fans of sufficient power to prevent the entrance of flies are in use at all ineffectively protected openings; or

(4) Flies are absent.

(5) Window and door screens must be tight-fitting and free of holes. This includes the screens for skylights and transoms.

CAFÉ FLOORS, WALLS, CEILINGS, VENTILATION, LIGHTING

FLOORS. The floors of all rooms where food or drink is stored, prepared, or served should be smooth, as seamless as possible, and easy to clean; they should be kept clean and in good repair. Kitchen floors should be impervious to water.

Public-health reason. Properly constructed floors in good repair can be more easily kept clean than improperly constructed floors. Kitchen floors having an impervious surface can be cleaned more easily than floors constructed of wood or other pervious or easily disintegrated material, will not absorb organic matter, and are therefore more likely to be kept clean and free of odors. Clean floors are conducive to clean food-handling methods.

Satisfactory compliance. The following conditions imply satisfactory compliance:

(1) The floors of all rooms where food or drink is stored, prepared, or served can be easily cleaned, are smooth, and are in good repair. Floors may be of concrete, terrazzo, tile, and so forth; or of wood covered with sheet vinyl; or of tight, finished wood construction.

(2) The floors of all rooms in which food is prepared or utensils are washed are constructed of concrete, terrazzo, tile, or other impervious material, in good repair and equipped with drains. However, where floors of such rooms are kept clean without flushing, sheet-vinyl or similarly impervious surfacing in good repair may be used and floor drains are not necessary. Wherever floor drains are used, they must be fitted with proper traps so constructed as to minimize clogging.

WALLS AND CEILINGS. Walls and ceilings of all rooms where food and drink is stored, prepared, or served should be kept clean and in good repair, should be finished in light color, and should have a smooth, washable surface up to at least the level that might be reached by splash or spray.

Public-health reason. Painted or otherwise properly finished walls and ceilings are more easily kept clean and are therefore more likely to be kept clean. A light-colored paint or finish aids in the even distribution of light and the detection of unclean conditions. Clean walls and ceilings are conducive to clean food-handling operations.

LIGHTING. All rooms in which food or drink is stored or prepared or in which utensils are washed should be well lighted.

Public-health reason. Ample light promotes cleanliness.

Satisfactory compliance. Satisfactory conditions prevail if artificial light sources provide 30 or more footcandles on all working surfaces in rooms where food or drink is prepared or utensils are washed, as measured by a light meter, and if such sources are in use except when equivalent natural light is present. Storage rooms are sufficiently well lighted at an illuminance of 5 footcandles, measured 30" from the floor.

VENTILATION. All rooms in which food or drink is stored, prepared, or served or in which utensils are washed should be well ventilated.

Public-health reason. Proper ventilation reduces odors, aerosol greases, smoke, and other air contaminants, and prevents condensation upon interior surfaces.

Satisfactory compliance. Satisfactory conditions prevail when all rooms are ventilated sufficiently to fulfill air-change-rate requirements, if any, and sufficiently to be reasonably free of disagreeable odors, smoke, and condensation. Ventilation equipment supplementary to windows and doors, such as adequate exhaust fans and stove hoods, must be provided as required.

CAFÉ LAVATORIES, TOILETS, WATER SUPPLY

LAVATORY FACILITIES.   Adequate and convenient hand-washing facilities must be provided, including warm running water, soap, and approved sanitary towels. The use of a common towel should not be permitted. No employee should be permitted to return from a toilet room without washing his or her hands.

Public-health reason. Washing facilities and sanitary towels are essential to the personal cleanliness of food handlers.

Satisfactory compliance. This item shall be deemed to have been satisfied if hand-washing facilities, including warm running water, soap, and individual cloth or paper towels are provided. Washing facilities must be adequate and convenient to the toilet rooms. Dish-washing vats shall not be accepted as washing facilities for personnel. Warm water must be on hand at all times or within a reasonable time after opening the faucets. Soap and towels should be provided by the management. No employee shall return from a toilet to a room where food, drink or utensils are handled or stored without first having washed his or her hands.

TOILET FACILITIES.   Every restaurant must be provided with adequate toilet facilities conveniently located and conforming with state and local ordinances and health regulations. Toilet rooms should not open directly into any room in which food, drink, or utensils are handled or stored. The doors of all toilet rooms should be self-closing and lockable from the inside. Toilet rooms should be constructed of hygienic, easy-to-clean materials, and should be kept clean, in good repair, well lighted, and well ventilated. Hand-washing signs should be posted in each toilet room used by employees.

Public-health reason. The need for toilet facilities and the necessity for protecting the food from toilet-contaminated flies are obvious.

Satisfactory compliance. The following requirements are implied conditions of satisfactory compliance:

(1) The toilet room, stool, etc., are kept clean, sanitary, in good repair, and free from flies.

(2) Durable, legible signs are posted conspicuously in each toilet room directing employees to wash their hands before returning to work. Such signs may be stenciled on the wall to prevent removal.

(3) A booth open at the top shall not qualify as a toilet room.

WATER SUPPLY.   A water supply should be easily accessible to all rooms where food is prepared or utensils are washed, and it must be of a safe sanitary quality.

Public-health reason. The water supply should be ample and convenient so as to encourage its use in cleaning operations and so that cleaning and rinsing will be thorough; and it must be of safe, sanitary quality in order to be suitable for drinking and to avoid the contamination of food and utensils.

CAFÉ EQUIPMENT, WASTE DISPOSAL, ETC.

CONSTRUCTION OF UTENSILS AND EQUIPMENT. All eating and cooking utensils and all show and display cases or windows, counters, shelves, tables, refrigerating equipment, sinks, and other equipment or utensils used in connection with the operation of a restaurant must be so constructed as to be easily cleaned and must be kept in good repair.

Public-health reason. If the equipment is not easy to clean and is not kept in good repair, it is unlikely ever to be cleaned properly.

Satisfactory compliance. The following requirements are implied conditions of satisfactory compliance:

(1) All surfaces with which food or drink comes in contact consist of smooth, not readily corrodible material.

(2) All surfaces with which food or drink comes in contact are in good repair, free of breaks, corrosion, open seams, cracks, and chipped places.

(3) All surfaces with which food or drink comes in contact are easily accessible for cleaning, and are self-draining.

(4) All display cases, windows, counters, shelves, tables, refrigeration equipment, stoves, hoods and other equipment are so constructed as to be easily cleaned, and are in good repair.

(5) The above requirement precludes the use of any type of equipment so designed as to permit food or drink routinely to come in contact with threaded surfaces.

(6) In all cases in which a rotating shaft is inserted through a surface with which food or drink comes in contact, the inspector shall make certain that the joint between the moving and stationary surfaces is close fitting.

(7) All mechanical equipment used in food storage, handling, and preparation must be of suitably hygienic construction and should be designed and approved for the restaurant trade. All such equipment is generally subject to inspection and approval by state and local authorities.

DISPOSAL OF WASTES. All wastes must be properly disposed of, and all garbage and trash must be kept in suitable containers in such a manner as not to become a nuisance or hazard.

Public-health reason. All garbage, refuse, and liquid wastes resulting from the normal operation of a food or drink establishment should be properly disposed of so as not to become a nuisance or a public-health menace.

REFRIGERATION. Waste water from refrigeration equipment should discharge into an open sink or drain, properly trapped and sewer-connected. However, in certain circumstances when sewer connections are not available, clean, adequate-sized watertight drip pans may be used.

MISCELLANEOUS. The surroundings of all restaurants should be kept clean and free of litter or rubbish. None of the operations connected with a restaurant should be conducted in any room used for domestic purposes. Adequate lockers or dressing rooms should be provided for employees' use. Soiled linens, coats, aprons, uniforms, and so forth should be kept in containers provided for the purpose.

LIQUOR BARS

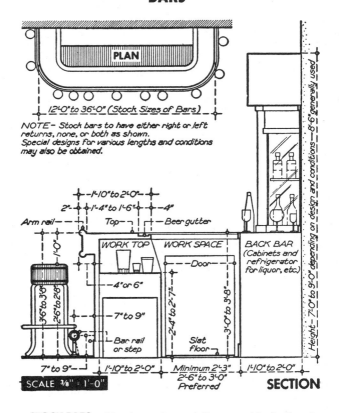

PLAN

12'-0" to 36'-0" (Stock Sizes of Bars)

NOTE— Stock bars to have either right or left returns, none, or both as shown.
Special designs for various lengths and conditions may also be obtained.

Arm rail

Top — Beer gutter

1'-10" to 2'-0"

2" — 1'-4" to 1'-6" — 4"

1'-0"

WORK TOP — WORK SPACE

BACK BAR
(Cabinets and refrigerator for liquor, etc.)

Door

4" or 6"

7" to 9"

Bar rail
or step

36" to 3'0"

2'-6" to 2'-8"

2'-4" to 2'-7"

3'-0" to 3'-8"

Slat floor

7" to 9"

1'-10" to 2'-0"

Minimum 2'-3"
2'-6" to 3'-0"
Preferred

1'-10" to 2'-0"

Height— 7'-0" to 9'-0" depending on design and conditions — 8'-6" generally used

SCALE ⅜" = 1'-0"

SECTION

STOCK BARS.   There is properly no such thing as a stock bar for dispensing liquor. The general measurements have been thoroughly established, and, if the section shown above is followed, such fittings as cabinets, refrigerators, sinks, and other similar equipment can be readily fitted into any special design. The handling of the bar front and the back bar offers wide scope for the imagination of the designer, and we find bars constructed of practically all decorative materials, such as glass blocks, plastics, structural glass, bricks, fieldstone, wood, and marble.

VARIATIONS.   One arrangement that can cause bar patrons serious discomfort arises when the worktop projects too little or when it has too deep an apron. The section above would provide considerably greater comfort to patrons using bar stools if the projection were increased from 4" or 6" to 8" or 10", and if the vertical thickness were reduced to a minimum. If more than one bartender is to work at a time, the aisle space between front and back bars should not be less than 3'-0".

SIZES OF
LIQUOR BOTTLES

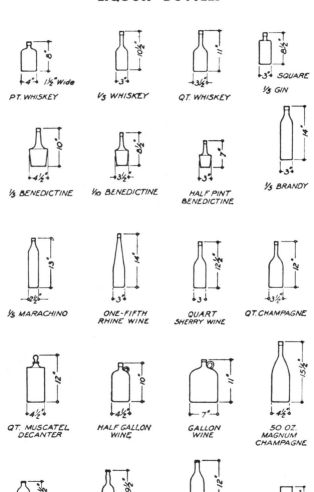

PT. WHISKEY

1/5 WHISKEY

QT. WHISKEY

1/3 GIN

1/3 BENEDICTINE

1/10 BENEDICTINE

HALF PINT
BENEDICTINE

1/5 BRANDY

1/5 MARACHINO

ONE-FIFTH
RHINE WINE

QUART
SHERRY WINE

QT. CHAMPAGNE

QT. MUSCATEL
DECANTER

HALF GALLON
WINE

GALLON
WINE

50 OZ.
MAGNUM
CHAMPAGNE

12 OZ.
BEER

12 OZ.
BEER

QUART
BEER

12 OZ. CAN
BEER

SIZES OF
LIQUOR GLASSES, ETC.

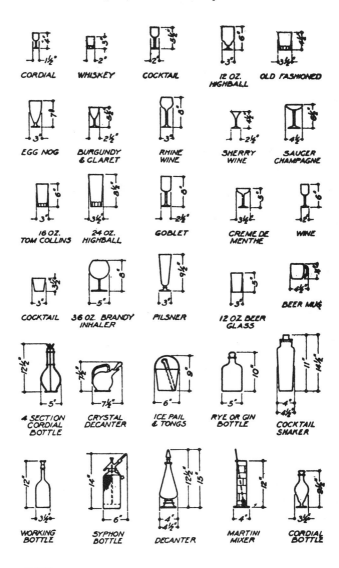

CORDIAL   WHISKEY   COCKTAIL   12 OZ. HIGHBALL   OLD FASHIONED

EGG NOG   BURGUNDY & CLARET   RHINE WINE   SHERRY WINE   SAUCER CHAMPAGNE

16 OZ. TOM COLLINS   24 OZ. HIGHBALL   GOBLET   CREME DE MENTHE   WINE

COCKTAIL   36 OZ. BRANDY INHALER   PILSNER   12 OZ BEER GLASS   BEER MUG

4 SECTION CORDIAL BOTTLE   CRYSTAL DECANTER   ICE PAIL & TONGS   RYE OR GIN BOTTLE   COCKTAIL SHAKER

WORKING BOTTLE   SYPHON BOTTLE   DECANTER   MARTINI MIXER   CORDIAL BOTTLE

218

SMALL STORE
PLANNING PRINCIPLES

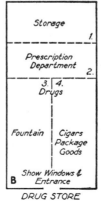

Since it is not required for selling space, this area is partitioned off from the rest of the store. It is available for storage, service department, living quarters, or for other purposes.
1.

This part of the selling space is generally allotted to that department of merchandise that does greatest amount of processing, weighing, cutting, transformation, etc.
2.

3. | 4.

These two areas in the front of the store are allotted to the balance of the merchandise, which is separated and placed on either side of the store according to relationship and similarity, and with due regard to promotional benefits.

1

Show Windows and Entrance

BASIC USE OF SPACE IN
RETAIL STORES

IN GENERAL. Store design is constantly adjusting itself to changing needs of the time and location insofar as merchants have been able to determine the wants of the customers and have been in a position to make desirable changes. To set down a specific plan for *any* kind of retail store is impossible. However, the basic principle of *departmentization* can be established to assist in planning.

SELLING AREAS. All merchants divide their stock and activities into departmental groups. Small businesses can be conveniently set up in 3 departments. The location of the departments will depend upon the special skill of the merchant and the local market demand for various goods. For example, the drawing shows the meat department to occupy the rear of the grocery store, yet a merchant wishing to lay heavier stress on meat because of profit potentialities or because of special skill in handling meat, might place this department forward at the side of the store.

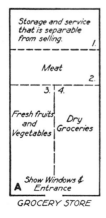

Storage and service that is separable from selling.
1.

Meat
2.

3. | 4.

Fresh fruits and Vegetables | Dry Groceries

Show Windows & Entrance

A GROCERY STORE

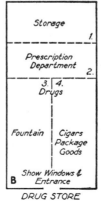

Storage
1.

Prescription Department
2.

3. | 4.
Drugs

Fountain | Cigars Package Goods

Show Windows & Entrance

B DRUG STORE

Storage
1.

Heavy hardware, bulk goods, wheel goods, and service.
2.

3. | 4.

Shelf Hardware and Tools | House Furnishings and Appliances

Show Windows & Entrance

C HARDWARE STORE

219

SHOW WINDOW
DESIGN PRINCIPLES

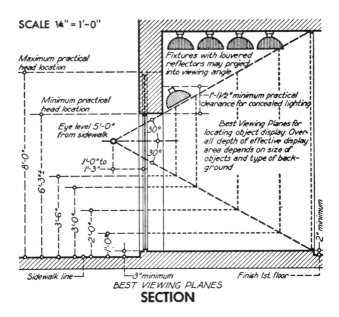

SCALE ¼" = 1'-0"

Maximum practical head location

Minimum practical head location

Eye level 5'-0" from sidewalk

Fixtures with louvered reflectors may project into viewing angle

1'—1½" minimum practical clearance for concealed lighting

Best Viewing Planes for locating object display. Over-all depth of effective display area depends on size of objects and type of back-ground

30°

30°

1'-0" to 1'-3"

8'-0"

6'-3¼"

3'-6"

3'-0"

2'-0"

1'-0"

2" minimum

Sidewalk line

3" minimum

Finish 1st. floor

BEST VIEWING PLANES
SECTION

In general, the smaller the objects are which must be displayed, the higher the bulkhead becomes and the shallower the display space becomes — to bring the objects closer to the observer's eye. Large objects such as automobiles and house furnishings will have a very low bulkhead and a relatively deep display area. In the diagram above is shown a method of locating the most favorable viewing plane for locating the objects on display. Obviously, the show-window back should be located sufficiently in the rear of this plane to furnish a proper background.

SIGHT LINES. The normal cone of human vision is approximately 60° — 30° in all directions from the optical center. Eye levels have been incorrectly suggested in various printed articles as 5'-3". Consumer Research says that women influence the majority of retail purchases so a 5'-3" sight line is incorrect as an average for prospective buyers. An eye level of 5'-0" or even 4'-10" more closely approximates true conditions. For bulkheads of various heights, the optimum viewing plane will be found at the intersection of the floor and the sight lines.

MULTIPLE SHOW
WINDOWS

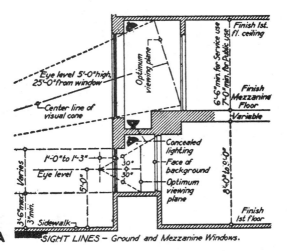

Finish 1st. fl. ceiling

6'-6" min. for Service use
7'-0" min. for Public use

Finish Mezzanine Floor

Variable

Eye level 5'-0" high,
25'-0" from window

Center line of visual cone

Concealed lighting

Face of background

Optimum viewing plane

Optimum viewing plane

1'-0" to 1'-3"

Eye level

30°

30°

5'-0"

8'-0" to 9'-0"

Finish 1st floor

Varies

3'-6" max.
3' min.

Sidewalk

A   *SIGHT LINES - Ground and Mezzanine Windows.*

Finish ceiling

Access door

8'-0" to 9'-0" with Mezzanine
(7'-6" minimum)
12'-0" min. without Mezz.

Finish 1st floor

Variable

1'-0" to 1'-3"

Eye level

30°

30°

5'-0"

Optimum viewing plane

Optimum viewing plane

Access door

Sidewalk

3" minimum

8'-0" to 9'-0"

SCALE
⅛" = 1'-0"

SECTIONS

Finish basement floor

B   *SIGHT LINES - Ground and Basement Windows.*

A comfortable viewing angle is 30° in all directions from the optical axis. Within this 60° cone, the eye sees quickly and without any appreciable physical effort of focusing. Thus it is practical to plan show windows which encompass two or more "viewing angle areas."

DIMENSIONS OF SHOW WINDOWS

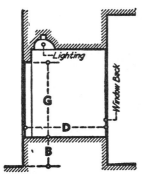

NO STANDARDS POSSIBLE. In analyzing the figures which are presented below, the designer is cautioned not to regard these as hard and fast standards which must not be varied. The figures given only represent reasonable averages for the types of stores listed. They will, however, provide a starting place since they take into consideration the basic principle of store front heights which dictates that the smaller the object displayed, the higher must be the display window floor.

Store	Bulkhead Height (B)	Glass Height (G)	Window Depth (D)	Lighting in Watts per Lin. Ft. Outlets 12"-15"	Window Backs
ARTISTS' MATERIALS	2'-2" to 3'-0"	4'-6" to 6'-0"	3'-0" to 4'-0"	100 to 200	Neutral color, suitable for tacking; no portion of window more than 3'-6" from access door
AUTOMOBILE MACHINERY (large)	0'-0" to 1'-0" (access window sometimes needed)	8'-0" to 10'-0"	6'-0" to 10'-0"	300 to 500 (special lighting effects; ceiling lights lowered, recessed spotlights	Open into store
MACHINERY (small)	0'-0" to 1'-0"	6'-0" to 10'-0"	5'-0" to 10'-0"	250 to 300	Closed
BAKERY CONFECTIONERY	2'-0" to 2'-6"	5'-0" to 6'-0"	2'-0" to 3'-6"	150	Glass or wood, closed. Screened vent ducts to outer air
BANKS (branch store type)	3'-0"	6'-0"	2'-6"	200	Preferably open or low; interior appearance important
BARS, CAFES, RESTAURANTS	1'-8" to 2'-4"	6'-0" to 8'-0"	0'-0" to 5'-0"		
REAL ESTATE AGENCIES	2'-0"	6'-0"	4'-0"		
BOOKS OR STATIONERY TOBACCO	2'-0" to 3'-0"	4'-6" to 6'-0"	2'-0" to 3'-6"	100	Closed or low railing, wood; possibly with shelving for displays
CAMERAS AND PHOTOGRAPHY	1'-8" to 3'-0"	4'-0" to 6'-0"	2'-0" to 3'-0"	200	Open or closed.
CHINA AND GLASSWARE	2'-0"	5'-6"	3'-0"	200	Closed
MUSICAL INSTRUMENTS PICTURES AND FRAMES	1'-4" to 2'-0"	5'-0" to 7'-0"	3'-0" to 5'-0"		
TOYS	1'-0"	7'-0"	6'-0"		
CLOTHING (Men's)	1'-4" to 2'-0"	6'-6" to 8'-0" Allow 4 sq. ft. per "torso" form; 3'-4" to 4'-2" high	3'-0" to 6'-0"	200 Additional spotlights and base outlets	Closed; partitions or screens often divide window into 4'-0" to 6'-0" units
CLOTHING (Women's)	1'-0" to 2'-6"	7'-0" to 9'-0" Allow 4 sq. ft. by 5'-10" per form	3'-0" to 6'-6"	200 Additional spotlights and base outlets	Closed
CUTLERY, NOVELTIES, SILVERWARE	1'-10" to 2'-6"	4'-6" to 6'-0"	2'-0" to 3'-0"	150	Closed, removable

DIMENSIONS OF
SHOW WINDOWS

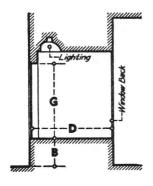

NO STANDARDS POSSIBLE.
In analyzing the figures which are presented below, the designer is cautioned not to regard these as hard and fast standards which must not be varied. The figures given only represent reasonable averages for the types of stores listed. They will, however, provide a starting place since they take into consideration the basic principle of store front heights which dictates that the smaller the object displayed, the higher must be the display window floor.

Store	Bulkhead Height (B)	Glass Height (G)	Window Depth (D)	Lighting in Watts per Lin. Ft. Outlets 12"-15"	Window Backs
DAIRIES, DELICATESSEN MEAT AND FISH	1'-8" to 2'-4"	5'-0" to 7'-0"	2'-6" to 4'-0"	150	Closed or partially open. Vent unless refrigerated
BIRDS AND PETS	1'-6"	7'-0"	4'-0"		
DEPARTMENT STORE	1'-0" to 2'-6"	8'-0" to 10'-0"	7'-0" to 10'-0"	250	Closed. Interior wall valuable
DRUG	1'-8" to 3'-0"	6'-0" to 8'-0"	2'-0" to 4'-0"	200	Partially closed or open; show interior
DRY GOODS HOSIERY AND LINGERIE	1'-4" to 2'-0"	6'-0" to 8'-0"	3'-0" to 5'-0"	200	Closed
FLOOR COVERINGS*					Open or closed.
ELECTRIC EQUIPMENT TYPEWRITERS	1'-8" to 2'-4"	6'-0" to 8'-0"	3'-0" to 4'-0"	200	Closed
FLORIST (General)	1'-0" Waterproof floor; drainage	6'-0" to 8'-0"	3'-0" to 6'-0"	150	Open or glass—additional glass and metal shelving—ventilated
FLORIST (Hotel, Cut Flowers)	3'-0"	4'-0" to 5'-0"	3'-0" to 4'-0"	100	Closed—additional glass and metal shelving—Vent unless refrigerated
FURNITURE	0'-0" to 1'-2"	9'-0" to 11'-0"	7'-0" to 12'-0" Room size, 9' x 12' rug, wall space	250 to 350 Convenience outlets	Closed; access doors, 4'-0"x 6'-8". Period background
FURRIER	1'-8" to 2'-4"	6'-0" to 8'-0"	3'-0" to 6'-0"	200 Spotlights and/or footlights; lenses necessary to protect furs	Semi-closed or closed, rich wood preferred
GROCERY LIQUOR	1'-8" to 2'-6"	5'-0" to 7'-0"	3'-0" to 6'-0"	150	Open or low rail—clear view into store
HABERDASHER (Varied Stock)	1'-4" to 2'-6"	6'-0" to 7'-0"	3'-0" to 5'-0"	200	Closed
HABERDASHER (Limited Stock)	2'-6" to 2'-8"	5'-0" to 6'-0"	2'-6" to 3'-0"	150	Closed

* Large enough for room setups and usually sold with furniture.

DIMENSIONS OF
SHOW WINDOWS

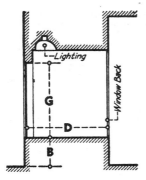

NO STANDARDS POSSIBLE.
In analyzing the figures which are presented below, the designer is cautioned not to regard these as hard and fast standards which must not be varied. The figures given only represent reasonable averages for the types of stores listed. They will, however, provide a starting place since they take into consideration the basic principle of store front heights which dictates that the smaller the object displayed, the higher must be the display window floor.

Store	Bulkhead Height (B)	Glass Height (G)	Window Depth (D)	Lighting in Watts per Lin. Ft. Outlets 12"–15"	Window Backs
HARDWARE OR PAINTS HOUSE FURNISHINGS	1'-0" to 2'-6"	6'-0" to 10'-0"	2'-6" to 6'-0"	200 Additional spotlights and outlets for mechanical contrivances	Closed
HATS (Men's)	1'-4" to 2'-4"	6'-0" to 8'-0"	3'-0" to 5'-0"	200	Closed
HATS (Women's Millinery)	1'-4" to 2'-8" 1 sq. ft. area x 1'-3" to 1'-8" height per hat	5'-0" to 7'-0"	3'-0" to 5'-0"	200	Closed
JEWELRY (Inexpensive)	2'-4" to 3'-0"	4'-0" to 6'-0"	2'-0" to 3'-6"	150	Low or closed, removable; provide access passage
JEWELRY (High Quality)	3'-2" to 4'-0"	3'-0"	1'-0" to 3'-0"	100 "Daylight" lenses preferred	Low or closed, removable; provide access passage. Miniature stage
LEATHER GOODS LUGGAGE	1'-4" to 2'-0"	6'-6" to 7'-6"	3'-0" to 8'-0"	200	Closed, provide shelves 1'-3" to 2'-0" apart for luggage displays
OPTICAL	3'-0" to 3'-6"	4'-0" to 5'-0"	2'-0" to 3'-0"	150	Closed or partially open; whole window free in design
ORGANS, PIANOS, STEREOS/TVS	0'-0" to 1'-0"	7'-0" to 10'-0"	5'-0" to 10'-0"	200	Open or closed
STEREOS/TVS ACCESSORIES (not many floor models) AUTOMOBILE ACCESSORIES REFRIGERATORS SPORTING GOODS	1'-4" to 2'-0"	6'-0" to 8'-0"	3'-0" to 6'-0"	200	Open or closed
SERVICE: BARBER SHOP, BEAUTY SHOP, CLEANER & DYER, LAUNDRY, TAILOR	1'-6" to 2'-4"	6'-6" to 8'-0"	1'-6" to 5'-0"	200	Preferably open; interior appearance important
SHOES (Men's)	1'-4" to 2'-2"	6'-0" to 7'-0"	2'-0" to 5'-0"	150	Closed
SHOES (Women's; Men's and Women's)	2'-0" to 2'-4" 4'-0" † 1½ sq. ft. per pair shoes	5'-0" to 6'-0" 3'-0"†	2'-0" to 5'-0"	150	Closed † (Exclusive shops may feature individual models in small windows)

STORE FRONT
SILL DETAILS

Typical details of store front construction are shown on this and the six following pages. They may be used as an aid in visualizing store front construction and as a guide for preliminary sketches. Details of the moldings and hardware to be used, of course, should be taken from the manufacturers' drawings of particular products or obtained in quantitative form from the manufacturer's representative before drawings are made.

Store front setting moldings are available in several materials and finishes and in numerous configurations. Specific dimensions and hardware vary widely.

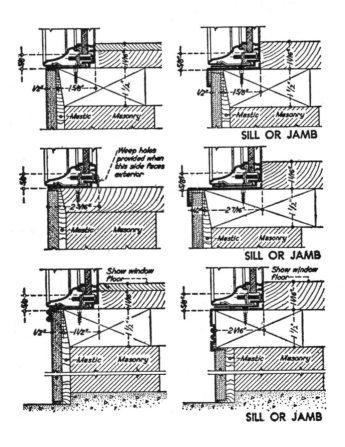

STORE FRONT
HEAD AND JAMB DETAILS

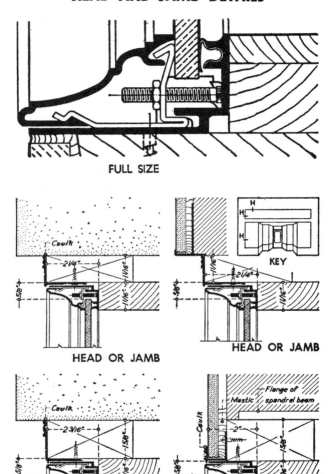

FULL SIZE

KEY

HEAD OR JAMB

HEAD OR JAMB

Scale 3"=1'-0"

HEAD OR JAMB

HEAD OR JAMB

STORE FRONT
TRANSOM BAR DETAILS

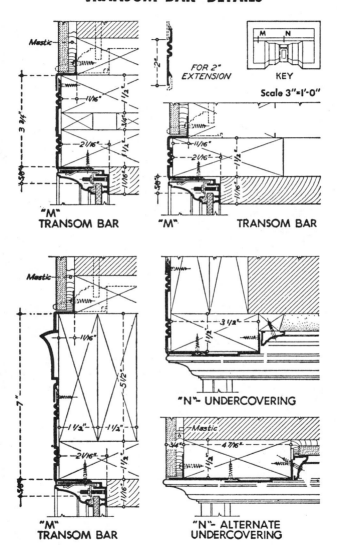

For 2" EXTENSION

KEY

Scale 3"=1'-0"

Mastic

"M" TRANSOM BAR

"M" TRANSOM BAR

"M" TRANSOM BAR

"N"- UNDERCOVERING

"N"- ALTERNATE UNDERCOVERING

STORE FRONT
DIVISION BAR DETAILS

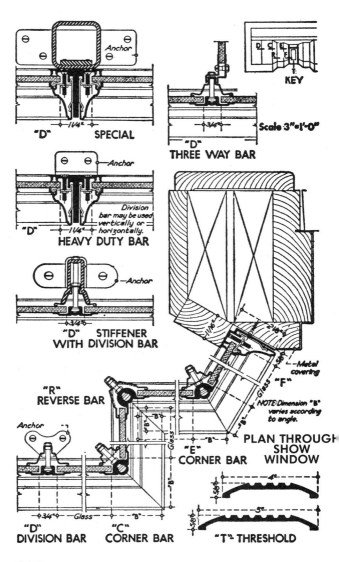

"D" SPECIAL — 1¼"

"D" THREE WAY BAR — ¾" — Scale 3"=1'-0"

KEY

"D" HEAVY DUTY BAR — 1¼"

Division bar may be used vertically or horizontally.

"D" STIFFENER WITH DIVISION BAR — ¾"

"R" REVERSE BAR

Anchor

"D" DIVISION BAR — ¾" — Glass

"C" CORNER BAR — "B"

"E" CORNER BAR

"F"

Metal covering

NOTE: Dimension "B" varies according to angle.

PLAN THROUGH SHOW WINDOW

"T" THRESHOLD — 4" — 5"

STORE FRONT
AWNING BAR DETAILS

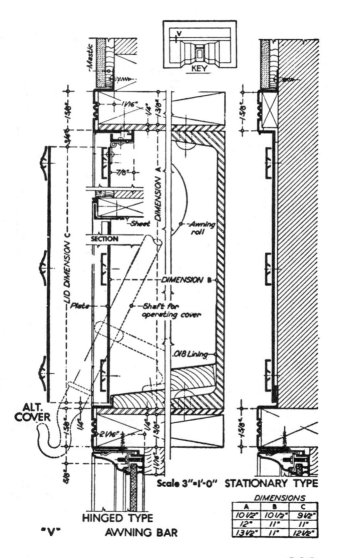

KEY

Mastic

1 1/16"   1/4"   3/16"

7/8"

SECTION

Sheet

DIMENSION A

Awning roll

DIMENSION B

LID DIMENSION C

Plate

Shaft for operating cover

.018 Lining

ALT. COVER

2 1/16"

Scale 3"=1'-0"   STATIONARY TYPE

"V"   HINGED TYPE
AWNING BAR

DIMENSIONS		
A	B	C
10 1/2"	10 1/2"	9 1/2"
12"	11"	11"
13 1/2"	11"	12 1/2"

STORE FRONT
AWNING BAR DETAILS

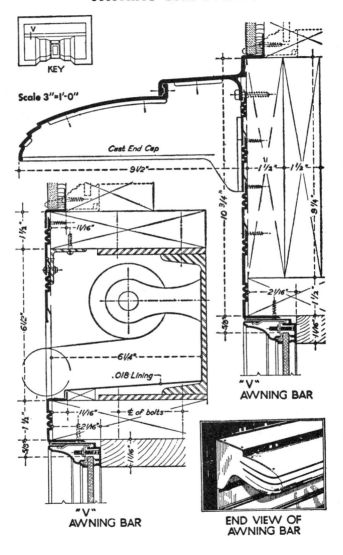

KEY

Scale 3"=1'-0"

Cast End Cap

9½"

10¾"

1½" | 1½"

9¼"

2¹⁄₁₆"

1½"

⅝"

"V"
AWNING BAR

1½"

1¹⁄₁₆"

6½"

6¼"

.018 Lining

1¹⁄₁₆" | ℄ of bolts

2¹⁄₁₆"

1¹⁄₁₆"

⅝" | 1½"

"V"
AWNING BAR

END VIEW OF
AWNING BAR

STORE FRONT
EXTRUDED MOLDINGS

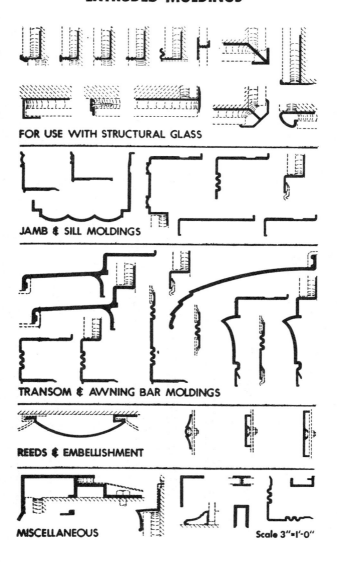

FOR USE WITH STRUCTURAL GLASS

JAMB & SILL MOLDINGS

TRANSOM & AWNING BAR MOLDINGS

REEDS & EMBELLISHMENT

MISCELLANEOUS

Scale 3"=1'-0"

ATTRACTING POWER
LIGHTED STORE WINDOWS

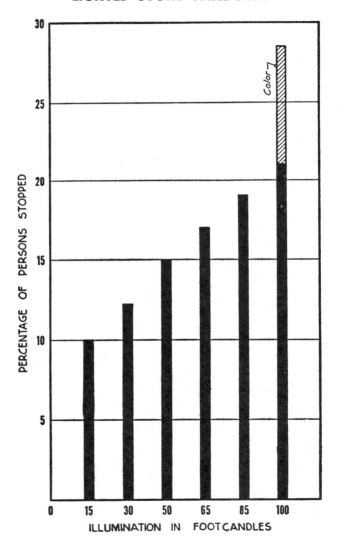

SHOW WINDOW LIGHTING

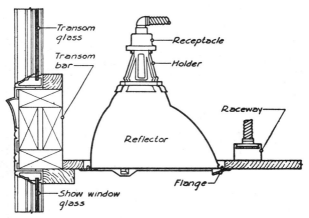

Transom glass
Transom bar
Receptacle
Holder
Raceway
Reflector
Flange
Show window glass

FLUSH CEILING REFLECTOR

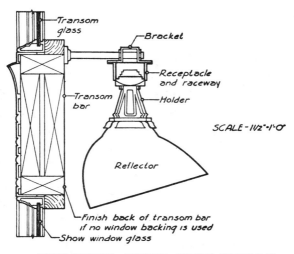

Transom glass
Bracket
Receptacle and raceway
Transom bar
Holder
SCALE - 1 1/2" = 1'-0"
Reflector
Finish back of transom bar if no window backing is used
Show window glass

BRACKET TYPE REFLECTOR

233

STORE FRONT
LIGHTING

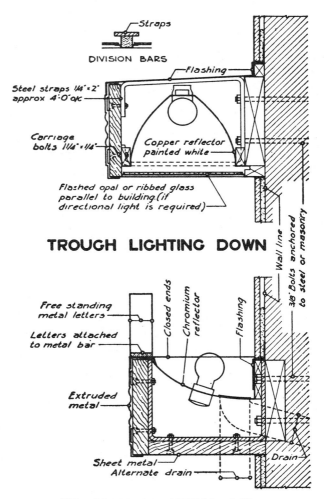

Straps

DIVISION BARS

Flashing

Steel straps 1/4"×2"
approx 4:0"o.c.

Carriage
bolts 1 1/4"×1/4"

Copper reflector
painted white

Flashed opal or ribbed glass
parallel to building.(if
directional light is required)

Wall line

3/8" Bolts anchored
to steel or masonry

TROUGH LIGHTING DOWN

Free standing
metal letters

Letters attached
to metal bar

Closed ends

Chromium
reflector

Flashing

Extruded
metal

Sheet metal
Alternate drain

Drain

TROUGH LIGHTING UP

SCALE-1 1/2":1'-0"

234

ILLUMINATED
STORE FRONT

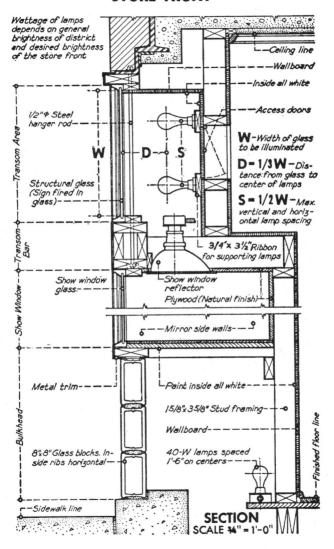

Wattage of lamps depends on general brightness of district and desired brightness of the store front

Ceiling line

Wallboard

Inside all white

Access doors

1/2"⌀ Steel hanger rod

Transom Area

Structural glass (Sign fired in glass)

W — Width of glass to be illuminated

D = 1/3 W — Distance from glass to center of lamps

S = 1/2 W — Max. vertical and horizontal lamp spacing

W   **D**   **S**

3/4" x 3 1/2" Ribbon for supporting lamps

Transom Bar

Show window glass

Show window reflector

Plywood (Natural finish)

Show Window

Mirror side walls

Metal trim

Paint inside all white

15/8" x 3 5/8" Stud framing

Wallboard

8" x 8" Glass blocks. Inside ribs horizontal

40-W lamps spaced 1'-6" on centers

Bulkhead

Finished floor line

Sidewalk line

SECTION
SCALE ¾" = 1'-0"

STORE FRONT
LIGHTING

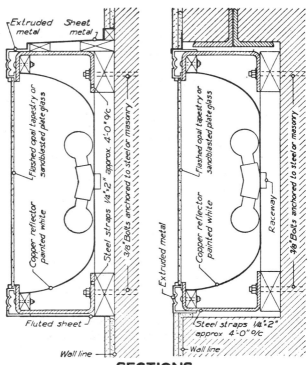

SECTIONS
SCALE-1½"-1'-0"

ELEVATION
SCALE-½"-1'-0"

STORE FRONT
LIGHTING

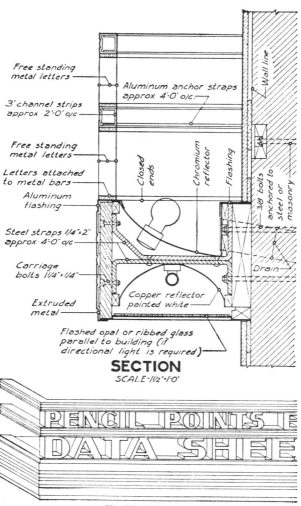

Free standing
metal letters

Aluminum anchor straps
approx 4'-0" o/c

3" channel strips
approx 2'-0" o/c

Free standing
metal letters

Letters attached
to metal bars

Aluminum
flashing

Steel straps 1/4"×2"
approx 4'-0" o/c

Carriage
bolts 1 1/4"×1/4"

Extruded
metal

Wall line

Closed
ends

Chromium
reflector

Flashing

3/8" bolts
anchored to
steel or
masonry

Drain

Copper reflector
painted white

Flashed opal or ribbed glass
parallel to building (if
directional light is required)

SECTION
SCALE-1 1/2"-1'-0"

PENCIL POINTS
DATA SHEE

ELEVATION
SCALE-1/2"-1'-0"

STORE FRONT
LIGHTING

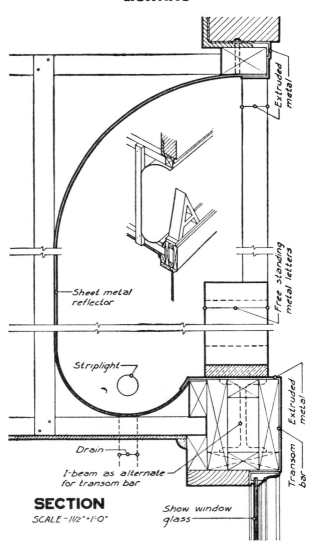

Extruded metal

Free standing metal letters

Sheet metal reflector

Striplight

Extruded metal

Transom bar

Drain

I-beam as alternate for transom bar

SECTION
SCALE - 1½" = 1'-0"

Show window glass

STORE FRONT
LIGHTING

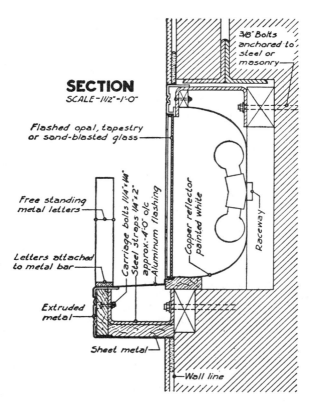

SECTION
SCALE-1 1/2"-1'-0"

3/8" Bolts anchored to steel or masonry

Flashed opal, tapestry or sand-blasted glass

Free standing metal letters

Letters attached to metal bar

Carriage bolts 1 1/4"x 1/4"

Steel straps 1/4"x 2" approx.-4'-0" o/c

Aluminum flashing

Copper reflector painted white

Raceway

Extruded metal

Sheet metal

Wall line

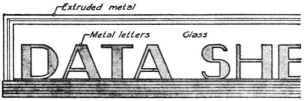

Extruded metal

Metal letters   Glass

DATA SHE

ELEVATION
SCALE- 1/2"-1'-0"

PORCELAIN ENAMEL
LETTERS FOR SIGNS

SIGN LETTERS. Letters may be composed either entirely of porcelain enamel or of porcelain faces inlaid in side flanges of aluminum, stainless steel, or other material. They can be used for identification on store fronts, roofs, marquees, or in any other position where permanent architectural lettering is required. Letters may be made to the architect's patterns or may be taken from stock alphabets. The letters can be produced in Gothic, Roman, thick and thin, modern, angular, script, and so forth. For custom designs, an accurate scaled drawing must be supplied. Simple or complicated trademarks or logotypes can also be supplied to the architect's design.

Permanent materials and finishes are used throughout. Many standard colors are available; special color matches often can be made, though some difficulty (as well as some delay and extra charges) may be encountered in making exact matches.

Letters of relatively large size (variable depending upon the manufacturer) can be fabricated as single units. Letters to unlimited size can be fabricated in sections. Backs can be furnished to enclose electrical work.

METHODS OF MOUNTING. The letters may be attached in a number of ways.
1. Attached to the face of a building.
2. Attached to the face of a building but set away from the wall.
3. Free-standing letters, base-attached on marquees, copings, or projecting parts.

For all-porcelain enameled letters, the attachment methods shown are applicable. For porcelain inlaid stainless steel letters, essentially the same methods are used, except that, where possible, the letters are supported from the heavy porcelain face rather than from the side flanges.

NEON LIGHTING. Letters can be provided with electrode holes and tube support holes. The neon tubing can be installed on the face of the letters to be directly visible at night, or can be placed in the back of freestanding letters to create a silhouette effect when illuminated. The local electrical sign contractor should be consulted as to code requirements. Numerous important restrictions and requirements, variable depending upon the exact nature of the lettering and the sign that it constitutes, must be taken into consideration. Neon tubing is set 1½" from the face or background on which it is mounted. Suitable approved bushings allow the passage of the tubing through any metal parts.

Transformers for letter installations are placed inside the building as close as possible to the letters (where accessible), or in a curb or transformer box under the letters, or in the letter itself if it is large enough. Transformer sizes vary with different manufacturers, and the capacity will vary with the length and diameter of the run of tubing. Several transformers are usually used on the average job.

BULB LIGHTING. Letters can be provided with lamp socket holes. Bulb lighting can be installed on the face of the letters, as shown in the detail; alternatively, through the use of an intermediate back to support the electrical work, the bulbs may be mounted on the backs of the letters to create a silhouette effect. Various sizes and types of lamps for either intermediate or medium bases may be used. As with neon tubing, various requirements of the National Electrical Code, and possibly local codes as well, must be met.

CAUTION. The method of attaching letter to background should produce a very sturdy connection—one that remains entirely concealed if this is at all possible. Attachments should be designed so as not to interfere with installations and servicing of electrical work.

240

TWO TYPES OF
PORCELAIN ENAMEL LETTERS

ALL-PORCELAIN LETTERS. Letters may be made with 14-gauge faces and 18-gauge side flanges (other combinations will work also, of course), completely welded and completely covered with porcelain enamel inside and out. Fold-back construction eliminates all exposed metal edges. Faces and flanges can be made in different colors. Porcelain letters may be made in several ways, three of which are shown below. A recessed face (about ¼") may be used for nonilluminated installations, and deep recess may be used for illuminated installations, if desired.

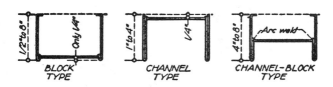

ALL-PORCELAIN LETTERS

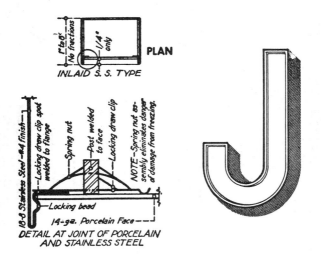

PORCELAIN INLAID STAINLESS STEEL LETTERS

PORCELAIN ENAMEL
LETTERS FOR SIGNS

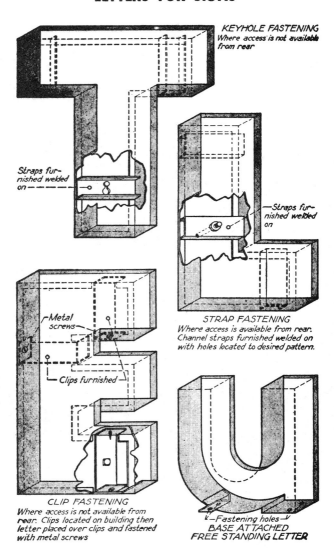

KEYHOLE FASTENING
Where access is not available from rear

Straps furnished welded on

Straps furnished welded on

STRAP FASTENING
Where access is available from rear. Channel straps furnished welded on with holes located to desired pattern.

Metal screws

Clips furnished

CLIP FASTENING
Where access is not available from rear. Clips located on building then letter placed over clips and fastened with metal screws

Fastening holes
BASE ATTACHED
FREE STANDING LETTER

PORCELAIN ENAMEL
LETTERS FOR SIGNS

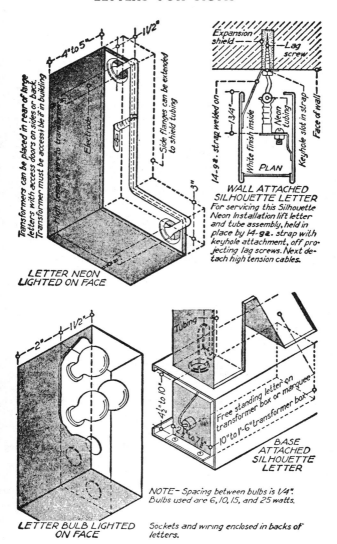

4"to 5"

1 1/2"

Transformers can be placed in rear of large letters with access doors on sides or back. Transformer must be accessible if in building

Electrode

High tension wire to transformer

Side Flanges can be extended to shield tubing

3"

**LETTER NEON
LIGHTED ON FACE**

Expansion shield

Lag screw

14-ga. strap welded on

1-3/4"

White finish inside

Neon tubing

Keyhole slot in strap

Face of wall

PLAN

*WALL ATTACHED
SILHOUETTE LETTER*
For servicing this Silhouette Neon Installation lift letter and tube assembly, held in place by 14-ga. strap with keyhole attachment, off projecting lag screws. Next detach high tension cables.

2"

1 1/2"

4 1/2" to 10"

Tubing

Free standing letter on transformer box or marquee

10"to 1'-6" transformer box

2 1/2" to 7 1/2"

**BASE
ATTACHED
SILHOUETTE
LETTER**

*NOTE— Spacing between bulbs is 1/4".
Bulbs used are 6, 10, 15, and 25 watts.*

**LETTER BULB LIGHTED
ON FACE**

Sockets and wiring enclosed in backs of letters.

BARBER SHOP
PLAN

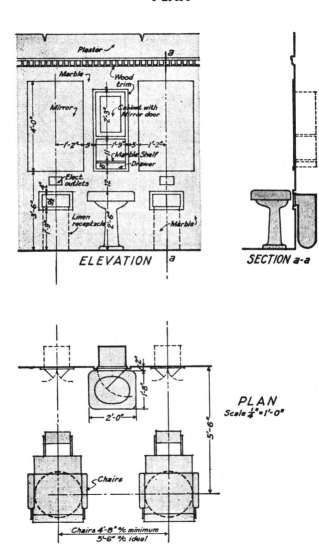

Plaster

Marble

Wood trim

Cabinet with Mirror door

2'-3"

Mirror

1'-2" 5 1'-9" 5 1'-2"

Marble Shelf

Drawer

Elect. outlets

4'-0"

3'-6"

7'-9"

12"

Linen receptacle

2'-6"

Marble

ELEVATION   a

SECTION a-a

2"

1'-8"

2'-0"

5'-6"

Chairs

PLAN
Scale ¼"=1'-0"

Chairs 4'-8" % minimum
5'-6" % ideal

TYPICAL OFFICE
BUILDING UNIT

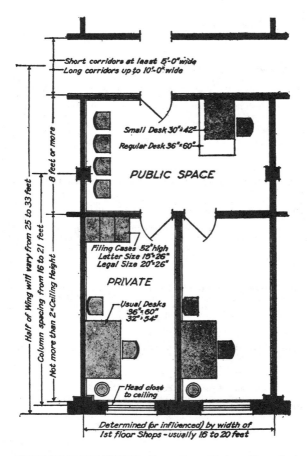

FLOOR TO FLOOR HEIGHT. From 10 to 13 feet. Floors having large undivided areas for general offices will require greater heights than small private units as shown here.

SIZE OF UNIT. Sizes recommended here are for usual city office buildings. They will vary with cost of land, kind of floor and beam construction adopted, whims of owner, shape of lot, etc. Structural requirements are extremely important if the office building is to be economically constructed.

PLANNING SCHOOL CLASSROOMS

The following Data Sheets on schoolhouse requirements are presented only as suggested practice for preliminary sketches. Regional and local variations make rigid standards on a national scale impossible to attempt.

REFERENCE. School design involves a good many variable problems and is circumscribed to a considerable extent by federal, state, and local regulations and requirements, as well as by local community desires and budgetary constraints. Prior to making preliminary designs, the designer should contact both state and local education authorities for current information. In addition, up-to-date design information should be obtained from the A.I.A. and from recognized authorities in the field. All such information should then be cross-checked with building and education laws, regulations, and codes in effect in the school location.

CONDITIONS ENTERING INTO THE PLANNING OF CLASS-ROOMS. In addition to mere classroom space, provide for heating and ventilation, chalkboards, bulletin boards, supply cabinets, book-cases, and means for the hanging of children's, students', and teachers' clothing.

Other desirable features, depending upon the character of the school, include provisions for room clock, temperature control, electric outlets for lighting, projection, and vacuum cleaning, interphone connections, radio connections, lavatory and drinking facilities, project lockers.

COLOR OF WALLS AND CEILINGS. All walls should be of a color that has a light reflecting factor of not less than 30% nor more than 50%. The ceiling should be ivory white or light cream, with a high reflecting factor of not less than 70%. Avoid glossy finish.

COLOR OF SHADES. Use translucent shades of a color that harmonizes with the walls.

COATROOMS, WARDROBES, AND LOCKERS. Provide each elementary classroom with suitable space for the children's outer garments in one of three ways:

(1) Ventilated coatrooms approximately 5′ wide, with an outside window having a glass area of not less than 1 sq. ft. to every 10 sq. ft. of floor area; also with two hook strips placed respectively 3′-6″ and 5′-0″ above the floor, each to be equipped with a sufficient number of hooks staggered 18″ apart on each strip. A pole equipped with hangers may be substituted for hook strips. Coatrooms as described above with a classroom wall in the form of a stationary screen are acceptable when the area behind the screen is properly ventilated.
(2) Ventilated wardrobes readily accessible and convenient for use, preferably opening into the classroom.
(3) Ventilated lockers in corridors, providing ample space for outer garments and placed so as to be convenient for use.

PLANNING SCHOOL CLASSROOMS

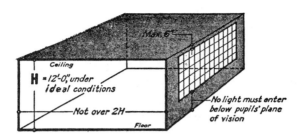

DIMENSIONS OF CLASSROOMS. Traditionally, the ceiling height of unilaterally lighted classrooms has been 12'. However, in recent years construction costs and other factors frequently have dictated ceiling heights in the 8' to 10' range.

Ideally, the width of a unilaterally lighted classroom should be no more than twice the height. This does not necessarily apply where ceiling heights are less than 12', but the ratio should be close to 2:1.

In the front end of the classroom provide approximately 8' between the first row of seats and the front wall. In the rear of the classroom, provide at least 3' between the last row of seats and the rear wall.

SEATING CAPACITY OF CLASSROOMS. The normal seating capacity of classrooms is determined by allowing 16 sq. ft. average, up to 30 sq. ft. for primary rooms, depending on the method of instruction. High school lecture rooms may need only 7 sq. ft. in theater arrangements; up to 20 sq. ft. with other types of seating and seats.

AISLES. For people's safety and convenience in passing up and down the classroom, aisles next to walls should be at least 30" wide. Intermediate aisles should be at least 18" wide.

DOORS TO CLASSROOMS. Classroom doors should be at least 3'-0" x 7'-0" x 1¾". A clear wireglass pane in the upper part of the door is desirable.

LOCATION OF WINDOWS. Bilateral daylight sources, such as clerestory windows and variations of skylights, may contribute materially to visual comfort and efficiency when properly controlled by shielding devices or orientation.

East and west fenestration is preferable to north and south. The top of the upper sash of windows should be 6" or less below the ceiling. No windowpanes should be placed so low that light enters the room below the plane of vision of pupils seated next to the windows.

GLASS AREA OF WINDOWS. Natural daylight illumination should be adequate at the desks of all pupils. The ratio of window area to floor area is governed by the intensity of illumination necessary for a proper distribution of a sufficient amount of light at each desk. For different regions of the country, the actual ratio of required glass area to floor area will vary between one-fourth and one-sixth.

ARTIFICIAL ILLUMINATION. Intensities of from 40 to 50 footcandles are satisfactory in a balanced brightness environment.

SCHOOL SEATING SCHEMES

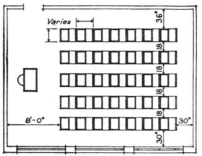

This is the usual arrangement of desks, especially the fixed type. The pupils at the rear and near the windows are facing a glare of light. The focus of attention on the teacher's desk is difficult for pupils at the sides near the front. It is the most economical arrangement. The local laws should govern the dimensions of aisles.

RECTANGULAR SCHEME

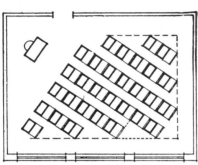

In this plan the desks are turned away from the glare of the windows, resulting in excellent lighting for all desks. The focus of attention is as difficult as in the rectangular scheme. Suitable to fixed or movable seating. Not particularly economical of floor space. Surrounding aisle sometime reduced in width.

OBLIQUE SCHEME

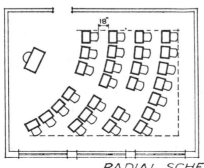

A scheme particularly adaptable to movable seating. Excellent light is secured on all desks, and the attention of the pupils is easily focused on the teacher's desk without turning in the seats. Difficult to lay out in plan. Somewhat wasteful unless the rear and window aisles are utilized for seating.

RADIAL SCHEME

PLAN OF ELEMENTARY
SCHOOL CLASSROOM

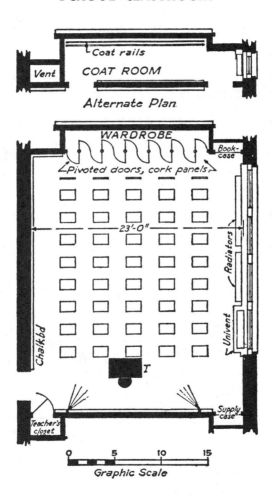

The alternate plan is for the rear of classrooms in schools where the State Laws require a coat room instead of the wardrobe unit. When the State Laws require two doors to each classroom the additional door should be placed near the rear of the room on the left-hand wall.

ELEVATIONS OF
ELEMENTARY CLASSROOM

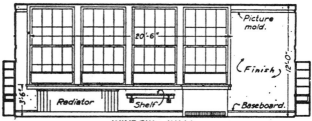

WINDOW WALL

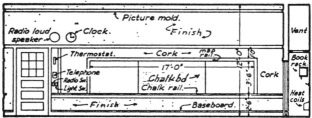

WALL OPPOSITE WINDOWS

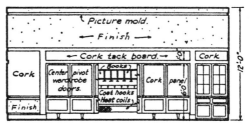

REAR OF ROOM

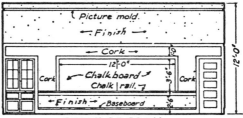

FRONT OF ROOM

PLAN OF JUNIOR OR SENIOR HIGH SCHOOL CLASSROOM

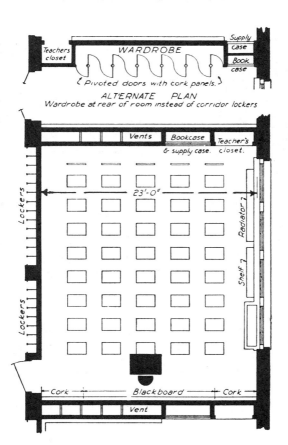

The alternate plan is for the rear of classrooms where lockers are not desired. When the state laws require two doors to each classroom, the additional door should be placed near the rear of the room on the left-hand wall.

ELEVATIONS OF JUNIOR OR SENIOR HIGH SCHOOL CLASSROOM

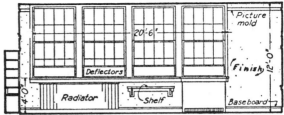

WINDOW WALL

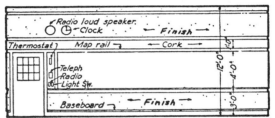

WALL OPPOSITE WINDOWS

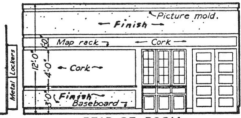

REAR OF ROOM

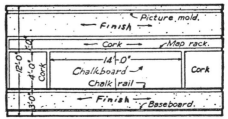

FRONT OF ROOM

Scale

DETAIL OF
SCHOOL WARDROBE

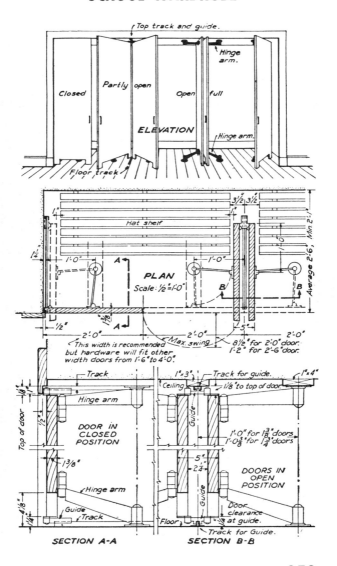

Top track and guide.

Hinge arm.

Closed   Partly open   Open   full

ELEVATION

Hinge arm.

Floor track

3½" 3½"

Hat shelf

A

PLAN

Scale: ½"=1'-0"

B   B

Average 2'-6", Min 2'-1"

1'-0"   1'-0"

½"

2'-0"   2'-0"   2'-0"

This width is recommended but hardware will fit other width doors from 1'-6" to 4'-0".

Max. swing

8½" for 2'-0" door.
1'-2" for 2'-6" door.

5"

Track   1"×3"   Track for guide.   1"×4"

Ceiling   1⅛" to top of door.

Hinge arm

Guide

**DOOR IN
CLOSED
POSITION**

1'-0" for 1⅜" doors
1'-0⅜" for 1¾" doors

1⅜"

5"
2¼"

**DOORS IN
OPEN
POSITION**

Hinge arm

Door
clearance
at guide.

Guide   Track

Floor   Track for Guide.

4⅛"

SECTION A-A   **SECTION B-B**

253

AREA AND
LOCATION OF CHALKBOARDS

Different locales may have specific requirements as to the position and area of the tackboards and light-colored or conventional chalkboards in the schoolroom. The total tackboard and/or chalkboard area, or the balance between the two, may vary with the type of classroom and the number of pupils to be served.

Kindergarten and primary school rooms should be equipped with enough tackboards to accommodate the display of pupils' work, as well as pictures and diagrams that are used extensively in these grades. Tackboards may be installed along the top of the chalkboard in elementary school classrooms so that work can be displayed without interfering with the chalkboard work. These tackboards should not replace lower-level tackboards, however, since young students tend to ignore the high ones completely as being above and beyond the youngsters' normal range of consideration.

An older extensive survey on the use of chalkboards indicates that 34 linear feet of board is a practical minimum, and that the ideal is a total length of 45.7 linear feet. This would take care of the teachers' activity with an average 17.5 linear feet, and the pupils' activity with a maximum of 28.2 linear feet. The survey figures are based on an analysis of 6,000 schools, grades 1 to 8 inclusive—some grades requiring less and some requiring more. However, in recent years the increased use of various audiovisual tools has reduced dependency on blackboards to some extent, with an attendant reduction in many cases in the needed linear footage of blackboard. Further, it may not always be possible to secure such large chalkboard areas because of window areas, doors, wardrobes, or in some cases because the budget has dictated that a smaller classroom be built. Swinging-leaf, rolling, multiple-sliding, or folding-screen chalkboards can be employed as necessary to make up whatever area is required.

Extensive tests made in schools where conditions were considered representative show the length of chalkboards to be as in the following table:

	Elementary Grades 1, 2, 3	Grade School Grades 4, 5, 6	7, 8	High School
Width of room	18'-6"	18'-9"	21'-0"	26'-0" to 33'-8"
Length of room	27'-4"	30'-0"	31'-4"	32'-0" to 43'-0"

In rural schools, a chalkboard 42" wide placed with its lower edge 26" above the floor, is serviceable. The top of such a board is 5'-8" above the floor. A board 48" wide would give a permanent writing space at the top that is very desirable in many cases.

Chalkboards are available in various light colors and with low reflectance factors. These are practical when the level of illumination is sufficiently high to overcome the loss in visibility due to the reduced brightness difference between white chalk and the light-colored board, as compared with white chalk on the conventional blackboard. The use of yellow chalk often mitigates such contrast problems. Where conventional blackboards are installed, a convenient means such as sliding panels (which can be tackboards) should be available for covering the blackboard with lighter-colored surfaces when it is not in use.

In some locales there is an increasing trend, especially in the upper grades, toward reducing the area of permanently installed chalkboard to an 8- to 12-foot panel on the front wall, to be used by the teacher for demonstrations. Additional chalkboard area can then be provided as needed with rolling or portable-screen chalkboards. Storage space should be provided for such chalkboards.

TYPICAL CHALKBOARD CONSTRUCTION

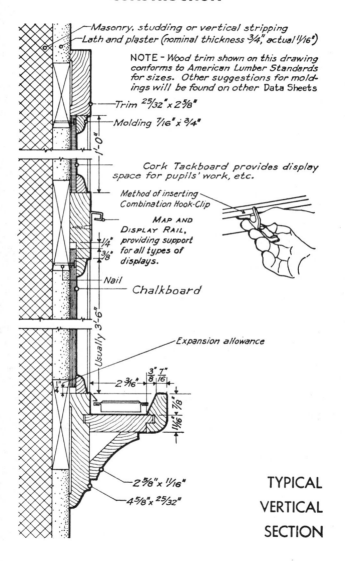

Masonry, studding or vertical stripping

Lath and plaster (nominal thickness ¾", actual 11/16")

NOTE – Wood trim shown on this drawing conforms to American Lumber Standards for sizes. Other suggestions for moldings will be found on other Data Sheets

Trim 25/32" x 2⅝"

Molding 7/16" x ¾"

1'-0"

Cork Tackboard provides display space for pupils' work, etc.

Method of inserting Combination Hook-Clip

MAP AND DISPLAY RAIL, providing support for all types of displays.

¼"
⅜"

Nail

Chalkboard

Usually 3'-6"

Expansion allowance

3" 7"
⅜ 16"

2 3/16"

7/8"

11/16"

2 ⅝" x 11/16"

4 ⅝" x 25/32"

TYPICAL VERTICAL SECTION

255

SUB-CHALKBOARD
CONSTRUCTION

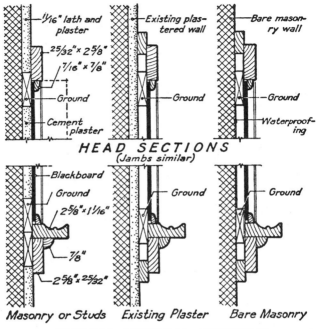

HEAD SECTIONS
(Jambs similar)

Masonry or Studs   Existing Plaster   Bare Masonry

TYPICAL VERTICAL SECTIONS

VERTICAL SECTIONS. Shown here are typical representations of traditional chalkboard/tackboard installations using wood moldings and trim. In more common use today are various systems of aluminum framing and chalkrails. Depending upon the type, these may be either integrally or surface mounted. In general, installation is much the same as for wood varieties. Details of the systems vary with different manufacturers.

FRAME/WOOD BACKING. Instead of installing chalkboards and tackboards over plaster or bare masonry as shown above, they may be secured to conventional wood-frame walls sheathed with plasterboard, usually by direct attachment to the studs. Alternatively, and irrespective of the construction of the wall itself, an excellent wood base can be installed. Nominal 1″ kiln-dried T&G flooring is excellent, but ⅝″ or ¾″ plywood, preferably exterior grade, is even better and is somewhat easier to install. Such backings are secure and stable, and no grounds are required for attachment of trim and chalkrail.

CHALK RAILS AND CHALKBOARD HEIGHTS

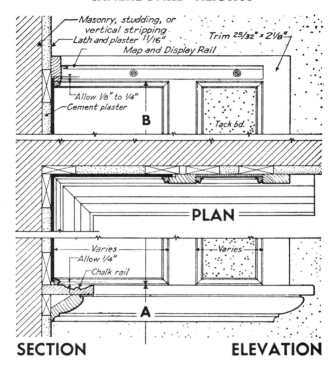

Masonry, studding, or vertical stripping
Lath and plaster 11/16"
Map and Display Rail
Trim 25/32" x 2⅛"

Allow ⅛" to ¼"
Cement plaster

B

Tack bd.

PLAN

Varies
Allow ¼"
Chalk rail

Varies

A

SECTION   **ELEVATION**

Chalkboards are manufactured in the generally recommended heights of 3'-0", 3'-6", and 4'-0". The 3'-0" height serves for kindergarten children, while the 3'-6" size suits children with varying arm reaches. The 4'-0" height is best for teachers' blackboards and for rural or ungraded systems where pupils of several or all grades must use the same blackboard. This height is also the best choice for school buildings where grades are apt to be shifted from room to room as enrollment demands vary.

Grade	**A** Chalk-rail Above Floor	**B** Height of Blackboard
Kindergarten	20"–24"	3'-0"
First and Second	24"–28"	3'-6"
Third and Fourth	27"–30"	3'-6"
Fifth and Sixth	30"–34"	3'-6"
Junior and Senior High	30"–36"	4'-0"
Back of Teacher, 1st to 8th	36"	to door height
Rural and Ungraded	26"	4'-0"

257

SUGGESTED
CHALK-RAIL DETAILS

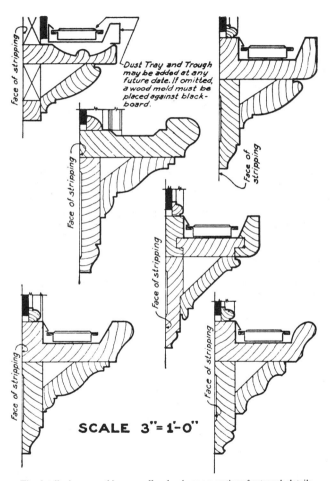

Dust Tray and Trough
may be added at any
future date. If omitted,
a wood mold must be
placed against black-
board.

SCALE 3"=1'-0"

The details shown on this page, offered only as suggestions for trough details, are typical of the more traditional wood framing styles for chalkboards. Other chalk rail details using lumber and moldings of standard dimensions and of simpler section are found on pages 255 and 256 of this book. In more common use today are the many configurations of aluminum chalkboard frames and trim, with either integral or separate troughs.

258

SCHOOL BUILDING CORRIDORS

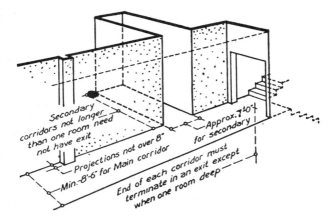

Secondary corridors not longer than one room need not have exit

Projections not over 8"

Min. 8'-6" for Main corridor

Approx. 7'-0" for secondary

End of each corridor must terminate in an exit except when one room deep

CORRIDOR CONSTRUCTION.   Corridors should be of "fireproof construction," with walls of approved masonry or reinforced concrete. Structural members should have a fire resistance of not less than a 4-hour rating for bearing walls, isolated piers, columns, and wall-supporting girders; a 3-hour rating for walls and girders other than those already specified and for beams, floors, roofs, and floor fillings; and a 2-hour rating for fire partitions. Construction must be in compliance with local and state codes.

WIDTH OF CORRIDORS.   The minimum clear passageway of the main corridor or corridors of any school building containing four or more classrooms should be 8'-6".

The minimum clear corridor width of secondary corridors may vary with the length of such corridors and the number of classroom doors leading to them, but such secondary corridors should be approximately 7'-0" wide. Check local codes.

TERMINATION.   Each end of each corridor should terminate on an egress or stairway, except that "pockets" not to exceed the length of one classroom may be planned when conditions dictate.

LIGHTING.   Corridors and passageways should be well lighted. Outside windows are always desirable. The recommended artificial illumination level for corridors in general is 20 footcandles.

WALL PROJECTIONS.   No projections should extend beyond the face of the corridor walls in excess of 8".

CORRIDOR EQUIPMENT.   No radiators, drinking fountains, wash basins, or other equipment should be placed on corridor walls unless the walls are recessed to receive them.

259

SCIENCE LABORATORIES

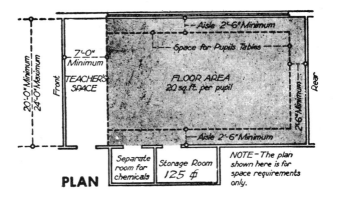

PLAN

NOTE—The plan shown here is for space requirements only.

DIMENSIONS OF LABORATORIES. Ideally, laboratories should be not less than 20'-0" wide and not more than 24'-0" wide. The length should be such that the room remains small enough to contain only a few students, allowing close control and supervision by the teacher.

DEPENDENCIES. Provide a separate, vented room for the storage of chemicals. One storage room for apparatuses and equipment and another for chemicals may be placed between two laboratories to serve both.

TEACHER'S TABLE. The teacher's table should have an acid-resisting top, acid-resisting sink, and drain. Each table should be provided with an electrical connection, and a direct current supply is desirable. Cold water is mandatory at the sink, and hot water is desirable. A gas connection should be provided.

STUDENTS' TABLES. Allow a minimum of 2'-6" of table length for each pupil's station. Gas, water, and electric connections should be convenient to each station. Table tops should be acid-resisting.

OTHER REQUIREMENTS. Provide both translucent and opaque window shades of approved type. Provide an electrical outlet suitable for a projection machine. A bulletin board of at least 15 square feet should be provided. A notebook case is desirable. A minimum of 20 linear feet of blackboard with at least 10 linear feet in the teacher's end of the room should be provided.

ART AND MECHANICAL DRAWING

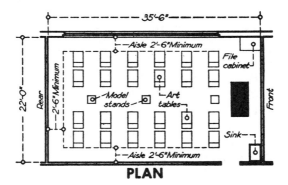

PLAN

ART ROOMS. Rooms devoted to art instruction should be so located as to receive north light. Not less than 30 square feet of floor area per student should be provided.

Provide adequate arrangements for electric connections, water supply, ventilation, display spaces, and storage facilities. A typical art room plan is shown above.

MECHANICAL DRAWING ROOMS. The provisions for mechanical drawing rooms are practically identical with those for art rooms. Where class schedules allow, the same room can be used for both purposes.

Rooms devoted to mechanical drawing should, if possible, be so located as to receive north light, and artificial lighting should be provided to correspond with standards established by the Illuminating Engineering Society or other appropriate authority. Current recommendations are for illuminances of 100 footcandles for general drafting and 150 footcandles for fine drafting. At least 30 square feet of gross floor area per student should be provided.

Special rooms suitably equipped should be considered for duplicating or blueprinting.

TABLES. Illustrated at left is a typical drawing table with adjustable top. Boards, instruments, and materials are stored in the compartments. A general drawer is provided for classroom equipment.

PHYSICAL EDUCATION

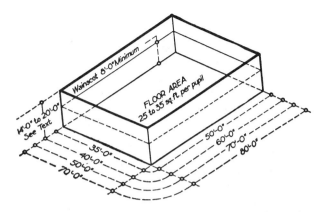

DIMENSIONS OF GYMNASIUMS. The floor dimensions should be computed on the basis of 25 to 35 square feet per pupil and will depend upon the enrollment, the school organization, and the age of the children. Recommended dimensions are shown on the drawing above.

In elementary schools, the ceiling should not be less than 14 feet for floor areas of 2,400 square feet or less; the ceiling should be 16 feet for floor areas over 2,400 square feet and less than 3,500 square feet.

High school gymnasiums should have a ceiling height of 18 to 20 feet.

SPECTATOR PROVISIONS. Seats along sides, beginning at or near the floor level, are preferable to other means of seating.

VENTILATION. Mechanical ventilation should be provided.

ORIENTATION. Southern exposure is desirable.

LIGHTING. Gymnasiums are sometimes designed with no windows. When windows are a part of the design, their sills should be located at least 6 feet from floor level. Bilateral lighting is preferred.

Equip windows with suitable shades and protective grilles. Windows may be glazed with "unbreakable" plastic or special strengthened or composite glass to avoid the use of grilles.

Artificial lighting should correspond to the standards established by the Illuminating Engineering Society or other recognized authority.

FLOORS, WALLS AND CEILING. Finish floor should be marked with painted or stained lines for games. Floors should be of such material and so installed as to provide suitable resiliency and freedom from slipperiness and splintering.

Provide a wainscot, to the height of at least 8 feet, of brick, block, glazed tile, wood, vinyl, or other suitable material. In elementary and middle school gymnasiums, and all gymnasiums where the playing field borders are close to the walls, provide heavy protective pads and means of attachment to a height of at least 6 feet, as a personal safety measure.

The ceiling and areas above the wainscoting should receive acoustical treatment to keep the time of reverberation to a reasonable limit.

DEPENDENCIES. Necessary rooms for instructors of both sexes should be provided. Storage and apparatus room of the dimensions required to carry out the physical education program should be provided.

ACCESSORY PROVISIONS. Provide one or more drinking fountains of the recessed wall type at convenient but out-of-the-way locations.

SCHOOL BUILDING
LIBRARIES

LIBRARY ROOM FOR ELEMENTARY SCHOOLS. The minimum desirable floor area is approximately the same as the size of a standard classroom. The pupil capacity of a library room is computed on a basis of from 15 to 25 square feet of net usable floor area per pupil.

LIBRARY ROOM FOR HIGH SCHOOLS. For high schools having an enrollment of less than 200 pupils, a separate classroom or an end of a study hall should be fitted with shelving, tables, and chairs.

For an enrollment of from 200 to 500 high school pupils, there should be a separate library room equipped with shelving, tables, and chairs. In addition thereto, there should be a charging desk, bulletin board, and other essential office equipment.

As the enrollment increases, workrooms should be provided. In all workrooms, lavatory facilities should constitute a part of the equipment. Whenever the enrollment and the use of a library justify it, separate small conference spaces should be added.

ACOUSTICS. Careful attention should be paid to the acoustics of the library, and a noiseless type of floor should be selected.

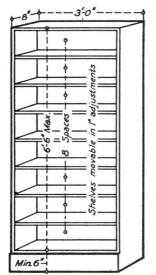

BOOKCASES. Book shelving in library rooms should be the open type, as shown in the illustration at the left. Shelves should be movable with 1-inch adjustments.

Provide one shelf for each 10 inches in height. Allow eight books to the shelf foot in computing capacity.

A limited number of sections should be provided that are 10 inches and 12 inches deep, in addition to the 8-inch-deep sections.

EQUIPMENT. Reading tables should not have glossy surfaces. Suitable chairs, librarian's desk, card catalog case, magazine rack, and closets should be provided as needed. A bulletin board at least 4'-0" x 3'-0" should be provided.

LIGHTING. Provision for natural lighting should be made on the same basis as for classrooms. Artificial lighting should correspond to standards established by the American National Standards Institute (ANSI), the Illuminating Engineering Society (IES), or other appropriate authority. Tables and desks should be so planned that pupils will not be obliged to face the windows.

CLASSROOM LIBRARY FACILITIES. In elementary school buildings in which a separate library room is provided, and when the educational program requires it, book reference facilities should be made available in each classroom. The minimum shelf space should be determined by the requirements. Such classroom shelving should be made readily accessible to pupils and should be of the closed, locking type. A reading table should be placed in the elementary classroom.

BOOKSHELVES

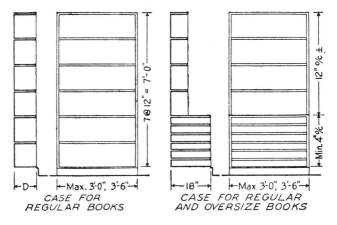

CASE FOR
REGULAR BOOKS

CASE FOR REGULAR
AND OVERSIZE BOOKS

SHELF DEPTH (D). Shelves deeper than necessary collect dirt and waste space. An analysis of a college library shows that the variation in book sizes requires shelving proportioned as follows:

$$85\% — 8'' \text{ shelves}$$
$$10\% — 10'' \text{ shelves}$$
$$5\% — 12'' \text{ shelves}$$

The following table gives recommended shelf depths for various types of books and data for estimating the capacity of shelving. In general libraries about one-third extra shelf space is usually allowed for expansion.

Kind of Books	Vols. per ft. of shelf	D
Fiction	10	8"
Economics	9	8
General Literature	8	8
History	8	8
Law	4½	8
Public Documents	6	8
Reference	8	10
Technical	7	10
Medical	6½	10
Bound Periodicals	5½	12

OVERSIZE BOOKS. Periodicals, atlases and other large books should preferably be shelved flat on account of their bulk and the consequent damage to their bindings in removing and replacing them on the shelves. Shelves 18" deep x 28" or more wide x 4" or more on center vertically are recommended. A collection of books on art or architecture will require a larger proportion of shelves for oversize books.

HEIGHT AND WIDTH OF UNITS. 7'-0" is a practical limit for height, and 3'-0" or 3'-6" for widths of units. All shelves should be adjustable, the number being based on a vertical spacing of 12" on center.

LIBRARY
STACKS

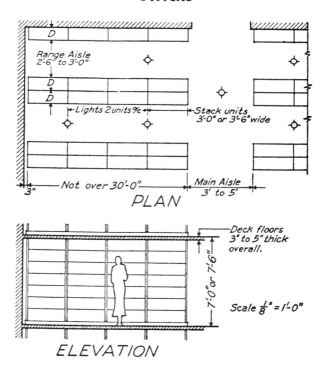

D

Range Aisle
2'-6" to 3'-0"

D

D

←Lights 2 units %→ ←Stack units
3'-0" or 3'-6" wide→

←3" ←――Not over 30'-0"――→ Main Aisle
3' to 5'

PLAN

Deck floors
3" to 5" thick
overall.

7'-0" or 7'-6"

Scale ⅛" = 1'-0"

ELEVATION

Depths (D) for books = 8", 9", 10", 12". For newspapers 18" and 22". No definite rules can be laid down for the depth of shelves required, as this dimension depends upon the method of classification, space available, and the nature of the library. Where economy and compactness of storage are important, 85% of the shelves could be planned for 8" depth in a general library. Most books are 6" or less in depth, and extra shelf width collects dust as well as wasting space which would be valuable if added to the width of range aisles.

For small multi-deck stacks a hand-power book dumb-waiter 16" x 20" x 30" is adequate. It may be built into a stack unit near the center of the stack room. For larger stack rooms an automatic push-button-control electric elevator car 3'-4" x 4'-4" (clear shaft 4'-9" x 5'-3") will accommodate a book truck and attendant.

No columns are necessary as the stack units are designed to act as supports for the floors and stacks above.

MINIMUM CEILING HEIGHTS— TYPICAL BUILDING CODE REQUIREMENTS

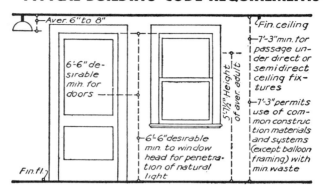

Aver. 6" to 8"

Fin. ceiling

6'-6" desirable min. for doors

7'-3" min. for passage under direct or semidirect ceiling fixtures

7'-3" permits use of common construction materials and systems (except balloon framing) with min. waste

5'-7½" Height of aver. adult

Fin. fl.

6'-6" desirable min. to window head for penetration of natural light

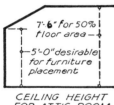

7'-6" for 50% floor area

5'-0" desirable for furniture placement

CEILING HEIGHT FOR ATTIC ROOM

All habitable spaces must have minimum ceiling heights of 7'-6", except for kitchens, bathrooms, and halls, which may have a minimum height of 7'-0" as measured to the lowest projection. Rooms with sloped ceilings are required to have the prescribed ceiling height over at least one half their area, and no part of such rooms measuring less than 5'-0" in height may be included in the minimum area calculation. Basements may or may not be classed as habitable spaces, but 7'-0" in the clear can be considered a minimum ceiling height.

HEIGHTS REQUIRED BY CONSTRUCTION
(to avoid waste of building materials)

Wall Construction	Ceiling Height[1]
MASONRY—	
Brick[2]	
35 courses	8'-2¾"
34 courses	8'-0"
33 courses	7'-9¼"
32 courses	7'-6½"
31 courses	7'-3¾"
30 courses	7'-1¼"
29 courses	6'-10¼"
Concrete Block[3]	
12 courses	8'-2½"
11 courses	7'-6½"
10 courses	6'-10½"
FRAME[4]—	
Platform	
8'-0 studs	8'-3¼"
7'-0 studs	7'-3¼"
Balloon	
18'-0 studs	7'-11¾"
16'-0 studs	6'-11¾"

[1] — Finish floor to finish ceiling.
[2] — Brick assumed 2¼" x 3¾" x 8" with ½" joints, top plates 3", ceiling ½", calculated from monolithic slab surface.
[3] — Concrete block assumed 7⅝" high with ⅜" joints, top plates 3", ceiling ½", calculated from monolithic slab surface.
[4] — Joists assumed 9¼", subflooring ⅝" plywood, finish floor ¾", ceiling ½", sole plate 1½", top plates 3".

MINIMUM WINDOW AREAS
FOR HOUSES AND HOUSING

SUGGESTED MINIMUMS. Suggested minimum window areas noted here are based on natural illumination requirements as determined by data on average daylight illumination and brightness of the sky for different regions of the United States. A minimum of 6 footcandles of natural light has been suggested in the past; this would require, at Washington, D.C., (lat. 39°N) a ratio (glass to floor area) of 15% or 1:6.7, if walls and ceilings are light in color. But since exacting eye work can usually be moved close to windows, or windows installed near obvious work-station locations, a general illumination of 5 footcandles or even less can generally be considered adequate. U.S. Public Health and Weather Bureau reports show that Great Plains states average 25% higher and Rocky Mountain states 46% higher daylight illumination than Northeastern states. To facilitate adaption of the table to local conditions, governing physical conditions for each region are listed.

MINIMUM RATIO — WINDOW AREA TO FLOOR AREA

Region	I Northeastern States	II Southeastern States	III Northwestern States	IV Southwestern States
Physical Condition — Latitude	high	low	high	low
Physical Condition — Altitude	low	low	high	high
Physical Condition — Air Pollution	high	moderate	low	low
Desirable Ratio Window Area to Floor Area	1:7	1:8	1:8	1:10
Minimum Openable Area Per Window	1/3	1/2	1/3	1/2

SPECIAL CASES

Location	Min. Ratio Glass to Floor Area
Bathroom and water closet compartments	1:8 (not less than 3 sq. ft.)
Kitchen	1:8 (not less than 9 sq. ft.)
Basement and Cellar	1:40
Stairways in multiple family buildings (more than 2 stories)	12 sq. ft. minimum per story height
Hallways in multiple family buildings	1:20

CURRENT PRACTICE. Building code requirements do vary to some extent throughout the country. However, the requirements of the Uniform Building Code are typical, and in the absence of other specifics they can be safely used as guidelines. The window area must be at least 1:10 in all habitable rooms of a dwelling, with a minimum glass area of 10 square feet per room; openable glass area in such rooms must be at least 1:20, with a minimum 5 square feet of openable glass. The former numbers represent lighting requirements, and the latter ventilation requirements.

The minimum window area for bathrooms, laundry rooms, and similar rooms is 1:20, with a minimum openable glass area of 1.5 square feet, unless a mechanical ventilation system is installed. (Lighting and ventilation requirements are the same.) Rooms and areas normally used or occupied by human beings in buildings of other occupancy types generally require a minimum glass/floor area ratio of 1:10.

LIVING ROOMS IN
MULTIPLE HOUSING

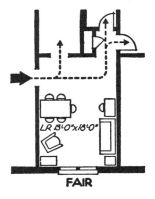

FAIR

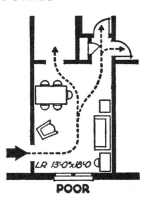

POOR

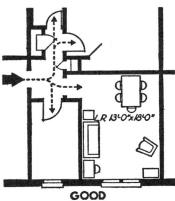

GOOD

USE OF DATA. The comments on this *Data Sheet* apply specifically to multi-family housing. Nevertheless, many of the suggestions are equally applicable to detached dwellings. The material has been taken in part from an FHA booklet entitled "Architectural Presentation and Desirable Physical Characteristics of Projects Submitted to the Rental Housing Division under Sections 207 and 210 of the National Housing Act."

SIZE. Size will vary with the size of the family, its economic status, and the contemplated use of the room. It should not be less than 11'-0" in width. If standard length lumber is used with 4" end bearing for joists, the width will usually work out 11'-4", 13'-4", 15'-4", etc., to avoid waste in cutting. The average living room size for detached bungalows has been found to be 12'-5" x 17'-4", equal to 216 square feet. In multiple housing, the rooms are usually smaller than in detached dwellings.

ASPECT. It is generally desirable for the living room to receive sun during the periods of the day when it is occupied, except in hot climates. For this reason south or west exposures are good. The living room should be located favorably with respect to attractive views. Cross-ventilation should be provided, if possible.

CIRCULATION. Living rooms should be entered through a small foyer where outer garments can be removed. The foyer acts as a buffer against direct intrusion into the living space. In no case should a design require people to pass diagonally through the living room to reach other rooms in the dwelling.

BEDROOMS IN
MULTIPLE HOUSING

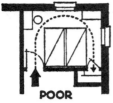

POOR **GOOD**

TWO SMALL BEDROOMS (13'-0"x10'-0")—illustrating the need of careful study of wall spaces.

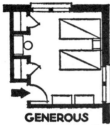
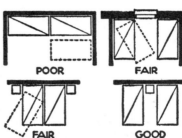

GENEROUS

A LARGE BEDROOM (16'-0" x 12'-0")—affording space for study, sewing, or play.

POOR **FAIR**

FAIR **GOOD**

Arrangements of TWIN BEDS rated for facility of housekeeping.

USE OF DATA. The comments on this *Data Sheet* apply specifically to multi-family housing. Nevertheless, many of the suggestions are equally applicable to detached dwellings. The material has been taken in part from the FHA booklet entitled "Architectural Presentation and Desirable Physical Characteristics of Projects Submitted to the Rental Housing Division under Sections 207 and 210 of the National Housing Act."

BEDROOMS IN LOW-INCOME HOUSING. It is desirable to have at least 1 bedroom that will accommodate twin beds. Where space-saving is essential to very low rentals, observance of this recommendation may not be imperative. Persons who must economize in the rent they pay, have the parallel problem of reduced household expense. A double bed costs less than a pair of twin beds and the recurrent cost of laundry is less.

SIZE AND SHAPE OF ROOM. Careful study of wall spaces for required furniture will largely determine the room size. It has been found that detached bungalows have an average bedroom size of 10'-5" to 10'-8" wide by 11'-8" to 12'-7" long.

PLANNING REQUIREMENTS. Privacy, ventilation, adequate storage space, quiet, and some sunlight during each day are desirable. Bedrooms must frequently serve as places for study, sewing or play, especially in small dwelling units, hence there should be adequate space for these activities. Facility of cleaning and bed-making is of special importance. Added time and labor are required to move beds or to make them up from 1 side only.

269

KITCHENS IN
MULTIPLE HOUSING

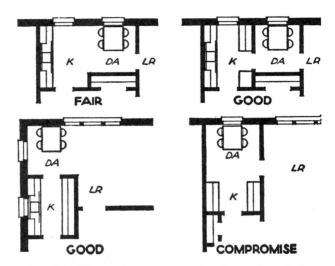

USE OF DATA. The comments on this *Data Sheet* apply specifically to multi-family housing. Nevertheless, many of the suggestions are equally applicable to detached dwellings. The material has been taken in part from the FHA booklet entitled "Architectural Presentatioin and Desirable Physical Characteristics of Projects Submitted to the Rental Housing Division under Sections 207 and 210 of the National Housing Act." Where FHA recommendations have seemed incompatible with good architectural planning, changes have been deliberately made.

KITCHENS IN LOW-INCOME HOUSING. In housing for low-income families, kitchen equipment must be adequate and must operate efficiently. Limited cupboard and storage space is allowable.

SIZE AND SHAPE OF ROOM. In general, an oblong room, wide enough to accommodate fixtures on both long sides, is more efficient than a square room. The minimum length is 6'-6" for kitchens with fixtures on both walls; 5'-6" for those with fixtures on one side only. The length should be greater if possible. In detached bungalows, the average kitchen was found to be 8'-3" x 11'-3" —equal to 92½ square feet. In large urban centers, there are a great many childless families in which both husband and wife work and usually "eat out." For such people, the small kitchen may be entirely appropriate.

STRIP KITCHENETTE. The kitchen installed in a niche or closet off the living room is to be condemned without exception. It is inadequate in equipment. It fills the room with cooking odors. If the living room is used for sleeping, the strip kitchenette can and has caused asphyxiation by the escape of cooking gas or refrigerants. Wall space is sacrificed.

LIGHT AND VENTILATION. The kitchen should face east or northeast, if possible. An east aspect is particularly desirable if a dining alcove is incorporated. Ample light and good ventilation for removing hot air and odors are important. An exhaust fan or range vent hood is highly desirable.

DINING SPACE. The dining alcove should be located where it will not interfere with food preparation.

AVERAGE
ROOM SIZES

Rooms	Bungalows	2-Story Detached	Row Houses
KITCHEN	8'3" × 11'3" 92½#	9'1" × 11'8" 107#	8'0" × 12'0" 96#
DINING ROOM	12'4" × 10'7" 131#	13'10" × 12'0" 166#	14'1" × 11'1" 156#
LIVING ROOM	17'4" × 12'5" 216#	21'4" × 13'2" 281#	16'0" × 13'6" 216#
BED ROOM #1	12'7" × 10'8" 134#	16'5" × 12'8" 208#	14'0" × 11'11" 166#
BED ROOM #2	11'8" × 10'5" 122#	13'5" × 11'5" 153#	12'11" × 9'10" 127#
BED ROOM #3		12'7" × 10'0" 126#	11'6" × 8'5" 97#

SOURCE OF INFORMATION. A number of years ago, houses were inspected in 31 cities during a survey conducted by staff members of the Division of Building and Housing of the U.S. Department of Commerce. The prices of the houses ranged from low to moderately high, encompassing numerous styles. In the period that followed, housing tastes changed somewhat, and the size of new houses increased. Now, however, the shift is back to smaller housing units. Thus, though overall housing designs have changed from time to time and doubtless will continue to do so, the dimensional details gained from this survey remain useful as a general guideline to livable, workable room sizes.

ROOM SIZE VARIATIONS. *Kitchens* were found more nearly alike in size than any other room. Most of them contained about 100 square feet with the width about three-quarters of the length, so that 8'-10" x 11'-8" would be typical. The variation in *Living Rooms* was from 11'-0" to 15'-0" in width, and 15'-0" to 22'-0" in length. The width was commonly about two-thirds of the length. The width of *Dining Rooms* tended to be about three-quarters of the length with an area about half again as large as the kitchen. *Bedrooms* ran noticeably larger in two-story houses than in one-story houses. The owner's bedroom in many two-story houses was over the living room and about the same size.

CEILING HEIGHTS. Ceiling heights were usually greater in the South than in the North, presumably because they are more comfortable in warmer climates and also on account of custom. In houses above the lowest price range there was an increasing tendency to obtain a higher living room ceiling by dropping its floor one or two steps lower than the rest of the first floor.

ASPECT OF
ROOMS

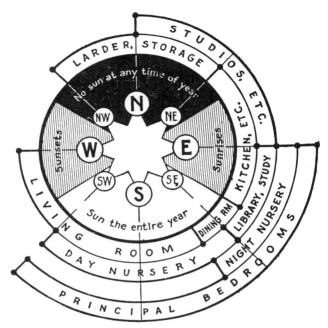

There are many other influences on the location of rooms besides their aspect with regard to the sun. Prevailing winds in both winter and summer, the range of temperature, natural obstructions influencing light and air, the number of sunny days per year, desirable vistas and personal preferences all have a bearing on the direction in which rooms face.

LIVING ROOM. The living room should generally receive sun during the periods of the day when it is occupied. For this reason the south or west (or both) exposures are desirable.

DINING ROOM. It is thought that the morning sun has a cheerful influence on the day's activities, so dining rooms should have an easterly exposure where possible.

KITCHEN. The housewife usually spends an appreciable part of her morning in the kitchen so that an easterly exposure is desirable. In the opinion of some experts, a northern exposure for the kitchen is desirable because of the diffused quality of the light to work by as well as the coolness.

BEDROOM. The location of the bedroom is largely a matter of personal preference, some people objecting to being awakened by the morning sun and its resulting heat in the summertime. A room exposed to the west receives the heat of the afternoon and is often unpleasantly warm at bedtime. A bedroom exposed to the north, if adequately heated, is preferred by many.

ORIENTATION CHART

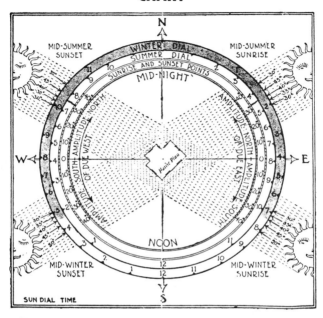

The chart is for midsummer and midwinter day on the 40th degree, north latitude, which is the best average line that runs midway of the country passing through or near New York, Philadelphia, Cincinnati, Columbus, Indianapolis, St. Louis, Kansas City, Denver, Salt Lake City, Sacramento.

The figures 30°-50°, alongside the suns, represent north latitudes, and the arrows show the direction of the sun's rays at sunrise and sunset. If your location is at or near any of the degrees of north latitude marked, consider the suns slipped around accordingly.

The inner circle represents the horizon, and the degrees upon it show the points of sunrise and sunset north or south of the direct east and west line. These amplitudes must not be confused with the latitudes on the earth's surface. The two outer circles are sun-dials for midsummer and midwinter day at 40°, north latitude. They show the direction, in plan, of the sun's rays during successive hours.

For seasons other than the solstices, move the suns down and up from the solstitial points about a third of the way to E for each succeeding month. At E the sun has reached either equinox and will rise directly east and set directly west.

At times other than the solstices, while you can thus get the direction of the rising and setting sun, the hourly dial will not apply. As the sun moves from the solstice to the equinox, the summer hour spaces grow more uniform, while the winter hour spaces lengthen. But with the general direction of the morning and afternoon light settled for the solstitial extremes, the hour position of the sun between seasons is not so important.

ANGLES OF
SUNLIGHT

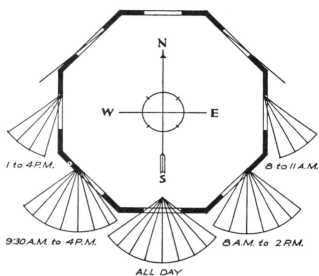

1 to 4 P.M.

8 to 11 A.M.

9:30 A.M. to 4 P.M.

8 A.M. to 2 P.M.

ALL DAY

WINTER EXPOSURE DIAGRAM
Latitude of Northern States

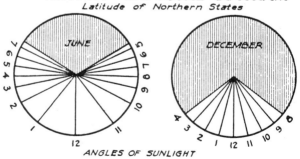

JUNE

DECEMBER

ANGLES OF SUNLIGHT

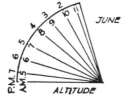

JUNE

ALTITUDE

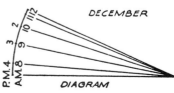

DECEMBER

DIAGRAM

KITCHEN PLANNING

POSITION OF KITCHEN IN THE HOUSE PLAN
Easy access to front and rear doors, telephone, toilet, stairs to second floor, stairs to basement, service yard, dining area.
Not a part of general household traffic pattern. Foot traffic should not have to pass through kitchen en route from house to garage, basement to outdoors, house to utility closet, or living room to bedrooms.
As few doors in kitchen walls as possible: two is minimum and ideal. Sunny exposure is not always desirable (in hot climates), north light is ideal. View over children's play area may be advisable.

SIZE AND SHAPE OF KITCHEN
Preferably oblong, U-shape or (less efficient) corridor.
Small as possible for efficiency kitchen; large and comfortable for "country" or "family" kitchen.

ROUTING
"Should proceed from right to left." Jury of a national competition.
"Should proceed from left to right." U.S. Gov't. Bul. 1315.

ELECTRICAL
Dedicated receptacles for refrigerator, freezer. Receptacles as required for other large appliances such as dishwashers and trash compactors. Ample receptacles for small appliances: toaster, mixer, microwave/convection oven, food processor, coffee-maker, crock-pot, and deep-fat fryer.
Bright general illumination from overhead fluorescent lighting fixtures. Task lighting at sink, range, work areas, serving plane, and so forth, directly over-head, on or in ceiling.

VENTILATION
Cross-draft important. Windows close to ceiling, 4'-0" to 4'-6' to stool. Ventilating hood over range, or self-venting range.

LAUNDRY
Kitchen plan efficiency, as well as hygiene, is impaired by inclusion of laundry; locate elsewhere. Built-in ironing board useful for occasional use.

UTILITY CLOSET
Under no circumstances should dirt-removing equipment and cleaning supplies be located in the kitchen.

STORAGE PANTRY
Materials and equipment are best stored where they are used, if possible— not in a separate room or closet. The doorway to a storage pantry often takes as much kitchen wall space as the required storage facilities would have taken.

SERVING PANTRY
A space waster that has no place in a small, servantless home.

BREAKFAST NOOK
Should not be tiny and cramped. Should be located so as not to increase the distance from the kitchen to the dining room, and be handy to the kitchen work area.

THE RANGE
Locate away from cross-draft, but handy to all major work areas. Working plane 34" to 36".

THE SINK
Not under a window, because of light contrast, glare, sun's heat in summer, cold air fall in winter. Wall space over sink is also very valuable for storing equipment needed there. Sink can be near dining room door or serving pass-through for easy clearing away of dishes. Rim from 34" to 39" above floor.

THE WORK COUNTERS
Toe space 4" to 5" high and 3" to 4" deep, cove or base shoe molding at the floor. Working plane 34" to 36" above the floor.
Cabinet doors removable for easy cleaning at sink. Sliding doors, so worker doesn't have to step away from them and to obviate cracking skulls in arising under swing door left open. Glass panel, so articles are visible without searching. No muntins, for easy cleaning and better vision. No open or curtained cabinets.
No live storage over 7'-0" from floor. No fixed shelves in base cabinets—pullout shelves and racks instead.

GREAT CONVENIENCES
Trash compactor; linen hamper; garbage disposal unit; desk for keeping kitchen records; cookbook shelving; storage for table linens, dish towels, and so forth; bulk storage bins/drawers for flour, sugar, dogfood, and so forth; built-in cutting board/chopping block; pullout marble pastry board; space-saver storage racks in cabinets; stereo radio/intercom station.

KITCHEN
ROUTING

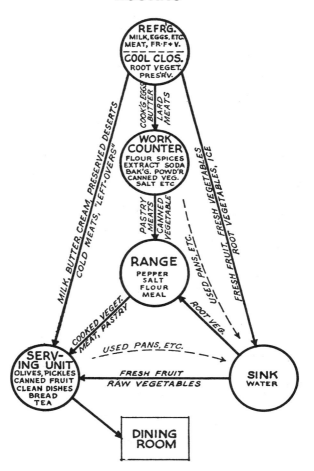

This diagram indicates the relation of the various units of kitchen equipment as used in the preparation of most frequently served foods. In each circle is given a few of the typical food materials stored at that location.

Clearing away process is: soiled dishes from dining room to sink, to dish storage; cream, butter, etc., from dining room to refrigerator.

KITCHEN CABINET
REQUIREMENTS

SOURCE OF INFORMATION. Several years ago a small group of men measured over 5,000 existing kitchens and pantries and ascertained whether or not sufficient storage space had been provided. From this mass of data it was possible to formulate simple rules for the amount of storage space necessary in a home or apartment.

NORMAL OCCUPANCY. The amount of kitchen storage space required is a function of the normal occupancy of the dwelling. Normal occupancy is determined by allowing 2 persons for the first or master bedroom, 1 person for each additional bedroom and 2 persons are added for accumulation and entertaining. In other words, the normal occupancy is taken as the number of bedrooms plus 3.

RESIDENCE STORAGE SPACE. Six square feet of shelf area should be allowed per person. Since the base storage cabinet may contain both drawers and shelves, this is measured in linear feet as shown in the diagram. The storage space over refrigerators and broom closets is disregarded in making calculations, as it is not particularly accessible and acts as a factor of safety.

APARTMENT STORAGE SPACE. Allow 10 to 14 square feet of upper shelves, and 3 to 4 linear feet of base storage for one-room studio apartments having alcove kitchens. Allow 14 to 20 square feet of upper shelves and 4 to 6 linear feet of base storage for apartments with 1 bedroom. Use the residence table for apartments of 2 or more bedrooms. It shows minimum requirements.

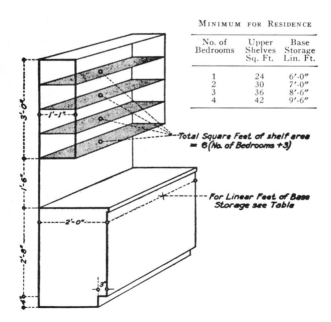

MINIMUM FOR RESIDENCE

No. of Bedrooms	Upper Shelves Sq. Ft.	Base Storage Lin. Ft.
1	24	6'-0"
2	30	7'-0"
3	36	8'-6"
4	42	9'-6"

Total Square Feet of shelf area
= 6 (No. of Bedrooms +3)

For Linear Feet of Base Storage see Table

KITCHEN CABINET,
USUAL DIMENSIONS

Both steel and wood kitchen cabinets are available in many styles, finishes, and sizes. Though there are some variations, manufacturers generally follow the dimensions given here. Notice that sizes of steel cabinets may vary from those of wood cabinets.

Usual widths for wall cabinets are: single-door – 12″, 15″, 18″, 24″; double-door - 24″, 30″, 33″, 36″, 42″, 48″; triple-door - 42″ to 60″. Corner wall units measure approximately 24″ x 24″ at the back. Peninsula overhead cabinets are from 18″ to 42″ wide.

Usual heights for wall cabinets are 15″, 18″, 21″, 24″, 30″. Peninsula overhead cabinet heights are 18″ and 30″.

Usual widths for base cabinets are: single-door - 12″, 15″, 18″, 21″, 24″; double-door - 24″, 27″, 30″, 36″, 42″, 48″. Corner base units measure approximately 36″ x 36″ at the back.

Sink cabinet widths are: single-door - 24″, 27″; double-door - 27″ to 48″; triple- or four-door - 48″ to 84″.

Standard height for base cabinets alone is 34½″, including standard 4″ base. Countertop thickness is 1½″, for a standard overall counter height of 36″.

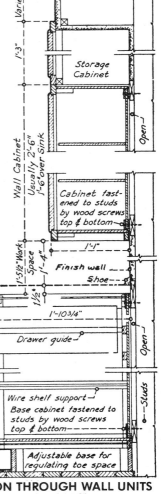

Storage Cabinet

Open

Varies

1′-3″

Wall Cabinet
Usually 2′-6″
1′-6″ over Sink

Cabinet fastened to studs by wood screws top & bottom

1′-1″

Finish wall

Shoe

1′-5½″ Work Space
1′-4″

½″

Countertop

1′-10¾″

Drawer guide

2′-10½″ Base Cabinet
Finish floor

Open

Studs

Wire shelf support

Base cabinet fastened to studs by wood screws top & bottom

Adjustable base for regulating toe space

SECTION THROUGH WALL UNITS
Scale - ¾″=1′-0″

Wall Cabinet

Open

½″ Plaster
½″ Mortar bed
½″ Tile
1½″

Studs

Base cabinet

ALTERNATE FOR COUNTER BACK

KITCHEN STORAGE
REQUIREMENTS

CANNED GOODS

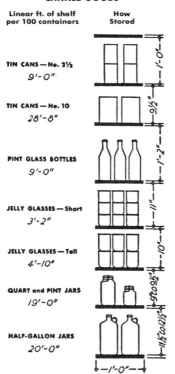

Linear ft. of shelf per 100 containers	How Stored

TIN CANS — No. 2½
9'-0"

TIN CANS — No. 10
28'-8"

PINT GLASS BOTTLES
9'-0"

JELLY GLASSES — Short
3'-2"

JELLY GLASSES — Tall
4'-10"

QUART and PINT JARS
19'-0"

HALF-GALLON JARS
20'-0"

The amount of canned goods stored differs so with family habits that it is impossible to determine the amount of storage space needed. For rural families extra storage space must be provided outside the kitchen. On the basis of studies of shelf space needed for different sized containers and an Indiana study of the average amounts of canned goods stored by rural families, 63 feet of shelving 12" wide, with shelves 9" apart, is needed for home canned foods, and 14 feet, with shelves 12" to 18" apart, for food in tin cans. With a ceiling 7'-3" high this requires a wall space 9 to 10 feet wide; a closet 4' x 4' with shelves on 3 sides is adequate.

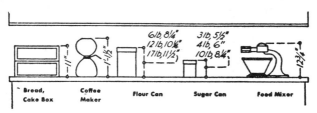

6 lb, 8¼"
12 lb, 10¼"
17 lb, 11½"

3 lb, 5½"
4 lb, 6"
10 lb, 8¼"

Bread, Cake Box   Coffee Maker   Flour Can   Sugar Can   Food Mixer

AVERAGE HEIGHTS OF ARTICLES STORED ON COUNTERS

CHECKLIST OF CULINARY EQUIPMENT

SINK EQUIPMENT

1 waste basket
1 towel rack
1 dishpan, about 12-qt. capacity
1 vegetable brush
1 garbage can
1 dish-drying rack (if no electric dishwasher)
6 dishcloths
12 dish towels
1 roll paper towels
2 pot scrubbers
1 soap dish
1 bar soap
1 hand lotion dispenser

GLASSWARE AND CHINA

8 dinner plates
8 dessert or salad plates
8 cereal dishes
8 breakfast or luncheon plates
8 soup plates or bowls
8 bouillon cups
8 sherbet glasses
3 vegetable dishes
1 sauce or gravy boat
16 glasses (water and iced tea)
Relish, candy, and nut dishes
8 cups and saucers
1 teapot and stand
1 cream pitcher
1 water pitcher
1 sugar bowl
1 salad bowl
1 large platter
1 medium platter
Salt and pepper shakers
Other glasses for juices, wines, beer, cocktails, and so forth, may be added according to need.

FOR KITCHEN CABINETS OR WORK COUNTERS

1 coffee-maker
1 set storage jars/cans
1 spice rack
5 mixing bowls, nested
2 measuring cups
1 grater
1 dough blender
1 set cookie cutters
1 set muffin pans (6- or 8-cup)
3 casseroles, 1-2 qt.
1 egg beater
1 set kitchen cutlery

4-6 wooden spoons
1 corkscrew/bottle opener
1 chopping bowl and knife
1 cake knife
1 breadboard
1 utility tray
1 colander
1 rolling pin
1 potato masher
1 flour sifter
1 bread box
1 cake box
1 pie box
2 wire strainers, 3" and 6"
2 sets measuring spoons

FOR STORAGE CABINETS

1 Dutch oven
1 cake pan 8" x 8"
1 cake pan 9" x 12"
1 loaf pan 5" x 9"
2 layer cake pans, 9"
2 cookie sheets
2 pie plates, 9" or 10"
2 wire cake coolers
1 steamer/waterless cooker
1 saucepan, 6-10 qt.
3 or 4 saucepans, 2-4 qt.
Saucepan covers
2 funnels (small and large)
1 whip or whisk
1 set refrigerator dishes
Assorted freezer containers
1 roll waxed paper
1 roll aluminum foil
1 box sandwich bags
1 double boiler
1 roasting pan, 11" x 16"
1 baster
1 griddle, 10"

TO KEEP NEAR RANGE

1 pair salt and pepper shakers
3 frying pans, 6"-12"
1 iron skillet, 10"
1 tea kettle
3 lipped saucepans, 1 pt., 1½ pt., and 1 qt.
6 pot holders
1 set spatulas

SIZES OF KITCHENWARE

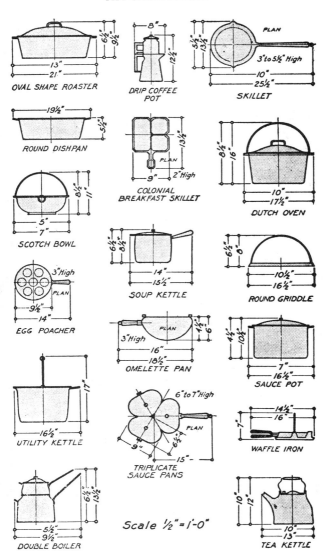

OVAL SHAPE ROASTER

DRIP COFFEE POT

SKILLET

ROUND DISHPAN

COLONIAL BREAKFAST SKILLET

DUTCH OVEN

SCOTCH BOWL

SOUP KETTLE

ROUND GRIDDLE

EGG POACHER

OMELETTE PAN

SAUCE POT

UTILITY KETTLE

TRIPLICATE SAUCE PANS

WAFFLE IRON

DOUBLE BOILER

Scale ½" = 1'-0"

TEA KETTLE

SIZES OF
KITCHENWARE

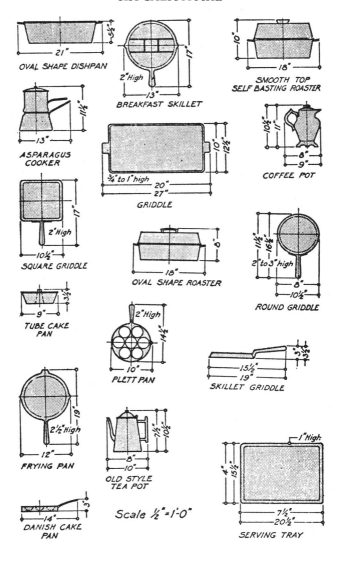

OVAL SHAPE DISHPAN
21" 5½"

BREAKFAST SKILLET
2" High 13" 17"

SMOOTH TOP
SELF BASTING ROASTER
10" 18"

ASPARAGUS
COOKER
13" 11½"

GRIDDLE
¾" to 1" high 20" 27" 10" 12½"

COFFEE POT
10½" 11" 8" 9"

SQUARE GRIDDLE
2" High 10½" 17"

OVAL SHAPE ROASTER
18" 6"

ROUND GRIDDLE
11½" 16½" 2" to 3" high 8" 10½"

TUBE CAKE
PAN
9" 3½"

PLETT PAN
2" High 14½" 10"

SKILLET GRIDDLE
15½" 19" 3" 3½"

FRYING PAN
2½" High 12" 19"

OLD STYLE
TEA POT
7½" 10½" 8" 10"

Scale ½" = 1'-0"

SERVING TRAY
1" High 4" 15½" 7½" 20½"

DANISH CAKE
PAN
14" 3"

SIZES OF
KITCHENWARE

2 QT. MILK

QT. MILK

PT. MILK

WATER JUG

REFRIG. DISHES

REFR. DISH

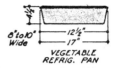

VEGETABLE
REFRIG. PAN

JUG WITH
ICE CUBE
INSERT

VITALIZER

2 QT.
PRESERVE
JAR

WATER
PITCHER

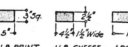

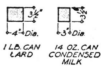

1 LB. PRINT
BUTTER

1 LB. CHEESE

1 DOZ. EGGS

1 LB. CAN
LARD

14 OZ. CAN
CONDENSED
MILK

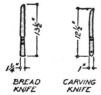

BREAD
KNIFE

CARVING
KNIFE

COOK'S FORK

PARING
KNIFE

SIZES OF
TABLEWARE

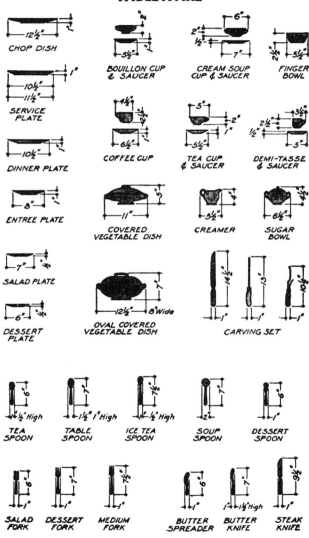

CHOP DISH — 12½"

BOUILLON CUP & SAUCER — 2", 5½"

CREAM SOUP CUP & SAUCER — 6", 2½", 4½", 7"

FINGER BOWL — 2½", 5½"

SERVICE PLATE — 1", 10½", 11½"

COFFEE CUP — 4½", 6½"

TEA CUP & SAUCER — 5", 2", 1", 5½"

DEMI-TASSE & SAUCER — 3½", 2½", 4½", 2"

DINNER PLATE — 10½"

ENTREE PLATE — 8"

COVERED VEGETABLE DISH — 5", 11"

CREAMER — 4", 5½"

SUGAR BOWL — 6½"

SALAD PLATE — 7"

OVAL COVERED VEGETABLE DISH — 7", 12½", 8" Wide

CARVING SET — 14½", 13", 10½", 1"

DESSERT PLATE — 6"

TEA SPOON — 6", ½" High

TABLE SPOON — 7", 1½", 1" High

ICE TEA SPOON — 7½", ½" High

SOUP SPOON — 7", 2"

DESSERT SPOON — 6", 1"

SALAD FORK — 6", 1"

DESSERT FORK — 7", 1"

MEDIUM FORK — 7½", 1"

BUTTER SPREADER — 6", 1"

BUTTER KNIFE — 7", 1½" High

STEAK KNIFE — 9½", 1"

284

SIZES OF
TABLEWARE

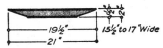

TURKEY PLATTER

OVAL PLATTER

ROUND PLATE

SOUP DISH

BERRY BOWL

COVERED
MUFFIN

CAKE PLATE

BAKER

SALAD PLATE

GRAVY BOAT

RAMEKIN

FRUIT SAUCER

CHEESE STAND

BREAD &
BUTTER
PLATE

DOUBLE
EGG CUP

CEREAL SAUCER

PITCHER

COFFEE POT

TEA POT

GRAVY BOAT
WITH PLATE

285

HOUSE CLEANING
EQUIPMENT

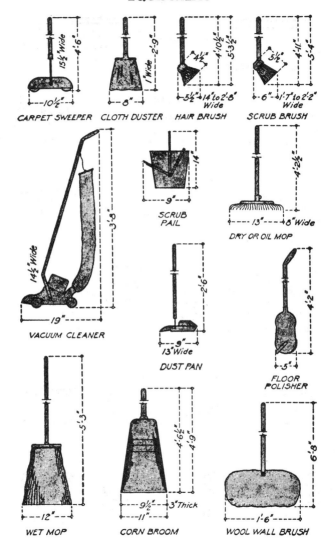

CARPET SWEEPER   CLOTH DUSTER   HAIR BRUSH   SCRUB BRUSH

SCRUB PAIL

DRY OR OIL MOP

VACUUM CLEANER

DUST PAN

FLOOR POLISHER

WET MOP   CORN BROOM   WOOL WALL BRUSH

BASIC REQUIREMENTS
FOR CLOSETS

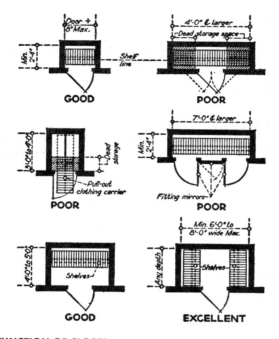

FUNCTION OF CLOSET. Bedroom closets should be classified as "live storage," not "dead." More erroneous and misleading information has been printed on closets than on any other space in the house. The following requirements are basic in the arrangement of desirable closet space:

1. Hanging space or shelving is useless unless it permits the easy removal and replacement of garments and stored items.
2. Excess front-wall space should be devoted to doors, so that garments are fully visible and accessible.
3. Mothproof bags for the protection of infrequently used garments are 2'-1½" to 2'-3" wide, making the usual closet recommendation of 24" barely adequate. No closet should be less than 2'-4" in the clear, if garments are not to rub and brush against the closet walls and doors. A man's overcoat will measure a minimum of 24" in width.
4. 1¼" steel pipe is infinitely preferable to wood rounds as hanging poles, and extendable, chrome-plated tubular steel closet rods are best for easy manipulation of hangers.
5. 5 linear feet of hanging space is a minimum for each person. Systems based on floor area or cubage for closets are not true indexes of the available hanging space and should not be employed in determining closet size.
6. The closet rod for adults' use should be 5'-8" above the floor to accommodate long coats and evening dresses—closet rods are invariably located too low. However, in a split-section closet, one section may have two rods—one located about 6'-6" high and the other about 39" below it—for shirts, blouses, and similar short garments. Rods for children should be low, but adjustable for height in future years.

WIDE, SHALLOW CLOSET, AND CLOSET DESIGN DATA

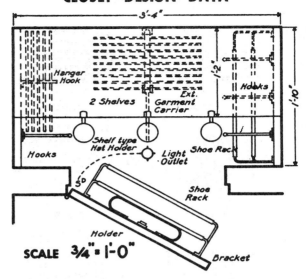

SCALE ¾" = 1'-0"

CAPACITY OF CLOSET ROD. Heavy garments for adults require 2'-0" clearance in width so that they will not brush against the wall—meaning that the closet rod should be 12" or more from the wall. Capacity or length of the rod will depend upon the type of garments hung. The following table gives the bar space required for various types of garments on hangers:

Men's suits	2"
Overcoats	5"
Women's dresses	1½"
Skirts	2"
Women's coats	5"
Fur & fur-collared coats	6"

SHOE RACK. The adjustable shoe rack may be extended from 1'-8" to 2'-4". At its 1'-8" position, it is suitable for a 24" closet door and will hold 2 pairs of men's shoes or 3 pairs of women's shoes. Width required for storing shoes is as follows:

Per pair men's shoes	8"
Per pair women's shoes	6"

CLOSET ROD HEIGHT. Extra cleats or brackets may be placed at a lower level for the installation of hooks and the closet rod, for the convenience of children. The following table gives the preferred height of the closet rod for children of various ages:

From 3 to 6 years	3'-0"
From 6 to 9 years	4'-0"
From 9 to 12 years	4'-6"
From 12 to 14 years	5'-0"
From 14 years up	5'-6"

288

MINIMUM PLAN FOR
DEEP, NARROW CLOSETS

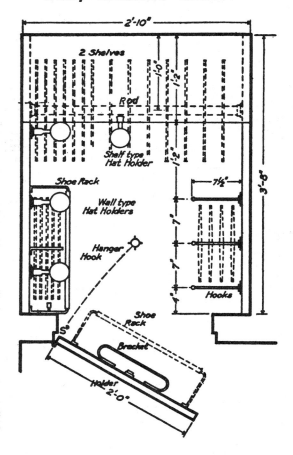

The plan shown above for a deep but narrow closet provides at least 3 times the accommodation in the limited space that can normally be obtained with the usual haphazard supply of coat hooks ordinarily provided. Yet easy access is provided to all parts of the closet.

The top shelf can be used for the storage of seasonal or seldom used articles. The lower shelf is used for hats and boxes, etc.

One or more shelf type hat holders can be used on the edge of the lower shelf without preventing easy access to the material behind.

INTERIOR VIEW
DEEP, NARROW CLOSETS

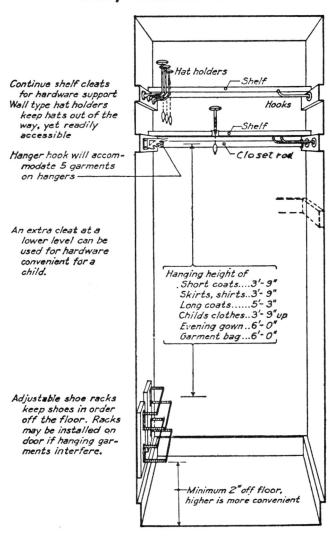

Continue shelf cleats
for hardware support
Wall type hat holders
keep hats out of the
way, yet readily
accessible

Hat holders

Shelf

Hooks

Shelf

Hanger hook will accom-
modate 5 garments
on hangers

Closet rod

An extra cleat at a
lower level can be
used for hardware
convenient for a
child.

Hanging height of
Short coats....3'- 9"
Skirts, shirts..3'- 9"
Long coats......5'- 3"
Childs clothes..3'- 9" up
Evening gown..6'- 0"
Garment bag...6'- 0"

Adjustable shoe racks
keep shoes in order
off the floor. Racks
may be installed on
door if hanging gar-
ments interfere.

Minimum 2" off floor,
higher is more convenient

INTERIOR VIEW

CEDAR CLOSETS

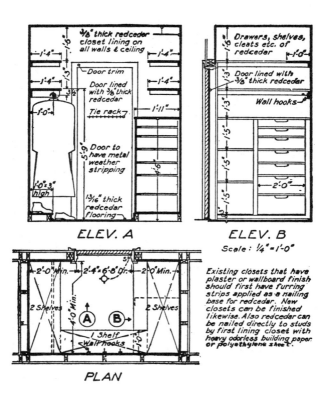

ELEV. A

ELEV. B
Scale: 1/4" = 1'-0"

PLAN

(Elevation A labels) 3/8" thick redcedar closet lining on all walls & ceiling — Door trim — Door lined with 3/8" thick redcedar — Tie rack — Door to have metal weather stripping — 13/16" thick redcedar flooring — 1'-0" x 3" high

(Elevation B labels) Drawers, shelves, cleats etc. of redcedar — Door lined with 3/8" thick redcedar — Wall hooks — 2'-0"

(Plan labels) 2'-0" Min. — 2'4" x 6'-8" Dr. — 2'-0" Min. — 2 Shelves — 4'-0" Min. — 2 Shelves — A — B — 1 Shelf — Wall hooks

Existing closets that have plaster or wallboard finish should first have furring strips applied as a nailing base for redcedar. New closets can be finished likewise. Also redcedar can be nailed directly to studs by first lining closet with heavy odorless building paper or polyethylene sheet.

The value of cedar in protecting clothing lies in the fact that it kills the newly hatched larvae of clothes moths. Well-made, tight chests or closets lined with aromatic eastern redcedar (*Juniperus virginiana*) heartwood can be depended upon for protection against clothes moths—provided the articles to be placed in them are first thoroughly brushed, combed, or otherwise treated to remove the older clothes-moth larvae. Southern redcedar (*Juniperus silicicola*) can be used for the same purpose.

The aroma, the persistent characteristic of redcedar heartwood, comes from a volatile oil present in the wood. As it is the aroma from this volatile oil that protects the clothing, the closet or chest should remain tightly closed at all times when clothing is not being removed from or placed in it. Any application of finish, such as wax, sealer, varnish, or paint, on the surface of the redcedar will destroy its effectiveness.

A lined closet or chest will give off its aroma indefinitely. When exposed to air for long periods of time, redcedar tends to seal itself on the surface. The lining can be kept efficient by lightly scraping, sanding, or steel-wooling the surface periodically at long intervals.

COAL STORAGE BIN
MADE OF CONCRETE

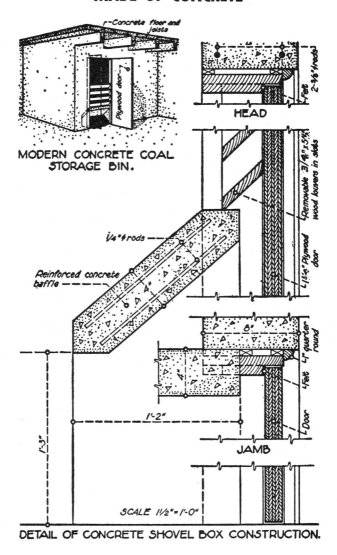

Concrete floor and joists

Plywood door

MODERN CONCRETE COAL STORAGE BIN.

HEAD

2-¾" rods

Felt

Removable ¾"x5¾" wood louvers in sides

¼"Ø rods

Reinforced concrete baffle

1¾" Plywood door

6"   4"

8"

¼" quarter round

Felt

4"

Door

JAMB

1'-2"

1'-3"

SCALE 1½" = 1'-0"

DETAIL OF CONCRETE SHOVEL BOX CONSTRUCTION.

292

COAL STORAGE BIN
OF WOOD

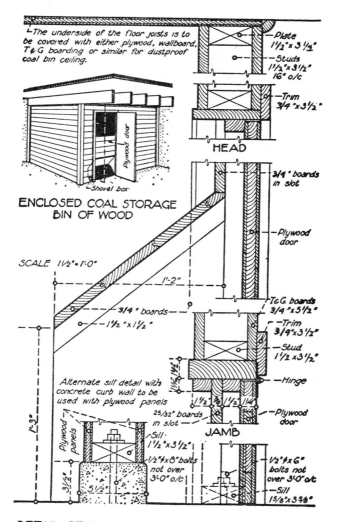

⌐The underside of the floor joists is to
be covered with either plywood, wallboard,
T&G boarding or similar for dustproof
coal bin ceiling.

Plywood door

Shovel box

ENCLOSED COAL STORAGE
BIN OF WOOD

SCALE 1½" = 1'·0"

Plate
1½" x 3½"

Studs
1½" x 3½"
16" o/c

Trim
¾" x 3½"

HEAD

3/4" boards
in slot

Plywood
door

1'-2"

3/4" boards

1½" x 1½"

T&G boards
¾" x 5½"

Trim
¾" x 3½"

Stud
1½ x 3½"

Hinge

1½   ⅝ 1½ ¼

Plywood
door

Alternate sill detail with
concrete curb wall to be
used with plywood panels

25/32" boards
in slot

Sill
1½" x 3½"

½"⌀x8"bolts
not over
3'·0" o/c

JAMB

½"⌀x6"
bolts not
over 3'·0" o/c

Sill
1⅝"x3⅜"

Plywood
panels

1'-3"

3½"

5½"

DETAIL OF WOODEN SHOVEL BOX CONSTRUCTION

COAL OR FIREWOOD CHUTE
MADE OF WOOD

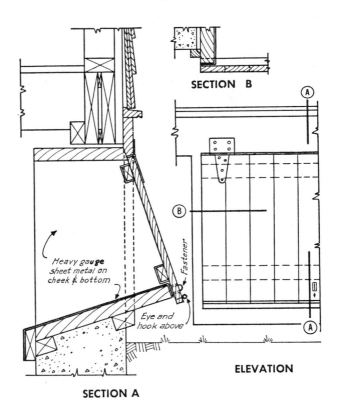

SECTION B

A

B

*Heavy gauge
sheet metal on
cheek & bottom*

Fastener

*Eye and
hook above*

A

ELEVATION

SECTION A

Any good planking can be used to construct the wood, coal, or firewood chute shown in the drawing. One of the best materials for the four sides of the chute is standard tongue-and-groove softwood decking in nominal 2″ x 6″ single T&G or, better yet, 3″ x 6″ double T&G size. The door can be made of the same material in the 2″ x 6″ size. The clear opening should not be less than 18″ x 24″ and can be up to 24″ x 30″. Heavy sheet metal lining should be secured inside the cheeks. Cracks can be filled with caulk to eliminate air infiltration. In cold climates the door can be made of 1″ T&G stock, doubled to sandwich a sheet of rigid thermal insulation.

294

LIQUEFIED PETROLEUM GAS INSTALLATION

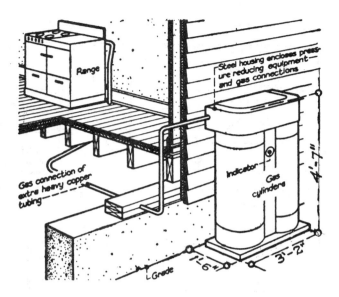

Range

Steel housing encloses pressure reducing equipment and gas connections

4'-7"

Indicator

Gas cylinders

Gas connection of extra heavy copper tubing

3'-2"

1'-6"

Grade

BOTTLED GAS. The term "bottled gas" usually means propane gas, though other gases are also "bottled." Propane (C_3H_8) is also called LP gas, or LPG; it has a heating value of 2,000-3,500 Btu per cubic foot. The "bottles" are steel or aluminum cylinders ranging from small cartridges through cylinders of 100-pound capacity to large storage tanks of various sizes. Propane is used in cold climates because it flows freely at subzero temperatures.

BUTANE. Butane (C_4H_{10}) is also referred to in some areas as bottled gas, and has a heating value of 3,200 Btu per cubic foot. It can be obtained in much the same fashion as propane, but its use is less common.

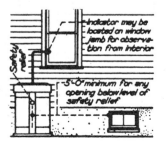

Indicator may be located on window jamb for observation from interior

Safety relief

5'-0" minimum for any opening below level of safety relief

USES. Equipment available for operation by bottled gas includes backpacking stoves, small blowtorches, emergency lighting, incinerators, barbecue units, laundry driers, kitchen ranges, heating units, water heaters, refrigerators, poultry brooders, orchard smudgepots, and internal combustion engines for trucks and motorhomes.

CODES. Local codes and supplier regulations should always be checked before designing or installing any bottled-gas equipment.

ECONOMICAL
AMATEUR DARKROOM

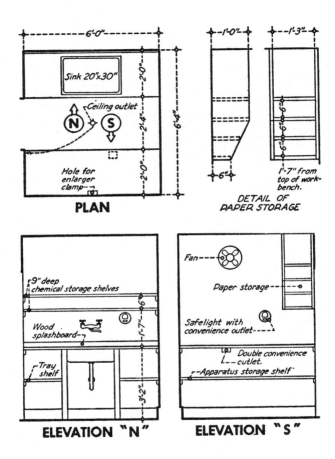

PLAN

Sink 20"x30"

Ceiling outlet

N   S

Hole for enlarger clamp

6'-0"

2'-0"

2'-4"

2'-0"

6'-4"

DETAIL OF PAPER STORAGE

1'-0"

1'-3"

6"

6"

6"

6"

6"

1'-7" from top of work-bench.

ELEVATION "N"

9" deep chemical storage shelves

Wood splashboard

Tray shelf

6"

1'-7"

3'-2"

ELEVATION "S"

Fan

Paper storage

Safelight with convenience outlet

Double convenience outlet

Apparatus storage shelf

In the plan the *wet* operations have been separated from the *dry*. The mixing of chemicals, development, washing, and fixing can all take place on the sink side of the room. The exposure in the enlarger or contact printer as well as the drying and trimming can be confined to the other side of the darkroom. The door should have a lock, and a ventilating fan is an absolute necessity. The door can be weather-stripped to make it light-tight.

AMATEUR DARKROOM

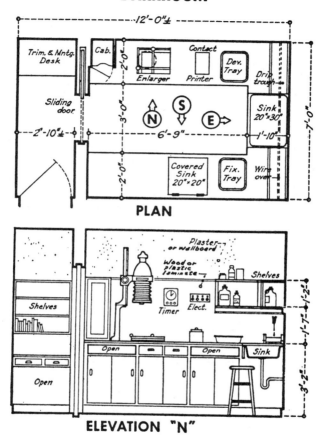

PLAN

ELEVATION "N"

For right handed persons, a clockwise sequence of operations will be most efficient. The sink at the end of the room serves as a rinse between development and fixing and will also be the location for mixing of chemicals. The covered sink is for the final washing of films or prints. The hinged cover, when closed, results in no loss of work space. Counter area is provided alongside this covered sink for ferrotyping, a print drier or other apparatus. The cabinet beside the enlarger is for printing paper up to 11" x 14", and other light-sensitive supplies.

AMATEUR DARKROOM

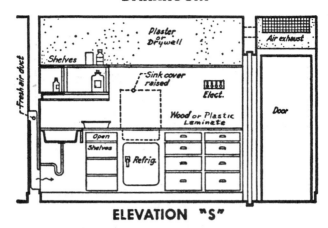

ELEVATION "S"

ELEVATION "E"

Use ¾″ interior plywood for countertops and cabinet structure; ¼″ plywood or tempered hardboard for sliding doors. Use dark, smooth, matte-finish plastic laminate for finish countertop, edging, and backsplash. Accommodate safelights for the various darkroom operations, a light-blocked ventilating system, and ample electrical receptacles (GFCI protected and away from sinks for safety) for darkroom equipment. Cover the floor with smooth, seamless vinyl flooring in a dark color, and install tall rubber or vinyl cove molding at all wall/floor joints, cemented in place. Paint walls and ceiling matte black, and eliminate or cover all light-colored or light-reflecting materials.

298

HOME VIDEO ROOM

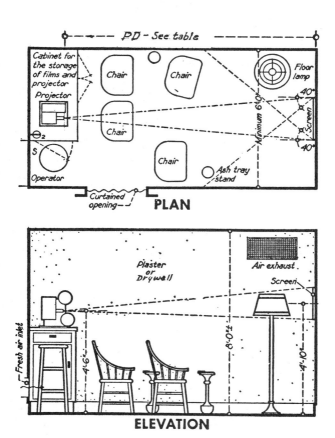

PLAN

ELEVATION

The plan and elevation above show a room arranged for an audience of four or five people and the operator. In general, principles of commercial movie theaters apply to the projection of home movies or slides, and indeed to the viewing of large-screen videotapes or television. The line of sight from the nearest viewer to the top of the screen should not exceed 30° from the horizontal. Side seats requiring the observer to view any part of the screen image at an angle greater than 40° from the line of projection are undesirable. Home projectors do not allow an angle of projection that deviates more than a few degrees from the horizontal without image distortion.

HOME
THEATER

PROJECTORS. Projectors are constructed to take 8-millimeter or 16-millimeter movie film and 35-millimeter or 2¼" x 2¼" slides for projection; less commonly, they may be able to take special projections such as stereo slides. Depending upon the size of the projection lamp used for illumination, the model of the projector, and the type of lens used, various screen-image sizes and projection distances or *throws* may be obtained to provide entertainment in rooms varying through a wide range of seating capacities. The values in the following table are average projection distances and will vary somewhat with specific projectors and lenses.

APPROXIMATE PROJECTION DISTANCES (PD)

Slides/ film	Lens focal length	Width of Screen (and image)					
		40"	50"	60"	70"	84"	96"
35mm 2x2	3"	7'	9'	11'	13'	16'	18'
	4"	10'	12'	15'	17'	21'	24'
	5"	12'	16'	19'	22'	26'	30'
	6"	15'	19'	22'	26'	31'	36'
	7"	17'	22'	26'	30'	37'	42'
	8"	20'	25'	30'	35'	42'	48'
16mm	1½"	13'	17'	20'	23'	28'	32'
	2"	18'	22'	26'	31'	37'	42'
	2½"	22'	27'	33'	38'	46'	53'
	3"	26'	33'	40'	46'	55'	63'
	3½"	31'	38'	46'	54'	64'	74'
	4"	35'	44'	53'	61'	73'	84'
8mm	¾"	14'	18'	22'	25'	31'	35'
	⅞"	18'	22'	26'	31'	37'	42'
	1"	19'	24'	29'	34'	41'	46'
	1½"	29'	36'	44'	53'	61'	70'

FITTINGS. The projection room should have a sturdy table or cabinet of appropriate height upon which the projector may be placed for use. The space underneath can be used to house the projector when it is not in use. Space for storage of films, slide trays, extra lamps, and tools may be provided in the same place or in a separate but convenient area of the room. Storage for films should be relatively light-tight, enclosed, and dust-free, and the temperature should be as cool as possible. Where two or more slide projectors are used in conjunction with dissolve and audio equipment, separate storage space of ample dimensions should be provided in some handy location; a large rolling cabinet works well.

MECHANICAL CONVENIENCES. The room should have facilities for changing the air by forced ventilation at least six times per hour. The light switches for the room illumination should be located near the operator's table, or controlled by an electronic sound-operated or other remote-control switch. Alternatively, room lights may be wired through a dimmer, so that the lights can be dimmed and brought up slowly to prevent ocular shock. An electrical outlet of ample capacity for the projector(s) and any ancillary equipment should be located for convenience and in a manner that avoids having coils of cord underfoot.

PLANNING
THE FARMSTEAD

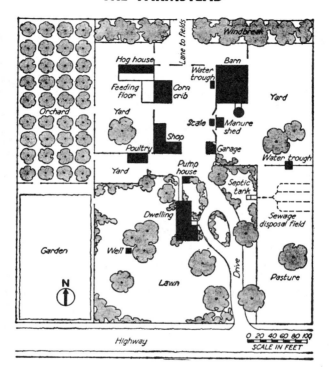

THE FARMSTEAD. Planning comprises the creation of a practical business establishment in combination with a home. The arrangement of buildings, yards, lots, and so forth, with relation to each other, to the fields, and to the highway should result in best efficiency for executing routine farmwork. The home must be attractive, fully habitable, and inspiring to its occupants. Pleasing architectural effects, tempered with economy in materials and construction, should be sought. The guideline dimensions listed in the following pages should be varied as required.

FARMSTEAD LOCATION. Easy access to fields is highly desirable. Proximity to the road eases travel to and from neighboring houses and towns, helps in daily errands and shipping, and promotes social intercourse and participation in community affairs—a source of considerable interest to most farm families. The directions in which the town, school, church, markets, and farm supply outlets lie will affect the farmstead location, especially for a large operation. Slope of land for good drainage, availability of public utilities, existence of a good water supply, direction of prevailing winds, and many other matters should be considered. See U.S. Department of Agriculture publications on farming and farmstead planning; see also some of the many good periodicals and books on the subject.

SPACE REQUIRED
FOR COW SHELTERS

Bull pens...................................10 x 10 to 12 x 12 or 100 to 150 sq. ft.
Calf pens, 4-6 calves...............8 x 10 to 10 x 10 or 75 to 100 sq. ft.
Maternity pens.........................10 x 10 or 100 sq. ft.
Box stalls, horse.....................10 x 10 to 10 x 12 or 100 to 120 sq. ft.
Brood sow pens........................8' x 8'
Sheep (ewe) pens....................4' x 4'
Pen barns..................................About 50 to 60 sq. ft. of floor space
    per cow.
Milking rooms...........................12' x 12' to 16' x 16' for each 4 milking
    stalls. (And 4 milking stalls for each
    20 cows.)
Milking parlors..........................Length 7'-6" per cow x 10'-0" wide.
Cow stanchions........................36", 42" and 48" widths.
Stanchions, heifer....................36" wide.
Stanchions, calf pens..............24" for calf pens.
Mangers....................................Bottom 1" above level of platform.
Manger widths..........................20, 24, 28 and 32 inches.
Toe hold....................................1" high and 16 to 18" from stanchion
    curb.
Horse stalls...............................5' wide by 9' to 10' long.
Bull yard...................................600 sq. ft. or more.
Stable heights..........................8'-0" to 8'-6".
Litter alleys..............................Wall alley 4'; drive alley 8'.
Cross alleys..............................3'-6" to 5' wide.
Feed Alleys..............................Wall alley 3'-6" to 4'-6"; center alley
    5' to 7'.
Gutters......................................16" wide x 8" deep on platform side.
Slope of litter alley.................¼" to the foot.
Slope of gutters or mangers...1" in 20'.
Hay and straw chutes.............4'-0" x 4'-0" to 4'-6" x 4'-6".
Doors for hay fork...................9'-0" to 10'-0" wide by 10'-0" to 12'-0"
    high.
Doors for hay slings................10'-0" to 12'-0" wide by 12'-0" to 15'-0"
    high.
Doors for mow floor drive......14' to 16' wide; 12' to 14' high.
Doors for straw carrier............5' wide x 10' high.
Doors for stock........................3'-6" to 4' wide x 7' high.
Doors for basement drive.......8' wide x 8' high.
Lighting......................................3 to 4 sq. ft. of glass per cow.
Ventilation.................................60 cu. ft. of air per cow per minute.

STANDARD STALL DIMENSIONS

		Length of Platform		
Breeds	*Width*	*Small*	*Medium*	*Large*
Holstein	3'-6" to 4'-0"	4'-10"	5'-2"	5'-8"
Shorthorn	3'-6" to 4'-0"	4'-8"	5'-0"	5'-6"
Ayrshire	3'-6" to 3'-8"	4'-6"	5'-0"	5'-6"
Guernsey	3'-4" to 3'-6"	4'-6"	4'-10"	5'-4"
Jersey	3'-4" to 3'-6"	4'-4"	4'-8"	5'-0"
Heifers	2'-9" to 3'-2"	3'-8"	3'-10"	4'-2"

DETAILS OF A MILK HOUSE

The milk house should contain the tank for cooling, racks for holding dairy utensils, and sufficient space for at least one worker. The plan shown is for a small, basic unit. If refrigeration, sterilization, and/or other equipment is to be installed, the building must be made at least 4' longer. Rigid insulation should have the highest practical R-value possible. A loading platform is a great convenience. An ample supply of water, both hot and cold, must be provided. Construction, materials, and interior detailing must conform to current local health and other regulations.

PLAN
SCALE ⅛" = 1'-0"

Cooling tank (Cover not shown)
Shelves: iron or cypress
6'-0"
13'-0"
9'-2"
Floor drain
1'-4"
4"
4"
1'-10" 8"
3'-2"
Sash door
10'-0"

SECTION A-A
SCALE ¼" = 1'-0"

Hip roof
Ventilating flue
1½" x 3½" rafters 3'-0" o.c.
1½" x 3½" joists, 3'-0" o.c.
¾" ceiling
¾" x 3½" brace
4'-0"
1'-0"
Damper
Studs 3'-0" o.c.
Guard rail
½" galv. pipe
¼" x 2" strap iron
Shelf for pails
T-hinge
3"∅ overflow pipe
Cover
Milk-cooling tank
8" channel iron
9'-0"
1'-3"
1'-4"
1'-6"
Shelf for cans
Concrete walls to extend 1'-6" above floor
4" concrete floor
3" concrete
2" thick waterproof insulation
1'-5½"
2'-6"

303

DETAILS OF AN
ICE HOUSE

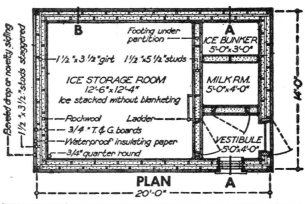

Beveled drop or novelty siding
1½" x 3½" studs staggered

B

Footing under partition

1½" x 3½" girt   1½" x 5½" studs

ICE STORAGE ROOM
12'-6" x 12'-4"
Ice stacked without blanketing

Rockwool   Ladder
3/4" T. & G. boards
Waterproof insulating paper
3/4" quarter round

A
ICE BUNKER
5'-0" x 3'-0"

MILK RM.
5'-0" x 4'-0"

VESTIBULE
5'-0" x 4'-0"

14'-0"

PLAN
20'-0"

Tons of Ice	Length	Width	Height	Tons of Ice	Length	Width	Height
10	10'-0"	7'-0"	7'-0"	30	14'-0"	10'-0"	10'-0"
20	14'-0"	8'-0"	8'-0"	40	18'-0"	10'-0"	10'-0"
25	14'-0"	10'-0"	8'-0"	50	16'-0"	12'-0"	12'-0"

INSIDE DIMENSIONS OF ICE HOUSES

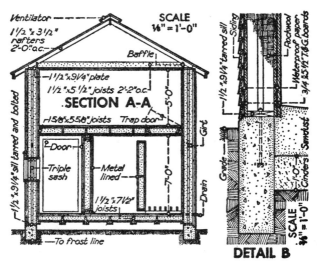

Ventilator
1½" x 3½" rafters 2'-0" o.c.

SCALE
¼" = 1'-0"

Baffle

1½" x 9¼" plate
1½" x 5½" joists 2'-2" o.c.

SECTION A-A

5/8" x 5⅝" joists   Trap door

1½" x 9¼" sill tarred and bolted

Door

Triple sash

Metal lined

1½" x 7½" joists

To frost line

Girt

5'-0"

7'-0"

Drain

1½" x 9¼" tarred sill   Siding

1½" x 9¼" tarred sill

Rockwool
Waterproof paper
3/4" x 5½" T. & G. boards

Grade

1'-0"   Cinders   Sawdust

SCALE
¾" = 1'-0"

DETAIL B

FARM SPRING
HOUSE

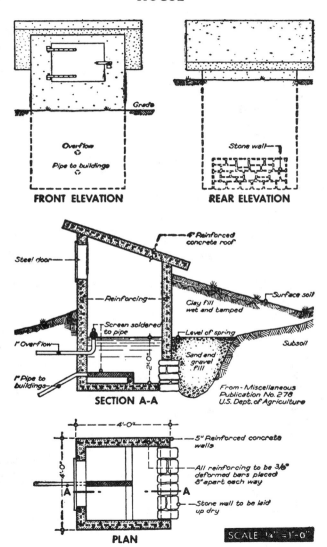

FRONT ELEVATION

Grade

Overflow

Pipe to buildings

REAR ELEVATION

Stone wall

4" Reinforced concrete roof

Steel door

Reinforcing

Surface soil

Clay fill wet and tamped

Level of spring

Screen soldered to pipe

1" Overflow

Subsoil

Sand and gravel fill

1" Pipe to buildings

SECTION A-A

From - Miscellaneous Publication No. 278 U.S. Dept. of Agriculture

4'-0"

4'-0"

5" Reinforced concrete walls

All reinforcing to be 3/8" deformed bars placed 8" apart each way

Stone wall to be laid up dry

PLAN

SCALE 1/4" = 1'-0"

305

BATHROOM PLANNING

POSITION OF BATHROOMS IN HOUSE PLAN. The placing of bathrooms over or adjacent to one another, and close by kitchen/laundry plumbing, results in the greatest economy. However, it is unwise to sacrifice convenience and utility for economy of piping. Usually good planning and plumbing economy are complementary.

It is best to locate bathrooms so that soil stacks do not run through partitions adjacent to rooms used for entertainment, since bathroom sounds may be heard. When this is not possible, piping should be surrounded by acoustic insulation and enclosed in a sound-wall, using studs wide enough that the pipe at no point touches the wall framing or covering.

SIZE AND SHAPE OF BATHROOM. Bathrooms larger than those ordinarily regarded as minimum are desirable, as they are disproportionately more convenient and workable. The care of children and/or invalids usually requires greater space than the minimum. Fully equipped luxury bathrooms often attain the size of a small bedroom. Small bathrooms are generally square or rectangular, but larger bathrooms are unrestricted as to design flow.

CHECKLIST OF EQUIPMENT

Lavatory	Hot tub/whirlpool spa
Toilet	Sauna
Bathtub	Separate tub compartment
Tub with shower over	Soaking tub
Separate shower compartment	Dressing table
Bidet	Manicure table
Separate toilet compartment	Double/triple basins
Shampoo basin	Exerciser
Sun lamp	Couch
Scale	Chair

BUILT-IN CONVENIENCES

Medicine cabinets	Full-length mirror
Medicine storage cabinet	General lighting fixture
Linen cabinet	Local lighting fixtures
Towel bars	Receptacles (GFCI protected)
Soap dishes	Exhaust fan
Paper holder	Auxiliary heater
Toothbrush/glass holder	Safety bars
Clothing hooks	Clothes hamper
Telephone	Stereo sound system

WINDOWS AND DOORS. Never locate a window over a bathtub, in a shower enclosure, or behind a toilet. Windows should be placed in the clear for convenient opening and closing. The stools of ordinary windows should be 4′ above the floor; sliding doors and full-height window units must be glazed with safety glass.

Usually a bathroom should have only one door; if it is to serve more than one bedroom, entrance should be from a hall. There are, however, exceptions, such as a bathroom opening onto a private terrace or patio.

TYPES OF FIXTURES. There are nine basic materials used in the manufacture of bathroom fixtures.
1. Porcelain
2. Vitreous china
3. Porcelainized cast iron
4. Enameled steel
5. Fiberglass
6. Stainless steel
7. Synthetic marble
8. Plastics
9. Pottery

BATHROOM HEATING. Radiation units are often placed under windows or along outside walls. Alternatively, radiant floor- or ceiling-panel heating may be used, or individual radiant panels may be placed nearly anywhere. The heating should be designed to take care of at least two air changes per hour and should be able to provide an inside dry-bulb temperature of 80°F.

TO DETERMINE FIXTURES
FROM FLOOR AREA

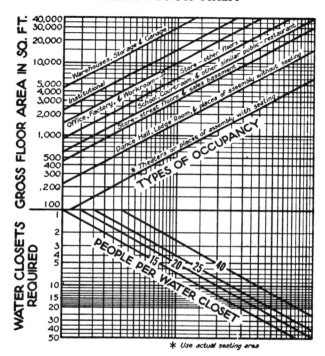

* Use actual seating area

First—read across from floor area to curve for type of building.

Second—at intersection read down to curve for number of persons per toilet as specified in the local code or as judged desirable. (15 persons per toilet represents generous conditions.)

Third—from this intersection read to the left for the number of water closets required.

Fourth—determine the probable sex proportion and divide the number of toilets found, in a suitable ratio, remembering that urinals in the men's toilet room augment the facilities offered by the water closets and makes relatively fewer water closets necessary for men than for women. For schools, allow 1 toilet to 25 girls, and 1 toilet to 40 boys.

Fifth—apportion urinals and lavatories as follows:

Type of Building	Urinals to 1 Closet	Lavatories to 1 Closet
Theaters	1½	½ to 1
Office Buildings	½ to 1	1 to 1½
Schools (boys' room)	1	1
Schools (girls' room)	—	½
Other Buildings	1	½ to 2

METAL
TOILET PARTITIONS

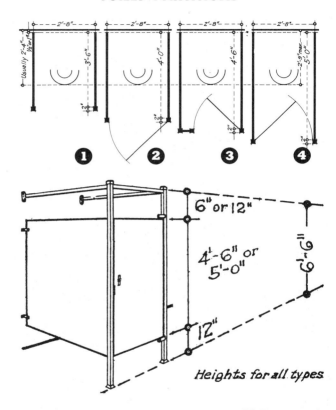

Heights for all types

The information on this page came originally from a U.S. Department of Commerce publication entitled *Simplified Practice Recommendation R101*, which is anything but. It is strongly suggested that the data required for any plans more definitive than rough sketches on the back of an old envelope be obtained from an informed representative of a manufacturer. Plans above show:

1. NO DOORS, OR THE INFORMAL TYPE
2. FULL-WIDTH DOORS SWINGING OUT, IN CASE OF FIRE
3. "L" FRONTS WITH NARROW DOORS SWINGING IN. Experience has shown that this type is fraught with the possibility of adventure, and these dimensions are recommended for use where special conditions do not make other dimensions preferable. "L" fronts are recommended for rigidity, permanence, space economy.
4. FULL-WIDTH DOORS SWINGING IN. Space wasting and completely disconcerting to a user entering or leaving; hence the type most used.

SIZES OF
SHOWER ENCLOSURES

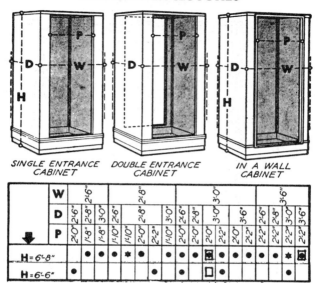

SINGLE ENTRANCE
CABINET

DOUBLE ENTRANCE
CABINET

IN A WALL
CABINET

	W	2'-6"	2'-6"	2'-6"	2'-8"	2'-8"	2'-8"	2'-8"	3'-0"	3'-0"	3'-0"	3'-0"	3'-6"	3'-6"	3'-6"	3'-6"
	D	2'-6"	2'-8"	3'-0"	2'-6"	2'-8"	2'-8"	3'-0"	2'-6"	2'-8"	3'-0"	3'-6"	2'-6"	2'-8"	3'-0"	3'-6"
⬇	P	2'-0"	1'-8"	1'-8"	1'-10"	1'-10"	2'-0"	2'-2"	1'-10"	2'-0"	2'-0"	2'-0"	2'-2"	2'-2"	2'-2"	2'-2"
H = 6'-8"			•	•	•	✱	•	•	•	•	■	•	•	•	•	✱	■
H = 6'-6"		•					•		•		□				•		

- *Manufactured standards*
- ✱ *Recommended*
- □ *Corner shower sizes*

The dimensions given above apply to factory-made types of job-constructed enclosures of tile or other material having a waterproof membrane or receptor. The following pages show typical details for adaptation to exposed or built-in installations. Specifics will vary depending upon the particular manufacturer of the materials and dictates of the job at hand. Notice that the 2'-6" size is an unusual one and really a bit small; 2'-8" should be considered a practical minimum.

Where space permits, the so-called combination shower consisting of valves and head over a bathtub, should be avoided because of:

1. Danger of slipping
2. Inadequate space for free movement
3. Discomfort from flapping curtain
4. Slopping of water around edges of shower curtain
5. Duplication of built-in wall accessories for standing and seated positions

The designer may choose between a curtain and a door glazed with glass or plastic; between factory-made enclosures in standard colors and selected colors in a job-constructed enclosure; between a dome light and no light (the switch should always be beyond reach of the bather's wet hands and grounded feet); between exposed and built-in construction; and from among various materials including enameled steel, fiberglass, and plastic.

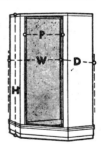

CORNER ENTRANCE
CABINET

SHOWER BATH CONSTRUCTION

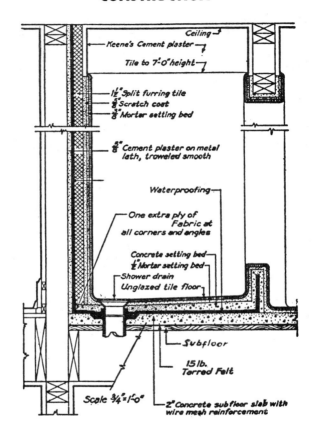

Ceiling

Keene's Cement plaster

Tile to 7'-0" height

1½" Split furring tile
⅜" Scratch coat
¾" Mortar setting bed

⅝" Cement plaster on metal lath, troweled smooth

Waterproofing

One extra ply of Fabric at all corners and angles

Concrete setting bed
½" Mortar setting bed
Shower drain
Unglazed tile floor

Subfloor

15 lb. Tarred Felt

Scale ¾"=1'-0"

2" Concrete subfloor slab with wire mesh reinforcement

Showers less than 36 x 36 in the clear should only be planned when space conditions make it mandatory. The door opening may be closed either with a water-proof curtain or any of the standard shower stall doors having ventilating panels.

The floor of the shower should be of an unglazed or abrasive tile, to prevent slipping. Glazed or unglazed tiles may be used for the walls at the discretion of the designer. Floor drains vary from 1½" to 3" wastes. The larger the better.

EXPOSED TYPE
SHOWER ENCLOSURES

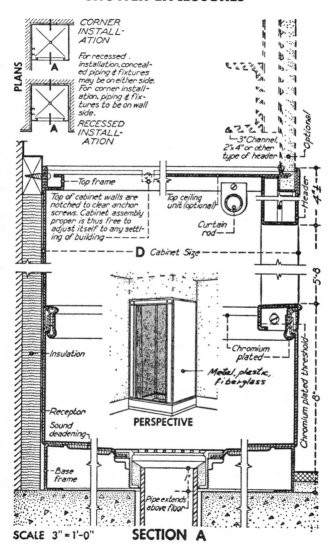

PLANS

CORNER INSTALL-ATION

For recessed installation, concealed piping & fixtures may be on either side. For corner installation, piping & fixtures to be on wall side.

RECESSED INSTALL-ATION

3"Channel, 2"x 4" or other type of header

Optional

Header

4'4"

Top frame

Top of cabinet walls are notched to clear anchor screws. Cabinet assembly proper is thus free to adjust itself to any settling of building

Top ceiling unit (optional)

Curtain rod

D Cabinet Size

5'-8"

Insulation

Chromium plated

Metal, plastic, fiberglass

PERSPECTIVE

Receptor

Sound deadening

Base frame

Chromium plated threshold

8"

Pipe extends above floor

SCALE 3" = 1'-0"   **SECTION A**

BUILT-IN TYPE
SHOWER ENCLOSURES

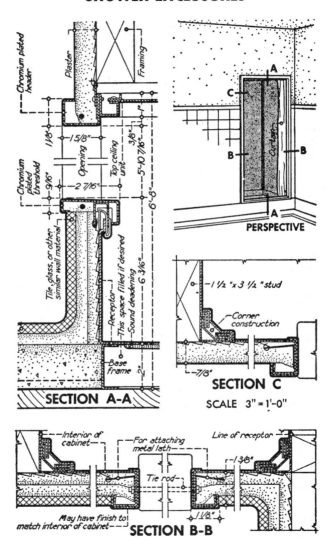

SECTION A-A

PERSPECTIVE

SECTION C

SCALE 3" = 1'-0"

SECTION B-B

RECEPTORS & CEILINGS FOR
SHOWER ENCLOSURES

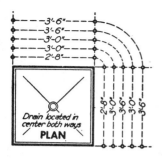

Drain located in
center both ways

PLAN

3'-6"
3'-6"
3'-0"
3'-0"
2'-8"

2'-8"
3'-0"
3'-6"
3'-0"
3'-6"

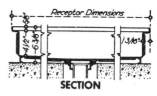

Receptor Dimensions

4 1/2"
6 3/8"

1 3/16"

SECTION

Top ceiling unit

Shower-lite

CEILING UNIT.
Should be used on all
models when built-in.

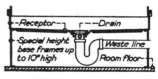

Receptor ————— Drain

Special height
base frames up
to 10" high

Waste line

Room floor

HIGH-BASE FRAME. Available to accommodate waste line or trap between receptor and room floor. This special frame is particularly useful in remodeling work.

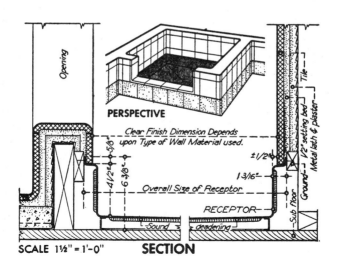

PERSPECTIVE

Opening

Clear Finish Dimension Depends
upon Type of Wall Material used.

Tile

± 1/2"

1/2" setting bed

Metal lath & plaster

Ground

5/8"
4 1/2"
6 3/8"

1 3/16"

Overall Size of Receptor

RECEPTOR

Sub floor

Sound — deadening

SCALE 1½" = 1'-0"   **SECTION**

WASHROOM PLANNING

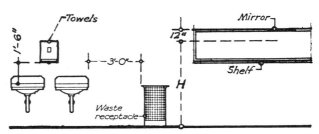

EFFICIENT PLAN-NING. Each fixture should be placed where it is handiest for the user and so that traffic moves rapidly at all times. This principle involves both the horizontal and vertical placement of equipment.

USED BY		H	H₁
MEN		5'-0"	2'-11"
WOMEN		4'-6"	2'-11"
CHILDREN	5 to 6 Years	3'-1"	2'-0"
	6 to 11 Years	3'-8"	2'-6"
	11 to 14 Years	4'-4"	2'-11"

Efficiency, however, can be carried to the point where (a) the person is inconvenienced, or (b) architectural dimensions become greater to attain the flow of traffic than a less "efficient" room would require. Lavatories spaced too close together will only be used alternately—half as many fixtures a few extra inches apart will take care of the same number of persons. No one who wants to wash his face should be forced to travel to another part of the room where the towels are located.

TOWEL CABINETS. There should be one towel cabinet above each alternate space between lavatories, located as low as possible so that water from wet hands does not run up the arm or sleeve.

LAVATORIES. The height of a lavatory rim should be ONE HALF of the users height. Invariably lavatories are placed TOO LOW in all types of buildings including residences. A six-foot man will find a 3-ft. height of rim most convenient, e.g.

The spacing of lavatories should be ONE HALF of the users height. A six-foot man will spread his elbows to about 2'-6" in washing so lavatories 3-ft. on center will allow clearance.

WASTE RECEPTACLE. Waste receptacle should not be located beneath the towel cabinet. Its position, as shown in the illustration, will lead the user away from the towel cabinet.

MIRRORS AND SHELVES. Mirrors over lavatories lead users to loiter in front of basins except where face-washing is essential to cleaning up as in some factories. Mirrors on towel cabinets create congestion and unnecessary use of more than 1 towel. In washrooms used by women, shelves need to be provided under mirrors for cosmetics, eyeglasses and handbags.

SOAP. Some states require the provision of soap. Liquid or powdered soap dispensers for correct types of soap to meet particular requirements are available from leading manufacturers.

OTHER EQUIPMENT. Consideration should be given to the necessity for and the placing of sand urns, cuspidors, hand lotion dispensers, sanitary napkin dispensers and receptacles, medicine or first-aid cabinets.

BUILT-IN TOWEL HOLDER RECESS

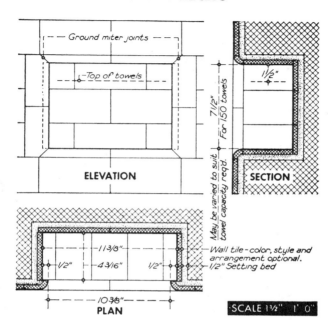

Where a recessed type holder is desirable for dispensing standard folded paper towels, it may be constructed as shown in the drawing above. Towels are retained in the wall recess by the ½″ lips projecting from the right and left sides. Replenishment of the supply is accomplished easily by adding towels to the pile to within 1½″ of the soffit of the niche.

Users simply remove towels as needed from the top of the pile. There is no restraint over the removal of towels, and this recessed type therefore will be most frequently found in buildings where a certain level of extravagance, or malicious mischief, is not a serious problem — such as clubs, homes, higher-class hotels and motels, and private office suites.

By following the dimensions shown in the drawings, details can be worked out for recessed holders in any type of wall finish, such as laminated plastic, vinyl-coated hardboard, sheet vinyl, marble, glass, or other interior wall materials. The recess shown is for 150-towel capacity. Greater or lesser capacity can be obtained by allowing about 1″ of towel height for every 25 towels.

The drawing showing the recess in tile makes use of 3″ x 6″ tile shapes. The only precaution that should be especially noted is that the miter joints at the four corners should be carefully ground from the standard tile stretcher cap member.

SIZES OF
STALL URINALS

GENERAL. Urinals are made of vitreous china, except the trough type which is made of enameled iron or steel, stainless steel, or high-impact plastic. The stall type is most commonly used; wall-hung urinals are second in popularity; the pedestal type third. In these pages, exclusive designs are not shown—the drawings and dimensions are representative of typical products. Specific manufacturers should be contacted for exact conformations and specifications.

SPACING. The usual spacing is 24″ o/c, which is entirely too little. Extended observation will reveal that, with the customary spacing, men will wait rather than crowd into a space between two urinals that are in use. Only alternate urinals will tend to be used when the spacing is less than about 30″ o/c. A 36″ o/c spacing may be regarded as ideal. It is believed that ten urinals on a 36″ spacing is much to be preferred to fifteen urinals on a 24″ spacing, under normal conditions and with the exception of toilet rooms in legitimate theaters where some inconvenience can be tolerated because of rush conditions between acts.

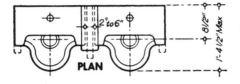

NOTE: The drawings shown are diagrammatic and do not represent the design of any manufacturer

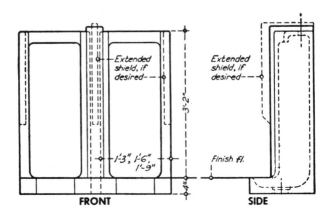

316

SIZES OF
WALL-HUNG URINALS

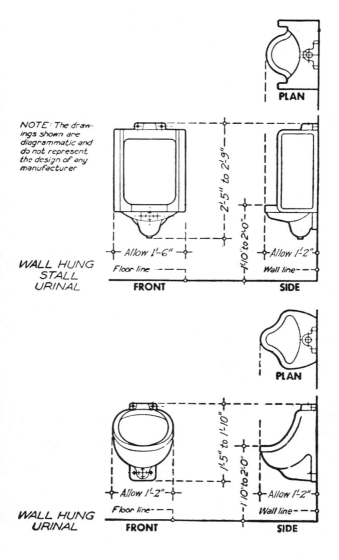

NOTE: The drawings shown are diagrammatic and do not represent the design of any manufacturer

PLAN

2'-5" to 2'-9"

1'-0" to 2'-0"

Allow 1'-6"

Allow 1'-2"

Floor line

Wall line

WALL HUNG STALL URINAL

FRONT

SIDE

PLAN

1'-5" to 1'-10"

1'-0" to 2'-0"

Allow 1'-2"

Allow 1'-2"

Floor line

Wall line

WALL HUNG URINAL

FRONT

SIDE

PEDESTAL AND TROUGH URINALS

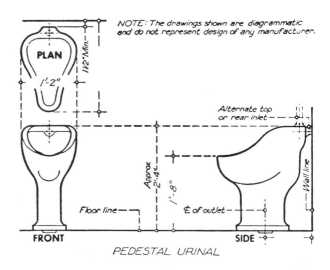

NOTE: The drawings shown are diagrammatic and do not represent design of any manufacturer.

PLAN

1'-2"

1½" Min.

FRONT

Alternate top or rear inlet

Approx 2'-4"

1'-8"

Floor line

℄ of outlet

Wall line

SIDE

PEDESTAL URINAL

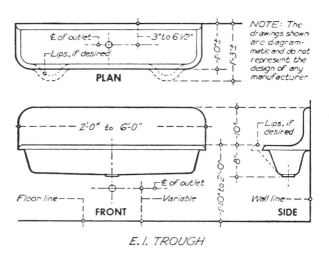

NOTE: The drawings shown are diagrammatic and do not represent the design of any manufacturer

℄ of outlet — 3" to 6½"

Lips, if desired

1'-0"±

1'-3"±

PLAN

2'-0" to 6'-0"

Lips, if desired

1'-0"

8"

℄ of outlet

Floor line

FRONT

Variable

1'-10" to 2'-0"

Wall line

SIDE

E. I. TROUGH

318

BATTERY
LAVATORIES

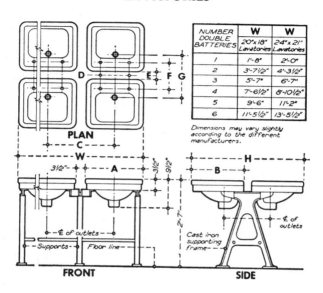

NUMBER DOUBLE BATTERIES	W 20"x18" Lavatories	W 24"x 21" Lavatories
1	1'-8"	2'-0"
2	3'-7½"	4'-3½"
3	5'-7"	6'-7"
4	7'-6½"	8'-10½"
5	9'-6"	11'-2"
6	11'-5½"	13'-5½"

Dimensions may vary slightly according to the different manufacturers.

PLAN

FRONT

SIDE

SIZE OF LAVATORY		DIMENSIONS									
A	B	C	D	E		F		G		H	
				Min.	Max.	Min.	Max.	Min.	Max.	Min.	Max.
20"	18"	1'-11½"	10"	2"	4"	8¼"	10¼"	1'-3¾"	1'-5¾"	3'-2"	3'-4"
24"	21"	2'-3½"	1'-0"	2"	4"	1'-0"	1'-2"	1'-7½"	1'-9½"	3'-8"	3'-10"

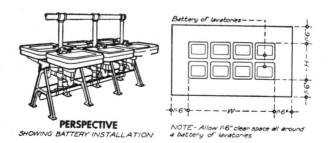

PERSPECTIVE
SHOWING BATTERY INSTALLATION

NOTE- Allow 1'-6" clear space all around a battery of lavatories

COLD WATER DISTRIBUTION

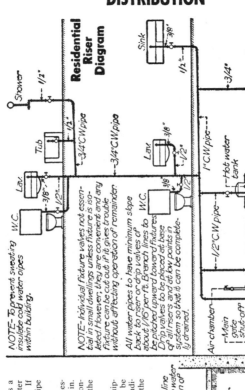

Residential Riser Diagram

SECTION

NOTE—To prevent sweating insulate cold water pipes within building.

NOTE—Individual fixture valves not essential in small dwellings unless fixture is isolated. However, they are convenient and any fixture can be cut out if it gives trouble without affecting operation of remainder.

All water pipes to have minimum slope back to riser or drip valves of about 1/16" per ft. Branch lines to be pitched upward toward fixtures. Drip valves to be placed at base of all risers and at low points in system so that it can be completely drained.

Shower — 1/2"
Tub — 1/2"
Lav. — 3/8" — 1/2"
W.C.
Sink — 3/8"
Lav. — 3/8" — 1/2"
W.C. — 3/8" — 1/2"
3/4" C.W. pipe
3/4" C.W. pipe
1/2"
Washer or Laundry tubs — 1/2"
3/4"
Drip valve
1" C.W. pipe
Hot water tank
1/2" C.W. pipe
Boiler
Air chamber
Main gate shut-off valve
Meter
Testing tee
Drip valves

Sidewalk
Service or curb box
2" Pipe sleeve
1" Service pipe (Brass, plastic, copper, galv. wrought iron or galv. steel pipe)
Gooseneck
Union
Corporation cock and tap
Street water main

Street

In North service pipe is laid below frost line from 4'-0" to 5'-0" and in South, to keep water cool, pipe is buried below heat penetration of sun (min. 3'-0").

For average one- or two-family dwellings a 3/4" service pipe will be satisfactory with water pressure about 40 lbs. per sq. in. or over. If water pressure is less, then the service pipe should be 1" if possible.

The most desirable water pressure for domestic service is between 30 and 50 lbs. per sq. in. The entire water supply system should be controlled by a main gate shut-off valve within the building.

Water hammer damages mechanical equipment and piping, so air chambers should be installed at the head of the line and/or at individual appliances and fixtures to prevent the problem.

320

INSIDE DIAMETERS OF PLUMBING PIPE

Nominal diameters in inches	Actual inside diameters in inches						
	Types of copper tubing			Steel [1]		Brass (I. P. S.)	
	K	L	M	Standard	Extra Strong	Standard	Extra Strong
3/8	0.40	0.43	0.45	0.49	0.42	0.49	0.42
1/2	.53	.55	.57	.62	.55	.63	.54
3/4	.75	.79	.81	.82	.74	.82	.74
1	1.00	1.03	1.06	1.05	.96	1.06	.95
1¼	1.25	1.27	1.29	1.38	1.28	1.37	1.27
1½	1.48	1.51	1.53	1.61	1.50	1.60	1.49
2	1.96	1.99	2.01	2.07	1.94	2.06	1.93
2½	2.44	2.47	2.50	2.47	2.32	2.50	2.32
3	2.91	2.95	2.98	3.07	2.90	3.06	2.89
3½	3.39	3.43	3.47	3.53	3.36	3.50	3.36
4	3.86	3.91	3.94	4.03	3.83	4.00	3.82
5	4.81	4.88	4.91	5.05	4.81	5.06	4.81
6	5.74	5.85	5.88	6.07	5.76	6.13	5.75

[1] For the most part, wrought-iron pipe corresponds in size to the corresponding weights of steel pipe, differing by not more than 0.01 inch in diameter for nominal diameters of 2½ inches and smaller and by not more than 0.02 inch for nominal diameters of from 3 to 6 inches.

CAST IRON
SOIL PIPE

Cast iron soil pipe and fittings suitable for installation and service in drainage, waste, vent, and sewer lines are manufactured of gray cast iron in three types: single-hub and spigot, double-hub, and hubless. The single- and double-hub types comprise one distinct system, the hubless another. Two different weights are available in the single- and double-hub types: service and extra-heavy. The extra-heavy weight is slowly being phased out, and now accounts for only about 3% of cast iron soil pipe installations. A complete line of fittings is available for all three types. Local building codes usually specify which of the pipe types and weights are permissible.

Cast iron soil pipe and fittings are furnished with a suitable protective coating that is nonadherent, not brittle, and free from any tendency to scale. The exact nature of the coating varies in different parts of the country, depending upon prevailing building codes. The purposes of the coating are to provide a smooth surface and to protect the iron from chemical action.

Dimensions of Hubs, Spigots, and Barrels for Extra-Heavy and Service Cast Iron Soil Pipe and Fittings, in.

NOTE—1 in. = 25.4 mm; 1 ft = 0.3 m throughout Tables.

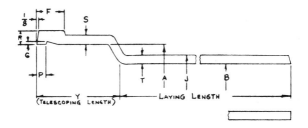

CAST IRON SOIL PIPE

Extra-Heavy Cast Iron Soil Pipe and Fittings

Size[A] Availability	Inside Diameter of Hub A	Outside Diameter of Barrel J	Telescoping length Y	Inside Diameter of Barrel B	Thickness of Barrel T	
					Nom	Min
2*	3.06	2.38	2.50	2.00	0.19	0.16
3*	4.19	3.50	2.75	3.00	0.25	0.22
4*	5.19	4.50	3.00	4.00	0.25	0.22
5*	6.19	5.50	3.00	5.00	0.25	0.22
6*	7.19	6.50	3.00	6.00	0.25	0.22
8*	9.50	8.62	3.50	8.00	0.31	0.25
10*	11.62	10.75	3.50	10.00	0.37	0.31
12*	13.75	12.75	4.25	12.00	0.37	0.31
15*	16.95	15.88	4.25	15.00	0.44	0.38

Size[A]	Thickness of Hub		Width of Hub Bead[B] F	Distance from Lead Groove to End Pipe and Fittings[B] P	Depth of Lead Groove	
	Hub Body S (min)	Over Bead R (min)			G (min)	G (max)
2	0.18	0.17	0.75	0.22	0.10	0.19
3	0.25	0.43	0.81	0.22	0.10	0.19
4	0.25	0.43	0.88	0.22	0.10	0.19
5	0.25	0.43	0.88	0.22	0.10	0.19
6	0.25	0.43	0.88	0.22	0.10	0.19
8	0.34	0.59	1.19	0.38	0.15	0.19
10	0.40	0.65	1.19	0.38	0.15	0.22
12	0.40	0.65	1.44	0.47	0.15	0.22
15	0.46	0.71	1.44	0.47	0.15	0.22

[A] Nominal inside diameter.
[B] Hub ends and spigot ends can be made with or without draft.

CAST IRON SOIL PIPE

Service Cast Iron Soil Pipe:

Size[A] Availability	Inside Diameter of Hub[a] A	Outside Diameter of Barrel[b] J	Telescoping length[a] Y	Inside Diameter of Barrel[a] B	Thickness of Barrel[a] T	
					Nom	Min
2	2.94	2.30	2.50	1.96	0.17	0.14
3	3.94	3.30	2.75	2.96	0.17	0.14
4	4.94	4.30	3.00	3.94	0.18	0.15
5	5.94	5.30	3.00	4.94	0.18	0.15
6	6.94	6.30	3.00	5.94	0.18	0.15
8	9.25	8.38	3.50	7.94	0.23	0.17
10	11.38	10.50	3.50	9.94	0.28	0.22
12	13.50	12.50	4.24	11.94	0.28	0.22
15	16.95	15.88	4.25	15.16	0.36	0.30

Size[A]	Thickness of Hub		Width of Hub Bead[B] F	Distance from Lead Groove to End, Pipe and Fitting[B] P	Depth of Lead Groove	
	Hub Body S (min)	Over Bead R (min)			G (min)	G (max)
2	0.13	0.34	0.75	0.22	0.10	0.19
3	0.16	0.37	0.81	0.22	0.10	0.19
4	0.16	0.37	0.88	0.22	0.10	0.19
5	0.16	0.37	0.88	0.22	0.10	0.19
6	0.18	0.37	0.88	0.22	0.10	0.19
8	0.19	0.44	1.19	0.38	0.15	0.19
10	0.27	0.53	1.19	0.38	0.15	0.22
12	0.27	0.53	1.44	0.47	0.15	0.22
15	0.30	0.58	1.44	0.47	0.15	0.22

[A] Nominal inside diameter.
[B] Hub ends and spigot ends can be made with or without draft.

CAST IRON SOIL PIPE

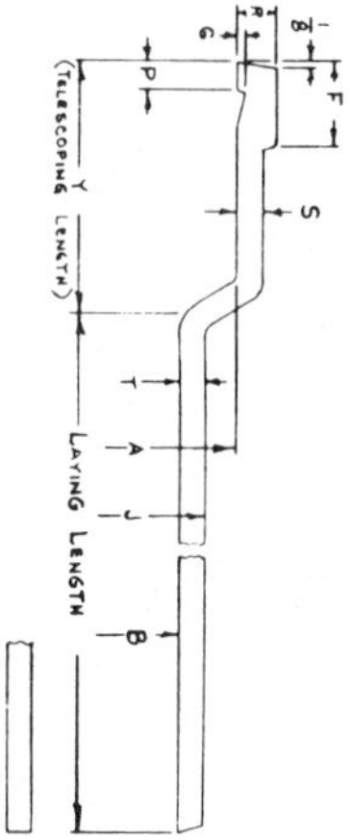

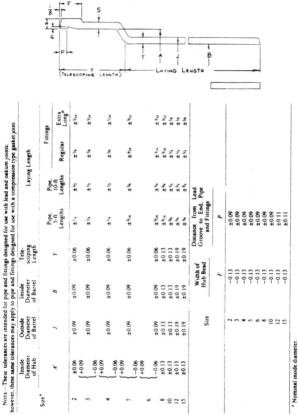

NOTE—These tolerances are intended for pipe and fittings designed for use with lead and oakum joints; however, these same tolerances may apply to pipe and fittings designed for use with a compression type gasket joint.

Size[A]	Inside Diameter of Hub	Outside Diameter of Barrel	Inside Diameter of Barrel	Telescoping Length	Laying Length			
					Pipe, 5-ft Lengths	Pipe, 10-ft Lengths	Fittings: Regular	Fittings: Extra Long[B]
	A′	J	B	Y				
2	±0.06	±0.09	±0.09	±0.06	±1/4	±1/2	±1/8	±1/16
3	+0.09 / −0.06	±0.09	±0.09	±0.06	±1/4	±1/2	±1/8	±1/16
4	+0.09 / −0.06	±0.09	±0.09	±0.06	±1/4	±1/2	±1/8	±1/16
5	+0.09 / −0.06	±0.09	±0.09	±0.06	±5/16	±5/8	±3/16	±3/32
6	+0.09 / −0.06	±0.09	±0.09	±0.06	±5/16	±5/8	±3/16	±3/32
8	±0.13	±0.13	±0.13	±0.13	±5/16	±5/8	±3/16	±3/32
10	±0.13	±0.13	±0.13	±0.13	±3/8	±3/4	±1/4	±1/8
12	±0.13	±0.19	±0.19	±0.19	±3/8	±3/4	±1/4	±1/8
15	±0.13	±0.19	±0.19	±0.19	±3/8	±3/4	±1/4	±1/8

Size	Width of Hub Bead	Distance from Lead Groove to End, Pipe and Fittings
	F	P
2	−0.13	±0.09
3	−0.13	±0.09
4	−0.13	±0.09
5	−0.13	±0.09
6	−0.13	±0.09
8	−0.13	±0.09
10	−0.13	±0.09
12	−0.13	±0.11
15	−0.13	±0.11

[A] Nominal inside diameter.
[B] These tolerances apply to each foot of extra-long fittings in excess of regular laying lengths specified herein.

CAST IRON
SOIL PIPE

RUBBER SLEEVE

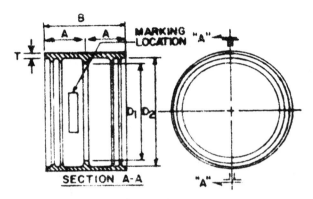

Dimensions for sizes 1 1/2" – 4"

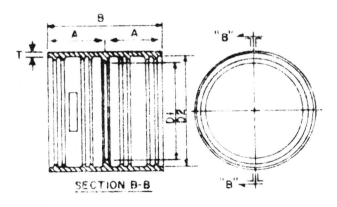

Dimensions for sizes 5", 6", 8" & 10"

CAST IRON SOIL PIPE

Dimensions in Inches

Dimensional Tolerances to be RMA Class 3

	1½"	2"	3"	4"	5"	6"	8"	10"
A	1.062	1.062	1.062	1.062	1.500	1.500	2.000	2.000
B	2.125	2.125	2.125	2.125	3.000	3.000	4.000	4.000
D1	1.531	1.968	2.968	4.000	4.968	5.968	7.968	9.975
D2	1.937	2.343	3.343	4.406	5.343	6.343	8.343	10.350
T	0.094	0.094	0.094	0.094	0.094	0.094	0.094	0.094

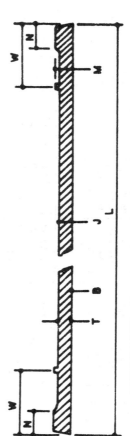

DIMENSIONS AND TOLERANCES (IN INCHES) OF SPIGOTS AND BARRELS FOR HUBLESS PIPE AND FITTINGS

CAST IRON SOIL PIPE

Size	Inside Diameter Barrel B	Outside Diameter Barrel J	Outside Diameter Spigot M	Width Spigot Bead N (± .13)	Thickness of Barrel		Gasket Positioning Lug W	Laying Length L[1]	
					T-Nom.	T-Min.		5 Foot (± .25)	10 Foot (± .50)
1½	1.50 ± .09	1.90 + .06 − .05	1.96 ± .06	.25	.16	.13	1.13	60	120
2	2.00 ± .06	2.35 + .09 − .05	2.41 ± .09	.25	.16	.13	1.13	60	120
3	3.00 ± .06	3.35 + .09 − .05	3.41 ± .09	.25	.16	.13	1.13	60	120
4	4.00 ± .06	4.38 + .09 − .05	4.44 ± .09	.31	.19	.15	1.13	60	120
5	4.94 ± .09	5.30 + .09 − .05	5.36 ± .09	.31	.19	.15	1.50	60	120
6	5.94 ± .09	6.30 + .09 − .05	6.36 ± .09	.31	.19	.15	1.50	60	120
8	7.94 ± .13	8.38 + .13 − .09	8.44 ± .09	.31	.23	.17	2.00	60	120
10[2]	10.00 ± .13	10.56 ± .09	10.62 ± .09	.31	.28	.22	2.00	60	120

[1] Laying lengths are for pipe only, and such pipe may be either 5'-0" long or 10'-0" long.
[2] O.D. of Barrel may have an out of round tolerance of .04.

CISTERNS
AND WELLS

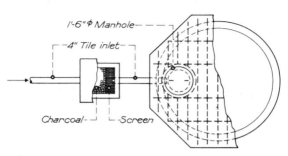

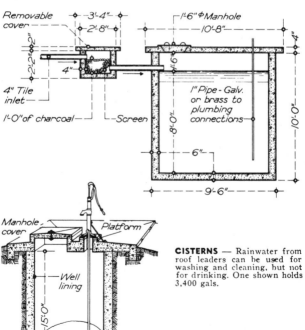

CISTERNS — Rainwater from roof leaders can be used for washing and cleaning, but not for drinking. One shown holds 3,400 gals.

WELLS — A natural spring should be protected by a well "curb" or lining to exclude seepage and burrowing animals. Diameter usually about 6 ft.

DETERMINING
WATER DEMAND

Selecting a water system that will operate satisfactorily must be based on certain variable factors. WATER DEMAND is the amount of water to be supplied. TYPE OF SOURCE may be *shallow*—if from a cistern, a lake, a stream, or a dug or driven well where the maximum lift from working water level to the pump inlet (*maximum suction lift*) does not exceed 20'-25' (depending upon exact pump specifications)—or *deep*, if the lift exceeds 20'-25'. YIELD of the well is the maximum potential flow of the well in gallons per hour (gph) or gallons per minute (gpm).

MAXIMUM WATER DEMAND occurs when all water-using devices flow at one time. Since this seldom occurs, it is ordinarily an impractical design basis, and short-cuts are commonly used.

1. RULE OF THUMB, based on accumulated experience, is embodied in Table 1. Requirements are approximate; in every case they should be checked to make certain that no unusually heavy demands will occur.

2. RATE OF FLOW method, more satisfactory when unusual conditions exist, is based on requirements of fixtures, livestock, and so forth, listed in Table 2. Minima shown are absolute; use of average requirements is preferred for calculations in order to maintain reserve water for fire protection and other abnormal demands. EXAMPLE: for residences, assume fixtures having greatest demand in every bath, kitchen, and laundry room are flowing concurrently. For 2 baths and kitchen:

$$[2 \times 200 \text{ (shower)}] + [5 \text{ (kitchen sink)}]$$
$$= 405 \text{ gph}$$

For farms, large herds of stock may determine maximum demand. However, peak loads occur at determinable intervals, and the supply can be replenished comparatively slowly. For a farm that must supply 10 cows in milk, 2 horses, 400 fowl, 2 hogs, and 5 people:

$$(10 \times 35) + (2 \times 10) + (4 \times 2) + (2 \times 2.5)$$
$$+ (5 \times 35) = 558 \text{ gph}$$

NOTE: To the above figures must be added special requirements such as water for heating system, irrigation, and swimming pool. Base calculations on water consumption in summer, and allow for dry spells or drought. Each water system is unique; installations vary accordingly.

TABLE 1 — WATER DEMAND BY RULE OF THUMB

TYPE OF INSTALLATION	PUMP CAPACITY (gph)*
Small home, cottage, small service sta'n	200 to 250
Large homes, more than 1 lavatory or bath	300 to 375
Average farm (normal quantity of stock)	300 to 400

*To accommodate normal demand

TABLE 2 — WATER DEMAND BY RATE OF FLOW

USE OR FIXTURE For Each:	DEMAND IN GAL. PER HOUR (H) OR PER DAY (D)	
	Min.	Average
Person (1)	25 D	35 to 50 D (100 D Max.)
Horse	5 D	10 D
Cow, dry	7 D	12 D
Cow, in milk	25 D	35 D
Hog	1 D	2 to 2.5 D
Sheep	1 D	2 D
100 Chickens		2 D
½" Hose Nozzle (2)	75 H	200 H
¾" Hose Nozzle	100 H	275 to 300 H
Lawn Sprinkler	125 H	200 H
Shower, hourly rate	100 H	200 H
Shower, per bath		30
Bath, Tub (1 filling)		30
Lavatory (1 filling)		1.5
Toilet (1 filling)		6
Sink, laundry (1 filling)		10
Sink, kitchen		75 D

Notes: 1. When using quantity demanded per person per day, it is not necessary to add individual fixtures listed below; these are for use when unusual conditions exist. 2. Eight gallons will sprinkle 100 sq. ft. of lawn; 16-20 gal. will soak it.

LOCATING THE WELL is not a precise matter, nor can its rate of flow (if indeed there is any) be predetermined. Advice should be obtained from reputable local well drillers, owners of adjacent wells, and appropriate local and state government agencies. If usual information fails, do not underestimate the powers of local "water-witchers" and dowsers of good repute. Unknown geological formations, drought, and number of adjacent wells affect the depth at which water may be found. Avoid locations close to barns, sewage disposal fields, and so forth; if possible, check rock formations to eliminate chances of penetrating strata that might carry polluted surface water into the well. Higher elevations are usually preferable. It is advisable, and often mandatory, to submit water samples to competent authorities to determine the water's purity. Well permits and/or testing (often periodic), as well as filtration and/or purification equipment may be required.

TYPES OF WATER
SOURCES AND SYSTEMS

SHALLOW WELLS (dug or driven wells, cisterns, lakes, streams, and so forth) are those in which the *working level* of water does not descend more than 20'-25' (depending upon the specific pumping system used) below pump inlet level when the pump operates—for example, during droughts. For these, *shallow well pumps* are available in two types: (1) *piston* and (2) *jet* or *ejector*. Both types operate by creating a vacuum in the well piping and thus enabling atmospheric pressure to raise water to pump level.

DEEP WELLS (usually drilled and cased, sometimes dug or driven) are those in which the working level of water is more than 20'-25' (depending upon the pumping system used) below pump inlet level. *Deep well pumps* are of two types: (1) *jet* or *ejector* and (2) *deep well submersible*. If the working level is within or close to the 20'-25' range, it is preferable to use a deep well pump.

SELECTING THE SYSTEM. 1. *The pump* would ideally have sufficient capacity, in gals. per hr., to enable it to pump a day's water demand in 1 hr. However, the yield of the well (rate of flow in gph) must be at least equal the pump capacity. If the yield is limited, a smaller pump and larger tank may be more satisfactory. *Peak water demand* (maximum demand occurring at any one period) may be used as the design basis.

2. *Tank size* should be sufficient to prevent too frequent starting and stopping of the pump, to avoid undue wear and excess current consumption, and to maintain adequate water reserves. For average homes, *available* tank capacity, shown diagrammatically, should be at least 5 to 10 gals. Average *nominal* tank capacity is 12.5% to 25% of the hourly pumping capacity. *Waterlogging* occurs when tank air is absorbed by water under pressure. To overcome this, an air control may be specified or a precharged or captured-air storage system may be used.

Storage tanks are available in a wide range of sizes; those normally found in residential service run from 2 to 120 gals. or more. Both vertical and horizontal tanks may be obtained in numerous configurations, either plain or containing a bladder for air-control purposes. Plain tanks may be epoxy-coated on the interior, or glass-lined. Dimensions and specifications should be obtained from individual manufacturers.

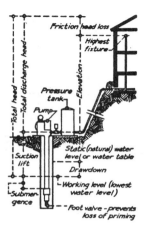

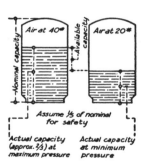

DEEP WELL DOUBLE PIPE JET PUMPS

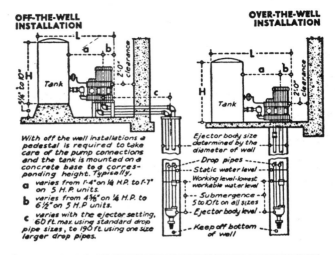

OFF-THE-WELL INSTALLATION

OVER-THE-WELL INSTALLATION

With off the well installations a pedestal is required to take care of the pump connections and the tank is mounted on a concrete base to a corresponding height. Typically,

a varies from 1-4" on 1/6 H.P. to 1-7" on 5 H.P. units.

b varies from 4⅝" on 1/6 H.P. to 6½" on 5 H.P. units.

c varies with the ejector setting, 60 ft. max. using standard drop pipe sizes, to 190 ft. using one size larger drop pipes.

Ejector body size determined by the diameter of well

Drop pipes
Static water level
Working level-lowest workable water level
Submergence 5 to 10 ft. on all sizes
Ejector body level
Keep off bottom of well

MAXIMUM CAPACITIES (gals. per hr.) at VARIOUS DEPTHS														
Maximum Depth* of Well										Tank Size in Gals.	Min. Well Dia.	Pump & Tank Space Required	
20'	40'	60'	70'	90'	110'	120'	130'	150'			L	W	H
480	295	185	....	....	....	....	....	....	18	4"	2'-9"x	1'-0"x	3'-2"
1110	780	410	350	240	....	....	....	....	42	4"	3'-1"x	1'-4"x	4'-2"
1110	810	500	440	320	....	....	....	....	82	4"	3'-7"x	1'-8"x	5'-2"
2820	2320	1980	1830	1320	960	870	760	540	220	6"	4'-7"x	2'-6"x	6'-5"

* From pump base to working water level (lowest level of water at which pump will produce flow).

The drawings and the figures shown in the table above are typical of the pump installations that are commonly made in new work and that satisfy most average conditions. Exact specifications, however, vary considerably.

DEEP WELL JET PUMPS. Used for water lift from 20'-25' to 100' or more in single-stage models, as measured from lowest anticipated water level in the casing to the pump. Multistage pumps are available for depths of 350' and more. Special pumps are available for greater depths and high-volume purposes, such as irrigation. Full particulars can be obtained from pump manufacturers.

Because there are no working parts in the well, the pump may be located away from the actual well. The pump works on the "suction" principle, but ejector or jet equipment increases lift. Installing the pump in a basement or pumphouse prevents freezing and allows easy access for maintenance and repairs. It may be installed in a well that is out of plumb. The only moving part is the impeller mounted on the motor shaft. Lack of gears, cylinders, springs, and leathers promotes quiet operation. With a pressure tank and air volume control, or a captive-air arrangement, jet systems are completely automatic.

DEEP WELL
SUBMERSIBLE PUMPS

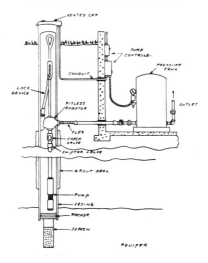

SUBMERSIBLE PUMPS. Submersible pumps are a type of centrifugal pump installed directly within drilled wells, typically of depths ranging from 20 to 550 feet, and often deeper. The centrifugal type of pump contains rotating impellers mounted on a shaft turned by the motor. The impellers increase the velocity of the water and force it into the surrounding casing, where the flow is slowed and converted to pressure. Each impeller is called a stage. As the required water-system operating pressure and/or the height that the water must be raised from the source's surface level increases, the number of stages is increased. As water passes through each stage successively, the pressure increases.

The submersible pump consists of a series of impellers and diffusers within a casing containing a screened inlet (liquid end) closely coupled to a special sealed, waterproof electric motor in its own casing (motor end). The whole constitutes a long cylinder with a drop-pipe connection at one end. The pump is suspended from the drop pipe (often in conjunction with a safety cable) within the well casing, below the maximum drawdown level, and is supported by the well cap or by the locking device of a pitless adapter. The electrical cable to the pump is likewise suspended in the well casing and must be of a special waterproof pump-cable type. The entire unit must be properly installed and grounded to minimize the possibility of short circuits and damage to the unit. Lightning protection should be provided.

Submersible pumps will deliver water across a wide range of volumes and pressures, limited only by the size and horsepower of the unit. Those for residential and general-purpose applications typically range from ½ to 5 horsepower, from 6 to 20 stages, and from 5 to 20 gallons per minute pumping capacity. Delivery of 1,500 gallons per hour or more from relatively shallow depths is easily possible.

Submersible pumps provide an even, smooth flow of pumped water under constant pressure. Installations are easily frost-proofed, straightness of the well casing is not critical, installation is relatively simple, and operation is silent. On the other hand, repair means withdrawing the pump from the well (provisions must be made for this). In addition, complete sealing of all electrical equipment is essential, and there is a definite susceptibility to excessive wear if sand or other abrasives are present in the water. In situations where the water is heavily mineralized, a three-wire pump is preferable to a two-wire pump, since the higher starting torque of the former is capable of breaking free any mineral deposits that may form on moving parts of the pump.

SHALLOW WELL SYSTEMS

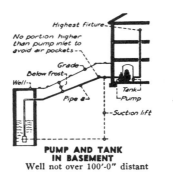

PUMP AND TANK IN BASEMENT
Well not over 100'-0" distant

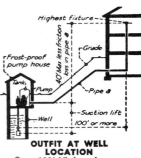

OUTFIT AT WELL LOCATION
Over 100'-0" from house

Shallow well pumps operate by lowering air pressure in the well piping and allowing atmospheric pressure at water level to force water up to the pump. The practical maximum vertical lift from working level to pump inlet is 22'-0" at sea level. As altitude increases, maximum lift decreases as shown in table below.

ALTITUDE, FT.	Sea	1000	2500	4000	5000	6500	8000
MAX. PRACTICAL SUCTION LIFT, FT.	22	21	20	18	17	16	15

Friction loss occurs in all pipe runs and, by reducing the head, may materially affect choice of type of system and of location of units. Amount of loss depends on: (1) pipe size; (2) length of run; (3) water pressure. Examples: In 100'-0" of ¾"-pipe the head is reduced 1'-11" at 2-lb. pressure, and 136'-0" at 20-lb. For 2" pipe, the head is reduced 6" per 100'-0" at 10-lb., and 6'-7" at 40-lb. Sharp elbows increase friction loss. If friction loss is unavoidable, any of several means of overcoming it may be used, depending on local conditions. Pipe sizes or pump size may be increased, or a high-pressure pump may be installed.

The casing is the lining of a driven well, or a supplementary lining in a dug well, which houses well piping. In old wells, its diameter may limit pump size. In new wells, 4" casings (minimum) are advisable; 6" diameter is preferred. Well seal is a sanitary ground level cap, required in some states.

PISTON PUMP. On its forward stroke, the piston creates a vacuum that draws in water; the backstroke forces accumulated water into a discharge chamber. Most are double-acting: a single stroke forces water out of one chamber while drawing fresh water into the other, thereby producing a more constant flow. Piston pumps deliver rated capacities at any stage of lift (0'-0" to 22'-0" approx.), at normal pressures (20 to 40 lb.). Piston pumps are slightly larger and noisier than jet (ejector) pumps.

JET OR EJECTOR PUMP. This pump is centrifugal: a motor-driven impeller scoops up water and forces it outward into discharge lines, thus creating a vacuum, which draws in more water. In addition, jet pumps have built into them a venturi, or device for increasing their capacity. Jet pumps for shallow well applications are generally designed for a maximum suction lift of 20' to 25' with a single 1¼" suction line, and they deliver normal pressures of up to 60 lbs.

HOT WATER STORAGE TANKS

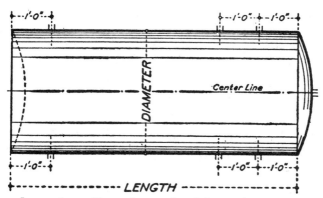

In accordance with unanimous action of 2 general conferences of manufacturers, distributors and users of hot water storage tanks, the U. S. Department of Commerce recommends that simplified dimensions and capacities of hot water storage tanks be established as follows: The tanks to be made in 2 working pressures; 65 pounds classified as *standard*, 100 pounds classified as *extra heavy*. Each tank is to be stenciled with its classification, working pressure, and name and address of its manufacturer. There are 6 tappings in each tank, placed as shown in the diagram above. 11x15-inch manholes may be placed either in the shell or the convex end. 4x6-inch hand holes may be located as desired. The tanks are interchangeable for either horizontal or vertical installation.

STANDARD TANK SIZES

Diameter	Length	Gallons	Diameter	Length	Gallons
20″	5 ft.	82	42″	7 ft.	504
24″	5 ft.	118	42″	8 ft.	576
24″	6 ft.	141	42″	10 ft.	720
30″	6 ft.	220	42″	14 ft.	1,008
30″	8 ft.	294	48″	10 ft.	940
36″	6 ft.	318	48″	16 ft.	1,504
36″	8 ft.	423	48″	20 ft.	1,880

MINIMUM SIZE HEATING COILS

Tank dimensions		Size of pipe	Minimum length of heating coil	Tank dimensions		Size of pipe	Minimum length of heating coil
Diameter	Length			Diameter	Length		
20″	5 ft.	1″	14 ft.	42″	7 ft.	1½″	22 ft.
24″	5 ft.	1¼″	14 ft.	42″	8 ft.	1½″	26 ft.
24″	6 ft.	1¼″	18 ft.	42″	10 ft.	1½″	34 ft.
30″	6 ft.	1¼″	18 ft.	42″	14 ft.	1½″	50 ft.
30″	8 ft.	1¼″	26 ft.	48″	10 ft.	2″	34 ft.
36″	6 ft.	1½″	18 ft.	48″	16 ft.	2″	58 ft.
36″	8 ft.	1½″	26 ft.	48″	20 ft.	2″	74 ft.

335

CAPACITY OF TANKS
IN GALLONS

CYLINDRICAL TANKS

Depth or Length	DIAMETER							
	18-in.	24-in.	30-in.	36-in.	42-in.	48-in.	54-in.	60-in.
1 Inch	1.10	1.96	3.06	4.41	5.99	7.83	9.91	12.24
1 ft.	13.	23.	37.	53.	72.	94.	119.	147.
1½ ft.	20.	35.	55.	79.	108.	141.	179.	220.
2 ft.	26.	47.	73.	106.	144.	188.	238.	294.
2½ ft.	33.	59.	92.	132.	180.	235.	298.	367.
3 ft.	40.	71.	110.	159.	216.	282.	357.	441.
3½ ft.	46.	82.	129.	185.	252.	329.	417.	514.
4 ft.	53.	94.	147.	211.	288.	376.	476.	587.
4½ ft.	59.	106.	165.	238.	324.	423.	536.	661.
5 ft.	66.	117.	183.	264.	360.	470.	595.	734.
5½ ft.	73.	129.	202.	291.	396.	517.	657.	808.
6 ft.	79.	141.	221.	317.	432.	564.	714,	881.
7 ft.	92.	164.	257.	370.	504.	658.	833.	1028.
8 ft.	106.	188.	294.	424.	576.	755.	952.	1175.
9 ft.	119.	212.	330.	476.	644.	846.	1071.	1322.
10 ft.	132.	235.	372.	530.	720.	940.	1190.	1475.
12 ft.	157.	282.	440.	634.	864.	1128.	1428.	1755.
14 ft.	185.	329.	514.	740.	1000.	1316.	1665.	2056.
16 ft.	211.	376.	587.	846.	1152.	1500.	1904.	2350.
18 ft.	238.	423.	661.	952.	1296.	1692.	2142.	2644.
20 ft.	264.	470.	734.	1057.	1440.	1880.	2430.	2940.

To find how many U. S. gallons a cylindrical tank will hold: Multiply the square of the inside diameter by 0.7854, which gives the area; multiply that result by the depth and this gives the cubic contents of the tank. If measurements are in inches, divide the cubic contents by 1728 and you then have contents expressed in cubic feet; then multiply by 7.4805 (U. S. gallons in each cubic foot of water) and the final result is the number of U. S. gallons the tank will contain.

RECTANGULAR TANKS

To find how many U. S. gallons any rectangular tank will hold: Multiply the inside length, depth and width, which gives the contents of the tank in cubic inches, or in cubic feet, as case may be. If in inches, divide by 1728 and you have the contents in cubic feet. Then multiply that result by 7.4805 (U. S. gallons in each cubic foot of water) and the final result is the number of U. S. gallons the tank will contain.

A gallon of water (U. S. standard) weighs 8 1/3 lbs. and contains 231 cubic inches.

A cubic foot of water contains 7 1/2 gallons, 1728 cubic inches, and weighs 62 1/2 lbs.

336

DESCRIPTION OF
SUMP PUMPS

GENERAL CHARACTERISTICS.

Sump pumps can be divided into two general types: upright, with the motor supported above water level on a column attached to a pedestal, as in the drawing; and compact and fully submersible, with the motor attached directly to the pedestal. They are classed as continuous-duty or intermittent-duty, and they may be intended for heavy-duty industrial purposes, light-duty general purposes, or occasional home and very light-duty purposes (with short, intermittent cycling). Special 12-volt direct-current standby models are also available for emergency use in power outages. Quality levels vary, and many different pump models are available.

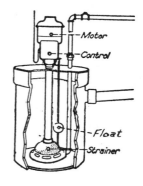

MOTORS.

Motors range from $\frac{1}{6}$ to $\frac{1}{2}$ horsepower as a general rule, though specialized heavy-duty pumps with larger motors are available. Continuous-duty models are often fan-cooled. Better motors are fitted with sealed double ball bearings requiring no lubrication; others are fitted with permanently lubricated sleeve bearings. Most motors are equipped with automatic thermal protection against overheating; such protection may be of either the automatic or the manual reset type. Most operate on 115 volts, 60 Hz alternating current. They can be direct-wired but are usually fitted with cord and plug. They do not cause radio or television interference.

ELECTRIC CONTROL.

A few small models of sump pumps (generally known as "portables"), as well as some industrial and contractors' dewatering pumps, are manually operated and start pumping as soon as plugged in. Most are fully automatic. Upright models are controlled by means of an adjustable copper or stainless steel ball mounted on a stainless steel rod. Rising water level raises the float and the rod, which in turn actuates a switch mounted on the motor, and starts the pump. The mechanical action is simple and virtually foolproof; water never reaches the switch, nor is clogging or interference likely. Submersible pumps may be operated in several ways: by a pressure-operated diaphragm switch located in the pump housing; by a small magnetic float that rises between reed switches; or by other types of enclosed, corrosion-resistant float-and-switch mechanisms within the pump housing.

CONSTRUCTION.

Pumps are constructed of various combinations of cast iron, stainless steel, bronze, and several types of plastics. Impellers may be of glass-filled nylon, glass-filled polypropylene, fiberglass-reinforced acetal plastic, nylon, or other plastics, cast bronze, or stainless steel. Choice of pump materials should depend upon desired service life, installation environment, and nature and temperature of liquids being pumped.

PUMP SIZE.

The discharge pipe size of most pumps is $1\frac{1}{4}''$ National Pipe Thread. Portable mini-pumps are fitted with standard $\frac{3}{4}''$ garden hose discharge. Large heavy-duty pumps have $2''$ National Pipe Thread discharge. The maximum head is generally 20 feet. Capacities range from 1,300 to 6,200 gallons per hour at a head of 5 feet to 130 to 3,000 gallons per hour at a head of 20 feet, including suction lift and friction loss. To determine the approximate total discharge head on an average job, measure the vertical distance or elevation in feet from the bottom of the sump to the highest point in discharge line, and add $\frac{3}{4}$ foot for each elbow. To select the proper pump size, estimate the approximate maximum inflow of liquid into the sump in gallons per minute, then multiply by 60 to get gallons per hour. Consult manufacturers' literature to determine which pump will readily discharge that inflow at the given discharge head.

SUMP PUMP
INSTALLATION

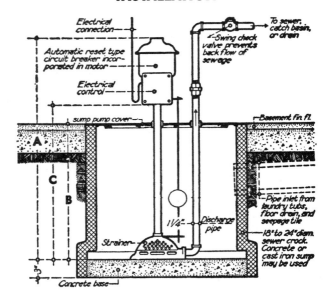

Mini-pumps, portable pumps, and some kinds of industrial and contractors' dewatering pumps require no installation. They are simply transported to the dewatering site, positioned, and put into operation. The only requirements are a power source and a power cable of sufficient capacity to handle the starting current of the motor and ensure no line loss.

Upright sump pumps are designed for permanent installation, typically as shown above. Dimension A generally is in the 30″ to 42″ range, with dimension C correspondingly less, depending upon motor size. The recommended depth of the sump (B) is usually 24″; this may be regarded as a minimum, and greater depth is sometimes desirable. However, in no case should the depth of the water exceed dimension C. Minimum diameter or width of the sump is 18″. Practically any sturdy material, such as concrete block, concrete drain tile, clay tile as shown, galvanized steel culvert, or even redwood or treated plywood may be used to construct the sump. Screening of sump inlets is not necessary.

For submersible sump pumps, dimension A ranges from approximately 6″ to 15″, depending upon model. Total width required runs from 6″ to 12″. They are permanently installed in the same manner as upright sump pumps, in sumps of the same size recommendations. Screening of sump inlets is not necessary, since the pump inlets are screened.

Sumps must be fitted with adequate covers for safety and to keep foreign objects and material out of the sump. Holes through which electrical cords run should be properly sized, and the cable should be secured with a proper clamp or protected by a substantial grommet. Covers should be readily removable for cleaning of the sump.

An electrical outlet must be located near the sump pump. Check the National Electrical Code and local codes for proper electrical installation.

DIAGRAMS OF SUMP
PUMP INSTALLATIONS

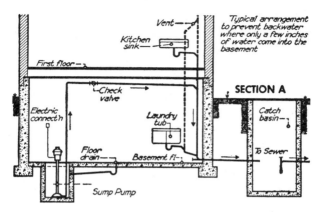

Typical arrangement to prevent backwater where only a few inches of water come into the basement

SECTION A

BACKWATER PREVENTION FOR ORDINARY CONDITIONS.

Overloading of the city sewer often results in flooding of basements. This diagram shows how a basement can be entirely cut off from drain connections through which backwater might otherwise enter. It will be necessary for the water in the sewer (and catch basin) to reach the height of the laundry tub rim to become a hazard. The catch basin is shown since certain localities make it mandatory. The catch basin acts as a large size grease trap, allowing only clear water to enter the sewer. It should be noted that fixtures discharging solid wastes are not drained to the sump pit nor to the catch basin. Therefore, they must be situated above the level of the sewer and connected directly to it through the usual house drain trap.

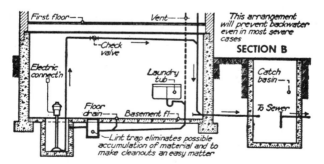

This arrangement will prevent backwater even in most severe cases

SECTION B

BACKWATER PREVENTION FOR SEVERE CONDITIONS.

If the laundry tubs are drained into the sump pit, through a lint trap, the height that is safe against backwater becomes sufficient to provide for the most severe conditions. Otherwise *Section B* corresponds to *Section A* above. The lint trap prevents the accumulation of material that might clog the sump pump strainer. See also *Section C* on the following page.

339

DIAGRAMS OF SUMP PUMP INSTALLATIONS

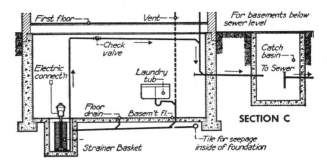

SECTION C

DRAINAGE FOR FIXTURES BELOW SEWER LEVEL. Even if the basement floor is considerably below the level of the city sewer, the occupant need not be denied the conveniences of plumbing in the basement. The floor drain for convenience in cleaning, laundry tubs, or other fixtures not discharging solid wastes, may be utilized in connection with a floatless sump pump. *Section C* shows such a piping diagram. The backing up of the sewer due to storms and other causes cannot flood a basement having such an arrangement of waste lines.

The catch basin is required in many localities, to prevent grease and soap from sinks and tubs from reaching the sewer.

In *Section C*, an alternate method of preventing lint from entering the sump pit (the use of a strainer basket) is shown.

Notice the foundation drain tile to pick up seepage of ground water from around footings. Ground water can be kept from entering the basement in this manner with any of the diagrams shown.

WHERE THERE IS NO SEWER. When no city sewer is available for wastes, buildings on level sites often make use of a septic tank disposal system. Since the drain field of such a system must be close to the surface of the ground to make use of the bacteria which attack and purify the outflow, the entrance to the septic tank system may be well above the basement level (on sloping sites this would not be true). The condition becomes, therefore, similar to that shown in *Section C* above, except that a grease trap might be substituted for the catch basin. The grease trap, if used, should not receive waste from cellar drains, or other waste.

BASEMENT WALL LEAKAGE. Occasionally, buildings are built in soil that does not allow ground water from storms to soak away, and a poor foundation permits this water to seep into the building periodically. Such conditions frequently occur during the spring thaws and rains. If the foundation is porous, it is practically impossible to render it water-tight from the inside—and it is too late to attack the problem from the exterior. Making use of a sump pump can often alleviate such conditions by removing the water as fast as it accumulates. This is done by trenching radially from a sump pit, under the basement floor. Open drain tiles are laid, and the trench is backfilled with coarse stones. Of course, the basement floor construction must be cut and patched for such an operation.

OTHER USES. Boiler pits, settling basins, flywheel and elevator pits, and similar places may develop seepage problems or may require means of removing drainage when the sewer is at a higher elevation.

SEPTIC TANK
SEWAGE DISPOSAL

HOUSE MAIN DRAIN. The 3″ or 4″ house main drain line should extend through the foundation, usually about 5′, where it is coupled to the house sewer line extending to the septic tank. If a well or other water supply is located nearer than 100′ from this point or from the septic tank or drain field site, a different location for one or the other must be selected. The water supply should not be on the downhill side of the sewage disposal system.

LAUNDRY BYPASS. In some installations a separate laundry bypass drain line, often 1½″ polyethylene pipe, is run directly from the laundry to the drain field, bypassing the septic tank entirely. This arrangement eliminates discharges of large quantities of water into the tank in short periods of time, which tends to disrupt proper tank action and can cause clogging solids into the drain field lines, and it also prevents a buildup of nonbiodegradable soaps in the tank.

GREASE TRAP. The septic tank may give trouble or a sewer line may clog from the collection of grease, most of which comes from the kitchen. There should be a grease trap in the kitchen line.

SEPTIC TANK ACTION. In a septic tank some of the solid matter floats on the surface as scum or "mat" and the heavier solids settle to form sludge. The septic tank causes the retained scum and sludge to decompose by biochemical action in the absence of oxygen, materially reducing the volume of the solids.

SEPTIC TANK. The septic tank should be watertight. Walls, top, and floor should be reinforced. The inlet and outlet of the first or settling chamber are arranged so as not to disturb the sludge or scum and carry solid particles to clog the following part of the system. An automatic siphon discharges the contents of its chamber, normally about every eight hours, flooding the drain field pipes. The tank should be tightly covered to prevent spread of odors, transmission of disease germs by flies, and accidents to children. It should be covered by a foot or two of earth to secure uniformity of temperature and warmth in winter to aid the biochemical action.

DISTRIBUTING BOX. This device ensures equal distribution of the outflow to each of the drain field branches; it is preferable to the scheme of branching directly from a main pipe with the leaching pipes.

DRAIN FIELD. The outflow of the septic tank, which contains disease germs and foul smelling matter in liquid form, soaks through the top soil from the subsurface drain pipes. The top 10″ or 20″ of soil contains beneficial bacteria that attack and purify the outflow. The siphon's intermittent action allows a rest period between discharges, better handling this process. The whole system should be watertight except for these drain field pipes, which are meant to leak.

MAINTENANCE. A few tanks need cleaning almost every year, while others may go 15 years or more without needing attention. Commercially available bacteria additives often help in keeping a tank properly functional and in a maximally liquid state. However, when scum and sludge together have accumulated to a thickness of about 2′ at the inlet area, the tank should be cleaned. This job must be left to a professional, who will pump the waste material into a tank truck and haul it away to be safely disposed of at a designated sanitary landfill site.

VARIATIONS IN METHODS. The details of septic systems and their installation vary widely throughout the country and nearly always the process is stringently regulated by municipal, county, or state government. Also, many types of tanks ranging from simple vaults to complex purification systems are in use today. In most cases the traditional federal guidelines used for so many decades are no longer followed, nor are there many new ones of universal consequence. The entire job of designing a septic system installation must be done in concert with local or state officials, with adherence to their recommended designs and even to specific approved products, and in light of the particular site conditions.

SEPTIC TANK
SEWAGE DISPOSAL

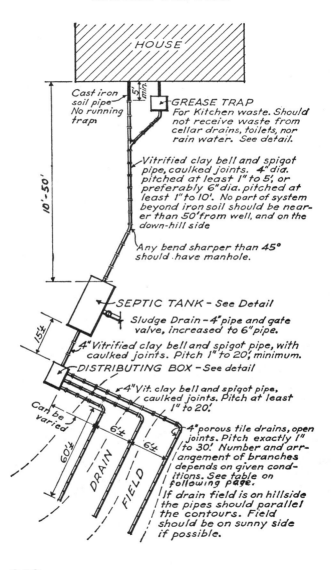

HOUSE

Cast iron soil pipe. No running trap.

5' min.

GREASE TRAP
For kitchen waste. Should not receive waste from cellar drains, toilets, nor rain water. See detail.

Vitrified clay bell and spigot pipe, caulked joints. 4" dia. pitched at least 1" to 5', or preferably 6" dia. pitched at least 1" to 10'. No part of system beyond iron soil should be nearer than 50' from well, and on the down-hill side

10'-50'

Any bend sharper than 45° should have manhole.

SEPTIC TANK - See Detail

Sludge Drain - 4" pipe and gate valve, increased to 6" pipe.

15'±

4" Vitrified clay bell and spigot pipe, with caulked joints. Pitch 1" to 20', minimum.

DISTRIBUTING BOX - See detail

4" Vit. clay bell and spigot pipe, caulked joints. Pitch at least 1" to 20'.

Can be varied

6'±

6'±

DRAIN FIELD

60'±

4" porous tile drains, open joints. Pitch exactly 1" to 30'. Number and arrangement of branches depends on given conditions. See table on following page.
If drain field is on hillside the pipes should parallel the contours. Field should be on sunny side if possible.

342

SEPTIC TANK
SEWAGE DISPOSAL

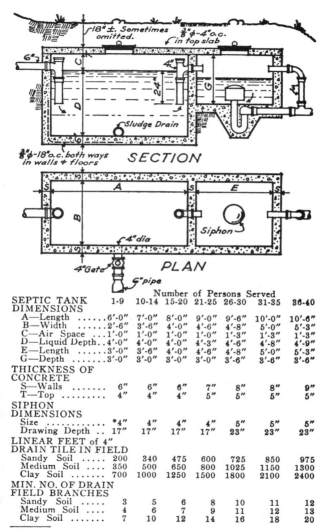

SECTION

PLAN

SEPTIC TANK DIMENSIONS	Number of Persons Served						
	1-9	10-14	15-20	21-25	26-30	31-35	36-40
A—Length	6'-0"	7'-0"	8'-0"	9'-0"	9'-6"	10'-0"	10'-6"
B—Width	2'-6"	3'-6"	4'-0"	4'-6"	4'-8"	5'-0"	5'-3"
C—Air Space	1'-0"	1'-0"	1'-0"	1'-0"	1'-3"	1'-3"	1'-3"
D—Liquid Depth	4'-0"	4'-0"	4'-0"	4'-3"	4'-6"	4'-8"	4'-9"
E—Length	3'-0"	3'-6"	4'-0"	4'-6"	4'-8"	5'-0"	5'-3"
G—Depth	3'-0"	3'-0"	3'-0"	3'-0"	3'-6"	3'-6"	3'-6"
THICKNESS OF CONCRETE							
S—Walls	6"	6"	6"	7"	8"	8"	9"
T—Top	4"	4"	4"	5"	5"	5"	5"
SIPHON DIMENSIONS							
Size	*4"	4"	4"	4"	5"	5"	5"
Drawing Depth	17"	17"	17"	17"	23"	23"	23"
LINEAR FEET of 4" DRAIN TILE IN FIELD							
Sandy Soil	200	340	475	600	725	850	975
Medium Soil	350	500	650	800	1025	1150	1300
Clay Soil	700	1000	1250	1500	1800	2100	2400
MIN. NO. OF DRAIN FIELD BRANCHES							
Sandy Soil	3	5	6	8	10	11	12
Medium Soil	4	6	7	9	11	12	13
Clay Soil	7	10	12	14	16	18	20

* In this smallest size the siphon is sometimes omitted.

343

SEPTIC TANK
SEWAGE DISPOSAL

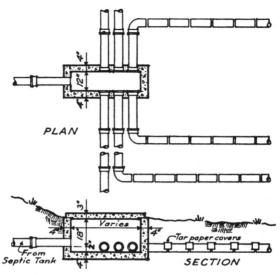

PLAN

SECTION

From Septic Tank

Varies

Tar paper covers

DISTRIBUTING BOX

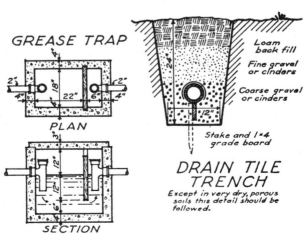

GREASE TRAP

PLAN

SECTION

Loam back fill

Fine gravel or cinders

Coarse gravel or cinders

Stake and 1×4 grade board

DRAIN TILE TRENCH

Except in very dry, porous soils this detail should be followed.

HEAT LOSS BY
Btu METHOD

Heat flows from substances of higher temperature to substances of lower temperature. To maintain the warmth of a room when the adjacent temperatures are lower, we must add to the air in the room an amount of heat equal to the amount that is constantly flowing away.

Calculation of a heating installation is divided into two parts. *First*, the heat loss must be determined. *Second*, the conditions required to balance that loss must be calculated. This may be simply expressed as:

$$Heat\ to\ Be\ Supplied = Heat\ Loss$$

Quantities of heat are measured in British thermal units (Btu). A Btu is the amount of heat required to raise the temperature of 1 pound of water $1°F$. Heating calculations are made on the basis of one hour as the unit of time.

Heat is lost from buildings in two ways: (1) by transmission through building sections (glass, doors, ceilings, floors, walls) to the outside, as well as to adjacent unheated spaces; and (2) by infiltration of outside air into the heated spaces and the concomitant exfiltration of heated air to the outside. This may be simply expressed as:

$$Heat\ to\ Be\ Supplied = HL_T = HL_I$$

The following pages will be devoted to the calculation of the two quantities involved.

To make heat-loss calculations it is necessary to arrive at the temperature at which the room is to be maintained, known as the "inside temperature," indicated as t_i. Current FEA/DOE recommendations are $65°F$ generally and $68°F$ for elderly or ill persons, while HUD recommends $70°F$ generally. Following are other possibilities.

TABLE A. t_i VALUES USUALLY SPECIFIED

Schools			**Theaters**	
Classrooms	70-72° F.		Seating Space	68-72° F.
Assembly Rooms	68-72° F.		Lounge Rooms	68-72° F.
Gymnasiums	55-65° F.		Toilets	68° F.
Toilets and Baths	70° F.		**Hotels**	
Wardrobe and Lock-			Bedrooms and Baths	70° F.
er Rooms	65-68° F.		Dining Rooms	70° F.
Kitchens	66° F.		Kitchens and Laun-	
Dining and Lunch			dries	66° F.
Rooms	65-70° F.		Ballrooms	65-68° F.
Playrooms	60-65° F.		Toilet Service Rooms	68° F.
Natatoriums	75° F.		Homes	70-72° F.
Hospitals			Stores	65-68° F.
Private Rooms	70-72° F.		Public Buildings	68-72° F.
Private Rooms (sur-			Warm Air Baths	120° F.
gical)	70-80° F.		Steam Baths	110° F.
Operating Rooms	70-95° F.		Factories, Machine	
Wards	68° F.		Shops	60-65° F.
Kitchens and Laun-			Foundries and Boiler	
dries	66° F.		Shops	50-60° F.
Toilets	68° F.		Paint Shops	80° F.
Bathrooms	70-80° F.			

It is also necessary to determine the "outside temperature," called t_o. The value usually selected is equal to a temperature that will be equaled or exceeded 97.5% of the time during the heating season, as taken from a standard table of weather data and design conditions for various locales. Tabulated figures should be modified as necessary to reflect localized conditions and microclimates. Record lows should not be used.

REPRESENTATIVE
OUTSIDE TEMPERATURES (t_o)

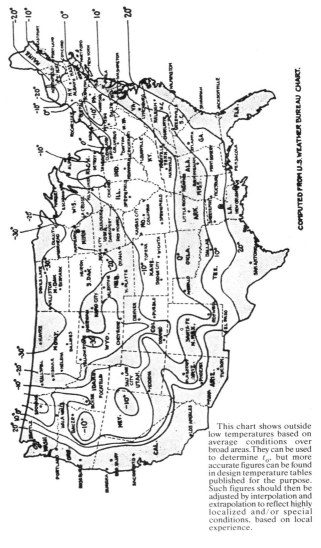

COMPUTED FROM U.S. WEATHER BUREAU CHART.

This chart shows outside low temperatures based on average conditions over broad areas. They can be used to determine t_o, but more accurate figures can be found in design temperature tables published for the purpose. Such figures should then be adjusted by interpolation and extrapolation to reflect highly localized and/or special conditions, based on local experience.

HL$_T$ = HEAT LOSS BY TRANSMISSION

The amount of heat in Btu that is lost through each square foot of building section (wall, ceiling, roof, doors, and so forth) per hour per degree of temperature difference between the warmer side and the cooler side of the section is designated by the symbol U, which stands for *thermal transmittance*. This value is called the *overall coefficient of heat transfer* through the section. It is often written in terms of its units: Btu/hr/sq ft/°F diff.

The values of the individual building materials that make up a building section are symbolized differently. If the material is homogeneous or of like composition throughout its entire thickness (concrete, wood), its capacity to conduct heat is symbolized as k, standing for *conductivity*. If the material is nonhomogeneous (heterogeneous) or is normally commercially available in thicknesses other than 1″ (hollow-core concrete block, cork tile), its capacity to conduct heat is symbolized as C, standing for *conductance*. For homogeneous materials, the value of k is listed per inch of thickness of the material. For heterogeneous materials the value of C is listed per indicated or per standard manufactured thickness of the material. These values are determined by laboratory tests, and are usually given in published tables as constants established at a mean temperature of 75°F.

When the U-value of any given building section is to be determined, the first step is to ascertain the k or C of each component material in the section. These values are then converted to *resistance* to heat flow, symbolized as R (these values are generally listed conjointly in published tables) by finding the reciprocals of k and C ($1/k$ and $1/C$).

The R-values are likewise listed per inch of thickness, or per stated or standard thickness. The second step is to add all of the R-values for the building section components. The total R-value, whether for one material or several, is then converted to U by finding the reciprocal of R ($1/R$).

The value of U must be determined for each different kind of building section in a structure (single glass; double glass; solid door; insulated steel door; each different wall, floor, and ceiling construction; and so on). Then, for each different building section, the particular U-value is multiplied by the area in square feet of that section to arrive at the total heat loss for the entire section. If two or more parts of a building have identical U-values, they can be combined during the calculations. The resulting figure must then be multiplied by the design temperature difference between the inside and the outside surfaces ($t_i - t_o$), both of which are selected as design constants according to certain criteria. The result is the total heat loss through the particular sections involved, per hour. This can be expressed as follows:

$$\text{Total Heat Loss through Section} = A_S \times U_S \times (t_i - t_o).$$

Assume that a room has five windows with a total area of 62 square feet of single glazing. The coefficient (U) of single glass in vertical position in wintertime is 1.10. With a difference in temperature of 70°F the calculation would be thus:

$$\text{Heat Loss through Glass} = A_G \times U_G \times (t_i - t_o) = 62 \text{ sq ft} \times 1.10 \text{ Btu} \times 70° \text{ diff.}$$
$$= 4,774 \text{ Btu per hour.}$$

This can expressed as:

$$HL_G = 4,774 \text{ Btu/hr.}$$

HL$_T$ = HEAT LOSS
BY TRANSMISSION

The coefficients that apply to other sections of the building, such as wall, ceiling, floor, or door, can be indicated in the same fashion as the glass section discussed on the previous page (U_W, U_C, U_F, and so on). If several different compositions of wall sections are involved, for example, they can be designated as U_{W1}, U_{W2}, and so on, which can ultimately be totaled as simply U_W. As in the glass example, the total heat loss through an outside wall, floor, or ceiling, or through a wall, floor, or ceiling adjacent to an unheated space, would be the area of the section *times* the coefficient *times* the difference in temperature. This can be expressed as follows:

$$HL_W = A_W \times U_W \times \Delta t$$

where HL_W = heat loss through wall, A_W = area of wall, U_W = U-value of wall section, and Δt = temperature °F difference. The answer is expressed in Btu/hr. The areas of walls, floors, and so forth may be computed from architectural drawings or by actual measurement. Typical coefficients for various composite constructions are shown as examples on pages that follow. However, for best accuracy the overall U-values should be calculated for the specific constructions involved in each case by totaling the R-values for each material and converting to U. Notice that in many cases the resistances for surface films and air spaces must be included in overall calculations, and often certain adjustments must be made for the position of sections, reflectivity, direction of heat flow, and so forth.

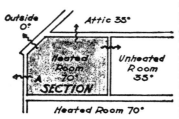

The temperature of unheated spaces in a building is usually taken as the mean between the room temperature and the outside design temperature. For instance, if the outside design temperature is 0°F, and the temperature of the room is taken as 70°F, the temperature of unheated spaces in the building can be assumed as 35°F. In a basement room that has part of its outside wall below grade and part above grade, the t_o of the ground can be taken at 50°F (assuming the other conditions of the above example are unchanged).

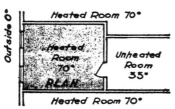

Figure 1

Reference to *Figure 1* will serve as an illustrated example. From the drawing, the total area of the wall exposed to the outside (*Arrow A*) may be calculated as 96 square feet, from which we subtract the area of the windows, 30 square feet, giving an actual wall area of 66 square feet. We will assume the coefficient for this type of wall construction to be 0.25 with a difference in temperature of 70°F. The calculation would be:

$$HL_W = 66 \text{ sq. ft.} \times 0.25 \text{ Btu} \times 70°F \text{ diff.} = 1,155 \text{ Btu/hr.}$$

A single room may have a wall that is exposed to the outside; a portion of the roof may be similarly exposed; the floor, ceiling, or inside partitions may be exposed to adjacent unheated spaces. Separate calculation is necessary for each condition, as marked by arrows on *Figure 1*. The sum of the losses thus obtained is considered the *heat loss through walls*.

HL₁ = HEAT LOSS
BY INFILTRATION

Cold air from the outside enters a building through cracks in the construction and sometimes through the walls themselves. Some engineers calculate the exact length of all cracks around windows, doors, etc., and arrive at a theoretical volume of air admitted. In most cases, however, it has become practice to assume a certain number of complete changes of air in a room per hour.

The usual number of air changes used in heating work are given in *Table C* below. Whether or not the actual infiltration of air amounts to the number of air changes shown in the table, it has been found that the heating system should allow for the number given. Doors and windows are opened for ventilation purposes, thus changing the air in the room, even though there is actually a very small infiltration. In rooms that have forced systems of ventilation, part of the heat loss may be supplied by the warmed air and part by the radiation in the room.

The cubic contents of the room multiplied by the number of air changes per hour gives the volume of air that must be heated. It requires 0.018 Btu to raise one cubic foot of air 1°F. Therefore, if we multiply the total volume of air by 0.018 we will have the number of Btu required to raise the temperature of this air by 1°F. Then, if this quantity is multiplied by the number of degrees required to raise the temperature, we will have the total heat required to balance the infiltration loss. This may be expressed as follows:

$$HL_I = 0.018 \times n \times C \times (t_i - t_o)$$

Assume a room having a floor area of 10 feet by 12 feet with a ceiling height of 9 feet. The cubic contents will be 1,080 cubic feet. If two air changes an hour are desirable, the calculation then becomes:

$$HL_I = 0.018 \times 2 \times 1,080 \times 70 = 2,720 \text{ Btu}$$

The loss thus obtained is considered the *heat loss by infiltration*.

TABLE OF *n* VALUES

Type or exposure	*Number*
Rooms, no windows or outside doors	½
Rooms, exposure 1 side	1
Rooms, exposure 2 sides	1½
Rooms, exposure 3 sides	2
Rooms, exposure 4 sides	2
Living Rooms in Residences	1 to 2
Stairways and Halls	½ to 1
Bedrooms	1½
Small Convention Halls	4
General Offices	3
Private Offices	4
Public Dining Rooms	4
Banquet Halls	5
Basement Restaurants	8 to 12
Hotel Kitchens	4 to 6
Public Libraries	3

U-VALUES
FOR FRAME WALLS

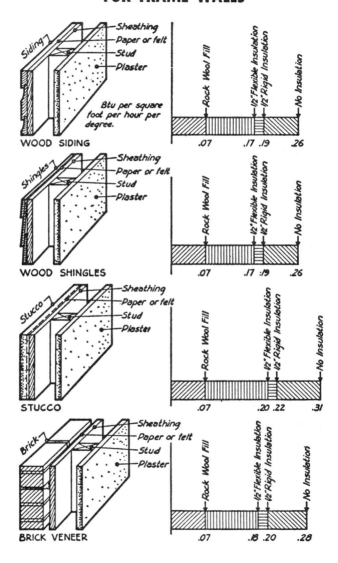

WOOD SIDING

Siding

Sheathing
Paper or felt
Stud
Plaster

Btu per square foot per hour per degree.

Rock Wool Fill   1/2" Flexible Insulation   1/2" Rigid Insulation   No Insulation

.07   .17 .19   .26

WOOD SHINGLES

Shingles

Sheathing
Paper or felt
Stud
Plaster

Rock Wool Fill   1/2" Flexible Insulation   1/2" Rigid Insulation   No Insulation

.07   .17 .19   .26

STUCCO

Stucco

Sheathing
Paper or felt
Stud
Plaster

Rock Wool Fill   1/2" Flexible Insulation   1/2" Rigid Insulation   No Insulation

.07   .20 .22   .31

BRICK VENEER

Brick

Sheathing
Paper or felt
Stud
Plaster

Rock Wool Fill   1/2" Flexible Insulation   1/2" Rigid Insulation   No Insulation

.07   .18 .20   .28

U-VALUES FOR
MASONRY WALLS

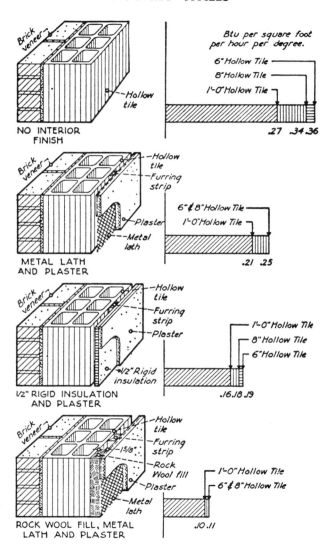

Btu per square foot
per hour per degree.

NO INTERIOR FINISH

Brick veneer
Hollow tile

6" Hollow Tile
8" Hollow Tile
1'-0" Hollow Tile

.27 .34 .36

METAL LATH AND PLASTER

Brick veneer
Hollow tile
Furring strip
Plaster
Metal lath

6" & 8" Hollow Tile
1'-0" Hollow Tile

.21 .25

1/2" RIGID INSULATION AND PLASTER

Brick veneer
Hollow tile
Furring strip
Plaster
1/2" Rigid insulation

1'-0" Hollow Tile
8" Hollow Tile
6" Hollow Tile

.16 .18 .19

ROCK WOOL FILL, METAL LATH AND PLASTER

Brick veneer
Hollow tile
Furring strip
1⅝"
Rock Wool fill
Plaster
Metal lath

1'-0" Hollow Tile
6" & 8" Hollow Tile

.10 .11

U-VALUES FOR
INTERIOR PARTITIONS

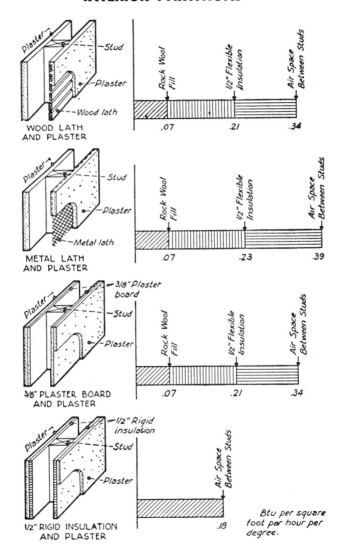

WOOD LATH
AND PLASTER

METAL LATH
AND PLASTER

3/8" PLASTER BOARD
AND PLASTER

1/2" RIGID INSULATION
AND PLASTER

Btu per square
foot per hour per
degree.

U-VALUES FOR
INTERIOR PARTITIONS

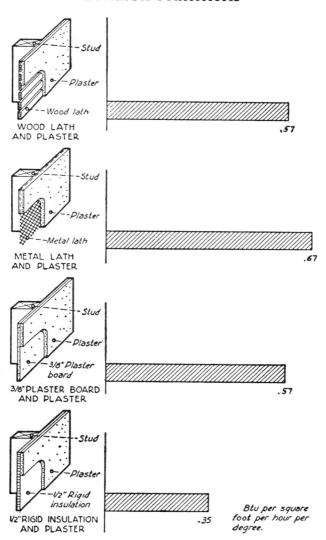

WOOD LATH
AND PLASTER — Stud, Plaster, Wood lath — .57

METAL LATH
AND PLASTER — Stud, Plaster, Metal lath — .67

3/8"PLASTER BOARD
AND PLASTER — Stud, Plaster, 3/8" Plaster board — .57

1/2"RIGID INSULATION
AND PLASTER — Stud, Plaster, 1/2" Rigid insulation — .35

Btu per square foot per hour per degree.

HOW TO FIGURE
HEAT LOSS

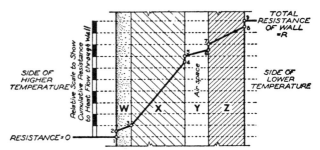

The heat loss through a wall, floor, or roof depends upon the overall resistance of the construction to heat flow.

In the drawing above, the way overall resistance of a heterogeneous wall equals the numerical sum of the resistances of the various parts is shown graphically. The total resistance to heat flow increases as we proceed through the wall from the side of higher temperature toward the side of lower temperature. Four types of resistance contribute to the total overall resistance (R) of a given wall.

F = The surface or film resistances.

M = The resistances of the solid materials.

A = The resistance of the air space or spaces.

I = The resistance of the insulating materials.

F = FILM RESISTANCE. The surface of a material exposed to air offers a resistance to heat flow that is called the film resistance. It is indicated here by the letter F, with a subscript for identification in case there are several. In the illustration, F_{1-2} is the film resistance of the wall face on the side of higher temperature. F_{4-5} and F_{6-7} are the film resistances of the surfaces that face air space; they are included in the value of A_{4-7} in making calculations. F_{8-9} is the resistance of the surface on the side of lower temperature.

In still air the film resistance of a vertical surface, such as the interior plastered wall of a house, would have a value of 0.68. The exterior film would be different and in practice is usually determined experimentally upon the assumption of an air movement of 15 miles per hour. The sum of F_{1-2} plus F_{8-9} can be indicated by the symbol ΣF.

Where two different materials are in contact, as at 3, there is no film resistance.

The film resistance varies depending upon whether the surface is vertical, horizontal, or at a 45° slope, and whether the heat flow is upward, downward, or horizontal. It also varies with the reflectivity of the surface. Comprehensive published tables are available; the following table presents some typical values for nonreflective surfaces.

Ordinary Surfaces	Values of F
Vertical, still air, horizontal heat flow	0.68
Horizontal, still air, upward heat flow	0.61
Horizontal, still air, downward heat flow	0.92
45° slope, still air, upward heat flow	0.62
45° slope, still air, downward heat flow	0.76
Vertical, outside, 15 mph wind	0.17

HOW TO FIGURE
HEAT LOSS

M = RESISTANCE OF SOLID MATERIALS. Different materials used for wall construction have different resistances. Material W in the illustration, for instance, might have considerably less resistance than material X. The resistance of the body of a material is indicated by the letter M, with a subscript for identification when several materials are used.

In the illustration, the resistance of material W is shown by the line 2-3, and the vertical distance indicates the amount of the resistance. The sum of M_W plus M_X plus M_Z can be indicated by the symbol ΣM. Thermal resistances are calculated by first determining the thermal conductivity (k) of homogeneous materials one inch thick and one foot square or the thermal conductance (C) of nonhomogeneous materials of standard unit thickness and one foot square, both in terms of the amount of heat that passes in one hour. This is then translated into the resistance, called the R or R-value, of the material by finding the reciprocal of the k or C. Most published tables list the heat transmission values of various common materials as R-values so that they may be easily used as design constants. The following table presents typical resistance values for several types of materials.

The resistance of heterogeneous materials, such as hollow tile or plasterboard made of gypsum between layers of heavy paper, do not vary directly with thickness—and values for each thickness have to be determined experimentally. However, when the resistance of a homogeneous material for 1″ thickness is known, the resistances for other thicknesses can be found by direct proportion; a 2″ thickness has twice the resistance of a 1″ thickness.

Type of Material	Thickness	R Value
Brick, common	1″	0.20
Brick, face	1″	0.11
Concrete, heavy aggregate	1″	0.08
Stucco	1″	0.20
Clay tile, hollow	8″	1.85
Concrete block	8″	1.11
Cinder block	8″	1.72
Stone	1″	0.08
Plasterboard	½″	0.45
Plywood	1″	1.25
Fiberboard sheathing	½″	1.22
Hardboard, medium density	1″	1.37
Particleboard, medium density	1″	1.06
Softwoods	1″	1.25
Hardwoods	1″	0.91
Carpet and rubber pad	av.	1.23
Terrazzo	1″	0.08
Floor tile, vinyl, etc.	av.	0.05
Plaster, cement sand	1″	0.20
Plaster, gypsum sand	1″	0.18
Asphalt shingles	av.	0.44
Asphalt roll roofing	av.	0.15
Wood shingles	av.	0.94
Wood drop siding	¾″	0.79
Aluminum siding over sheathing	av.	0.61

HOW TO FIGURE
HEAT LOSS

A = AIR SPACE RESISTANCE. Heat is conducted across an air space by a combination of radiation, conduction, and convection. The resistance of an air space increases with the air space width, but after ½″ has been reached the increases are small. However, some practical gains in resistance can be made up to a width of about 3½″. A ¾″ air space is perhaps the most common of intentionally designed spaces.

In the illustration the resistance of air space Y is shown by the line 5-6 and the vertical distance indicates the air space resistance, which is identified by the symbol A_Y. In a wall having several air spaces the total resistance of the spaces can be identified by the symbol ΣA. The resistance of air spaces varies with the reflectivity (or effective emittance E) of the bounding surfaces, the temperature difference between the surfaces, the mean temperature within the air space, the relative position of the air space (vertical, horizontal, or 45°), and the direction of heat flow (up, down, or sideways). The range of values is substantial. For example, a ¾″ space with two nonreflective surfaces may have a resistance as low as 1.70 or as high as 4.70. Comprehensive published tables of these values should be consulted in the interest of accuracy.

I = RESISTANCE OF INSULATION. Resistance values per inch of thickness do not afford a true basis for comparison between insulating materials as applied, although they frequently are used for that purpose. The value of an insulating material is measured in terms of its resistance to the transmission of heat, which depends not only upon the resistance per unit of thickness but also upon the actual thickness as installed, upon the presence of air spaces that provide film resistances, and upon the ambient temperature.

In the illustration, no insulation material is shown. For purposes of this explanation the symbol I is used to designate the resistance of an insulation material; in the case of several occurring in the same construction, the symbol ΣI would be employed. In usual practice, however, the resistance of an insulating material is designated in the same fashion as any other building material (M = resistance of solid materials, on at least one previous page of this book) as R, and they are conjoined when making calculations—the insulation is merely another part of the building section. The following table gives typical resistance values for I.

Type of Insulation	Thickness	Value of I
Fiberglass batt	2″	6.3
Fiberglass batt	3½″	11.0
Fiberglass batt	6″	19.0
Fiberglass batt	8″	25.3
Fiberglass loose fill	1″	1.1
Rock wool batt	1″	3.7
Cellular glass	2″	5.9
Cellulosics	1″	3.5
Polystyrene, extruded, cut cell	1″	4.0
Polystyrene, extruded, smooth skin	1″	5.0
Urea-Formaldehyde	1″	4.2
Polyurethane, expanded, aged	1″	6.3
Polyisocyanurate, smooth skin	1″	7.7
Perlite, loose fill	1″	2.7
Vermiculite, loose fill	1″	2.2
Sawdust/shavings	1″	2.2

HOW TO FIGURE HEAT LOSS

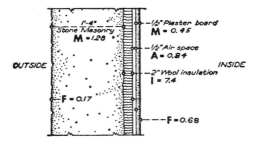

TYPICAL EXAMPLE.   In the illustration is shown a 16″ masonry wall with 2″ rock wool insulation, an air space, and ½″ plasterboard. It is desired to find the transmission heat loss for this building section. From our original statement we find that:

$$\text{Total resistance } (R) = \Sigma F + \Sigma M + \Sigma A + \Sigma I$$

An examination of the drawing shows two film resistances (F), one in still air, the other in moving (outside) air. Consulting the foregoing tables, we find that the still value is 0.68. The outside film resistance is 0.17.

$$\Sigma F = 0.68 + 0.17 = 0.85$$

The resistance of the masonry and the ½″ plasterboard are found in the foregoing tables, and we have:

$$\Sigma M = 1.28 + 0.45 = 1.73$$

In a similar manner we have the resistance of the air space:

$$\Sigma A = 0.84$$

In a similar manner we find:

$$\Sigma I = 7.4$$

The total overall resistance of the wall section is:

$$R = 0.85 + 1.73 + 0.84 + 7.4 = 10.82$$

THERMAL TRANSMITTANCE.   The heat loss through the building section is expressed as the overall coefficient of heat transfer, which is its U, or U-value, the thermal transmittance through a combination of materials. This is the equivalent of the heat transmission through a square foot of the building section and is equal to the reciprocal of the resistance of the materials—which is a fancy way of saying:

$$U = \frac{1}{R}$$

or:

$$U = \frac{1}{10.82} = 0.092$$

U-value is expressed in terms of Btu per square foot, so this means that each square foot of the wall section will transmit 92-thousandths of a Btu per degree of temperature difference between the inside (or warmer surface) and the outside (or cooler surface), per hour.

PRINCIPLES OF
HEAT TRANSMISSION

Heat may be transmitted by three means—conduction, convection, and radiation. (1) *Conduction.* If one part of a body is at a higher temperature than another part, there will be a flow of heat toward the part at lower temperature. For example, if one end of a metal is held in a fire, heat will flow to the opposite end; or, if a body is in contact with another of a different temperature, heat will flow to the cooler body. (2) *Convection.* When a body is in contact with a cooler fluid (liquid or gas), heat leaves the warmer body by *conduction* from its surface to the fluid in contact with it. The warmed particles of fluid will rise, giving place to cooler particles of fluid from above. The essential characteristic of *convection* is this continuous conveyance of heat and renewal of the fluid layer at the contact surface. (3) *Radiation.* Radiant energy is conveyed by electromagnetic waves through air or a vacuum, not affecting either. It is mostly invisible—a part of the infrared spectrum. When the waves contact another object they immediately transform into heat energy, warming that object and in turn warming others in the vicinity by conduction and convection, and by reradiation. The intensity of radiant energy at a particular point is inversely proportional to the square of the distance of that point from the energy source. While gases are transparent to radiant energy, most solids *are almost perfectly opaque to it.*

Heat is lost through a building section (wall, floor, door, or other section) in the following manner: the inside surface of the section becomes warmed by its contact with the warm air of the room. The heat in the materials composing the inside face of the section tends to flow to the colder outside face by conduction. If the section contains an air space or is transparent (glazing), part of the heat will cross or bridge the space, or simply flow outward through the glazing, by radiation. The remaining part of the heat is carried through air spaces by convection—that is, by the fluid motion of the air in the spaces. All three processes occur simultaneously and continuously, in a constant attempt at temperature equalization between cooler and warmer bodies. The relative extent to which this occurs is dependent upon the ease or difficulty with which the materials involved are able to transmit heat.

The ability to minimize all kinds of heat transmission is the basis of an efficient thermal insulation. Effective thermal insulation greatly retards heat loss by interposing a barrier material of low thermal conductivity between the warmer and cooler sides of building sections. This effectiveness can be proved conclusively by comparing the heat loss through building sections that are fitted with various types of thermal insulation, using any standard text giving heat transmission coefficients of materials. In addition to having good thermal qualities, however, a thermal insulation must be resistant to decomposition, fire-retardant or fireproof, vermin-proof, odorless, noncorrosive, moisture resistant, and economical.

HEAT LOSS COEFFICIENTS IN Btu/hr/sq ft

Type of Section	Uninsulated	Insulated
Frame wall (wood siding and sheathing, plasterboard)	0.206	0.128
Masonry cavity wall (8″ brick, 2.5″ air space, plasterboard)	0.204	0.128
Solid masonry wall (4″ brick, 8″ block, 0.75″ air space, plasterboard)	0.171	0.140
Flat frame roof (ceiling tile, plasterboard, 2 x 8 rafters, plywood deck, built-up surface)	0.159	0.046
Pitched frame roof (plasterboard, 2 x 8 rafters, plywood deck, underlayment, asphalt shingles)	0.442	0.127

WHOLE-HOUSE FAN COOLING

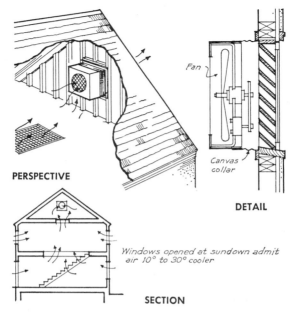

PERSPECTIVE

Fan

Canvas collar

DETAIL

Windows opened at sundown admit air 10° to 30° cooler

SECTION

The drawings above show a typical arrangement of a long-popular method of whole-house cooling, the attic fan. This arrangement works well when windows and doors are opened throughout the house, allowing cooler outside air to enter, circulate throughout the house, and exhaust through the attic as warmer air. A considerably more efficient arrangement makes use of a large plenum chamber mounted over the attic floor grille and extending at least halfway along the attic floor, with the exhaust fan mounted vertically at the far end of the chamber. Warm air is exhausted through sizable gable or roof vents. Alternatively, the plenum can be extended all the way to a gable wall. In either case, little or no hot attic air is moved by the fan, allowing it to do a better cooling job.

The recent advent of the whole-house fan, designed to operate continuously at a low noise-level in a horizontal position, has made possible the most practical arrangement for the average homeowner. The unit combines fan and vent grille, and is mounted directly in the attic floor at a central location, providing gentle air circulation, controllable by opening various windows throughout the house.

Any type of whole-house fan operating *in concert with* air conditioning does not effect any cooling cost savings, but when operated *instead of* air conditioning can produce substantial savings. In a house without air conditioning, the whole-house fan greatly increases hot-weather comfort.

To calculate the size of fan that is required, first find the volume (height x length x width = cubic feet) of the rooms to be ventilated; exclude closets, storage areas, basement, pantry, and attic. Then estimate how often a complete air change is desired. A common recommendation is once every minute in the South and parts of the Southwest, and once every 1½ minutes elsewhere. Divide the ventilated volume by this number. For example, to change 10,000 cubic feet of air every 1½ minutes would require a fan capacity of 6,666 cubic feet per minute (CFM).

The net free open area of the exhaust vents located at the attic gable end or in the roof should equal 1 square foot for each 750-CFM capacity of the fan. Thus, the fan in the example above would require 20 square feet of net free vent area. Screening and/or louvers reduce the net free vent area of vents, and this space reduction must be taken into consideration. The amount of reduction depends upon the size and number of the obstructions.

1-PIPE GRAVITY SYSTEM

The 1-pipe gravity steam system is the simplest and most economical to install for small and medium-size buildings. It requires less piping, fewer fittings, and less labor to install than other systems. The system would operate equally well on large installations, except that the largeness of the piping needed would not be economical.

The system features a main pipe situated above the waterline of the boiler; the pipe extends horizontally from the boiler to the most remote radiator and is known as the *supply main*. The radiators are generally either finned tube or convector units in new work, although the obsolete upright cast iron types still remain in many existing installations. From the main there are branch pipes called *risers*, extending vertically to the radiators above. In the 1-pipe system, steam travels up and the condensate from the radiators travels down the same pipe.

From the end of the supply main, a pipe is brought back to the boiler; this is called the *return*. The return may be above the waterline until it connects to the boiler, in which case it is known as a *dry return*. If the return is below the waterline, it will contain water and is known as a *wet return*. There are variations in return piping, such as the Hartford loop. The type or design of return does not affect the operating principle of the system.

To be satisfactory, the 1-pipe gravity system must perform three functions:

1. Carry steam uniformly to all radiators.
2. Return the condensate properly.
3. Vent the air in the piping and radiators.

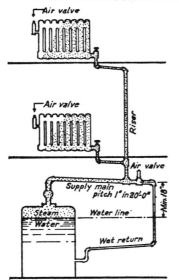

The successful performance of the first two functions depends upon proper design and installation. The third function depends upon efficient air valves.

Air must be eliminated from the radiators and piping to allow the entrance of steam. Air valves allow air to be pushed out by the entering steam, but do not allow steam or condensate to pass. As the radiators cool, air is again drawn into them through the valve. This is one disadvantage of the system. The inflowing air causes a more rapid cooling of the radiators, militating against uniform temperatures.

A valve for ridding the piping of air is placed near the end of the supply main, at the point where the main is "dripped" into the return.

No modulation is possible with a 1-pipe system.

1-PIPE GRAVITY
VACUUM SYSTEM

The piping of this system is the same as that of the 1-pipe gravity system. The difference consists in the use of *air-and-vacuum valves* on the radiators and mains to correct one of the basic shortcomings of the simple gravity system.

The air-and-vacuum valve may be described as a "one-way" valve. It allows the elimination of air (but not the escape of steam or condensate) and does not allow the air to reenter the radiation units as they cool. When the steam pressure drops due to a lowering of the fire, the condensing steam in the radiator forms a partial vacuum in the radiator—permitting the steam to reenter the radiators while encountering practically no resistance. Thus, the radiators will reheat quickly, will remain heated with a subatmospheric boiler pressure, and will maintain a much more even heat.

In order to retain the vacuum in the system as long as possible, all fittings and connections must be tight against the in-leakage of air. Not only the piping, but also the radiators and boiler must be made airtight. Because it is difficult to prevent some in-leakage of air around the stem of the ordinary radiator supply valve, the bellows type of packless supply valve is recommended.

All boilers with this type of system require a *compound pressure and vacuum gauge.* Such a gauge is necessary to tell the operator how the system is operating so that the greatest efficiency and comfort may be obtained from it. Certain movements of the gauge indicator signal air in-leakage.

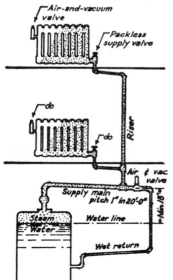

The installation cost of this system is slightly more than the 1-pipe gravity system— and is fully justified by the increased economy and comfort that result.

Any 1-pipe gravity system can easily be converted into a vacuum system by substituting air-and-vacuum valves for the air valves and substituting packless supply valves for the ordinary type used to control the radiators. At the same time, the system must be made tight against air in-leakage.

361

2-PIPE GRAVITY SYSTEMS

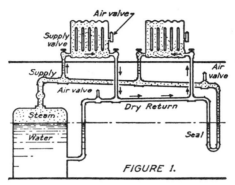

FIGURE 1.

The obvious cure for the defects of the 1-pipe system was to provide separate paths for the steam supply and the return of the condensate. This is the basis of all 2-pipe systems.

If a *dry return* were used in a 2-pipe system, the steam would flow completely through the radiator and into the return piping. Unless prevented, the steam might thus enter a following radiator from both the supply and return ends, trapping air in the center of the radiator. (See arrows on Figure 1.) This air, being unable to reach the air valve, lessened the heat output of the radiator. A water seal protected the return against the entrance of steam from the supply.

If a *wet return* were used, the return main would be below the waterline. This sealed each return pipe so that steam could not enter any radiator from the return pipes. The low point of the supply main was dripped into the return as shown by dotted lines on Figure 2.

Both the supply and the return in such 2-pipe systems had to have control or shutoff valves at each radiator. It was found that occupants of rooms tried to control the radiators by the operation of only one valve. If the supply valve (A) was closed, the condensing steam in the radiator would fill the radiator with air sucked in through the air valve (or full of water if an air-and-vacuum valve were being used). If the return valve (B) was closed, the radiator would operate on a 1-pipe system—the condensate having been forced to run back through the supply riser, causing hammer.

To overcome the difficulty of having two shutoff valves, *the radiator trap* was used at the return end of each radiator. The trap allowed air and condensate to pass but blocked steam.

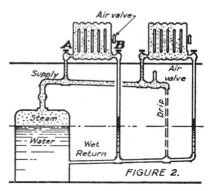

FIGURE 2.

2-PIPE GRAVITY SYSTEM
WITH CONDENSATION PUMP

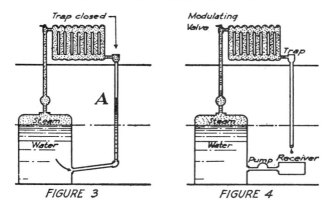

FIGURE 3

FIGURE 4

To overcome the inconvenience of having two shutoff valves at each radiator, the *radiator trap* may be used at the return ends. The trap allows air and condensate to pass but blocks steam. Only the supply valve need be operated to control the radiator with such an arrangement. Since no condensate exits through this orifice, it is possible to use a modulating valve that controls the supply of steam to the radiator.

The radiator trap closes against approaching steam. When it has closed, it shuts off the boiler pressure through the supply main and the radiator—pressure otherwise exerted on the return. Therefore, the water in the return would rise as at *A* in Figure 3, until its weight balances the pressure in the boiler exerted through the return main as indicated by the arrow in Figure 3. This "reversed" pressure might force water from the boiler into the return piping, lowering the level of the water in the boiler and thus causing serious damage.

There are three devices used to overcome this defect of the 2-pipe system using radiator traps. They are:
1. Differential loop (used on small systems with a boiler pressure of 8 oz. or less).
2. Boiler return trap (used on small systems up to approximately 8,000 square feet of radiation).
3. Condensation pump (has its widest application on relatively large installations).

Methods 1 and 2 are based on the principle of equalizing the pressure exerted on the return water to prevent forcing water from the boiler into the return piping.

Method 3 is illustrated diagrammatically in Figure 4. When the radiator trap closes, there is no tendency for any condensing steam in the return to suck water up into the pipe because the pipe is open to the air. An automatic float in the receiver actuates the water pump, returning the condensate to the boiler.

No air valves are used, since the air is eliminated through the condensation pump and receiver.

363

2-PIPE VACUUM
RETURN SYSTEM

Steam traveling in the supply pipes loses some of its initial pressure because of pipe friction and fittings. In a large steam system, this loss of pressure results in sluggish circulation if gravity alone is used to distribute steam to the radiators. Consequently, the boiler pressure must frequently be increased to unreasonably high pressures if the steam is to circulate against the pipe friction and the air resistance.

A method of overcoming this difficulty was sought and found in the creation of a vacuum in the return line of 2-pipe systems, increasing the pressure differential between the supply and the return to such an extent that the system operates satisfactorily on pressures of no more than 2 pounds. The circulation of the 2-pipe system thus becomes a positive mechanical operation that eliminates the forcing, noise, and difficulty of obtaining equal distribution of steam to the radiators.

The arrangement of boiler, piping, and radiators in the 2-pipe vacuum return line system is practically identical with the 2-pipe gravity system with radiator trap except for the addition of the vacuum pump. The steam enters the radiators through modulating supply valves. At the return end of the radiators, the steam is prevented from passing into the return piping by radiator traps. The traps allow air and condensate to escape. The return leads to a vacuum pump, which vents the air from the system, pumps the condensate back into the boiler, and creates a vacuum in the return piping. Steam is prevented from entering the return piping by drip traps.

For the system to operate successfully, the radiator traps must function properly. Any leakage of steam through the traps into the return pipes makes it difficult to maintain the proper vacuum—and proper vacuum is a vital prerequisite to maintaining positive circulation.

Quicker warming-up of the system, better removal of air from it, and better circulation in return lines having air or water pockets—are advantages of the 2-pipe vacuum return line system. Radiators may be located below the waterline of the boiler. Vacuum systems are somewhat more economical to operate because of the lower radiator temperatures that can be carried in mild weather when little heat is needed. Vacuum return line systems are best suited to large buildings, where the advantages to be gained will justify the slightly higher initial cost.

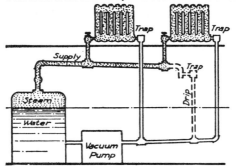

2-PIPE VAPOR-VACUUM SYSTEM

The boiling point of water varies with pressure. Water boils at 212°F under atmospheric pressure. As the pressure is reduced below atmospheric, the boiling point of water becomes lower. For instance, the temperature of steam at 20 inches of vacuum is 161°F. The vapor vacuum system utilizes this principle of physics by circulating vapor, which is steam at a pressure at or below atmospheric. Steam, at this lowered pressure, is lower in temperature and gives the vapor system an increased flexibility to handle very mild weather.

First, enough steam is produced by firing the boiler to fill the entire system with steam above atmospheric pressure. This steam enters the supply pipes and the radiation. The thermostatic traps at the return ends of the radiators close at the approach of steam.

The supply and the radiation being full of steam above atmospheric pressure, the fire is now reduced or turned off and steam is produced more slowly. This system is airtight so that, as the steam in the radiation condenses, the pressure is lowered below atmospheric. As the pressure is lowered, the boiling point of the boiler water is lowered, and steam continues to be formed. This process will continue at constantly lower temperatures and pressures as long as there is sufficient heat in the boiler water to generate steam at the pressure thus created, or until the fire is again accelerated. When the temperature of the steam falls somewhat below 212°F the radiator traps open, allowing the escape of water and air.

A device called a *boiler return trap* is used in connection with check valves. This arrangement is used to return the boiler water to the boiler, and to prevent the boiler water from entering the return piping when the thermostatic radiator traps are closed during the pressure period of the cycle.

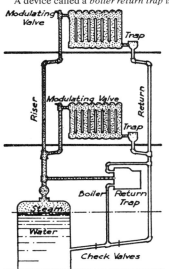

The cycle consists of alternate operation under pressure above atmospheric and under pressure below atmospheric. It is repeated as the heat demands of the building are met through manual or thermostatic operation of the fuel supply and dampers.

This system is well adapted to hand or stoker firing, but of course it is most commonly used with gas or oil firing.

DISTRIBUTING STEAM
EVENLY TO RADIATORS

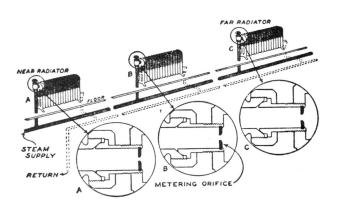

UNEVEN STEAM DISTRIBUTION. When the boiler is fired, a steam pressure is built up in the supply header of the boiler. The steam flows to the various radiators in the building against the resistance offered by the supply piping and fittings. The nearest radiator will offer comparatively little resistance against the flow of steam to it. The most remote radiator, because of the length of piping, the number of elbows, etc., will offer comparatively great resistance against the flow of steam to it. The steam will take the easiest path, so that the nearest radiators will be filled most quickly.

In extremely mild weather when the boiler pressure is low, the steam may never reach the most remote radiators. If the steam pressure is increased to a point which will overcome the resistance to the most remote radiators, overheating will result.

HOW EVEN DISTRIBUTION IS ACCOMPLISHED. Metering orifices, when properly selected and installed, effect even distribution of steam to all parts of the heating system. In the illustration it can be seen that the metering orifice in near radiator A is quite small. Radiator B, being more remote, offers comparatively great resistance against the flow of steam to it, so the metering orifice is slightly larger. In other words, the resistance against the flow of steam to each radiator has been equalized or balanced by the installation of the correctly sized metering orifice. The result is that each radiator receives a proportionate amount of steam regardless of pressure. In large buildings, intermediate metering orifices are used where needed in the branch mains to assist in primary distribution. They are placed in pipe lines between union flanges or between companion flange and gate valve.

SUPPLY VALVES. Supply valves are furnished in both the wheel and lever handle type. Steam flow to the radiators can be varied from an *off* position to a full *on* position, but cannot be increased over the amount determined by the metering orifice.

366

PRINCIPLES OF
RADIANT HEATING

DESIGN PRINCIPLES.   The heat loss from a room must be balanced by the heat supplied to that room. The ratio of radiant to convected heat in any given installation involves so many unpredictable variables that it excites the minds of physicists and advertising executives. The question is prudently left to them.

Burning one pound of 12,500-Btu coal in a system operating at 60% efficiency will deliver 7,500 Btu to a room, and it makes not the slightest difference what *species* of heat it is. A radiator full of steam at 215°F delivers 240 Btu per hour per square foot, which in a properly sized installation offers a completely satisfactory comfort level. In such systems, the occupant cares very little about how much of the heat is radiant and how much is convected. Experience has shown that humidity, air motion, and location of the heat sources should be considered carefully—but 240 Btu are 240 Btu.

Any "panel" performs largely as an air heating device. In the design of a radiant system, the SQUARE FEET, LOCATION, and the EMISSIVITY of the wall, floor, and ceiling area are the basis of the design. Find out what heat loss must be balanced by supply, divide this by the emissivity of the heated surface, and you will know how many square feet are required. Locate this area so as to produce "even" heating.

EXTERIOR WALLS.   Radiation of body heat to masses at lower temperature is a primary cause of discomfort. The idea of locating radiant surfaces in interior partitions as a method of "saving" fuel is ridiculous. The loss from radiant panels in outside walls to the outside can be reduced with proper insulation to a value consistent with that to be found if the panels were located elsewhere. The extra cost of such insulation is probably not excessive if comfort is the objective. For a surface temperature of 120°, the heat transfer to the room will be 90 heat units per hour.

CEILINGS.   Ceilings adjacent to the outside or to unheated spaces follow the same principle proposed above, unless they are so high as to have no noticeable radiant cooling effect. Using the ceiling-floor construction to heat the rooms both below and above may lead to serious control problems unless the loads are practically constant. For a surface temperature of 120°, the heat transfer to the room will be 70 heat units per hour.

FLOORS.   Floor slabs on fill are often assumed to reach temperature equilibrium but the locality and slab construction should be analyzed to establish the probable need for insulation. A floor slab at 85° will transfer 31.5 heat units per hour.

SOURCES OF HEAT.   In the following pages are shown some methods of installing pipes or tubing for steam or hot water. In localities where the cost of electricity is favorable, the type of rubber, plastic, or glass panels with resistance wire embedded in them makes the heating system as simple as turning on the lights. Hot air may be used as a medium in floor, wall, or ceiling channels. Baseboards, with and without convective paths, provide another method that in many cases is adaptable to remodeling as well as for use in new work where the requirement of simple construction disallows built-in piping.

WALL AT 120°	56% R **10**	RELATIVE PANEL AREAS FOR SAME HEAT DELIVERY
CEILING AT 120°	72% R **13**	
FLOOR AT 85°	48% R **29**	

In the bar chart above, the percentage figures for radiant heating effect show that no panel system does more than provide an approach to radiant heating. The dark part of the bars represents the air heating effect, which is in all locations a considerable part of the total.

RADIANT HEATING CONSTRUCTION

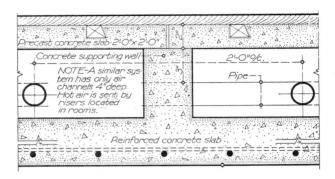

A special case of concrete floor slab construction where the heating effect is attained either by pipe coils in the air spaces or by circulation of hot air through these spaces. Note that this construction permits access to the pipes without cutting major structural parts. This construction can also be used where the floor is on the ground.

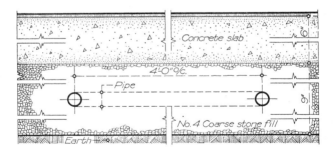

Typical concrete floor on ground with the coils located below the slab in the stone fill. This method can only be used where the fill is always dry. Note that no insulation is indicated below the coils as dry earth is a fairly good insulator but will absorb large quantities of heat at the start and return it at the end of the season, thus introducing considerable time lag when any attempt is made to change the floor temperature.

RADIANT HEATING
CONSTRUCTION

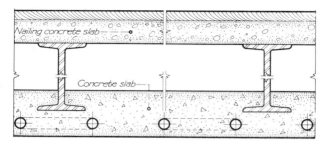

A special case of concrete ceiling slab construction in which the air space acts as insulation and materially limits upward heat flow to the floor above. Note that no plaster is indicated in this figure and that radiation will take place directly from the concrete ceiling surface.

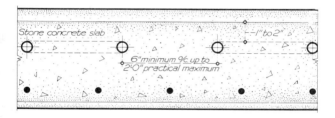

Typical concrete floor construction with the coils in top of slab for greater heating effect to the floor above.

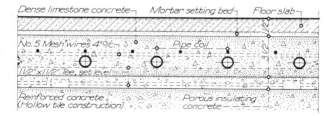

Typical fireproof floor construction using floor fill and coils in the fill. Note the use of insulating concrete to prevent heat flowing to the space below. This construction can also be used where the floor is on the ground.

RADIANT HEATING
CONSTRUCTION

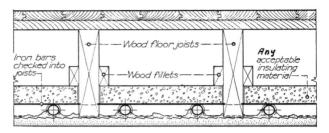

Typical wood joist ceiling construction with coils in plaster ceiling and with insulation to prevent heating the floor above. Scrim should be worked into the finished white plaster coat to prevent cracks. Pipes are generally spaced 6″ to 9″ apart although if uniform surface temperatures are not important, the spacing may be as much as 24″.

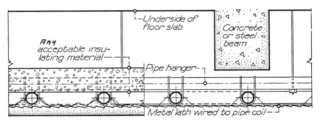

Typical fireproof ceiling construction with coils in plaster hung ceiling with insulation, scrim and pipe spacing as noted above.

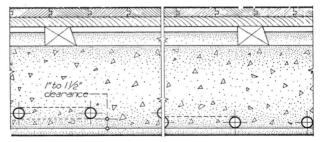

Typical floor-ceiling concrete construction. While wood flooring is indicated this method may be used with practically any floor finish such as cement, terrazzo, tile, linoleum, etc. Note that no insulation is indicated and therefore the heating effect is both up and down in proportion to the thermal values above and below the coils. The left hand section of this figure indicates the pipe coil being used as slab reinforcement.

METAL GRILLES FOR HEATING/COOLING

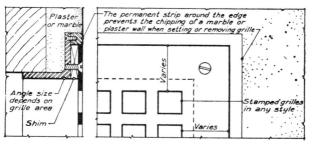

GRILLE IN PLASTER OR MARBLE FINISH

Plaster or marble

The permanent strip around the edge prevents the chipping of a marble or plaster wall when setting or removing grille

Angle size depends on grille area

Shim

Varies

Varies

Stamped grilles in any style

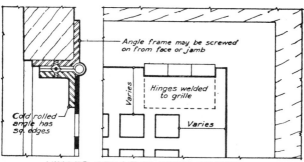

HINGED GRILLE IN WOOD FINISH

Angle frame may be screwed on from face or jamb

Hinges welded to grille

Cold rolled angle has sq. edges

Varies

Varies

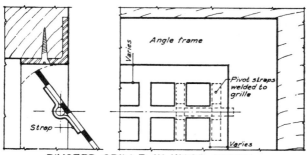

PIVOTED GRILLE IN WOOD FINISH

Scale Half Size

Varies

Angle frame

Pivot straps welded to grille

Strap

Varies

METAL GRILLES FOR HEATING/COOLING

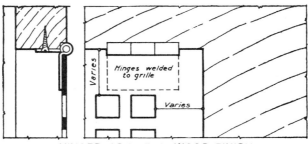

HINGED GRILLE IN WOOD FINISH

HINGED GRILLE IN WOOD FINISH

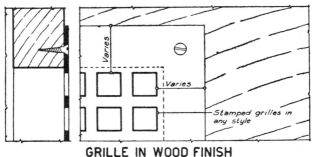

GRILLE IN WOOD FINISH
Scale Half Size

METAL GRILLES FOR
HEATING/COOLING

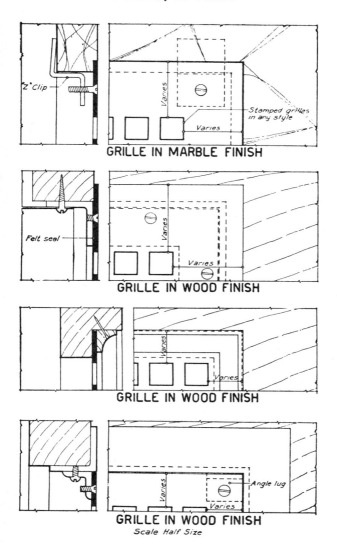

"Z" Clip

Varies

Varies

Stamped grilles
in any style

GRILLE IN MARBLE FINISH

Felt seal

Varies

Varies

GRILLE IN WOOD FINISH

Varies

Varies

GRILLE IN WOOD FINISH

Varies

Varies

Angle lug

GRILLE IN WOOD FINISH
Scale Half Size

COST OF
100,000 Btu

The charts on this page and the page following will permit rapid comparisons of heating costs.

FUEL COMPARISON.   Chart 2 shows that, at $100.00 a ton, 12,000 Btu coal produces 100,000 Btu for 70 cents. To obtain the same cost using 110,000 Btu oil, it would have to be available for 50 cents per gallon.

An analysis of local fuel costs and the calorific value of these fuels will provide a useful method of determining possible economies. It should be remembered, however, that each fuel has distinctive characteristics and advantages, consideration of which must form a part of any such analysis.

EFFICIENCIES.   The efficiency assumed in preparing the charts is noted on each one. If a different efficiency is to be used, the cost per 100,000 Btu will be equal to:

cost per 100,000 Btu shown on chart $\times$ $\dfrac{\text{efficiency shown on chart}}{\text{revised efficiency}}$

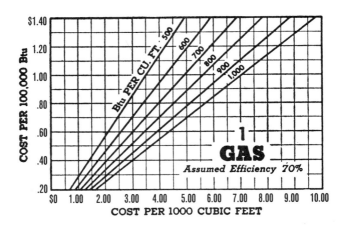

COST OF
100,000 Btu

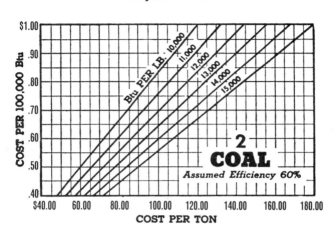

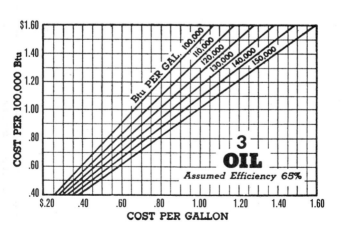

CHIMNEY SIZE FROM
HEAT DEMANDS OF BUILDING

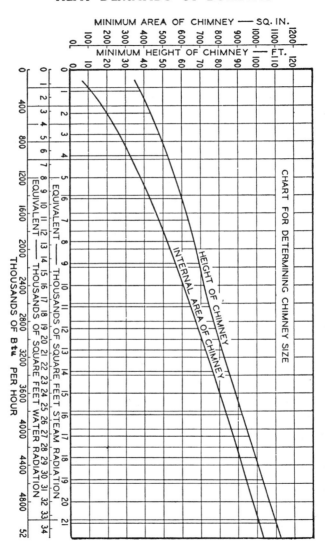

CHART FOR DETERMINING CHIMNEY SIZE

GRATE REQUIRED
FROM HOUSE VOLUME

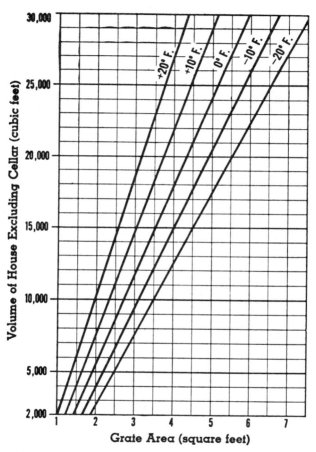

This chart may be used to determine the approximate sizes of coal-fired boilers or heaters from sketch drawings, before plans have progressed to a stage at which accurate heating calculations are possible. The table should be used with discretion, since a poorly built house will require a larger heating unit than that indicated, and a well built and well insulated one less. The chart is inapplicable to superinsulated houses.

Example: The chart shows that to heat a house of 19,000 cubic feet to 70° in −10°F. weather, a grate area of 4¾ square feet is required.

SIZE OF CHIMNEY FLUE
FROM GRATE AREA

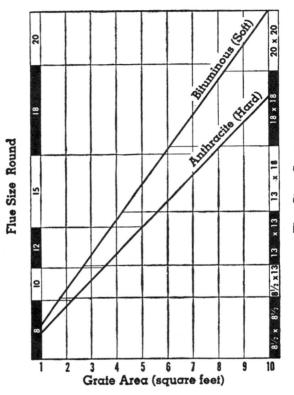

This chart can be used to determine approximate chimney sizes from sketch drawings when the plans have not progressed to a stage at which accurate heating calculations are possible. The chart is calculated from the formula:

$$A = \frac{182G}{\sqrt{H}}$$

in which:

    A = the area of the flue in square inches
    G = Grate area in square feet
    H = Height of chimney above the grate level in feet

The chart has been calculated for a chimney 36' high—an average height for most two-story residential construction.

Example: A coal-fired furnace, boiler, or heater having a grate area of 6½ square feet would require a 15"-diameter round flue or a 16" x 20" rectangular flue for anthracite; an 18"-diameter round or a 20" x 20" rectangular for bituminous.

378

STANDARD SIZES OF
CLAY FLUE LININGS

RECTANGULAR

Outside Dimensions of Flue Linings, Inches	Inside Dimensions of Flue Linings, Inches	Inside Cross Sectional Area of Flue Linings, Sq. Ins.	Thickness of Shell, Inches	Length, Feet
4½x 8½	3¼x 7¼	23.56	⅝	2
4½x13	3¼x11¾	38.19	⅝	2
8½x 8½	7¼x 7¼	52.56	⅝	2
8½x13	7 x11½	80.5	¾	2
8½x18	6¾x16¼	109.69	⅞	2
13 x13	11¼x11¼	126.56	⅞	2
13 x18	11¼x16¼	182.84	⅞	2
18 x18	15¾x15¾	248.06	1⅛	2
20 x20	17¼x17¼	297.56	1⅜	2
20 x24	17 x21	357.0	1½	2
24 x24	21 x21	441.0	1½	2

ROUND

Outside Diameter of Flue Linings, Inches	Inside Diameter of Flue Linings, Inches	Inside Cross Sectional Area of Flue Linings, Sq. Ins.	Thickness of Shell, Inches	Length, Feet
7¼	6	28.27	⅝	2
9½	8	50.26	¾	2
11¾	10	78.54	⅞	2
14	12	113.0	1	2
17¼	15	176.7	1⅛	2
20½	18	254.4	1¼	2
22¾	20	314.1	1⅜	2
27¼	24	452.3	1⅝	2
31	27	572.5	2	2½
34¼	30	706.8	2⅛	2½
37½	33	855.3	2¼	2½
41.0	36	1017.9	2½	2½

FIREPLACE
FLUE SIZES

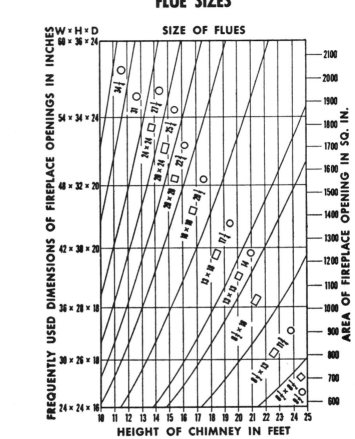

The commonly used rules of thumb for proportioning fireplace flues are very inaccurate methods since the draft of a flue may be said to vary inversely as the square root of the height. If we take a chimney 25'-0" from the top of the fireplace opening to the top of the flue as being satisfactory, on the basis of flue area equal to 1/12th the opening area we can derive the following formula from which the above chart has been plotted:

$$\text{Flue area in sq. ins.} = \frac{.41 \times \text{opening width}'' \times \text{opening height}''}{\sqrt{\text{chimney height}'}}$$

This chart should provide proper flue area for fireplaces having less than usual height.

FIREPLACE
CONSTRUCTION

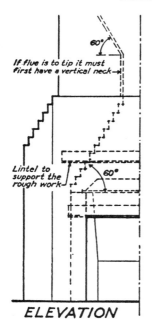

If flue is to tip it must
first have a vertical neck

Lintel to
support the
rough work

60°

60°

ELEVATION

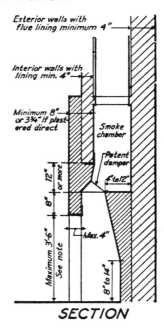

Exterior walls with
flue lining minimum 4"

Interior walls with
lining min. 4"

Minimum 8"
or 3¾" if plast-
ered direct

Smoke
chamber

Patent
damper

4"to12"

12"
or more

8"
or more

Maximum 3'-6"   See note

Max. 4"

8' to 14"

SECTION

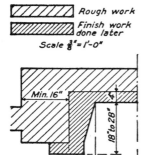

Rough work

Finish work
done later

Scale ¾" = 1'-0"

Min. 16"

4"

18 to 28"

5"

Slope of side same
as selected damper   PLAN

A good maximum height is 3'-6"
for openings up to 6' wide. For
openings over 6', a height of 4'
should not be exceeded. The higher
the opening is made, the greater
the chance of smoking becomes.

The width is usually greater than
the height of the opening. 30" is
a practical minimum width.

The method shown here is for
the finished work to be done after
the rough construction is completed. Occasionally it is all done
at one time.

Notice that fireplace constructions and dimensions vary widely.
Current references and code requirements should always be
checked before final designs are
drawn up.

381

SIZE OF CHIMNEY FLUE
BASED ON HEATING SYSTEM

| Warm air sq. in. leader pipe | RADIATION | | Round lining inside dimen. | Rectangular lining outside dimen. |
	Steam sq. ft.	Hot Water sq. ft.		
790	590	973	10″	* 8 1/2″x13″
1000	690	1140	*10″	13″ x13″
	900	1490	12″	*13″ x13″
	1100	1820	*12″	13″ x18″
	1700	2800	15″	*13″ x18″
	1940	3200	*15″	18″ x18″
	2130	3520		18″ x18″
	2480	4090	18″	20″ x20″
	3150	5200	*18″	20″ x24″
	4300	7100	20″	24″ x24″
	5000	8250		24″ x24″

*These sizes produce the exact minimum required areas of flue for heights of 25 ft. or greater. The other sizes without the asterisk will furnish slightly in excess of the required minimum.

As guides in selecting the size of a flue for a furnace, boiler, or heater, the recommendations of the manufacturer generally can be followed. Otherwise, 8″ x 12″ (nominal) should be considered the minimum. Chimneys for small units such as a wood stove, kitchen range, or freestanding fireplace often require an 8″ x 8″ or an 8″-diameter (nominal) round flue, but the requirements for a few are as small as 4″ x 12″ or 6″ diameter (nominal).

The most common error found in chimney construction is in the relation of the sectional area to the height. A chimney may be high enough and yet have too small an area to carry off the necessary volume of gases; or, the area may be sufficient but the height of the chimney too little to produce a draft that will draw enough air through the fuel bed for correct combustion. For chimneys less than 25′ high, the following formula can be used to get the approximate sectional area of the shorter flue:

$$A_L = \frac{5 A_H}{\sqrt{H_L}}$$

in which:

A_L = Section area of low flue

A_H = Sectional area of 25′-high flue

H_L = Height of low flue

No flue should ever be built without flue linings. It is best to have a width of 4″-thick brickwork between adjoining flues in the same chimney. Every flue in a chimney must be separate and have separate ash pits and clean-outs.

Even though a house is to be heated with gas or oil, a flue large enough for coal and wood should be installed. The uncertainty of supply of all types of fuels, as well as considerations of economy, may dictate a conversion later. Multiple-fuel boilers or furnaces should have their flues sized to accommodate maximum combustion gas volume.

The best location for the chimney is near the center of the house. It will not be cooled by outside temperatures and will aid in heating the building.

CHIMNEY CONSTRUCTION

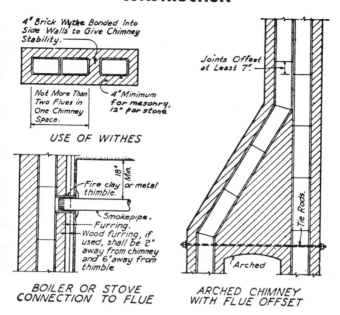

4" Brick Wythe Bonded Into Side Walls to Give Chimney Stability.

Not More Than Two Flues in One Chimney Space.

4" Minimum for masonry, 12" for stone

USE OF WITHES

Joints Offset at Least 7".

18" Min.

Fire clay or metal thimble.

Smokepipe.

Furring.

Wood furring, if used, shall be 2" away from chimney and 6" away from thimble

Tie Rods.

Arched

BOILER OR STOVE CONNECTION TO FLUE

ARCHED CHIMNEY WITH FLUE OFFSET

Masonry chimneys for residences must be constructed of solid masonry units, reinforced concrete, or refractory cement at least 4" thick, or of rubble stone at least 12" thick.

When two flues in the same chimney are separated only by fire-clay flue liners the joints of the liners must be staggered by at least 7".

Separate flue spaces within the same chimney must be separated by a masonry wythe at least 4" thick, bonded into the chimney structure. There can be no more than two flues in any series of adjoining flues in the same chimney without such separation.

Residential masonry chimneys must be lined with fire-clay flue lining at least ⅝" thick or with other approved flue lining. The lining must be separate from the chimney walls, with the gap unfilled.

Connectors must enter the chimney through a fire-clay or metal thimble or a masonry flue ring. The top of the connector must be 18" or more below any combustible materials, including lath and plaster or plasterboard. Neither the connector nor the thimble may protrude into the chimney. There may be no combustible materials within 6" of a ventilated metal thimble, or within 8" of a metal or fire-clay thimble.

Flues should be constructed so that they are as nearly vertical as possible to allow good draft and eliminate lodging places for soot and creosote. Preferably, direction changes should not depart more than 30° from the vertical, and in no case more than 45°.

Chimneys that have openings within their width or depth should have tie rods located over those openings to relieve thrust.

CHIMNEY
CONSTRUCTION

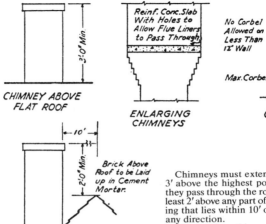

CHIMNEY ABOVE
FLAT ROOF

ENLARGING
CHIMNEYS

CORBEL

CHIMNEY ABOVE
SLOPING ROOF

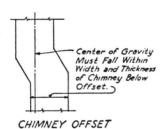

CHIMNEY OFFSET

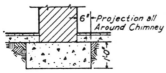

CHIMNEY FOOTING

Chimneys must extend at least 3' above the highest point where they pass through the roof, and at least 2' above any part of the building that lies within 10' of them in any direction.

Where the chimney passes through floors or roof, a clearance of at least 2" must be maintained between the chimney face and all wood framing members and any combustible materials. All spaces between chimneys and floors/ceilings must be firestopped with a noncombustible material 1" thick.

There must be no change in the size or shape of a chimney flue within 6" above or below the roof rafters or beams.

The total offset, overhang, or corbel of a chimney should not exceed ⅜ths of the width of the chimney in the direction of offset.

Corbeled chimneys should not be supported by hollow walls or walls of hollow units. Solid walls supporting corbeled chimneys must be at least 12" thick, and corbeling must not project more than 1" per course and not more than 6" in any case.

Chimneys must be built upon solid masonry, reinforced portland cement concrete, or refractory cement concrete foundations properly proportioned to carry the imposed weight without settling or cracking. The footing for an exterior chimney must start below the frost line.

ELECTRIC WIRING
ADEQUACY

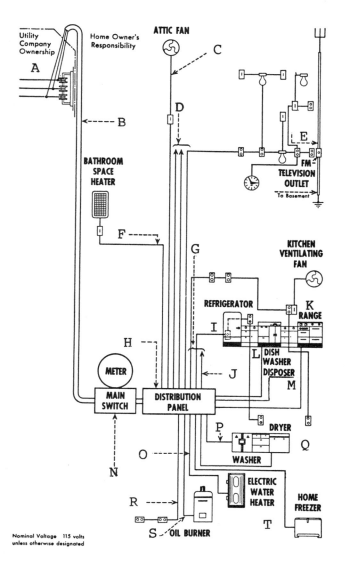

Nominal Voltage 115 volts
unless otherwise designated

ELECTRIC WIRING
ADEQUACY

The layout on the preceding page is representative only. Details are subject to change according to revisions in the National Electrical Code and may vary with requirements of local electrical codes.

A. Utility company ownership ends at different points. In an overhead service, ownership may end at the termination of the service drop or at the meter box. In an underground service, it may end at the connection point at the top of the service raceway or at the meter box. Other termination points are possible, as well.

B. All wiring beyond the point of the end of utility company responsibility is the responsibility of the building owner. Service is nominal 115/230-volt, 3-wire. For residential service, copper service-entrance conductors may be used as follows: #4, 100 amperes; #3, 110 amperes; #2, 125 amperes; #1, 150 amperes; #1/0, 175 amperes; #2/0, 200 amperes. Aluminum or copper-clad aluminum may be employed as follows: #2, 100 amperes; #1, 110 amperes; #1/0, 125 amperes; #2/0, 150 amperes; #3/0, 175 amperes; #4/0, 200 amperes.

C. An attic fan or whole-house ventilating fan may require or be most conveniently connected to a separate 20-ampere circuit with an automatic control or a switch located at a convenient point.

D. General-purpose branch circuits for lighting and occasional appliance use should be calculated according to current National Electrical Code provisions. Avoid overlong circuits and consequent voltage drop in large houses by using feeders and distribution panels. All receptacles installed in bathrooms, garages, and outdoors (where accessible from grade level) must have ground-fault circuit-interrupter protection.

E. Radio, television, and amateur radio/CB antenna systems must be installed in accordance with Article 810 of the National Electrical Code.

F. Bathroom space heater should be placed on a separate 20-ampere circuit, 115 or 230 volts as required.

G. In addition to other branch circuits, a minimum of two small-appliance branch circuits must be run to serve receptacles in the kitchen. A pantry, breakfast room, dining room, or similar area must also be served by required small-appliance branch circuits, which may be in addition to or a part of the two required kitchen circuits. These must be 20-ampere circuits, for small-appliance (including refrigeration equipment) use only. Use of 3-wire circuits is recommended to balance load and improve appliance operation.

H. Branch circuit protection for circuits is as follows: general purpose, 15 amperes; small-appliance and laundry, 20 amperes. Protection required for individual circuits with copper conductors is as follows:

ELECTRIC WIRING
ADEQUACY

#14, 15 amperes: #12, 20 amperes; #10, 30 amperes; #8, 40 amperes; #6, 60 amperes. For the same protection categories, aluminum or copper-clad aluminum conductors must be one size larger than copper (i.e., #10, 20 amperes).

I. Refrigerator may be connected to one of the two required small appliance circuits, but a separate dedicated circuit is recommended.

J. Though only two small-appliance circuits are required, in most homes adequacy demands at least one and often two more.

K. Range requires a separate 3-wire circuit, sized according to the provisions of Article 220 of the National Electrical Code.

L. In most cases the dishwasher should be connected to a separate 20-ampere circuit.

M. In most cases the garbage disposal unit should be connected to a separate 20-ampere circuit and controlled by a conveniently located switch as required.

N. Main disconnect may be a single fused switch or circuit breaker, or a set of two to six fused switches or circuit breakers, grouped. The main disconnect for a single-family dwelling with either an initial computed load of 10kw or more or an initial installation of six or more 2-wire branch circuits, must have a minimum rating of 100 amperes. Otherwise, the minimum service disconnect rating is 60 amperes.

O. An electric water heater usually requires a separate 20-ampere or 30-ampere circuit, and occasionally an even larger one. Most heaters operate on 230 volts. Consult local utility for wiring needs; sometimes water heaters are put on separate meters.

P. One 20-ampere small appliance circuit is required to serve each laundry room or laundry area; the clothes washer may be connected to this circuit, which must feed only receptacles in the laundry area.

Q. A clothes dryer must be served by a separate 230-volt circuit, sized according to Article 220 of the National Electrical Code.

R. A workshop generally requires at least one separate 20-ampere circuit to provide adequacy, and often it requires two or more.

S. A fuel-fired boiler or furnace generally requires a separate 20-ampere circuit with appropriate controls and safety switches. An electric furnace, heat pump, air conditioning unit, or electric space heating system requires different circuitry, generally 230-volt.

T. A freezer unit should be connected to a dedicated 20 ampere 115- or 230-volt circuit, as required.

ELECTRIC WIRING SYMBOLS

CEILING OUTLETS

Outlet for light _ _ _ _ _ _ _ _ _ _ _ _ _ _ _ _ _ ○

Blanked outlet _ _ _ _ _ _ _ _ _ _ _ _ _ _ _ _ _ Ⓑ

Drop cord _ _ _ _ _ _ _ _ _ _ _ _ _ _ _ _ _ _ Ⓓ

To indicate electric outlet when circle used alone might
be confused with columns or other symbols _ _ _ _ _ Ⓔ

Fan outlet _ _ _ _ _ _ _ _ _ _ _ _ _ _ _ _ _ _ Ⓕ

Junction box _ _ _ _ _ _ _ _ _ _ _ _ _ _ _ _ _ Ⓙ

Lamp holder _ _ _ _ _ _ _ _ _ _ _ _ _ _ _ _ _ Ⓛ

Lamp holder with pull switch _ _ _ _ _ _ _ _ _ _ _ ⓁPS

Vapor discharge lamp outlet _ _ _ _ _ _ _ _ _ _ _ Ⓥ

WALL OUTLETS

Outlet for light _ _ _ _ _ _ _ _ _ _ _ _ _ _ _ _ ⊦○

Blanked outlet _ _ _ _ _ _ _ _ _ _ _ _ _ _ _ _ _ ⊦Ⓑ

To indicate electric outlet when circle used alone might
be confused with columns or other symbols _ _ _ _ _ ⊦Ⓔ

Fan outlet _ _ _ _ _ _ _ _ _ _ _ _ _ _ _ _ _ _ ⊦Ⓕ

Junction box _ _ _ _ _ _ _ _ _ _ _ _ _ _ _ _ _ ⊦Ⓙ

Lamp holder _ _ _ _ _ _ _ _ _ _ _ _ _ _ _ _ _ ⊦Ⓛ

Lamp holder with pull switch _ _ _ _ _ _ _ _ _ _ ⊦ⓁPS

Vapor discharge lamp outlet _ _ _ _ _ _ _ _ _ _ _ ⊦Ⓥ

Exit light outlet _ _ _ _ _ _ _ _ _ _ _ _ _ _ _ _ ⊦Ⓧ

Clock outlet (*specify voltage*) _ _ _ _ _ _ _ _ _ _ _ ⊦Ⓒ

388

ELECTRIC WIRING
SYMBOLS

SWITCHES

Single pole switch ─ ─ ─ ─ ─ ─ ─ ─ ─ ─ ─ ─ ─ ─ ─ ─ ─ S

Double pole switch ─ ─ ─ ─ ─ ─ ─ ─ ─ ─ ─ ─ ─ ─ ─ ─ S_2

Three-way switch ─ ─ ─ ─ ─ ─ ─ ─ ─ ─ ─ ─ ─ ─ ─ ─ ─ S_3

Four-way switch ─ ─ ─ ─ ─ ─ ─ ─ ─ ─ ─ ─ ─ ─ ─ ─ ─ S_4

Automatic door switch ─ ─ ─ ─ ─ ─ ─ ─ ─ ─ ─ ─ ─ ─ S_D

Electrolier switch ─ ─ ─ ─ ─ ─ ─ ─ ─ ─ ─ ─ ─ ─ ─ ─ S_E

Key-operated switch ─ ─ ─ ─ ─ ─ ─ ─ ─ ─ ─ ─ ─ ─ ─ S_K

Switch with pilot lamp ─ ─ ─ ─ ─ ─ ─ ─ ─ ─ ─ ─ ─ ─ S_P

Circuit breaker ─ ─ ─ ─ ─ ─ ─ ─ ─ ─ ─ ─ ─ ─ ─ ─ ─ S_{CB}

Weatherproof circuit breaker ─ ─ ─ ─ ─ ─ ─ ─ ─ ─ ─ S_{WCB}

Momentary contact switch ─ ─ ─ ─ ─ ─ ─ ─ ─ ─ ─ ─ S_{MC}

Remote control switch ─ ─ ─ ─ ─ ─ ─ ─ ─ ─ ─ ─ ─ ─ S_{RC}

Weatherproof switch ─ ─ ─ ─ ─ ─ ─ ─ ─ ─ ─ ─ ─ ─ ─ S_{WP}

Fused switch ─ ─ ─ ─ ─ ─ ─ ─ ─ ─ ─ ─ ─ ─ ─ ─ ─ ─ S_F

Weatherproof fused switch ─ ─ ─ ─ ─ ─ ─ ─ ─ ─ ─ ─ S_{WF}

Pull switch from ceiling ─ ─ ─ ─ ─ ─ ─ ─ ─ ─ ─ ─ ─ ⓢ

Pull switch from wall ─ ─ ─ ─ ─ ─ ─ ─ ─ ─ ─ ─ ─ ─ ⊢ⓢ

SPECIAL PURPOSE OUTLETS

Special purpose, convenience outlet described in the plans or specifications. ▲

Any standard symbol with the addition of a subscript letter may be used to designate some special variation of standard equipment, and should be explained in the Key or Legend of Symbols.   a, b,

ELECTRIC WIRING SYMBOLS

CONVENIENCE OUTLETS

Duplex outlet (without subscript numeral) _ _ _ _ _

Outlet other than duplex: 1 = single, 3 = triplex _

Weatherproof outlet _ _ _ _ _ _ _ _ _ _ _ _ _

Range outlet _ _ _ _ _ _ _ _ _ _ _ _ _ _ _ _

Switch and outlet _ _ _ _ _ _ _ _ _ _ _ _ _ _

Radio and outlet _ _ _ _ _ _ _ _ _ _ _ _ _ _

Floor outlet _ _ _ _ _ _ _ _ _ _ _ _ _ _ _ _

PANELS, CIRCUITS, AND MISCELLANEOUS

Lighting panel _ _ _ _ _ _ _ _ _ _ _ _ _ _ _

Power panel _ _ _ _ _ _ _ _ _ _ _ _ _ _ _ _

Branch circuit concealed in ceiling or wall _ _ _ _ _

Branch circuit concealed in floor _ _ _ _ _ _ _ _

Branch circuit exposed _ _ _ _ _ _ _ _ _ _ _ _

Home run to panel board with number of circuits indicated by number of arrows _ _ _ _ _ _ _ _ _ _ _

Generator _ _ _ _ _ _ _ _ _ _ _ _ _ _ _ _ _

Motor _ _ _ _ _ _ _ _ _ _ _ _ _ _ _ _ _ _

Instrument _ _ _ _ _ _ _ _ _ _ _ _ _ _ _ _

Power transformer (*or draw to scale*) _ _ _ _ _ _

Controller _ _ _ _ _ _ _ _ _ _ _ _ _ _ _ _

Isolating switch _ _ _ _ _ _ _ _ _ _ _ _ _ _ _

ELECTRIC WIRING SYMBOLS

AUXILIARY SYSTEMS

Push button _____

Buzzer _____

Bell _____

Annunciator _____

Public utility telephone _____

Interconnecting telephone _____

Telephone switchboard (*or draw to scale*) _____

Bell-ringing transformer_____

Electric door operator _____

Fire alarm bell _____

City fire alarm station _____

Fire alarm central station _____

Automatic fire alarm device._____

Watchman's station _____

Watchman's central station _____

Horn _____

Nurse's signal plug _____

Maid's signal plug _____

Signal central station _____

Interconnection box _____

Battery _____

Special auxiliary outlet _____

CHECKLIST FOR
RESIDENCE ELECTRIC OUTLETS

FRONT AND OTHER ENTRANCES

⊦O **OR** O To conform with architectural style.

⊦⊘**WP** Near front entry door for decorative lighting. GFCI protection required.

⊦⊙ At front or other entrances.

⬜ᗡ Bell or chimes at interior location, usually kitchen or other central spot.

⬤ Special purpose outlet, such as illuminated house number, intercom station (indicate in subscript).

S One or more just inside door to control entrance lighting.

⬜ᵢ Intrusion alarm sensor.

RECEPTION HALL

⊦O **OR** O Central ceiling light, or wall, cove, or valance lighting, which should have switch.

⊦⊝ One in each usable wall space 2' or more in length, spaced so that no point along the floor line in any usable wall space is more than 6' horizontally from an outlet.

LIVING ROOM, LIBRARY, SUNROOM, BEDROOMS

O Usually one required; long and narrow rooms, rooms over 400 square feet, or rooms with low ceilings may require two or more. Should be switched.

⊦O For better decorative effect, valance, cove, or wall lighting may replace or augment ceiling lighting.

⊦⊝ One in each wall space 2' or more in length, with others located so that no point in any usable wall space (unbroken by doors, fireplaces, and so forth, but including sliding glass panels) is more than 6' horizontally along the floor line from an outlet. At least two such outlets should be switch controlled.

CHECKLIST FOR
RESIDENCE ELECTRIC OUTLETS

LIVING ROOM, LIBRARY, SUNROOM, BEDROOMS (Continued)

R   Radio and outlet (or TV, stereo, FM; indicate).

$\square_i$ Intrusion alarm sensors.

FA   Fire/intrusion alarm central or master station.

FS   Fire alarm sensors.

$\bigcirc_s$ Smoke detectors.

   Telephones.

$\bigcirc$ F   Ceiling fan for winter heat distribution and summer cooling.

CLOSETS

OR $\bigcirc$ Usually necessary in most closets and storage spaces, but not permitted by code in some.

S_D Automatic door switches are an added convenience.

   May be required in large closets.

HALLS

OR $\bigcirc$ One per 15′ of length of hall, and at turns in direction, intersections, stair headings, or stair landings. Use multiple switching for convenience.

   Same as living room, for vacuum cleaners, console table lights, and so on.

$\bigcirc_s$ Smoke detectors in vicinity of bedrooms.

FS   Fire alarm sensors.

CHECKLIST FOR
RESIDENCE ELECTRIC OUTLETS

RECREATION ROOM

At least one for every 150 square feet or major fraction thereof. Should be switched.

Valence, cove, or wall bracket lighting to supplement or take the place of ceiling lighting, depending upon use of room. Should be switched.

Same as for living room.

Radio and outlet (or TV, stereo, FM).

Telephone.

Intercom station.

Ceiling or wall fan for ventilation, summer cooling, and winter heat distribution.

Fire alarm sensors.

Intrusion alarm sensors.

LAUNDRY OR LAUNDRY SPACE

One for general background illumination.

One for each work center.

For hand iron.

One for each work center. The 6' rule applies as for living rooms.

For dryer.

For washer.

Fire alarm sensor.

Intrusion alarm sensors.

Intercom station.

394

CHECKLIST FOR
RESIDENCE ELECTRIC OUTLETS

GARAGE

(L)   One over hood location at each car space.

⊦O   Exterior light or lights to illuminate path to house for detached garages, and/or to flood driveway for either detached, attached, or integral garages, with multiple switch control from garage and house.

⊦⊝   At least one for every two car spaces, more preferable. Must be GFCI protected if readily accessible.

S WP3   Exterior switch to turn on garage lights, with 3-way or multiple low voltage control switching from house.

⬤ D   Automatic garage door opener(s).

◁   Intercom station.

FS   Fire alarm sensors.

☐ i   Intrusion alarm sensors.

KITCHEN, KITCHENETTE, PANTRY

O   Centrally located for general illumination. Fluorescent lamps in troffers or luminous ceiling most effective.

⊦O OR O   Over sink, range, work stations, choice depending upon window and cabinet arrangement. Additional undercabinet lights may be desirable.

⊦⊝   One for each separate counter space wider than 12″; the 6′ rule applies as for living rooms.

⊦⊝ R   Dedicated, for refrigerator.

⊦⊝ F   Dedicated, for freezer.

©   At high-visibility location, for clock.

⊦⊝ R   For electric range, and separate broilers or ovens.

395

CHECKLIST FOR
RESIDENCE ELECTRIC OUTLETS

KITCHEN, KITCHENETTE, PANTRY (Continued)

For special equipment, such as dishwasher, garbage disposal, and trash compactor; provide individual outlets as necessary.

Telephone.

Intercom master station.

Fire alarm sensor.

Intrusion alarm sensors.

For kitchen ventilating fan or range hood.

DINING ROOM, DINETTE, OR NOOK

One over table center.

Valance or cove light with switch may replace or supplement ceiling lighting.

No point of wall periphery at floor line should be more than 6' from an outlet, as for living room. Wall spaces 2' or more to have outlet. Outlet at 36" from floor in any wall space accommodating serving table or buffet, or any table used against wall. Outlet above any built-in serving counter or buffet top.

TERRACES AND PATIOS

One for each 15' or major fraction thereof of adjoining house wall. May be switched from inside door. GFCI protection required.

Exterior switch at convenient location to control patio, terrace, yard, walkway, garden, or architectural lighting.

At least one post, flood, or other type of light.

CHECKLIST FOR
RESIDENCE ELECTRIC OUTLETS

BATHROOMS AND WASHROOMS

○   One if floor area is 60 square feet or more. Enclosed shower compartment requires a ceiling lighting fixture approved for use in damp locations, switched from outside the shower compartment.

⊢○   One on each side or a single light above mirror.

⊢⊖   One near and at level of bottom edge of mirror for electric razors, hair dryers, etc. GFCI protection is required.

◉   Sunlamp or overhead auxiliary heater at convenient point.

$S_{.T}$ Timer switch to control sunlamp.

Ⓕ   Ventilating fan.

☐ᵢ   Intrusion alarm sensors.

BOILER OR UTILITY ROOM

Ⓛ$_{PS}$ One in each enclosed space, one over workbench, one in front of heating/cooling equipment. Additional lights for a minimum of one to each 150 square feet of open space.

⊢⊖   At workbench and/or other convenient location.

◉$_a$ For heating/cooling equipment.

◉$_b$ For electric water heater.

[FS]   Fire alarm sensor.

☐ᵢ   Intrusion alarm if windows or outside doors are present.

CHECKLIST FOR
RESIDENCE ELECTRIC OUTLETS

BASEMENT (UNFINISHED)

Ⓛ   One in each enclosed space. Additional lights for each 150 square feet of open space. Switch control, part of 3-way system, at interior stairhead, with pilot.

S WP 3   Exterior switch, part of 3-way system, beside outside door or bulkhead.

At least one in convenient location.

FS   Fire alarm sensors.

Intrusion alarm sensors.

ATTIC (UNFINISHED)

Ⓛ   One in each enclosed space. One or two additional lights to illuminate open spaces.

At least one for general use.

S P   One at foot of attic stairs to control stairway and attic lights, with pilot.

FS   Fire alarm sensors.

STAIRWAYS

OR   One at head and foot of each flight between active floors, or a single light on straight, short flights.

One at each blind or semienclosed landing.

S 3   Multiple control switches to provide convenient switching of lights on two floors from either floor.

CHECKLIST OF
ELECTRIC EQUIPMENT

FRONT HALL

Doorbell Chimes	
Lighted House Number	20
Ceiling Fixture	
Bracket Fixtures	
Table Lamps	
Vacuum Cleaner	600
Telephone	

LIVING ROOM OR LIBRARY

Ceiling Fixture	
Bracket Fixtures	
Floor Lamps	
Table Lamps	
Electric Fan	200
Vacuum Cleaner	600
Unit Air Conditioner	1,500
Humidifier/Dehumidifier	700
Stereo System	100
Electric Clock	2
Television	125
Telephone	

DINING ROOM

Ceiling Fixture	
Bracket Fixtures	
Vacuum Cleaner	600
Electric Fan	200
Toaster	1,000
Coffeemaker	1,650
Chafing Dish	700
Warming Tray	400
Electric Clock	2
Telephone	

KITCHEN

Ceiling Fixture	
Bracket Fixtures	
Kitchen Desk Lamp	
Undercounter Lamps	
Electric Range	7,000–17,000
Refrigerator	650
Freezer	700
Dishwasher	1,600
Garbage Disposal Unit	1,500
Electric Clock	2
Ventilating Fan	50
Juicer	150
Toaster	1,000
Coffeemaker	1,650
Grill	1,100
Griddle	1,500
Blender	800
Microwave Oven	650
Food Processor	1,000

KITCHEN (Continued)

Single Burner Hotplate	750
Double Burner Hotplate	1,100
Waffle Iron	1,500
Electric Frying Pan	1,500
Meat Slicer	800
Deep Fat Fryer	1,400
Slow Cooker	800
Dehydrator	900
Electric Mixer	300
Corn Popper	1,200
Trash Compactor	500
Broiler-oven	1,500
Telephone	
Intercom Master	100

BASEMENT

Ceiling Fixture	
Oil Burner/Furnace Blower	800
Electric Water Heater	5,500

LAUNDRY

Clothes Washer	800
Clothes Dryer	6,000
Flat or Steam Iron	1,100
Ventilating Fan	50

BEDROOM

Ceiling Fixture	
Bracket Fixtures	
Floor Lamps	
Table Lamps	
Clock Radio	35
Electric Blanket	180
Vacuum Cleaner	600
Sun/Heat Lamp	800
Electric Fan	200
Unit Air Conditioner	1,500
Medicinal Vaporizer	
Heating Pad	100
Hair Blow Drier	1,200
Hair Curler	45
Telephone	

BATHROOM

Ceiling Fixture	
Bracket Fixtures	
Ventilating Fan	50
Fan Heater	1,450
Hair Blow Drier	1,200
Hair Curler	45
Heat/Sun Lamp	800
Electric Razor	10

NOTE: This checklist may be used to ensure completeness of plans, as a questionnaire for the owner to determine the electrical conveniences desired, and as a guide in determining the number and capacities of circuits required. The figures given are for the wattages of the equipment. They are subject to some variation either up or down, but are representative.

WATTAGE OF
ELECTRIC OUTLETS

(1.) Locate the outlets in plan, as nearly as possible according to the following rule: using direct lighting, the units should be spaced not to exceed the distance from floor to outlet; using indirect or semi-indirect lighting, the spacing of the units should not exceed the ceiling height; from wall to unit should not exceed half the regular spacing.

(2.) Determine the number of square feet of floor to be lighted by each unit.

(3.) Select the proper wattage per square foot for the given class of occupancy from the following table.

Watts per Square Foot for General Illumination

Class of occupancy		Good practice	Min.
Auditoriums		4.0	1.0
Auto Parking Spaces		2.0	.25
Banking Room, w/added local illum		4.0	3.5
Churches	Auditorium	2.0	1.0
	Sunday School Room	5.0	3.0
Drafting Room, no local illum.		7.0	3.5
Factory, w/added bench illum.		4.0	2.0
Garage	Dead storage	2.0	.25
	Filling station yards	2.0	.50
	Live storage	3.0	.50
	Repairs, wash rack	3.0	1.0
Hospitals	Laboratories	5.0	3.0
	Private rooms	5.0	2.0
	Wards	3.0	2.0
Hotels	Bedrooms	3.0	2.0
	Dining room	4.0	2.0
	Lobby	5.0	2.0
Library	Reading room	6.0	3.5
	Stacks	12.0 per lin. ft.	
Offices, general		5.0	3.5
Recreation	Gymnasium	5.0	3.0
	Indoor swimming pool	5.0	3.0
	Shower rooms	2.0	1.0
Restaurants		3.0	2.0
Schools	Classrooms, man. train.	5.0	3.0
	Sewing rooms	7.0	5.0
Stores		6.0	3.0

(4.) Find wattage of outlet required (= area per outlet × watts per square foot).

It is doubtful that a more accurate method than the above would be consistent with the indeterminate factors of maintenance, adherance by the tenant to the design assumptions, safety factor in wiring installation, variation in footcandle intensities recommended by various authorities, and so forth. The foregoing method will assuredly provide sufficient capacity at the outlets for the service required. However, greater accuracy can be obtained (if desired) by first determining the lighting installation design by one or another of the usual methods (zonal cavity, lumen-per-foot, etc.) as required, and then determining the most appropriate wattage factors.

ELEMENTS OF
FLUORESCENT LIGHTING

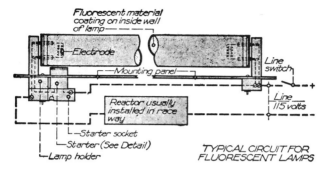

Fluorescent material coating on inside wall of lamp

Electrode

Line switch

Mounting panel

Reactor usually installed in race way

Line 115 volts

Starter socket

Starter (See Detail)

Lamp holder

TYPICAL CIRCUIT FOR FLUORESCENT LAMPS

FLUORESCENT PRINCIPLE. Fluorescence is a natural phenomenon by which short wavelengths of radiant energy are converted to longer waves. The term is applied to a group of light sources first made available in 1938, in which invisible ultraviolet radiations are changed to visible light. By coating the inside of low-pressure mercury lamps with materials known as *phosphors*, a large percentage of the energy input of the lamps is radiated as visible light.

The phosphors used are of many types, many hundreds being known. The actual choice of a phosphor depends on the color of light desired and the range of ultraviolet they utilize.

OPERATION. At each end of the lamp there is an electrode in the form of a small coil of wire, coated with a material that freely emits electrons when heated. Electrons are necessary to carry the arc current that passes through the vaporized mercury. Since mercury is a liquid at normal temperatures, a slight amount of argon gas is used to facilitate starting.

1. THE STARTER. A self-timing device in the starter preheats the lamp electrodes and then automatically switches the circuit in such a way as to provide a high-voltage surge to start normal lamp operation. If the lamp arc fails to strike, the cycle is repeated. The starter is in the form of a small aluminum cylinder that has bayonet-type contacts and is readily replaceable.

2. REACTOR. This prevents the arc current from increasing beyond the limit set for each size of lamp. Essentially, it is a choke. The reactor is also called the *Ballast* or *Current Limiting Device*.

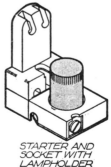

STARTER AND SOCKET WITH LAMPHOLDER

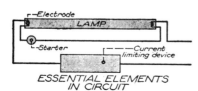

Electrode

LAMP

Starter

Current limiting device

ESSENTIAL ELEMENTS
IN CIRCUIT

MULTIOUTLET ASSEMBLIES AND PLUG-STRIPS

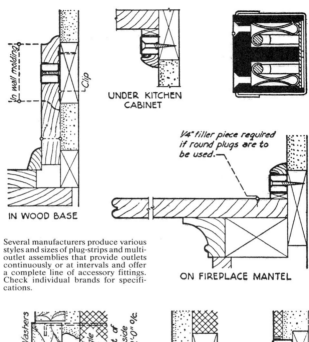

UNDER KITCHEN CABINET

IN WOOD BASE

Several manufacturers produce various styles and sizes of plug-strips and multioutlet assemblies that provide outlets continuously or at intervals and offer a complete line of accessory fittings. Check individual brands for specifications.

¼" filler piece required if round plugs are to be used.

ON FIREPLACE MANTEL

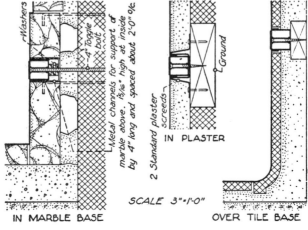

IN MARBLE BASE

Metal channels for support of marble above. 15/16" high at inside by 4" long and spaced about 2'-0" o/c.

IN PLASTER

2 Standard plaster screeds

SCALE 3"=1'-0"

OVER TILE BASE

FLOODLIGHTING
OF BUILDINGS

BUILDINGS AND MONUMENTS

Representative Building Materials	Approx. Reflection Factors Percent	Surroundings	
		Bright	Dark
White terra cotta White plaster Cream terra cotta Light marble	70-85	15	5
Light gray limestone Concrete, tinted Buff limestone Smooth buff face brick	45-70	20	10
Briar Hill sandstone Smooth gray brick Medium gray limestone Common tan brick	20-45	30	15
Dark field gray brick Common red brick Brownstone Stained shingles/siding	20	50	20

Utilitarian and Protective Purposes

Construction Work 2-10
Dredging 2
Gasoline Service Stations
    Buildings and Pumps 20-30
    Yard and Driveways 1.5-5
Parking Spaces 5
Protective Industrial 2-20
Quarries 5
Shipyards (construction) 5-30

Special Applications

Trees 5-20
Flags 20-100
Loading Docks 5
Loading Platforms 5-10
Signs 20-100
Smokestacks 15
Art Glass Windows 20-200
Waterfalls 10
Water Tanks 15

NUMBER OF PROJECTORS. Use the following formula to determine the number of projectors required to produce the required level of illumination:

$$\text{Number of projectors} = \frac{(\text{Area in Square Feet}) \times (\text{Footcandles})}{0.7 \times (\text{Beam Lumens}) \times \text{CBU}}$$

Area = area of surface to be lighted, in square feet.

Footcandles = level of illuminance, from table above.

0.7 = the *Maintenance Factor*, representing an allowance of 30% for depreciation in service.

Beam lumens = level of luminous flux, obtained from manufacturers' catalogs for the specific equipment under consideration.

CBU = *Coefficient of Beam Utilization*, representing the percentage of the beam lumens striking the lighted area (should be between 60% and 90%).

ROOM COLORS
AFFECT LIGHTING

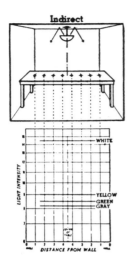

Indirect

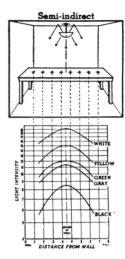

Semi-indirect

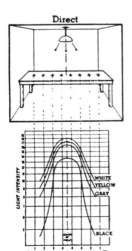

Direct

It must be noted that these charts are to indicate the principle involved rather than to serve as quantitative guides for design.

All surfaces absorb some of the light that strikes them. Consequently, the architect will take this factor into consideration when selecting a surfacing material. A black surface absorbs practically all the light that strikes it while a white surface absorbs practically none. An ideal white would reflect 100% of all the light that is directed upon it. Magnesium oxide, reflecting about 98%, generally is used as a *standard white* by scientific investigators. As this condition is not obtainable in common practice, maximum efficiency must be tempered with a consideration of commercial availability.

The charts in *Figures 1* and *2* are indicative of the wide variation resulting from the use of different colors in reflecting surfaces of a room. The colors selected for test are those having an adaptability to business and industrial use under various types of illumination. Basis for the diagrams is a series of tests conducted by the New Jersey Zinc Company technologists, contained in a 16-page booklet titled *Using Paint As Light.*

IMPORTANCE OF
LIGHTNING PROTECTION

The fundamental objective of lightning protection for buildings is to provide a means by which a discharge may enter or leave the earth without passing through a nonconducting part of the structure such as wood, brick, tile, or concrete. Damage is caused by the heat and mechanical force generated in such nonconducting portions by the discharge; in metal parts, the heat and mechanical forces are of negligible effect if the metal has sufficient cross-sectional area. There is a strong tendency for lightning discharges on structures to travel on metal parts that extend in the general direction of the discharge. Hence, if metal parts of proper proportions and distributions are provided and adequately grounded, damage can be prevented.

REASONS FOR LIGHTNING PROTECTION. That lightning involves a very real personal hazard and is the cause of tremendous financial loss is conclusively proved by statistics compiled by numerous agencies, organizations, and businesses that have an interest, for various reasons, in this phenomenon. The reasons for installing lightning protection equipment on any given building may be outlined as follows:

1. LOSS OF LIFE AND PHYSICAL INJURY. The lives of the occupants of many buildings are in danger during every electrical storm. The degree of danger depends upon the type of building, the amount of protection it affords, and its geographical and topographical location. Each year some 150 persons are killed by lightning, and many others are injured. Though most casualties occur in the open and few occur in modern structures, a substantial number take place in older buildings. The danger is considerable, and the aftereffects of even a minor shock are extremely unpleasant.

2. TERROR OF LIGHTNING. It is said that Napoleon feared lightning more than he did enemy fire. A great many persons experience real terror during an electrical storm, which can be eliminated by the feeling of security resulting from lightning protection.

3. DAMAGE TO BUILDINGS FROM DIRECT LIGHTNING STROKES. Many costly buildings have been entirely demolished as a result of a direct lightning stroke. In other cases the resultant fire has finished the job started by the lightning. In total, some 30,000 structures are destroyed or badly damaged every year in the United States alone, accounting for an estimated annual loss of $20,000,000 or more. Though many modern buildings, especially large ones, are quite lightning-proof because of their design, many others of recent construction include numerous disconnected metal parts that actually increase the lightning hazard to property.

4. DAMAGE TO BUILDINGS FROM "FLICKERS." An enormous amount of lightning damage done to buildings occurs without the knowledge of the building owners. When lightning strikes a building directly, plenty of evidence is left of its visit; but often only a minor part of a nearby stroke "flickers" onto the building. These minor strokes injure gutters, downspouts, and flashings, crack tile, slate, and other roofing materials, and cause rapid disintegration of chimney walls, foundation walls, and ceilings. Signs of damage from flickers are often misinterpreted as signs of settling by the building.

5. DAMAGE TO BUILDING CONTENTS. Furniture, equipment, personal belongings and treasures, valuable papers, and other building contents having value far in excess of the cost of an adequate lightning protection installation may be destroyed by a lightning stroke.

6. ELECTRICAL INSTALLATION AND EQUIPMENT. Cost of replacement in case of damage is high today. As a rule, insurance is inadequate because in many cases there is no fire or prompt action reduces the extent of the fire loss to a negligible proportion. The only method of insuring against such a loss is by eliminating the hazard through installation of proper lightning protection.

405

LIGHTNING PROTECTION
FOR TREES

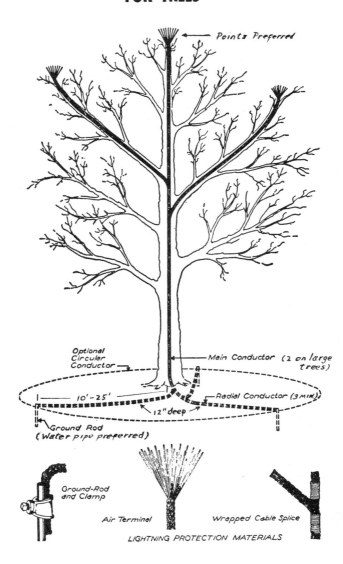

Points Preferred

Optional
Circular
Conductor

Main Conductor (2 on large trees)

10'-25'

Radial Conductor (3 min.)

12" deep

Ground Rod
(Water pipe preferred)

Ground-Rod
and Clamp

Air Terminal

Wrapped Cable Splice

LIGHTNING PROTECTION MATERIALS

LIGHTNING PROTECTION
FOR TREES

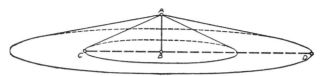

Minimum Cone BC = 2AB   Maximum Cone BD = 4AB.

CONE OF INFLUENCE. The 1975 National Fire Protection Association *Lightning Protection Code* (NFPA 78-75) points out that experiments have shown that tall conducting masts "will afford protection to a cone-shaped volume the height of the mast and with a ground area having a radius twice the height of the mast."

Thus a lightning-protected tree will tend to protect nearby structures or trees that are totally within the cone-shaped space represented above. However, as Dr. M. G. Lloyd has shown, the *cone of influence* is not a zone of complete protection, and lightning occasionally strikes within such a zone.

EFFECT OF LOCATION. Trees standing alone or above their neighbors and trees along avenues, streams, and lakes, are struck more frequently than others.

SPECIES SUSCEPTIBILITY. There is considerable difference in susceptibility to lightning attack among trees of different species. Studies made abroad tend to show the following are relatively free of lightning attack:

beech	birch
horse chestnut	holly

The same studies indicate that the following are struck frequently:

oak	maple
elm	ash
pine	spruce
poplar	

As a general rule decayed or rotten trees are greater sufferers from lightning than sound, undecayed specimens. Deep-rooted trees are generally believed to be more liable to lightning injury than those with shallow and widespreading root systems.

PRINCIPLES OF TREE PROTECTION. Air terminals should be placed at the highest point or points in a tree. It is unnecessary to place air terminals on lateral branches where such terminals would fall within the cone of influence. Copper cables may be attached to trees with copper nails—never with steel nails—in order to avoid electrolysis, or other approved fasteners may be employed. The use of insulated fasteners is not recommended. Three or more ground terminals should be provided for each conductor.

LOCATION OF
AIR TERMINALS

Prominent dormers, cupolas, wood flag poles, chimneys, spires, monitors and other similar projections, and all gables should have air terminals.

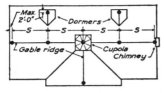

Gable Roof with Cupola and Dormers

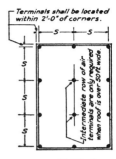

Terminals shall be located within 2'-0" of corners.

Intermediate row of air terminals are only required when roof is over 50ft wide

Flat Roof or low pitch roof

Mansard roofs should have air terminals at each corner of deck and also intermediates if necessary.

Max. 2'-0"

Mansard Roof

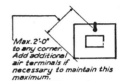

Max. 2'-0" to any corner. Add additional air terminals if necessary to maintain this maximum.

Chimney Cap

(S) Spacing of air terminals on roofs and roof projections should not exceed 20 feet.

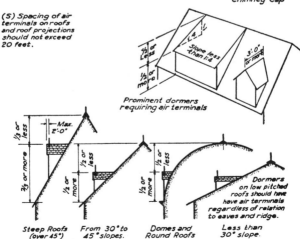

Slope less than 1:4

½ or Less

½ or more

3'-0" or more

Prominent dormers requiring air terminals

Max. 2'-0"

⅓ or less

⅔ or more

½ or less

½ or more

½ or less

½ or more

Dormers on low pitched roofs should have air terminals regardless of relation to eaves and ridge.

Steep Roofs (over 45°) *From 30° to 45° slopes.* *Domes and Round Roofs* *Less than 30° slope.*

HEIGHT OF DORMER RIDGE IN RELATION TO EAVES AND RIDGE OF ROOF WILL DETERMINE WHETHER AIR TERMINALS ARE NEEDED

LIGHTNING PROTECTION

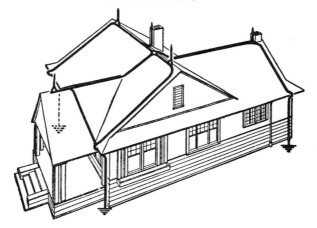

EFFICACY OF LIGHTNING RODS. An analysis of lightning fires showed that only 5 out of 100 occurred in protected structures, some of which had old or defective systems.

AIR TERMINALS. Rods are single-pointed and composed of solid or tubular copper, cast or tubular aluminum, cast bronze, or nickel-plated bronze, and obtainable in various finishes. Rods should be placed on upward projections such as chimneys and towers, on flat roofs 50′ o/c, and on the edges of flat roofs and the ridges of pitched roofs 20′ o/c. Rods should be from 10″ to 36″ above the object to be protected; above 24″ in height the rods should be braced.

CONDUCTORS. There must be two paths from each rod to ground, both *outside* the building (though in certain instances individual rods may be "dead-ended" with only one path to the main conductor). Conductors may be stranded copper cable, either plain, lead-coated, or tinned; plain or tinned solid copper strip; or pure aluminum stranded cable. Conductors should be in straight runs without sharp bends. Painting conductors above the ground does not affect their value.

FASTENERS. Only approved fasteners, selected for compatibility with the mounting surface, should be used; there are variations for all purposes. Insulators are not required. Holes made in roofing or walls by the fasteners should be made watertight.

GROUND. A solid, permanent electrical connection must be made from conductors to moist earth. The connection point should be made not less than 1′ below grade, and not less than 2′ from the foundation. There must be ample contact area; a resistance of no more than 50 ohms to ground is desirable. The metal used in the grounding electrodes must have little or no susceptibility to corrosion. A metal rod extending 10′ vertically into the earth is satisfactory where the earth is wet and clayey. Where the topsoil is shallow, or the earth is sandy, gravelly, or rocky, other approved grounding configurations must be used in special arrangements. Connections to a water pipe outside the building makes one of the best (and simplest) grounds available, and is used in addition to other required groundings. Only approved electrodes and connectors may be used.

409

CONCEALED LIGHTNING PROTECTION SYSTEM

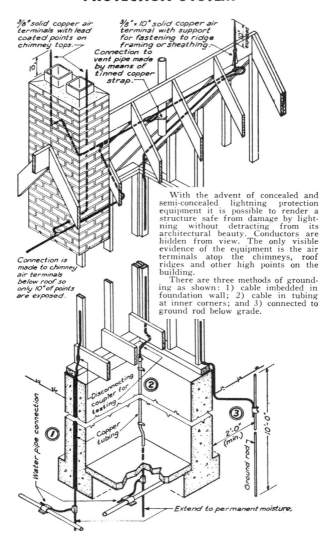

3/8" solid copper air terminals with lead coated points on chimney tops.

3/8" x 10" solid copper air terminal with support for fastening to ridge framing or sheathing.

Connection to vent pipe made by means of tinned copper strap.

Connection is made to chimney air terminals below roof so only 10" of points are exposed.

With the advent of concealed and semi-concealed lightning protection equipment it is possible to render a structure safe from damage by lightning without detracting from its architectural beauty. Conductors are hidden from view. The only visible evidence of the equipment is the air terminals atop the chimneys, roof ridges and other high points on the building.

There are three methods of grounding as shown: 1) cable imbedded in foundation wall; 2) cable in tubing at inner corners; and 3) connected to ground rod below grade.

Disconnecting coupler for testing

Water pipe connection

Copper tubing

2'-0" (min.)

Ground rod

10'-0"

Extend to permanent moisture.

410

OPEN SPECIFICATION FOR LIGHTNING PROTECTION

Architects will find that it best serves the interests of their clients to exclude lightning protection from the general contract and to take separate bids. The following short form specification is suggested as a guideline. The science of lightning protection and its correct installation is a highly technical and specialized field. The Architect's best guarantee of an effective and honest installation is to deal with a manufacturer whose integrity and record are beyond reproach and with installation specialists whose knowledge, expertise, and workmanship can be depended upon.

Lightning protection is an absolute necessity for many types of buildings and a desirable precaution for buildings of all kinds. A lightning stroke invariably deadens telephones and renders private electric pumping systems inoperative on country estates. Churches, schools, hospitals and public or semipublic buildings with spires, chimneys, domes, or projecting gable ends almost invariably require lightning protection. Low, flat buildings may be in greater need of lightning protection than higher buildings in more favorable locations are. Metal-roofed or metal-clad buildings used for manufacturing where explosive gases, dangerous fumes, or dust are created require an "excess" lightning protection system. Country club buildings as a rule occupy exposed locations and make ideal lightning targets.

1. WORK INCLUDED. This Contract includes the furnishing of all labor and materials for complete protection of the building from lightning damage and the installation of lightning protection equipment on trees that have been noted on the plot plan.

> NOTES: Valuable trees should be protected since it takes a generation to grow a beautiful shade tree. Each tree adds hundreds of dollars to the value of grounds. Trees overhanging the building itself or in close proximity to it may attract lightning to the building even though the building is properly rodded. Trees can be protected with a simple, inexpensive, and practically invisible system.
>
> The Owner should be advised by the Architect that cooperation is required for continued protection in the future. The location of a new building in close proximity to the protected structure, the growth of a tree so that it commands or overhangs the protected building, the addition of metal vents and pipes that project upward from the roof's surface, the addition of dormers or wings and porches, the installation of radio antennas or electrical and telephone wiring, the addition of a flagpole, the reroofing of a building in which the conductor system is reapplied by the roofers in a haphazard manner—all these things affect the efficiency of the lightning protection system as originally installed. Such changes may require the installation of additional lightning protection equipment to render the altered building lightning-proof.

2. SHOP DRAWINGS. Furnish a complete layout of the lightning protection system for the Architect's approval before starting the installation of materials.

411

OPEN SPECIFICATION FOR LIGHTNING PROTECTION

3. COOPERATION OF OTHER TRADES. Instruct the general contractor, and any subcontractors whose cooperation may be required, on the installation details so that chimney anchors, cable fastenings, concealed conductors, or any other devices may be placed at the appropriate time during construction without the need for rework or changes.

4. MATERIALS AND WORKMANSHIP. Furnish and install materials in accordance with the most recent revision of the "Lightning Protection Code" as promulgated by the National Fire Protection Association (NFPA No. 78) or the same as adopted by the American National Standards Institute (ANSI C5.1), and in accordance with the provisions of the "LPI Installation Code" of the Lightning Protection Institute. Use no materials or devices that do not bear Underwriters' Laboratories Labels. Full compliance with the manufacturer's rules and regulations for the installation system on this particular building is also a requirement. Upon completion of the installation of the lightning protection equipment, make successful application for award of Underwriters' Laboratories Master Label plate and furnish same to Owner. In addition, successfully file for LPI Systems Certification, including Form A, "Pre-Installation Witness of Grounding Report and Certification" and the "LPI Code Compliance Agreement," and Form B, "Post-Installation Inspection and Maintenance Procedure Agreement."

5. SAMPLES AND SCHEDULE OF MATERIALS. Submit samples of terminals, anchors, conductors, and other visible parts of the system to the Architect for selection and approval at the time and place the Architect designates. At the same time submit a typewritten schedule of the materials to be used, giving catalog numbers and complete description, to the Architect for approval.

6. INSTALLATION. Employ only specially trained and thoroughly competent workmen who are experienced in the installation of lightning protection equipment. Make the entire installation in an inconspicuous manner so as not to mar the architectural design of the structure. Provide an adequate number of air terminals. Firmly anchor all air terminals. Course the conductors properly and run them straight where they are supposed to be straight; make proper bends where bends are required. Use the proper attachment for each building or building surface. Attach conductors to the building firmly so that they cannot and will not come loose. See that all joints and connections are well made and will stay that way. Make all required metalwork connections in such a way that they are permanent and durable. The course of all conductors must be horizontal or downward, never upward. No branch leads may be longer than 16 feet, and in some cases 8 feet, without an additional ground.

> NOTE: Lightning protection systems may be planned in such a way that they do not detract from the appearance of the building. Conductors may be run from roof to ground along corners, behind downspouts, or inside downspouts. A fully concealed system may also be specified whose only visible parts are the air terminals— the conductors being run inside the building during construction.

7. PROMINENT PARTS. Spires, cupolas, ventilators, chimneys, high dormers, gable ends, water tanks, flagpoles, stair and elevator penthouses, and other vertical projections, must be protected by air terminals.

412

OPEN SPECIFICATION FOR LIGHTNING PROTECTION

8. AIR TERMINALS FOR PITCHED ROOFS. On pitched roofs install air terminals not more than 20' on centers along all ridges. There must be an air terminal within 2' of the ends of all ridges, whether they occur on the main roof or on protected dormers. On gently sloping roofs where the span is 40' or less and the pitch 1/8 or less, or where the span is 40' or more and the pitch is 1/4 or less, install terminals at the corners and edges so that they are spaced no more than 20' on centers. Provide two conductors to ground for the first 200 linear feet of protected roof perimeter, as measured along the eave line or equivalent, and provide one additional conductor to ground for each additional 100' of perimeter. Provide whatever additional conductors to ground are required to relieve low-positioned air terminals or to avoid a branch lead of over 16' on pitched roofs of irregular arrangement.

9. AIR TERMINALS FOR FLAT ROOFS. Install air terminals at the corners of all flat roofs and not more than 20' on centers around the entire perimeter. Install an additional row of air terminals spaced 20' on center for each 50' of roof width over 50'. For the first 200' of perimeter of flat or flat pitched roofs, install two conductors to ground. Install one additional conductor to ground for each additional 100' of perimeter or fractional part thereof.

10. CHIMNEYS. Provide lead-covered air terminals on chimneys, so located that no chimney corner is more than 2' distant from an air terminal. Air terminals must extend at least 10" above the highest part of the chimney construction.

11. GROUNDING. Provide an adequate number of effective grounds. For the purposes of estimating and bidding, it will be assumed that the earth is permanently moist to within 3' from finish grade. If, during the excavation, conditions are encountered that are at variance with this assumption, an adjustment will be made between the Owner and this Contractor for the greater expense that is involved in establishing the proper ground connections for the lightning protection system, each as a rule extending into the earth to a depth of 10' or equivalent. Install ground conductor guards where necessary to prevent mechanical injury.

> NOTE: Terminating a conductor in a few feet of dry, nonconducting earth so limits its capacity as to greatly interfere with, if not totally destroy, the protective power of the lightning protection system. Groundings in sand, gravel, or stony soil should be made by adding metal in the form of cable, driven rods, strips, plates, and so forth, buried in trenches extending away from the building, or by other approved means suitable to the installation.

12. BONDING METALWORK. Connect metal ventilators, stacks, vent pipes, and other metallic objects that project above the rodded structure to the system so that they serve as additional terminals. Connect metal roofing, ridge rolls, valleys, guy wires, and other metal bodies of conductance, as well as all bodies of inductance that might be subject to induced charges, to the grounding system as required. Electric wires, radio and television leads, telephone wires, and similar conductors entering buildings must be properly protected so that lightning cannot enter the building by these means.

GENERAL REQUIREMENTS
OF EMERGENCY LIGHTING

Emergency lighting systems are required by law to be installed in many occupancies and are equally valuable in those occupancies (such as residences and small stores) where they are not required. Most systems operate automatically, and many kinds of systems are available to satisfy various needs. Models range from individual storage-battery operated units, through battery systems, to motor-generator sets; they may be used to provide illumination, power, or both in whatever degree is needed.

Emergency lighting protection should be installed in all places where continuous light and/or power must be ensured to prevent:

1. Danger to workers, occupants, employees, or patrons.
2. Damage to property.
3. Theft.
4. Interruption of business activities or industrial processes.
5. Loss of goodwill.
6. The fire hazard attendant upon use of substitute lighting.

HOSPITALS.   The same storm, fire, or accident that causes a power outage may cause injuries requiring the immediate use of operating and emergency rooms. A power failure in a modern hospital would have extremely serious consequences and must be protected against at all times.

THEATERS, AUDITORIUMS.   Building codes require emergency lights in the interest of life safety. An independent power supply and automatic operation are a must.

INDUSTRIAL PLANTS.   Many industries conduct dangerous or delicate processes, control over which might be lost if power failed. In many instances, serious safety and health hazards can result from power loss. Immediate restoration of power is necessary to avert difficulties and danger.

BANKS.   Emergency lighting permits the continuation of business during power failures and is important in the prevention of thefts and holdups.

STORES AND MARKETS.   Emergency lighting protection allows the store to continue business, prevents shoplifting, protects the cashiers and cash registers, maintains the safety of patrons, eliminates the difficulties and hazards of makeshift substitute lighting, and prevents loss of goodwill.

ENGINE ROOMS.   Power outages are apt to originate in engine, boiler, transformer, or equipment rooms. Lighting is necessary in order to find the problem quickly and make repairs.

HOMES.   Homes large and small, urban and rural are all subject to short outages, long blackouts, and brownouts. An emergency lighting/power system, in the form of either a bank of rechargeable storage batteries or a motor-generator set, provides worthwhile security and convenience.

SUITABILITY OF WOODS
FOR FLOORING

SUBFLOORS (HOUSE)

Usual requirements: Requirements are not exacting, but high stiffness, good strength, medium or less shrinkage and warp, and ease of working are desired.

Highly suitable: Most commonly used—softwood plywood. Excellent but much less often used—Douglas-fir, western larch, southern pines.

Good suitability: Hemlocks, ponderosa pine, spruces, white fir, eastern and western white pine, sugar pine, red pine. Also useful but not readily available, expensive, and hard to work—beech, birches, elms, hackberry, maples, oaks, tupelo.

Grades used: Plywood is generally engineered C-D interior grade, sometimes with interior glue and often with exterior glue, or a combination subfloor-underlayment grade. For softwood boards, No. 2 grade is used in the more expensive constructions, while No. 3 serves in more economical projects. No. 4 is available and usable but entails waste in cutting and in time. Commercial hardwoods are almost never employed.

LIVING ROOM AND BEDROOM FLOORING

Usual requirements: High resistance to wear, attractive figure, color, and grain, minimum warp and shrinkage.

Highly suitable: Most commonly used hardwoods—hard maples, red and white oaks. Not commonly used—white ash, beech, birches, walnut. Not commonly available and hard to work and to nail—hickory, black locust, pecan.

Good suitability: Baldcypress, western hemlock, Douglas-fir, western larch, redwood, southern pines, western and eastern white pine (quite soft), cherry, sweetgum, sycamore (all quartered). The last three species are not commonly available but are highly decorative and suitable where wear is light and maintenance good.

Grades used: In commercial hardwood flooring grades Clear or First are ordinarily used for the better class of homes. Grades Select or Select & Better, Second, or Second & Better are used in more average work. NOFMA No. 1 Common may be used in some lower-cost oak floors. Third grade makes an economical but attractive and serviceable flooring. Softwood finish flooring grades are generally Superior Finish, Prime Finish, or B&B in better class homes; C, C&Btr, C Select, or D Select stock is employed for more economical work.

KITCHEN FLOORING (UNCOVERED)

Usual requirements: Resistance to wear, fine texture and close grain, ability to withstand washing and wear without discoloring and slivering, minimum warp and shrinkage.

Highly suitable: Fine textured—beech, birches, hard maples. Open textured—ash, red and white oaks, soft maples. (Also excellent are some of the imported hardwoods, such as teak.)

Good suitability: Baldcypress, Douglas-fir, western hemlock, western larch, redwood, southern pines, red pine (vertical grain preferred), elms, hackberry, sycamore.

Grades used: The grades used are the same as for living rooms and bedrooms. However, because of the hard use and wear that a kitchen floor receives, it is suggested that only the best quality of clear, defect-free woods be employed, whether hardwood or softwood.

PORCH AND DECK FLOORING

Usual requirements: Medium to good decay resistance, medium wear resistance, nonsplintering, freedom from warping.

Highly suitable: Baldcypress, Douglas-fir (vertical grain), western larch (vertical grain), southern pines (vertical grain), redwood, white oak. (If full drainage is not available, only the heartwood of white oak, redwood, and baldcypress can be given a high rating.) Black locust, walnut (impractical unless cut from home-grown timber). Most softwood species that are commonly available as pressure-treated lumber are also satisfactory.

STANDARDS FOR
SOFTWOOD FLOORING

Softwood flooring, which may be in strip form (widths to 3¼″) or in plank form (widths greater than 3¼″) is manufactured in accordance with the grading standards promulgated by the Southern Pine Inspection Bureau (SPIB), the West Coast Lumber Inspection Bureau (WCLIB), and the Western Wood Products Association (WWPA). The terminology and grading specifics differ somewhat among the three organizations.

SPIB.   Flooring manufactured from the several species of southern pines is graded according to SPIB rules. It is available side- and end-matched or side-matched only, and either hollow-backed or scratch-backed. The stock may be flat-grain or edge-grain, or near-rift, which is a compromise between the two. The grades are: B&B, supreme quality and generally clear; C, choice quality and quite clear but with minor surface defects and little permissible warp; C&Btr, a mix of the two above grades; D, good quality and fully usable; and No. 2, low cost but serviceable after some cutting and waste. All grades are available in flat-grain stock, and all except No. 2 are available in edge-grain and near-rift stock. Sizes that can be produced (but not all of which may be on hand at any given time) are:

$\frac{5}{16}″ \times 1\frac{1}{8}″, 2\frac{1}{8}″, 3\frac{1}{8}″, 4\frac{1}{8}″, 5\frac{1}{8}″$
$\frac{7}{16}″ \times 1\frac{1}{8}″, 2\frac{1}{8}″, 3\frac{1}{8}″, 4\frac{1}{8}″ 5\frac{1}{8}″$
$\frac{9}{16}″ \times 1\frac{1}{8}″, 2\frac{1}{8}″, 3\frac{1}{8}″, 4\frac{1}{8}″ 5\frac{1}{8}″$
$\frac{3}{4}″ \times 1\frac{1}{8}″, 2\frac{1}{8}″, 3\frac{1}{8}″, 4\frac{1}{8}″, 5\frac{1}{8}″$
$1″ \times 1\frac{1}{8}″, 2\frac{1}{8}″, 3\frac{1}{8}″, 4\frac{1}{8}″, 5\frac{1}{8}″$
$1\frac{1}{4}″ \times 2\frac{1}{8}″, 3\frac{1}{8}″, 4\frac{1}{8}″, 5\frac{1}{8}″$

Lengths range from 1′ to 20′, and standard practice is to include certain percentages of various lengths in each standard bundle of flooring, those percentages depending upon the grade and whether the material is end-matched or plain-end.

WCLIB.   Flooring manufactured of Douglas-fir, white fir, Sitka spruce, western redcedar, and western hemlock is graded under WCLIB rules. Stock is available in vertical-grain (synonymous with edge-grain in southern pines), flat-grain, or mixed grain cuts. All flooring stock is kiln-dried, and may be had with partially-surfaced backs, scratch-backed, or hollow-backed. The edges are tongued and grooved; ends are squared. There are three grades: C&Btr, fine appearance and mostly to completely clear, some light warp allowable; D, serviceable with reasonably good appearance, some surface defects and some warp; and E, with numerous nonweakening defects, intended for subfloor use. Sizes available are:

$\frac{9}{16}″ \times 3\frac{1}{8}″$
$\frac{3}{4}″ \times 2\frac{1}{8}″, 3\frac{1}{8}″, 5\frac{1}{8}″$
$1″ \times 2\frac{1}{8}″, 3\frac{1}{8}″, 5\frac{1}{8}″$

Lengths range from 4′ to 16′, occasionally longer.

WWPA.   Flooring manufactured from a wide variety of western softwoods is graded under WWPA rules. Stock is available in vertical-grain, flat-grain, or mixed grain cuts. The flooring is kiln-dried and processed hollow-backed, scratch-backed, or with partially surfaced backs. The edges are tongued and grooved; ends are squared. The standard grades are: Superior Finish, top-grade finish lumber with many pieces completely clear, very light cup, crook, or warp, limited minor surface defects; Prime Finish, very good appearance with some crook and warp, somewhat more minor surface defects than the top grade; C Select, good appearance and capable of taking fine finish, some staining and some surface checks and other defects; and D Select, similar to but of slightly lower quality than C Select on one face, with more numerous defects on the reverse face. Sizes available are:

$\frac{3}{4}″ \times 2\frac{1}{8}″, 3\frac{1}{8}″, 5\frac{1}{8}″$
$1″ \times 2\frac{1}{8}″, 3\frac{1}{8}″, 5\frac{1}{8}″$
$1\frac{1}{2}″ \times 5″, 6\frac{3}{4}″, 8\frac{3}{4}″$
$2\frac{1}{2}″ \times 5″, 6\frac{3}{4}″, 8\frac{3}{4}″$

Standard lengths are 4′ and longer, with certain percentages of various lengths permissible in standard bundles, depending upon the grade.

FINISH FLOORING
OF SOUTHERN PINE

PINE FLOORING. The best flooring stock is cut from the heavier butt logs in order to take advantage of their denser growth which, in the finished product, will stand up under hard wear. Heart-face, edge-grain southern pine flooring is practically indestructible.

ARCHITECT'S SPECIFICATIONS. The specifications, to be complete, must clearly state a choice in the following items:

1. Plain end or end-matched.
2. Grade.
3. Method of sawing.
4. Proportion of heart face.
5. Face width and thickness.

GRADES OF PLAIN END FLOORING. The grade usually specified in good construction is B&B, which consists of pieces practically free of defects on the face side. C Flooring grade admits a limited number of minor defects on the face side that do not detract from a smooth, well-groomed surface. C&Btr grade is a combination including pieces of both B&B and C Flooring. D grade admits slight defects that are visible but do not affect the soundness of the wood. It is suitable under carpets, in closets, and so forth. Moisture content in D or higher grade kiln-dried (KD) flooring does not exceed 12% in 90% of the pieces delivered. No. 2 grade is suitable for low-cost, utility construction. It can be trimmed as laid, to eliminate most defects without loss of more than 10% of the length of any piece. Moisture content in No. 2 grade kiln-dried (KD) flooring does not exceed 15%. Standard lengths in all grades are 4' to 20'.

GRADES OF END-MATCHED. In using end-matched material, the carpenter does not need to trim except upon reaching the side of a room. Thus, practically 100% of the flooring material is used, and substantial waste labor is eliminated. Where end joints occur, they are permanently maintained flush with each other and adjoining pieces. Grades are the same as in plain end flooring, above. Standard lengths are 2' to 16', nested in bundles 8' and longer in multiples of 1'.

EDGE GRAIN FLOORING. *Edge Grain, Rift Grain, Vertical Grain,* or *Quarter Sawn* flooring receives painter's finish evenly and is most durable.

FLAT GRAIN FLOORING. *Flat Grain, Plain Sawn,* or *Slash Grain* flooring is suitable for general flooring use where strict economy is necessary.

HEART FACE FLOORING. If unusual durability and uniform color are required, *Heart Face* flooring should be specified. *Heart Face* flooring is free from sapwood on the face side. It is unusually decay-resistant.

NEAR-RIFT FLOORING. Almost edge-grain, but has fewer rings per inch across the face, or the rings form an angle of less than 45° to the face. Durability is less than edge-grain flooring, greater than flat-grain.

OAK FLOORING
STANDARDS

Oak flooring is primarily manufactured as strip material. Both red oaks and white oaks are used, and the flooring stock is available either unfinished or factory prefinished. Most oak flooring is made from plain-sawed stock (equivalent to flat-grain in softwoods), where the angle of the annual rings is less than 45° from the strip face. Quarter-sawed stock (equivalent to edge-grain in softwoods), where the angle of the annual rings to the strip face is greater than 45°, is also obtainable, usually on special order. Quarter-sawed oak flooring is the more expensive of the two, but exhibits a striking flake and grain pattern and is subject to only minimal shrinking and swelling because of its greater dimensional stability. Stock is kiln-dried, and is grade- and trade-marked if of approved manufacture.

GRADES. Oak flooring materials are manufactured under the grading rules promulgated by the National Oak Flooring Manufacturers Association (NOFMA). There are three top grades covering both plain-sawed and quarter-sawed flooring material:

Clear. This is the top grade, presenting the finest appearance. Color consistency or matching is not considered, but it is naturally relatively uniform. Faces of the strips are virtually clear, but tiny amounts of visible sap are allowable.

Select. Appearance remains very good, but clearness is somewhat less than in the Clear grade because of the presence of tiny streaks, pinworm holes, very small knots or burls, and similar imperfections. Such defects, however, must average no more than one to every three linear feet of stock.

Select & Better. This grade consists of a balanced combination of the above two grades.

In addition, there are two lesser grades that cover only plain-sawed oak flooring:

No. 1 Common. This flooring is sound and solid but may have numerous surface imperfections, wormholes, streaks or stains, knots or burls, and the like, as well as possible minor defects caused by the machining processes during manufacture. It makes a very presentable floor, and is often used in residential applications.

No. 2 Common. White and red oak may be mixed in this grade. The wood is perfectly sound, but numerous imperfections and manufacturing defects are allowable. On the other hand, it is an economical flooring and when properly finished has an attractive and interestingly varied appearance.

In addition, there are several grades of prefinished oak flooring: *Prime, Standard, Standard and Better,* and *Tavern.* These grades approximate Clear, Select, Select and Better, and No. 1 Common in the unfinished flooring. Prime, however, is generally only available on special order.

SIZES. Oak flooring is made in a limited number of sizes. They are:

⅜″ x 1½″, 2″

½″ x 1½″, 2″

¾″ x 1½″, 2″, 2¼″, 3¼″

The 2″ width in the ⅜″ and ½″ thicknesses is not generally available, but can be had on special order; other sizes can also be made up to order. The ¾″ thickness is commonly used for residential and similar service, laid upon a subfloor. The thinner types are best suited for light-duty applications and are generally laid upon a doubled or heavy subfloor or an existing finish floor.

Lengths of oak flooring, whether red or white, range upward from 1¼′. In the top three grades, the average length of the strips in a bundle is 3¾′, while in the bottom two grades the average length is 2¾′. Most flooring stock is side-matched and end-matched, but square-edged and square-ended stock is available. Backs may be hollowed or flat.

FLOORING STANDARDS:
MAPLE, BIRCH, BEECH, AND PECAN

Maple, beech, birch, and pecan all make excellent hardwood flooring, as they are particularly dense, hard, and wear-resistant, and all take excellent finishes. All four are commonly produced as plain-sawed stock. Quarter-sawed flooring is available as well, generally on special order; however, neither the appearance nor the durability of the material as flooring is much different in plain-sawed than it is in quarter-sawed stock. In view of its added expense and limited availability, quarter-sawed stock is seldom used and no differentiation between the two is made in the grading rules.

Flooring manufactured of these woods is side- and end-matched normally, but square-edged and square-ended stock is available. Most is hollow-backed. All approved flooring is grade- and trade-marked, and the stock is kiln-dried.

GRADES. Maple, birch, and beech flooring is manufactured under grading rules of the Maple Flooring Manufacturers Association (MFMA). Flooring made of these three woods, as well as that made of pecan, is also produced under grading rules of the National Oak Flooring Manufacturers Association (NOFMA). There is, except for the grade names themselves, substantial similarity between the two sets of rules insofar as requirements are concerned; sizing differs somewhat. Neither association considers coloration of the wood in the grading process. However, color in all four species is relatively uniform, often completely so. Notice that under MFMA rules, quarter-sawed wood is defined as having annual rings at an angle of at least 30° to the strip face, rather than 45° as required by other grading rule sets. The grades for maple, birch, and beech are as follows:

First. Top grade, excellent appearance, virtually no defects, clear wood.

Second (NOFMA only). Perfectly serviceable, but small knots and other imperfections allowable; can be laid without any waste.

Second and Better (MFMA only). As above.

Third. Will make a serviceable floor, but imperfections may be fairly numerous, and some cutting and waste are involved.

The grade requirements for pecan are slightly different:

First. The face is almost free of small defects, but some will be encountered.

Second. Defects consisting of small, tight knots, wormholes, streaks or stains, and small machining imperfections are allowable; makes serviceable flooring with no waste.

Third. Imperfections and manufacturing defects are relatively numerous, and cutting and waste are involved in laying, but makes a serviceable floor and takes good finish.

SIZES. NOFMA sizes are as follows:

$\frac{3}{8}$" x 1$\frac{1}{4}$", 2", 2$\frac{1}{4}$", 3$\frac{1}{4}$"

$\frac{1}{2}$" x 1$\frac{1}{2}$", 2", 2$\frac{1}{4}$", 3$\frac{1}{4}$"

$\frac{3}{4}$" x 1$\frac{1}{4}$", 2", 2$\frac{1}{4}$", 3$\frac{1}{4}$"

Notice that the $\frac{3}{8}$" and $\frac{1}{2}$" thicknesses are usually available only on special order. Lengths are from 2' up for First and Second grades of maple, birch, and beech, and for First grade of pecan; all others are 1$\frac{1}{4}$' and up. Certain minimum percentages of shorts appear in full bundles, depending upon grade.

MFMA sizes are as follows:

$\frac{25}{32}$" x 1$\frac{1}{2}$", 2$\frac{1}{4}$", 3$\frac{1}{4}$"

$\frac{33}{32}$" x 1$\frac{1}{2}$", 2$\frac{1}{4}$", 3$\frac{1}{4}$"

$\frac{41}{32}$" x 1$\frac{1}{2}$", 2$\frac{1}{4}$", 3$\frac{1}{4}$"

Lengths of First grade are 2' to 8'; lengths in Second and Third grades are 1$\frac{1}{4}$' to 8'. Certain percentages of various lengths are contained in standard bundles, the percentages depending upon grade.

WOOD FINISH
FLOORING

Finish flooring laid either parallel or at right
angles to joists. Thickness varies.
Building Paper lapped 4"
Subflooring 3/4" thick S4S 4"or 6" wide, laid
diagonally with 1/8" spaces, or plywood.
Joists spaced 12" to 18"o.c.-designed to
limit deflection. Bridging at least
every 8'-0"of span, and two rows at
least in spans of 12'-0"or over.

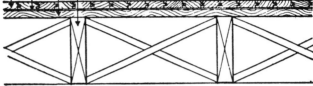

OVER WOOD JOISTS

Finish flooring laid either parallel or at right
angles to sleepers. Thickness varies.
Subflooring 3/4" thick S4S 4"or 6" wide, laid
diagonally with 1/8" spaces, or plywood.
Building paper lapped 4".
Sleepers 2"x3"or 2"x4" ripped to
give beveled edge. Space 16"o.c.
or 18"o.c. If subfloor is omitted
(not advised) space sleepers 12"o.c.
Space of 1/2" for ventilation.
Cinder concrete fill.

*OVER CONCRETE

SHOE
Correct
nailing is
to subfloor.

*Good Nailing Concrete permits
omission of both sleepers and
subfloor, when properly applied and given several weeks
to dry. Concrete floors on
earth should have an effective
damp-prfg. applied under wood

WOOD FINISH FLOORS
OVER GYPSUM PLANK

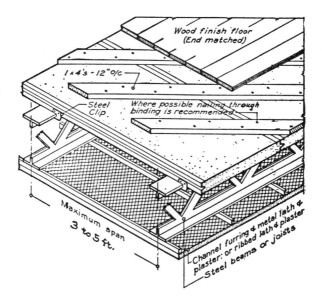

Wood finish floor
(End matched)

1 x 4's - 12" o/c

Steel
Clip

Where possible nailing through
binding is recommended

Maximum span
3 to 5 ft.

L Channel furring & metal lath &
plaster; or ribbed lath & plaster

Steel beams or joists

WOOD FINISH OVER WOOD STRIPS. A satisfactory and economical method is shown in the drawing above. 1" x 4" boards spaced 12" o/c may be laid either diagonally or at right angles to the direction of the gypsum plank. These strips are face-nailed to the plank, about 8" o/c on opposite edges, with the nails driven on the slant. The finish must be end-matched and is nailed to the strips at each bearing.

WOOD FINISH OVER SUBFLOORING. A finished wood floor can be installed over gypsum plank in the conventional manner. Subflooring may first be laid, consisting of an appropriate plywood or underlayment plywood, or of nominal 1" x 4" or 1" x 6" S4S boards positioned either diagonally or at right angles to the plank with ⅛" spacing between the boards. Plywood should be secured in accordance with the standard nailing schedule for the type used, with the nails slanted slightly; boards should be face-nailed on opposite edges 12" o/c, driving the nails to slant toward each other.

WOOD PARQUETRY BLOCK FLOORS. Ordinary wood parquetry block floors are laid directly over gypsum plank as they would be over concrete.

THICK FLOOR FINISHES. For terrazzo, granolithic, ceramic tile, or other poured finishes, gypsum plank should first be coated with tar, asphalt, or gypsum sealer. From the top of the plank to the finish floor line should be not less than 1½" thick. For best results, a minimum of 2" is recommended. The poured finish should be provided with adequate expansion joints and mesh reinforcing.

421

WOOD FINISH FLOORS
OVER CONCRETE

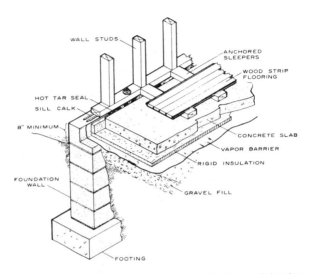

In basements or other below-grade locations, as well as in on-grade locations, where it is desired to install a wood finish floor over a concrete slab, precautions must be taken to prevent moisture from reaching the wood. Whether or not the ground appears damp at the time of construction, the foundation walls should be properly dampproofed. If there is evidence of either periodic or permanent ground moisture, the foundation should be waterproofed. A 4" to 6" layer of gravel or crushed rock should be laid in the slab area; footings should have a gravel-bedded footing drain system around them. A layer of heavy-duty polyethylene sheet should be laid as a vapor barrier beneath the concrete slab. Rigid perimeter insulation should be installed beneath the slab if and as required.

If the finish flooring is to be wood tile or parquet, either can be laid by first applying a coating of mastic, of a type recommended by the flooring manufacturer, directly to the clean concrete surface and then setting the tiles or pieces in place. The concrete slab must be clean, dust-free, and fully cured, and the instructions of the materials manufacturers must be exactly followed. Wood strip or plank floorings can be laid in the same manner as wood tile or parquet flooring, with mastic. In most cases, however, the first step with such commercial floorings is to secure pressure-treated wood 2" x 4" sleepers to the concrete slab, 12" or 16" o/c. This can be done with lags or other concrete anchors or with concrete nails (nail anchoring is most easily done when the concrete is only a few days old and still green). Then the wood strip flooring is nailed to the sleepers at right angles or diagonally in the usual fashion, after the concrete has fully cured and dried (which can be several months).

If the concrete slab already exists, a different arrangement must be used. First, if free moisture is present in any appreciable quantity, even if only periodically, finish wood flooring may be completely impractical. If conditions are normal, the first step is to apply a high-quality waterproof coating to the concrete slab surface; an asphaltic compound is one possibility. Pressure-treated 1" x 4" wood sleepers are then anchored to the slab, 12" or 16" o/c. A sheet of heavy-duty polyethylene is then placed over the sleepers, lapped 8" or so at joints and lapped up the walls. This should be stretched tight and stapled to the sleepers. A second set of untreated 1" x 4" sleepers is then nailed to the first. Wood strip or plank flooring can then be installed, at right angles or diagonally, on the sleepers in the usual fashion. Wood tile or parquet, however, should be laid on a subfloor of plywood. If desired, rigid insulation can be installed between the first-layer sleepers.

PARQUETRY
PATTERNS

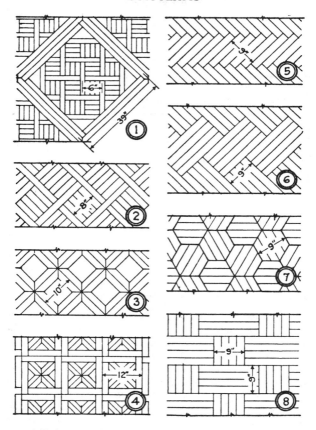

1. Design used in Palace at Fontainebleau.
2. From the Library of Marie Antoinette, Versailles.
3. From Thos. Jefferson's own house at Monticello.
4. In Melbury House, Dorset.

5. Butted Herringbone.

6. Clustered Herringbone.

7. Hexagons.

8. Basket Weave.

All drawings are made to the same scale of $\frac{3}{8}'' = 1'\text{-}0''$.

Common thicknesses for parquetry floors are $\frac{15}{32}''$ and $\frac{3}{8}''$; others also appear. The sizes of the patterns may be varied to suit the dimensions of the room and the standard sizes of wood flooring to be used.

CAUSE OF CRACKS
IN WOOD FLOORS

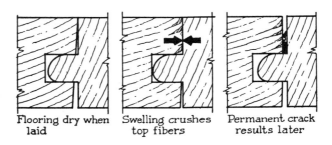

| Flooring dry when laid | Swelling crushes top fibers | Permanent crack results later |

When flooring with a high moisture content is laid tight, it dries out with the building, and as it shrinks from drying, cracks appear. Why cracks appear when the flooring is dry and carefully put down is not so well understood, however. These diagrams show what happens.

A succession of damp days, high humidities from the drying of wet plaster, or other cause, will allow the dry flooring to absorb moisture. The swelling of the flooring often causes a perceptible bulging of some boards. A crushing of the wood fiber on the sloped female edge is bound to take place—and we are then face to face with the most common cause of all cracking, known as "compression set." The subsequent drying of the house to a state of moisture equilibrium simply causes each board to shrink away from its neighbor, and the width of the crack is roughly equal to the amount of crushing or set that has taken place. Foreign matter or particles in such cracks will still farther open them during subsequent damp and dry cycles of swelling and shrinking.

Cracking can be prevented if the following precautions are carefully observed: 1) Use dry flooring to begin with, 2) Edge-grain flooring, even in lower grades, is to be preferred to mixed or flat-grain, 3) Building must be dried out, 4) Apply painter's finish to floors as soon as laid, and 5) Maintain a dry temperature until building is occupied.

PROTECTING CONSTRUCTION
AGAINST CONDENSATION

HOW CONDENSATION OCCURS. The average relative humidity in a house during the heating season is about 20%. As moisture is added to the air for comfort and health by deliberate humidification and/or inadvertently by laundry operations, (cooking, bathing, and so forth), the relative humidity is increased and may reach 40% or more. This humidified air can readily pass through unprotected walls and ceilings in an attempt to reach the outside atmosphere where the humidity is lower. The temperature of the wall construction decreases from the room temperature on the inside surface to the outside temperature on the exterior surface. If the humidified air, during its passage outward, reaches a point within the insulation or on the inside surface of the wall construction that is at or below dewpoint temperature, condensation will occur.

RESULT OF CONDENSATION. Some tests of sheathing have shown a moisture content as high as 35%. Instances of ice or frost in the walls have been found and are not uncommon. Thermal insulation of various types has been reported as wet or frosted in a number of cases. Condensation is frequently found on roof sheathing, forming as frost during cold weather. Warmer weather causes the frost to melt and drop off, soaking through the finish ceiling and producing the effect of roof leaks. Condensation can be retained in sheathing and/or thermal insulation, introducing the hazard of decay.

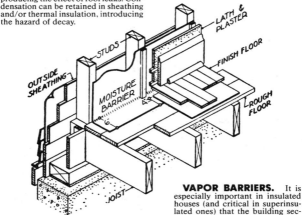

VAPOR BARRIERS. It is especially important in insulated houses (and critical in superinsulated ones) that the building sections be protected from condensation by the installation of effective vapor barriers. Polyethylene film has qualities that make it highly suitable for this use, though other materials with a permeance rating of 1 perm or less can also be used. PE film is rated at 0.08 perm in the 4-mil thickness, and 0.06 in the 6-mil thickness; either serves well as a vapor barrier. The material is inexpensive, easy to handle and install, available in large unbroken sheets, and readily repairable if punctured or slit. In addition, it makes an admirable air infiltration barrier.

APPLICATION OF BARRIER. The vapor barrier must be applied on the warm (heated) side of the building section. Common practice is to fasten it to the inner faces of wall studs, ceiling joists, roof rafters, and between subfloor and underlayment or finish floor. Polyethylene sheet can be stapled to the framework, leaving a 6″ lap at joints and overlaps onto floor or wall. All openings (electrical boxes, vents, windows, and so forth) should be covered over, and lap joints should be sealed with construction adhesive or duct tape. Each room or area should be completely sealed into a plastic envelope; openings are slit out and sealed in as windows, electrical devices, and other details are installed.

AIR INFILTRATION

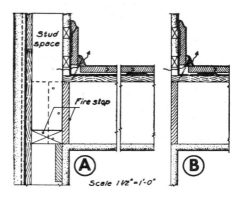

Stud space

Fire stop

Ⓐ Ⓑ

Scale 1 1/2" = 1'-0"

AIR LEAKAGE PROBLEM.   Inward infiltration of cold air through build-ing sections and at the joints between different types of sections constitutes a major heat-loss factor in buildings, especially those of a residential type. It has been found that the total infiltration in houses has a typical upward range that exceeds the infiltration that would occur through a 1.5 foot square opening to the outdoors, per 10,000 cubic feet of volume of the interior space of the building. Some 20% of the infiltration takes place around windows and doors, while the remainder is through walls and ceilings and, to some extent, through floors. The leakage is due to imperfect sealing around window and door frames, around the units themselves, at the floor/wall joints, at wall/ceiling joints, at wall corners, and to the lack of an impervious barrier over the wall, floor, and ceiling expanses. Proper sealing is extremely important, both as an infiltration barrier and as a vapor barrier.

STOPPING LEAKAGE.   Air infiltration can be stopped by sealing off the joints and cracks in a building by enclosing the entire interior in an envelope of 6-mil polyethylene sheeting (and by concurrently maintaining a high quality level of construction). One critical area in balloon-framed structures and floor constructions of any type built on or over a cold surface or space is the wall/floor joint. Where the finish floor is laid before the baseboards are installed, a layer of polyethylene sheet should be laid over the subfloor with 6″ laps and turned up the full length of the base as shown in A. When the baseboards are first installed, a strip of polyethylene should be secured at the wall/floor joint, lapping up onto the wall and out onto the floor as shown in B. In either case, the entire subfloor area should eventually be covered and the underlayment and/or finish floor installed on top of it.

The wall/ceiling joints are sealed in much the same way. The entire wall surface should be covered with polyethylene sheeting stapled to the inside faces of the studs, lapped 6″ or more and lapped out onto both floor and ceiling. The ceiling is similarly covered and lapped down onto the walls. Window, door, and other rough openings should be covered over, and the openings later cut out so as to leave 3″ or 4″ of polyethylene lapping out into the openings. This material is then slit diagonally at the corners and folded under the interior trim, providing a full seal across the rough-opening framework joints.

AIR CONDITIONING.   An air conditioned room usually has a higher humidity than the outside framing spaces. Escaping moisture will condense on the cold side of the thermal insulation, lowering its efficiency and possibly damag-ing both it and the structure. This is prevented by the polyethylene infiltration barrier, since it also serves as a vapor barrier.

PERMANENCE AND ECONOMY.   Polyethylene sheeting installed as explained above affords an excellent measure of infiltration and migrant moisture protection for several reasons. The material is inexpensive, very easy to apply, fully flexible, and workable with few tools; it is also unaffected by moisture, normal hot/cold cycles, rot, fungus, or insects and is available in large, unbroken sheets (roll form); finally, it is labor-saving compared to other types of barriers. The savings in heat loss alone will pay for the whole-house installation in a heating season or two.

TILE
FLOORS

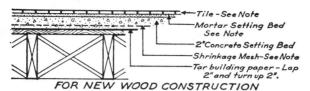

Tile – See Note
Mortar Setting Bed See Note
2" Concrete Setting Bed
Shrinkage Mesh – See Note
Tar building paper – Lap 2" and turn up 2".

FOR NEW WOOD CONSTRUCTION

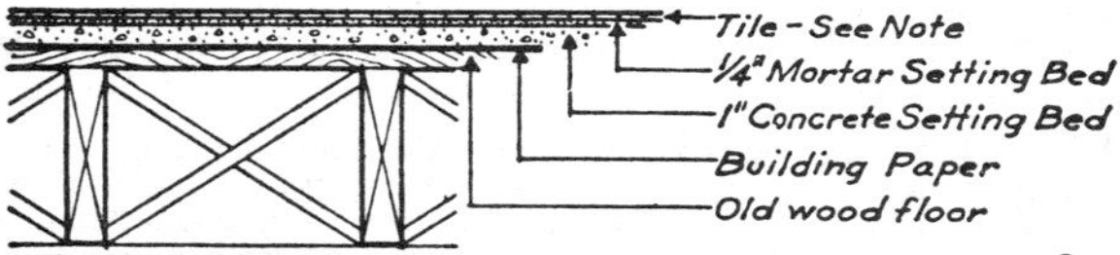

Tile – See Note
¼" Mortar Setting Bed
1" Concrete Setting Bed
Building Paper
Old wood floor

OLD WOOD CONSTRUCTION – MAX. AREA 100 φ

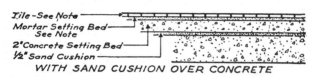

Tile – See Note
Mortar Setting Bed See Note
2" Concrete Setting Bed
½" Sand Cushion

WITH SAND CUSHION OVER CONCRETE

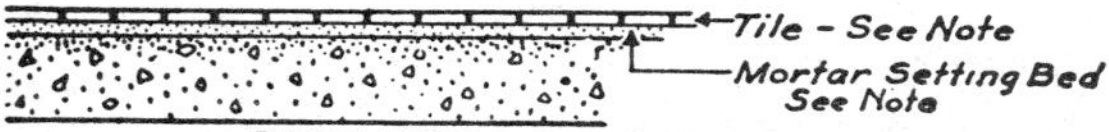

Tile – See Note
Mortar Setting Bed See Note

DIRECTLY ON CONCRETE

THICKNESS OF TILES

Ceramic Mosaic	¼"
Glazed Wall	¼"–5/16"
Glazed Floor	3/8"–½"
Quarry	3/8"–½"
Paver	3/8"–½"
Patio	½"–1"
Packinghouse	¾"–1 3/8"
Faience	½"–¾"
Faience Mosaic	½"

SETTING BEDS

Portland Cement	¾"–1¼"
Dry-set Mortar	1/16"–¼"
Latex-Portland Mortar	1/8"–¼"
Epoxy Mortar	1/16"–1/8"
Epoxy Adhesive	Surface Coat
Furan Mortar	Surface Coat
Organic Adhesive	1/16"

REINFORCEMENT (For Thick Mortar Bed Only)

Welded Wire	2" x 2" 16/16 ga.
	3" x 3" 13/13 ga.
	1½" x 2" 16/13 ga.
Expanded Metal Lath	2½ lb./sq. yd.
Sheet Metal Lath	4½ lb./sq. yd.

VINYL FLOOR
COVERINGS

Vinyl, in combination with other materials, has completely supplanted linoleum floor coverings in this country. Vinyl floor coverings are available in two forms: sheet and tile. There are four principal types and several subgroups.

VINYL SHEET. Vinyl sheet products are manufactured in two forms. The first is rug-sized sheets, generally 9' x 12', 12' x 12', and so forth, wherein a very thin layer of vinyl (usually less than 0.014") covers a colored imprint on a thin, asphalt-saturated felt backing. These sheets are of relatively low quality and wear-resistance, and are not meant for permanent installation; they are simply laid loose.

The second form consists of standard-width bulk sheeting manufactured in long rolls, from which the desired lengths for installation are cut by the dealer or installer. Standard widths are 6' and 12', and a special 4'-6" size is available from some sources. A very wide selection of colors and patterns is offered. Two types of vinyl surfacing are made: *filled* vinyl, and *clear* or *unfilled* vinyl. Filled vinyl consists of polyvinylchloride (PVC) resins, fillers, pigments, vinyl chips of a decorative nature, and stabilizers. These ingredients, in various combinations, are mixed and rolled into sheets. The sheets are chopped into tiny pieces, more vinyl chips and resins are added, and the mixture is spread over a backing and bonded to it under heat and pressure in the desired pattern and decorative aspect. Clear vinyl sheets, on the other hand, are made by first applying a decorative pattern to the backing with vinyl inks and then covering it with a coating of clear vinyl. Or, the decorative pattern may be applied to the back of the clear vinyl wear surface, which is then bonded to the backing.

Both filled and clear vinyl sheeting are made in thicknesses (gauges) of 0.065" to 0.160", with wear surface thicknesses ranging from 0.010" to 0.050". The minimum recommended thickness for residential applications is 0.065", with a minimum wear surface of 0.020" for the filled variety, and a minimum thickness of 0.010" of clear vinyl wear surface for that type.

Backings for vinyl sheet products are made from several materials, including resin- or asphalt-saturated felt, asbestos fibers impregnated with polymers, or vinyl. The backing material determines where in a building the vinyl sheeting may be installed: *below grade, on grade,* or on *suspended* construction. Typically, asbestos-backed and vinyl-backed products may be installed in all three locations, while felt-backed products are suitable only for suspended locations—that is, for subfloor

VINYL FLOOR
COVERINGS

assemblies that are not in any way in contact with the ground. Manufacturers' recommendations should always be followed in this respect. In some cases a thin layer of vinyl foam is sandwiched between the wear surface and the backing or a thicker layer is bonded to the underside of the backing. In both cases the foam serves to increase resiliency and comfort.

Sheet vinyl products are installed by cutting the material to fit and then laying it on a bed of special adhesive. Manufacturers' recommendations as to the type of adhesive and the method of application should be followed.

VINYL TILE. Vinyl tile may be composed of solid vinyl, mixed and formed under heat and pressure into sheets and then cut into tile sizes, or of vinyl sheet products with various backings, as explained above, cut into tile sizes. Standard sizes are 9″ x 9″ and 12″ x 12″ in squares, as well as various interlocking patterns, rectangular shapes, and large squares. Standard gauges of vinyl tile are $1/16″$, 0.080″, $3/32″$, and $1/8″$ for solid vinyl and 0.050″ to 0.095″ for backed tile. Installation requirements are the same as for vinyl sheet, except that some tile is available with self-adhesive backing for installation on hardboard or plywood underlayment or subfloor.

VINYL-ASBESTOS TILE. Vinyl-asbestos tile is composed of polyvinylchloride (PVC) resin binders, asbestos fibers, pulverized limestone, and various pigments and plasticizers. Standard gauges are $1/16″$, 0.080″, $3/32″$, and $1/8″$. Vinyl-asbestos tile is suitable for installation in all three subfloor locations and is installed in the same manner as vinyl tile. Manufacturers' instructions should be followed.

VINYL-CORK TILE. This tile is composed of ground bark from the cork oak tree and of various synthetic resins. The ingredients are mixed and then pressed into sheets and baked. A layer of polyvinylchloride (PVC) is bonded to the top surface, and the material is cut into tiles of standard sizes, in gauges of $1/8″$ and $3/16″$. Therefore, this is actually a *vinylized* product, rather than a vinyl product throughout. Its use is not widespread, nor is it readily available. It does, however, make a fine, resilient flooring that may be applied to either on-grade or suspended subfloor locations.

THIN FLOOR FINISHES
OVER GYPSUM PLANK

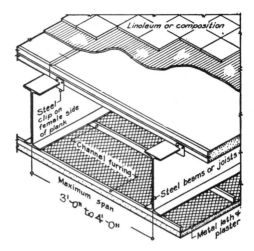

Linoleum or composition

Steel clip on female side of plank

Channel furring

Steel beams or joists

Maximum span 3'-0" to 4'-0"

Metal lath + plaster

THIN FLOOR FINISHES. When a thin finish flooring, such as linoleum, rubber tile, or asphalt tile, is to be applied over gypsum plank, a leveling coat should first be used. Spacing of joists or beams should be limited to 3' or 4', depending on the manufacturer's recommendations, in order to maintain a ratio between the overall thickness of the floors and the spans that will assure adequate stiffness.

LEVELING COAT. This composition bonds with gypsum, sets quickly and firmly, and is easily troweled to a smooth, hard finish. The first step in applying it is to sweep the plank floor thoroughly to remove any loose material or debris.

Instructions for applying the leveling coat should be followed closely. Generally, the plank is first dampened with water, and the material is applied in two coats. It sets hard in about one to three hours and must not be retempered in mixing. The finish flooring should not be laid until after the leveling coat is thoroughly dry.

OVER LIGHT STEEL JOISTS. Clips are attached alternately to the opposite flanges of the joists, thus allowing the gypsum plank itself to act as a series of struts. Clips are used at every intersection of the plank with the joists. Steel clips are securely nailed to the female side of the plank with two 4d galvanized slater's nails.

430

TERRAZZO FLOORS
ON SAND CUSHION

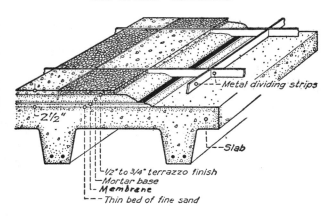

Metal dividing strips

2½"

Slab

½" to ¾" terrazzo finish
Mortar base
Membrane
Thin bed of fine sand

CRACKS IN TERRAZZO FLOORS. Shrinkage cracks are largely eliminated or localized by the brass dividing strips that form the pattern of the floor.
Structural cracks are usually caused by cracking of the base slab. Structural
cracks may be eliminated by constructing the floor finish without bond with
the base. This is accomplished by separating the base slab from the finish with
a layer of sand, covered with a membrane of reinforced building paper or polyethylene sheeting. The sand provides a cushion, and cracks originating in the
base slab from settlement, contraction, or vibration do not appear on the surface.
An alternative to the sand layer is a double layer of heavy polyethylene sheeting,
provided the base surface is smooth.

WHY USE A MEMBRANE? In the case of the double polyfilm mem-
brane substituting for the sand layer, the reason is obvious; it is the slipperiness
of the sheeting that allows movement and forms the isolation barrier between
the base and the finish. A single layer of polyethylene or building paper over
the sand layer prevents the wet mortar mixture from absorbing the sand and
bonding to the base, negating the whole procedure. Reinforced paper or heavy-
duty polyethylene must be employed (and protected from abuse after installation) to prevent ruptures or splits that might allow the wet mix to creep through
the barrier and form "points of support." This too would destroy the function of
the sand cushion.

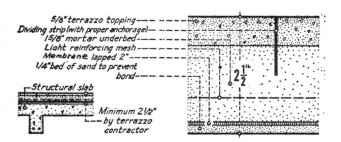

5/8" terrazzo topping
Dividing strip (with proper anchorage)
15/8" mortar underbed
Light reinforcing mesh
Membrane lapped 2"
1/4" bed of sand to prevent
bond

Structural slab

Minimum 2½"
by terrazzo
contractor

2½"

TERRAZZO
FLOORS

5/8" terrazzo topping
Dividing strip (with proper anchorage)
1 1/8" mortar underbed
Structural slab
Minimum 1 3/4"
by terrazzo
contractor

$1\frac{3}{4}$"

BONDED TO CONCRETE

5/8" terrazzo topping
Dividing strip (with proper anchorage)
1 7/8" mortar underbed
Temperature and reinforcing
bars or wire mesh
Precast concrete, steel or
wood joists
Minimum 2 1/2"
by terrazzo
contractor

$2\frac{1}{2}$"

DIRECTLY OVER JOISTS

5/8" terrazzo topping
Dividing strip (with proper anchorage)
1 1/8" mortar underbed
Minimum 3" concrete by terrazzo contractor or others
Light reinforcing mesh.
Minimum 1 3/4"
by terrazzo
contractor

$1\frac{3}{4}$"

DIRECTLY OVER EARTH

Terrazzo topping—200 lbs. marble granule to not less than 94 lbs. white or gray portland cement. Mortar underbed (cement: 4 sand 1½ to be indicated on plans and/or specified to show: 1). Depth: (1¼" is standard) 2). Gauge or face thickness
Dividing strips to be indicated on plans and/or specified to show: 1). Depth: (1¼" is standard) 2). Gauge or face thickness
3). Material: Brass, White Metal (zinc or nickel silver) or composition 4). Spacing and arrangement.

432

TERRAZZO
FLOORS

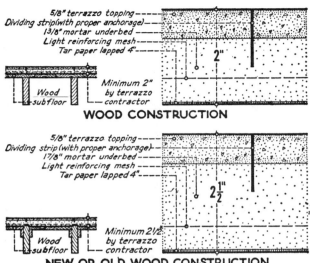

5/8" terrazzo topping ----
Dividing strip (with proper anchorage) ----
13/8" mortar underbed ----
Light reinforcing mesh ----
Tar paper lapped 4" ----

2"

Wood
subfloor

Minimum 2"
by terrazzo
contractor

WOOD CONSTRUCTION

5/8" terrazzo topping ----
Dividing strip (with proper anchorage) ----
17/8" mortar underbed ----
Light reinforcing mesh ----
Tar paper lapped 4" ----

$2\frac{1}{2}$"

Wood
subfloor

Minimum 2½"
by terrazzo
contractor

NEW OR OLD WOOD CONSTRUCTION

Terrazzo topping-200 lbs. marble granule to not less than 94 lbs. white or gray portland cement. Mortar underbed 1 cement:4 sand. Dividing strips to be indicated on plans and/or specified to show: 1). Depth (1¼" is standard). 2). Gauge or face thickness 3). Material: Brass, White Metal (zinc or nickel silver) or composition. 4). Spacing and arrangement.

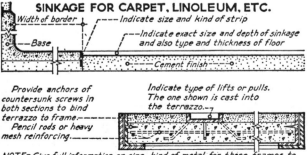

SINKAGE FOR CARPET, LINOLEUM, ETC.

Width of border
---- Indicate size and kind of strip

---- Indicate exact size and depth of sinkage and also type and thickness of floor

Base

Cement finish

Provide anchors of countersunk screws in both sections to bind terrazzo to frame. ----
Pencil rods or heavy mesh reinforcing. ----

Indicate type of lifts or pulls. The one shown is cast into the terrazzo. ----

NOTE- Give full information on size, kind of metal for these frames to be furnished and set by the terrazzo contractor.
TRENCH COVER

433

TERRAZZO BASES
AND SHOWER STALLS

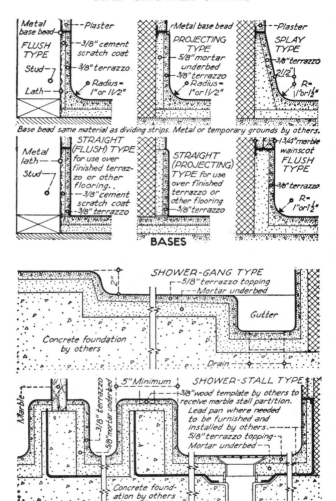

Metal base bead | Plaster
FLUSH TYPE
—3/8" cement scratch coat
Stud
—3/8" terrazzo
Lath
Radius = 1" or 1½"

Metal base bead | Plaster
PROJECTING TYPE
5/8" mortar underbed
—3/8" terrazzo
Radius = 1" or 1½"

Plaster
SPLAY TYPE
—3/8" terrazzo
2½"
R= 1" or 1½"

Base bead same material as dividing strips. Metal or temporary grounds by others.

Metal lath
Stud
STRAIGHT (FLUSH) TYPE for use over finished terrazzo or other flooring.
—3/8" cement scratch coat
—3/8" terrazzo

STRAIGHT (PROJECTING) TYPE for use over finished terrazzo or other flooring
—3/8" terrazzo

—1¾" marble wainscot
FLUSH TYPE
—3/8" terrazzo
R= 1" or 1½"

BASES

SHOWER-GANG TYPE
2" | 5/8" terrazzo topping
Mortar underbed
Gutter
Concrete foundation by others
Drain

SHOWER-STALL TYPE
Marble
5" Minimum
—3/8" terrazzo
5/8" mortar underbed
3/8" wood template by others to receive marble stall partition.
Lead pan where needed to be furnished and installed by others.
5/8" terrazzo topping
Mortar underbed
Concrete foundation by others

SHOWER FLOORS
SCALE OF DETAILS - 1½" = 1'-0"

434

TERRAZZO STAIR CONSTRUCTION

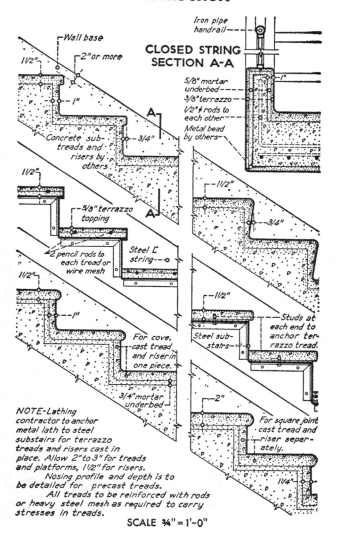

Wall base

1 1/2"

2" or more

1"

Concrete sub-treads and risers by others

3/4"

Iron pipe handrail

CLOSED STRING SECTION A-A

5/8" mortar underbed

3/8" terrazzo

1/2" ⌀ rods to each other

Metal bead by others

1"

A

A

1 1/2"

5/8" terrazzo topping

2 pencil rods to each tread or wire mesh

Steel ⌶ string

1 1/2"

3/4"

1 1/2"

1"

For cove, cast tread and riser in one piece.

3/4" mortar underbed

1 1/2"

Steel sub-stairs

Studs at each end to anchor terrazzo tread.

2"

For square joint cast tread and riser separately.

1 1/4"

NOTE—Lathing contractor to anchor metal lath to steel substairs for terrazzo treads and risers cast in place. Allow 2" to 3" for treads and platforms, 1 1/2" for risers.

Nosing profile and depth is to be detailed for precast treads.

All treads to be reinforced with rods or heavy steel mesh as required to carry stresses in treads.

SCALE 3/4" = 1'-0"

TERRAZZO DOOR TRIM
AND WINDOW STOOL

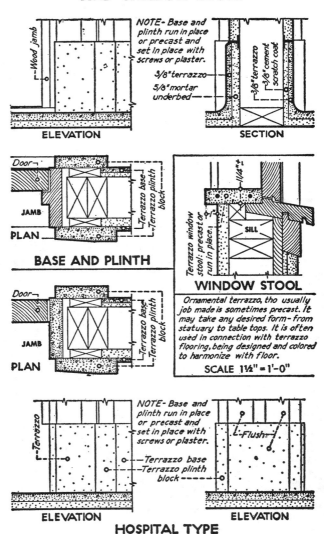

NOTE- Base and
plinth run in place
or precast and
set in place with
screws or plaster.

Wood jamb

3/8"terrazzo
5/8"mortar
underbed

3/8"terrazzo
3/8"cement
scratch coat

ELEVATION

SECTION

Door

JAMB

Terrazzo base
Terrazzo plinth
block

PLAN

BASE AND PLINTH

1/4"±

Terrazzo window
stool; precast or
run in place.

SILL

WINDOW STOOL

Ornamental terrazzo, tho usually
job made is sometimes precast. It
may take any desired form- from
statuary to table tops. It is often
used in connection with terrazzo
flooring, being designed and colored
to harmonize with floor.

SCALE 1½" = 1'-0"

Door

JAMB

Terrazzo base
Terrazzo plinth
block

PLAN

NOTE- Base and
plinth run in place
or precast and
set in place with
screws or plaster.

Terrazzo

Terrazzo base
Terrazzo plinth
block

Flush

ELEVATION

ELEVATION

HOSPITAL TYPE

436

TERRAZZO
WAINSCOTS & PARTITIONS

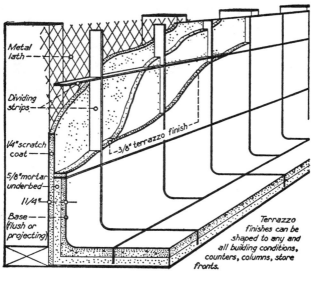

Metal lath

Dividing strips

1/4"scratch coat

5/8"mortar underbed

1 1/4"

Base — flush or projecting

— 3/8" terrazzo finish

Terrazzo finishes can be shaped to any and all building conditions, counters, columns, store fronts.

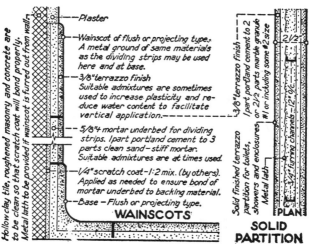

Hollow clay tile, roughened masonry and concrete are to be clean so that scratch coat will bond properly. Metal lath to be provided if wainscot is furred out from walls.

—Plaster

—Wainscot of flush or projecting type. A metal ground of same materials as the dividing strips may be used here and at base.

—3/8" terrazzo finish
Suitable admixtures are sometimes used to increase plasticity and reduce water content to facilitate vertical application.

—5/8"4 mortar underbed for dividing strips. 1part portland cement to 3 parts clean sand-stiff mortar. Suitable admixtures are at times used.

—1/4" scratch coat-1:2 mix. (by others). Applied as needed to ensure bond of mortar underbed to backing material.

—Base — Flush or projecting type.

WAINSCOTS

Solid finished terrazzo partition for toilets, showers and enclosures.
Metal lath

—3/8"terrazzo finish
1part portland cement to 2 or 2 1/2 parts marble granule #1 or including some #2 size

—1/4"furring channels-12"o/c

2 1/2"

PLAN

SOLID
PARTITION

437

TERRAZZO
SPECIFICATIONS

AVAILABLE REFERENCE MATERIAL. The product information, applications, specifications, design data, and installation of terrazzo and mosaic materials constitute a specialized field of some complexity. Contact the National Terrazzo & Mosaic Association, Inc. (NTMA), 3166 Des Plaines Avenue, Suite 15, Des Plaines, Illinois 60018 (toll-free telephone number: 1-800-323-9736) for complete and accurate, up-to-date general and technical information on the subject, as well as design assistance when problems are encountered. NTMA offers an extensive *Terrazzo Design/Technical Data Book* that may be used for guidance in selecting colors and designs and in outlining specifications. In addition, it can supply current *Technical Data* sheets containing complete specifications, architectural details, and product information about terrazzo/mosaic materials and their proper installation.

NTMA makes available scaled and traceable architectural details together with technical data and complete guide specifications ready for copying, as follows:

> Sand Cushion Terrazzo
> Bonded Terrazzo
> Monolithic Terrazzo
> Rustic Terrazzo (Structural System over Granular Fill)
> Rustic Terrazzo (Bonded System with Setting Bed)
>   (Over Concrete Slab)
> Rustic Terrazzo (Direct Bond to Concrete Slab)
> Rustic Terrazzo (Structural System over Granular Fill
>   on Insulated and Waterproofed Base
>   Slab over Heated Space)
> Palladiana
> Terrazzo over Permanent Metal Forms
> Precast Terrazzo Base
> Poured in Place Terrazzo Base
> Precast Terrazzo Stairs
> Poured in Place Terrazzo Stairs
> Vertical Terrazzo
> Polyacrylate Modified Terrazzo
> Epoxy Terrazzo
> Polyester Terrazzo
> Mosaics
> Textured Epoxy Mosaics
> Textured Polyacrylate Mosaics
> Textured Polyester Mosaics
> Conductive Epoxy Terrazzo
> Conductive Polyester Terrazzo
> Fine Aggregate Epoxy Floors (Interior Use Only)
> Fine Aggregate Polyester Floors (Interior Use Only)

Should information not covered in these guide specifications be required, the Director of Architectural Services at NTMA may be contacted.

MAINTENANCE OF TERRAZZO AND MOSAIC. Consult your terrazzo contractor concerning locally available dressings, cleaners, and sealers suitable for terrazzo, and/or about special staining or damage problems. Proper minimum care of terrazzo flooring includes: daily removal of spillage with clean cloths, removal of stains and scuff marks with neutral cleaner in warm water, and removal of gum or asphalt by applying dry ice and then lifting with a putty knife, all followed by dust-mopping with a non-oily dressing; twice weekly damp-mopping with a neutral cleaner; weekly scrubbing with a neutral cleaner, followed by buffing; monthly application of floor dressing followed by buffing (necessary only if a fine appearance is required); and finally, twice-yearly stripping of old sealer and dressing, followed by resealing.

Complete details on proper maintenance of terrazzo and mosaic may be obtained from NTMA.

FINISH FLOORS AT OR BELOW GRADE

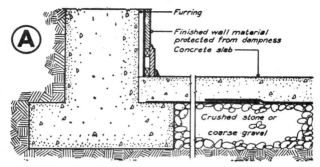

MONOLITHIC FLOORS. In clay, heavy loam, or other soils that hold moisture, special precautions must be taken to protect the basement rooms from dampness. A bed of crushed stone or coarse gravel is commonly set under the floor in such situations, providing a layer of material that does not readily hold moisture. Over this, 6-mil or 8-mil polyethylene sheeting should be laid, as shown in Figure A above, lapping the joints at least 9″. The film provides a nearly impervious vapor barrier and also prevents the wet concrete mixture from flowing into and filling the voids in the layer of crushed stone, which depends for its value on the existence of such voids. Care should be taken to prevent puncturing the film during placement of reinforcing bars or mesh, through excessive foot traffic across it, or through careless use of tools. In some cases, a primary layer of reinforced kraft paper may be advantageous; any damage to the film can be repaired immediately prior to pouring the concete.

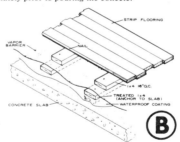

WALL PROTECTION. In such cases, plaster, drywall, wood paneling, or finished walls of any type should be protected from dampness by a vapor barrier of appropriate type, applied between the foundation wall and the finished wall. In addition, the exterior of the foundation wall should be properly dampproofed or waterproofed, depending upon the severity of the moisture problem, and a footing drain (not shown here) installed around the perimeter of the foundation.

FLOORS HAVING A FINISH. Wood, sheet vinyl flooring, carpeting, and most other floor coverings should not be laid on concrete if dampness is present. A very dry-mix cinder concrete slab should first be poured. Roofing felt can then be laid, with the joints overlapping by 6″ and sealed with asphalt. Wood sleepers can then be bedded in asphalt on 12″ or 16″ centers, and a plywood subfloor laid on them. Alternatively, as in Figure B, the concrete slab can be waterproofed and 1″ x 4″ sleepers nailed to the concrete. A layer of 6-mil polyethylene sheeting is stretched across the sleepers and stapled in place, then a second set of 1″ x 4″ sleepers is nailed to the first. Either a finish flooring or a plywood subflooring can then be laid.

CONCRETE FLOOR RESURFACING

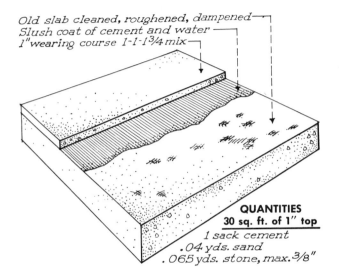

Old slab cleaned, roughened, dampened
Slush coat of cement and water
1" wearing course 1-1-1¾ mix

QUANTITIES
30 sq. ft. of 1" top
1 sack cement
.04 yds. sand
.065 yds. stone, max. ⅜"

Old concrete slab that is to be resurfaced must be clean of loose particles, grease, oil, paint, and any other material that might interfere with the bonding of the new top.

Saturate the slab with water overnight; then allow it to dry for 2 hours. No pools should be left standing.

Brush on a thin coat of cement mixed with water to the consistency of heavy cream or thick paint.

Place the wearing surface before the slush coat has dried or set.

Screed to proper true level, float with a wood float, and trowel to desired smoothness.

Careful curing will determine the strength of the new top and the amount of wear it will withstand. Protect it thoroughly with wet sand, wet burlap, or polyethylene sheeting, or by ponding, as soon as the new surface can be sprinkled and walked on. Keep the top completely moist for 4 to 7 days minimum.

Not more than 5 gallons of water should be used in the mix for each sack of cement. Screeding, floating, and troweling should not bring free water to the surface. Do not dust the top with dry cement, or sand and cement, to take up excess water.

CURING AND PROTECTION
OF CONCRETE AND TERRAZZO FLOORS

EFFECTIVENESS OF CURING AND PROTECTION WITH IMPERVIOUS COVERING. Moisture is a prime requisite for proper curing of concrete; if allowed to dry rather than cure, concrete will never gain its full strength. The chart below shows the effectiveness of covering the concrete surface with waterproof and airproof fiber-reinforced paper or polyethylene sheeting to retain the original moisture automatically—entirely eliminating the human element. In addition to this, the covering protects the floor against stains and construction dirt right up to the completion of the building. However, polyethylene film can present a safety hazard if wet, or especially if snow-covered, since it is extremely slippery.

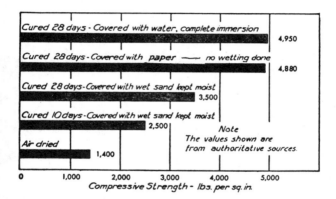

Cured 28 days - Covered with water, complete immersion — 4,950
Cured 28 days - Covered with paper ——— no wetting done — 4,880
Cured 28 days - Covered with wet sand kept moist — 3,500
Cured 10 days - Covered with wet sand kept moist — 2,500
Air dried — 1,400

Note
The values shown are from authoritative sources.

Compressive Strength - lbs. per sq. in.

CURING MONOLITHIC CONCRETE FLOORS. Monolithic slabs are generally placed before the building is closed in, and sometimes before any other construction takes place. Planks or other weights are used to hold the edges of the covering in place on the slab.

CURING GRANOLITHIC FLOORS. Concrete slabs that receive a granolithic finish are not ordinarily poured until the building is closed in. For floors that are to have a granolithic cove, the covering should be stopped 3″ from the wall to allow the placing of the cove without disturbing the covering of the floor. If no cove is to be used, the covering should be run right up to the wall.

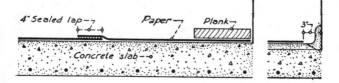

4″ Sealed lap — Paper — Plank — 3″
Concrete slab —

BASIC RULES
FOR MASTIC

1. Mastic will not withstand oils, greases, gasoline, animal or butter fats, solvents such as naphtha, carbon tetrachloride, etc.

2. Mastic will not stand temperatures above the normal atmospheric, unless special construction and special mixes are used.

3. Mastic will not remain on vertical surfaces over 3″ high unless reinforced with expanded metal.

4. Mastic will support the heaviest type of *moving* load, but is incapable of sustaining exceptionally heavy *standing* loads over prolonged periods without indenting, unless a special mix or metal floor grids are used.

5. Mastic must be applied over a firm, suitable base as it possesses little structural strength.

6. Edges of mastic, such as at elevator wells, stair treads, etc., must be protected by metal strips, angle irons or other means to prevent the mastic from fraying.

7. Mastic must be troweled on—it cannot be screeded unless an unusually soft mix is used.

8. Do not rely upon mastic to form a bond with the base over which it is applied. This is the reason for establishing a minimum thickness of 1″. Mastic will creep or shove under trucking when applied less than 1″ thick.

9. If a mastic floor is to be installed above the first floor, it is advisable to allow the mastic contractor the exclusive use of an elevator. The mastic mixture must not be allowed to cool as might be the case if delay were caused from a joint use of hoist or elevator.

10. Red rosin sized paper or insulating paper must be applied over boards before receiving mastic. Resinous or green unseasoned boards must not be used.

11. Mastic is impractical for roof gardens unless special precautions are taken by an experienced contractor.

12. Where mastic is to be applied over a waterproofing membrane, apply 2 layers of dry waxed craft paper before applying mastic.

13. Mastic mixtures must be applied while hot and cannot be applied over wet or damp surfaces without blistering.

Mastic properly used, has many advantages for use in flooring. It is resilient and decreases fatigue due to standing at work; it is quickly and easily laid; it is sanitary, odorless, non-dusting, non-absorbent and non-glaring; it is easy to maintain; it is waterproof and can be washed as often as desired without injury to the floor or danger of leakage to floors below. It is easily installed over any old floor that is solid; mastic floors are ready for use 3 hours after being laid.

Mastic is an ideal industrial flooring and is also used in the construction of sidewalks, tennis courts, swimming pools. It is suitable for cold storage floors and floors subjected to acid liquors.

MASTIC FLOORING FOR ORDINARY CONDITIONS

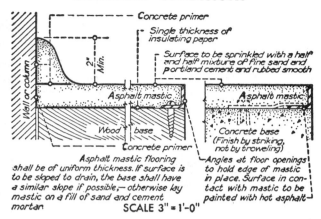

Concrete primer

Single thickness of insulating paper

Surface to be sprinkled with a half and half mixture of fine sand and portland cement and rubbed smooth

Asphalt mastic

Asphalt mastic

Wall or column

2" Min.

Wood base

Concrete primer

Concrete base (Finish by striking, not by troweling)

Asphalt mastic flooring shall be of uniform thickness. If surface is to be sloped to drain, the base shall have a similar slope if possible;— otherwise lay mastic on a fill of sand and cement mortar.

Angles at floor openings to hold edge of mastic in place. Surface in contact with mastic to be painted with hot asphalt.

SCALE 3" = 1'-0"

Location	Base	Mastic Thickness (Min.)
Outdoor foot and light traffic	Concrete or firm base	1"
	Wood	1"
Indoor foot and light traffic	Concrete or firm base	1"
	Wood	1¼"
Cold storage spaces	Concrete or firm base	1"

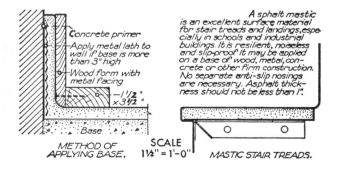

Concrete primer

Apply metal lath to wall if base is more than 3" high

Wood form with metal facing

—1½" × 3½"

Base

METHOD OF APPLYING BASE.

SCALE 1½" = 1'-0"

A sphalt mastic is an excellent surface material for stair treads and landings, especially in schools and industrial buildings. It is resilient, noiseless and slip-proof. It may be applied on a base of wood, metal, concrete or other firm construction. No separate anti-slip nosings are necessary. Asphalt thickness should not be less than 1".

MASTIC STAIR TREADS.

MASTIC FLOORING
FOR SEVERE CONDITIONS

Mastic is a bituminous mixture of asphalt, asphalt flux, filler, sand and gravel. When hot it is sufficiently plastic for spreading with wooden trowel or float. It hardens as it cools and is ready for use two or three hours after laying. This mastic mixture is waterproof, acid-resisting, non-dusting, sanitary, slip-proof, sound absorbent and noiseless. It is also resilient and therefore less tiring to workers. These features make it an excellent flooring material for all types of industrial buildings, canneries and bottling plants, chemical and acid plants, railroad platforms, loading platforms, sidewalks, roofing, tennis courts and other outdoor game areas. It can be made exceptionally hard to withstand heavy trucking and heavy loads by the addition of a special hardener.

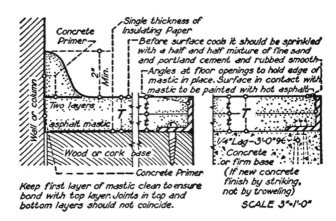

Keep first layer of mastic clean to ensure bond with top layer. Joints in top and bottom layers should not coincide.

(If new concrete finish by striking, not by troweling)

SCALE 3"=1'-0"

Location	Base	Mastic Thickness	
		t	T (Min.)
Outdoor heavy trucking and traffic	Concrete or firm base	¾"	1½"
	Wood	¾"	1½"
Indoor heavy traffic	Concrete or firm base	¾"	1½"
	Wood	¾"	1½"
Cold storage spaces	Wood or cork	⅝"	1¼"
Plating rooms, acid tank rooms and floors subject to liquid acids	Concrete or firm base	⅝"	1¼"
	Wood	⅝"	1¼"

UNDERLAYMENT
ON ROOFS

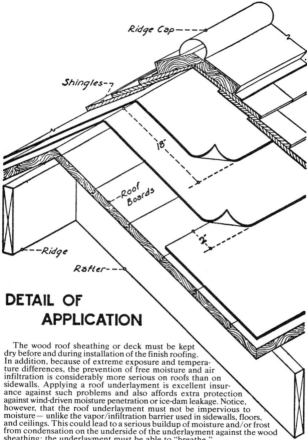

Ridge Cap

Shingles

18"

Roof Boards

2"

Ridge

Rafter

DETAIL OF
APPLICATION

The wood roof sheathing or deck must be kept
dry before and during installation of the finish roofing.
In addition, because of extreme exposure and temperature differences, the prevention of free moisture and air
infiltration is considerably more serious on roofs than on
sidewalls. Applying a roof underlayment is excellent insurance against such problems and also affords extra protection
against wind-driven moisture penetration or ice-dam leakage. Notice,
however, that the roof underlayment must not be impervious to
moisture— unlike the vapor/infiltration barrier used in sidewalls, floors,
and ceilings. This could lead to a serious buildup of moisture and/or frost
from condensation on the underside of the underlayment against the wood
sheathing; the underlayment must be able to "breathe."

The most commonly used and probably most cost-effective underlayment
material is 15-pound roofing felt, a roll-form felt 36" wide saturated with asphalt,
weighing 15 pounds per square (100 square feet) and universally available. On
roofs with a slope of 4" or more per foot of run the felt should be applied as
shown above, with 2" top laps (most felts have a white guideline for this purpose)
and 4" side laps. If the slope is less than 4" per foot of run, the felt should be
applied in double coverage by starting at the eave with one strip 19" wide, and
then applying a full-width strip over the starter strip, lapping the second full-
width strip 17" over the first, and lapping all subsequent strips 19" over the
preceding strip. The felt should be carried 18" beyond all hips, valleys, and
ridges from both directions, as shown in the detail. If stained wood shingles are
to be laid as a finish roofing, they must be thoroughly dry before application.

WOOD SHINGLE
ROOFING

Shingles covered by this standard are known as "Grade No. 1 (Blue Label)" and are chiefly cut from western redcedar (*Thuja plicata*). This wood has an extremely fine grain, low expansion and contraction from moisture, very high strength-to-weight ratio, and a high impermeability to water, making it an ideal roof shingle material.

WIDTH.   Maximum width is 14″, and minimum width is 3″. The shingles have parallel sides.

THICKNESS.   Shingles are measured for thickness at the butt ends and designated according to the number of pieces necessary to make up a specific unit of thickness. For example, 4/2 indicates that four butts measure 2″ in thickness.

Length in inches	*Thickness in inches*	*Maximum exposure to weather on roofs*	*Maximum exposure to weather on walls*
16″	5/2	5″	6 3/4″
18″	5/2¼	5 1/2″	7 1/2″
24″	4/2	7 1/2″	10″

Adapted from *Commercial Standard CS 31*

ROOF PITCH.   The standard exposures for wood shingles are 5″, 5½″, and 7½″ for shingle lengths of 16″, 18″, and 24″ respectively. These exposures are used on slopes of 5″ of run in 12″ of rise, or greater. For a slope of 4″ of rise in a run of 12″, the exposures should be decreased to 4½″, 5″, and 6¾″ for 16″, 18″, and 24″ shingles respectively. For a slope of 3″ of rise in 12″ of run, the exposure should be further decreased to 3¾″, 4¼″, and 5¾″ for 16″, 18″, and 24″ shingles respectively. Wood shingles should not be applied to slopes of less than 3″ of rise in 12″ of run.

ROOF CONSTRUCTION.   Wood shingles may be applied to either wood sheathing or strips. When 1″ x 3″ or 1″ x 4″ strips are used, they are spaced the same distance apart on centers as the shingles are exposed to the weather, but the spaces between the strips should be no more than the width of the strips themselves. Such construction without tight sheathing should be used only when heating costs are not a consideration (as in a garage or shed, or in a warm climate) or when special precautions have been taken to insulate the building fully.

Roof boarding of nominal 1″ x 6″ T&G laid tight, or sheathing of an appropriate grade and thickness of plywood, is considered a better base for wood shingles and is necessary in most parts of the country. Underlayment is usually not required with wood shingles, but it does afford extra protection to the sheathing and acts as a barrier against air infiltration. No. 15 asphalt-saturated roofing felt is most often used. In any case, a double course of shingles should start the roofing at the eaves.

CHARACTERISTICS OF WOOD SHINGLE ROOFS.   Wood shingles are lightweight, have good insulating value, are easily applied, and result in pleasing architectural effects. Top-grade shingles properly applied are long-lived—35 years or more on slopes of 8″ in 12″, and 20 years or more on slopes of 4″ in 12″. Initial cost is relatively high. The main objection to their use is the fire hazard, and for this reason they are disallowed in some locales and restricted in use in some ways by model building codes.

SLATE ROOFING

Slate roofing is available in numerous lengths and widths, all of which are generally considered standard, as shown in the table below. Not all sizes are obtainable from every manufacturer.

SLATE SHINGLES FOR SLOPING ROOFS

Lengths	Widths										Exposure
10	6	7	8								3½
11		7	8								4
12	6	7	8	9	10	11	12				4½
14		7	8	9	10	11	12		14		5½
16			8	9	10	11	12		14		6½
18				9	10	11	12	13	14		7½
20				9	10	11	12	13	14		8½
22					10	11	12	13	14		9½
24						11	12	13	14	16	10½
26									14		11½

To carry out a desired design on special roofs, it is sometimes necessary to make shingles longer than 24″, in which case the thicker slates are used. Smaller shingles, such as 12 or 14″ lengths, should be used for pents, porch and dormer roofs and cheeks, garages, and other low buildings—even when the main roof is composed of larger slates as a means of maintaining proper scale.

Slate shingles most commonly are available in a standard thickness of $\frac{3}{16}$″, plus or minus small tolerable variations. Other readily available thicknesses, often in stock, are ¼″, ⅜″, ½″, and ¾″. While the standard thickness is more than adequate for ordinary roofing purposes, the heavier slates are often used for extraordinary service requirements, special purposes, and various architectural reasons. Grades are not formally standardized, but some terms in common use are: Standard ($\frac{3}{16}$″), Rough Texture ($\frac{3}{16}$″–⅜″), Heavy (⅜″–½″), and Architectural (¾″).

Roofing slate is most commonly seen in a typical slate-gray color. Other colors available are green, red, variegated purple, charcoal gray, and black.

Roofing slates are relatively uniform in shape, size, and thickness, and are provided with prepunched nail holes. In most cases they are laid on roof slopes of 6″ or more per foot of rise, upon an underlayment of 30-pound roofing felt. The roof frame must be specially sized and constructed to bear the considerable weight of the slate. A minimum 3″ headlap should be allowed. Attachment of the roofing slates is made with slater's hard copper wire nails. Nail size depends upon the thickness of the slates and the nature of the roof deck. When making coverage calculations, allow for a 2″ projection at all eaves and a 1″ projection at the rakes.

GRADUATED SLATE ROOFS

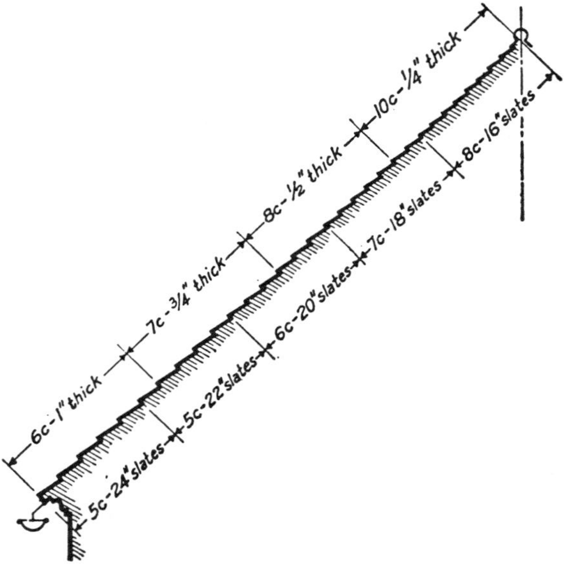

To lay out a graduated slate roof, first divide the rafter length from ridge to eaves into the same number of equal parts as there are different *thicknesses* of slate to be used.

Next divide the distance again into the same number of equal parts as there are *lengths* of slate to be used. A greater number of lengths should be used than thicknesses.

Then lay out the courses to correspond as nearly as possible with the divisions made, as shown in the drawing. The exposure for each length is found by subtracting 3" for the "head lap" from the length, and dividing the remainder by two.

With a graduated slate roof random widths should always be used.

SUGGESTED GRADUATIONS

Thicknesses	*Lengths*
¼, ⅜, ½, ¾, 1	14, 16, 18, 20, 22, 24
3/16, ¼, ⅜, ½, ¾	14, 16, 18, 20, 22, 24
3/16, ¼, ⅜, ½	12, 14, 16, 18, 20
3/16, ¼, ⅜	12, 14, 16, 18, 20
½, ¾, 1, 1¼, 1½	16, 18, 20, 22, 24, 26, 28, 30

GENERAL INFORMATION ON
MINERAL FIBER SHINGLES

PROCESS OF MANUFACTURE. Mineral fiber shingles, once known as asbestos shingles, are made of a combination of asbestos fiber, portland cement, and (occasionally) other fibers, formed under great pressure. A variety of colors, obtained by the addition of the highest quality pure mineral pigments, is possible. The range of colors enables the planner to obtain an effective harmony between the house, the roof, and the surrounding landscape. These shingles offer a carefree permanence and attractive appearance.

ACCESSORIES. Starter units, eave and gable trim, ridge and hip shingles, ridge and corner roll, fasteners, and adhesives are available for various methods of application.

APPLICATION. Roof boarding should be of narrow-width, well seasoned lumber, preferably tongue-and-groove. The boards should be laid with staggered joints, with at least two nails at each rafter. Unless end-matched lumber is used, the ends of every board should be securely nailed at a bearing. Plywood sheathing of an appropriate grade to suit the roof framing and loads makes an excellent roof deck. Before applying the shingles, the roof deck should be covered with an underlayment. The recommended material is 15-pound or 30-pound roofing felt, laid in single coverage on decks having a slope or more than 5″ for every foot of run, and in double coverage on decks with a slope of 3″ to 5″ for every foot of run. Instructions for proper application are included in every bundle of shingles.

FASTENERS. On new roofs, concealed roofing nails should be needle- or diamond-pointed and made of aluminum or hot-dipped galvanized steel, 1¼″ long. If the roof deck is plywood, threaded nails should be used; 1¼″ annular threaded if steel, and 12½° screw thread if aluminum.

FLASHING. The materials used for flashing and their method of application for a mineral fiber shingle roof are identical to those for a roof covered with slate or wood shingles.

ESSENTIALS OF MINERAL FIBER SHINGLE CONSTRUCTION

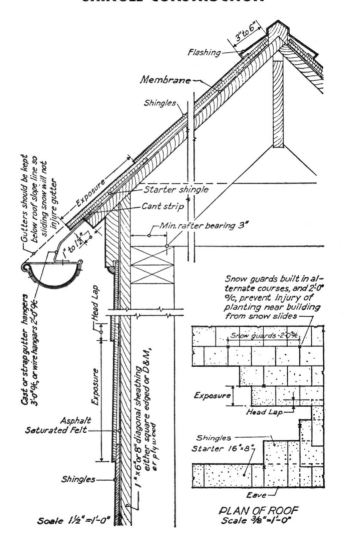

Flashing

3" to 6"

Membrane

Shingles

Exposure

Starter shingle

Cant strip

Min. rafter bearing 3"

Gutters should be kept below roof slope line so sliding snow will not injure gutter

1" to 1¾"

Cast or strap gutter hangers 3'-0"%, or wire hangers 2'-0"%.

Head Lap

Exposure

Asphalt Saturated Felt

Shingles

1"x6"or 8" diagonal sheathing either square edged or D&M, or plywood

Scale 1½"=1'-0"

Snow guards built in alternate courses, and 2'-0" %, prevent injury of planting near building from snow slides

Snow guards -2'-0"%

Exposure

Head Lap

Shingles
Starter 16"x8"

Eave

PLAN OF ROOF
Scale ⅜"=1'-0"

450

KINDS OF
BUILT-UP ROOFING

Built-up roofing is either smooth-surfaced or finished with an aggregate (slag or gravel). Roofs may be classified as follows:

1. Hot smooth—hot asphalt or coal tar pitch and saturated organic, asbestos, or fiberglass felt (use of coated felts is also possible), with pitch, asphalt, saturated felt, or mineral surfaced roll roofing as surfacing.
2. Hot aggregate—hot asphalt or coal tar pitch and saturated organic, asbestos, or fiberglass felt, with aggregate topping.
3. Cold smooth—cold asphalt cutback or asphalt emulsion and coated organic, asbestos, or fiberglass felt (use of saturated felts is also possible), with asphaltic coating, coated felt, or mineral surfaced roll roofing as surfacing.

SLAG OR GRAVEL. The gravel or slag protects the pitch or asphalt from drying out through the evaporation of natural oils. It also prevents ignition from burning embers and it provides a wearing surface and permits the application of a much heavier coat of bitumen (which is permanently anchored in place) than would otherwise be possible. Choosing between gravel and slag depends upon their price and availability in any given locality.

COAL TAR PITCH. This is a hydrocarbon obtained by the distillation of coal or from blast furnaces. It is loosely referred to as "tar" or "pitch." Coal tar pitch has high resistance to the penetration of water and is self-sealing.

COAL TAR PITCH ROOFING. Because of the relatively low melting point of coal tar pitch, it is generally recommended for use on slopes of no more than approximately ½″ per foot. On steeper slopes the pitch may flow off or the membranes loosen and slip. Water and moisture have a preservative effect on coal tar pitch, which also favors the use of relatively flat or almost dead-level slopes (perfectly dead-level roofs are not recommended) for this material. Coal tar pitch roofing has high resistance to acid fumes and corrosive gases. Experience indicates that coal tar pitch roofs are extremely long-lived under conditions not antagonistic to their use.

ASPHALT. Asphalt is a bitumen—a natural mixture of hydrocarbons. It is found in superficial deposits in various parts of the world and is obtained as a by-product in the distillation of petroleum, refined for commercial use.

ASPHALT ROOFING. Asphalt has a higher melting point than coal tar pitch and is available in several different melting-point consistencies. It is used on slopes of up to 6″ per foot. Asphalt is less impervious to moisture than coal tar pitch and hence is not normally recommended for use on slopes of less than about ½″ per foot, to ensure proper drainage. In climates having exceptionally hot sunshine, asphalt is particularly suitable because of its high melting point.

SMOOTH FINISH ROOFINGS. Smooth finish roofings have the advantage of light weight, compared to aggregate roofings. They may be used on slopes of from 1″ to 6″ per foot, and on steeper slopes under certain conditions. Saturated asbestos or fiberglass felt is recommended for the plies for its fire-resisting qualities and since it has the property of preventing the rapid drying out of the asphalt, which is unprotected by slag or gravel. Asphalt retains its oils better than coal tar pitch does, so it is frequently employed for smooth surfaced roofing (though coal tar pitch can be employed as well in some circumstances). Mineral surfaced roll roofing is also used as a final course in certain specifications on steep slopes.

CONSOLIDATED ROOFING TABLE

	Wood				Poured Concrete or Gypsum					
Slag, Lbs. per Sq.	400	400	:	:	400	400	400	:	:	:
or Gravel, Lbs. per Sq.	300	300	250	:	300	300	300	250	:	:
Asphalt, Lbs. per Sq.	:	:	40-50	60	:	:	:	:	90	:
Pitch, Lbs. per Sq.	150	125	140	:	200	200	175	160	:	200
Asphalt Primer	:	:	:	:	:	:	:	:	Yes	:
Weight of Mineral Surfaced Roofing, Lbs. per Sq.	:	:	:	110	:	:	:	:	110	:
Mineral Surfaced Roofing, No. of Plies	:	:	:	2	:	:	:	:	2	:
Weight of Tarred Felt, Lbs. per Sq.	75	60	75	30	60	60	45	60	15	75
No. of Plies, Tarred Felt	5	4	5	2	4	4	3	4	1	5
Sheathing Paper No. of Plies	1	1	1	:	:	:	:	:	:	:
Insulation	Optional	Optional	Optional	Optional	Optional	Optional	Optional	Optional	Optional	Optional
Surface Finish	S. or G.	S. or G.	Slag	Mineral Surfaced	S. or G.	S. or G.	S. or G.	Slag	Mineral Surfaced	Promenade Tile
Maximum Slope of Deck, In. per Ft.	2	2	{5 max. 2 min.}	{Over 3 in.}	2	¼	2	{5 max. 2 min.}	{Over 3 in.}	1
No. of Plies	5	4	5	4	4	4	3	4	3	5
Bond Years	20	15	20	10	20	20	15	20	10	:

CONSOLIDATED ROOFING TABLE

	Precast Concrete		Precast Gypsum			Spray Pond Over Concrete or gypsum		Steel		
Slag, Lbs. per Sq.	400	400	400	400	...	700	700	400	400	...
or Gravel, Lbs. per Sq.	300	300	300	300	250	500	500	300	300	250
Asphalt, Lbs. per Sq.	...	...	...	...	...	...	...	35	35	35
Pitch, Lbs. per Sq.	200	175	150	125	140	300	275	{200/35}	{175/35}	{60/35}
Asphalt Primer	...	...	...	...	...	...	...	Yes	Yes	Yes
Weight of Mineral Surfaced Roofing, Lbs. per Sq.	...	...	...	...	...	...	...	...	...	...
Mineral Surfaced Roofing, No. of Plies	...	...	...	...	...	...	...	...	...	...
Weight of Tarred Felt, Lbs. per Sq.	60	45	75	60	75	60	45	60	45	60
No. of Plies, Tarred Felt	4	3	5	4	5	4	3	4	3	4
Sheathing Paper No. of Plies	...	...	...	...	...	...	...	...	...	...
Insulation	Optional	Optional	Optional	Optional	Optional	No	No	Yes	Yes	Yes
Surface Finish	S. or G.	S. or G.	S. or G.	S. or G.	Slag	S. or G.	S. or G.	S. or G.	S. or G.	Slag
Maximum Slope of Deck, In. per Ft.	2	2	2	2	5 max. 2 min.	1	1	1	1	4 max. 2 min.
No. of Plies	4	3	5	4	5	4	3	4	3	4
Bond Years	20	15	20	15	20	20	15	20	15	10
Type of Deck	Precast Concrete		Precast Gypsum			Spray Pond Over Concrete or gypsum		Steel		

5-PLY 20 YEAR ROOF
OVER WOOD DECK

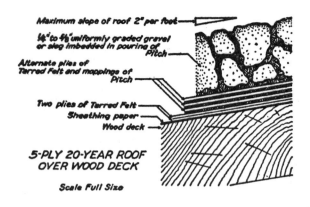

Maximum slope of roof 2" per foot

¼" to ⅝" uniformly graded gravel
or slag imbedded in pouring of
Pitch

Alternate plies of
Tarred Felt and moppings of
Pitch

Two plies of Tarred Felt
Sheathing paper
Wood deck

5-PLY 20-YEAR ROOF
OVER WOOD DECK

Scale Full Size

DECK.   The roof deck is made of clean, smooth lumber that is free from knotholes, large cracks, or loose boards. The lumber should be well seasoned, kiln-dried, or treated.

CONSTRUCTION.   One thickness of sheathing paper is laid over the deck with 1″ laps. Over this are laid two thicknesses of tarred felt, each 36″ strip or "course" overlapping the preceding one 19″ in clapboard fashion, nailed where necessary to hold the plies in place.

Over this surface a mopping of pitch is applied. Another thickness of tarred felt is laid with laps of 24⅔″, leaving 11⅓″ to the weather. Each lap is mopped so that felt never touches felt. Two more plies are added in the same way, alternating with moppings over the entire surface. Each strip of felt is nailed every 2′ along the upper edge. All nails must be covered by two plies of felt.

Over the last ply is poured a uniform coating of pitch. While the pitch is hot, gravel or slag is embedded in it.

INSULATION.   When insulation is applied on the wood deck, it must be thoroughly dry and of approved type. It is nailed to the deck and must be able to retain the roofing nails used in applying the roofing. When one layer of insulation is used, the roofing is applied as above. When more than one layer is used, the sheathing paper may be omitted.

ROOFS OVER WOOD OR GYPSUM BLOCK, 0" TO 3" SLOPE

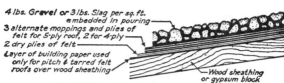

4 lbs. Gravel or 3 lbs. Slag per sq. ft. embedded in pouring
3 alternate moppings and plies of felt for 5-ply roof, 2 for 4-ply
2 dry plies of felt
Layer of building paper used only for pitch & tarred felt roofs over wood sheathing

Wood sheathing or gypsum block

GRAVEL OR SLAG SURFACING

Total Plies	Bond Years	Base	Type of Roofing	Roof Slope	Wt. Per Sq. Ft.*
5	20	Wood	Pitch and Tarred Felt	0" to 2"	6.32
5	20	Gyp. Bl.	Pitch and Tarred Felt	0" to 2"	6.32
4	15	Wood	Pitch and Tarred Felt	0" to 2"	5.91
4	15	Gyp. Bl.	Pitch and Tarred Felt	0" to 2"	5.91
5	20	Wood	Asphalt and Asphalt Felt..............	0" to 2"	5.91
4	15	Wood	Asphalt and Asphalt Felt..............	1/2" to 3"	6.07
				1/2" to 3"	5.66

4 lbs. Gravel or 3 lbs. Slag per sq. ft. embedded in pouring.
4 alternate moppings of plies and felt.
Base
Gypsum block

GRAVEL OR SLAG SURFACING

Total Plies	Bond Years	Base	Type of Roofing	Spec. No.	Roof Slope	Wt. Per Sq. Ft.*
4	20	Gyp. Bl.	Asphalt and Asphalt Felt..............	205	1" to 4"	6.20

Weight given is in pounds per square foot, with gravel surfacing; for slag surfacing the weight is 1# less. All roofs in table are Underwriters Class A.

CHOICE OF ROOFING. The selection of the proper specification will depend upon the slope of the deck, the number of plies desired, and whether a pitch or asphalt type is preferred, as shown in the table. See page 447.

CONSTRUCTION OF WOOD DECKS. The roof deck should be built of nominal 1" x 6" boards (preferably T&G) laid diagonally or of plywood of an appropriate grade and thickness. All boards must have a bearing on rafters at each end and must be nailed securely at each bearing. Plywood edges must bear on rafters at each end, and side-edge blocking may be required. The deck must not deflect perceptibly under an average person's weight. Knotholes and cracks wider than 1/4" must be covered with sheet metal, securely nailed. If boards are used, they must be thoroughly seasoned in order to avoid tearing the roofing plies by their movement. The deck must be smooth, dry, carefully graded to drains, and swept clean of all loose material.

CONSTRUCTION OF GYPSUM BLOCK DECK. The blocks must be dry. If pronounced ridges or depressions are present, they must be leveled off before the roofing operation is begun.

ROOFS FOR WOOD DECKS
SLOPE 1″ TO 6″ PER FT.

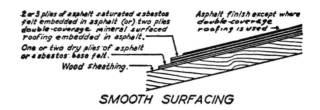

2 or 3 plies of asphalt saturated asbestos felt embedded in asphalt (or) two plies double-coverage mineral surfaced roofing embedded in asphalt.

One or two dry plies of asphalt or asbestos base felt.

Wood sheathing.

Asphalt finish except where double-coverage roofing is used

SMOOTH SURFACING

Total Plies	3	4	3	3	4	4
Bond Years	10	10	15	15	20	20
Roof Slope, per ft.	1″-6″	over 3″	1″-6″	1″-6″	1″-6″	1″-6″
Wt. in lbs. per sq. ft.	1.37	1.92	1.87	1.60	2.07	1.95
Dry Plies Asphalt Felt	1	1	1	1	1	1
Mopped Plies Asb. Felt	2	—	2	2	3	3
Mopped Plies Double-Cov.	—	2	—	—	—	—
Total Moppings	3	2	3	3	3	4
Cold Coating Surface Finish	x	—	x	x	x	x
Surface Fin. Double-Cov.	—	x	—	—	—	—

CHOICE OF ROOFING. The selection of the proper specification will depend upon the slope of the deck, the number of plies desired, and whether an asphalt surface finish or mineral-surfaced roll roofing is preferred. See page 447.

CONSTRUCTION OF WOOD DECKS. The roof deck should be built of nominal 1″ x 6″ boards (preferably T&G) laid diagonally or of plywood of an appropriate grade and thickness. All boards must have a bearing on rafters at each end and must be nailed securely at each bearing. Plywood edges must bear on rafters at each end, and side-edge blocking may be required. The deck must not deflect perceptibly under an average person's weight. Knotholes and cracks wider than ¼″ must be covered with sheet metal, securely nailed. If boards are used, they must be thoroughly seasoned in order to avoid tearing the roofing plies by their movement. The deck must be smooth, dry, carefully graded to drains, and swept clean of all loose material.

ROOFS OVER CONCRETE, GYPSUM, BOOK TILE, OR INSULATION, 0" TO 3" SLOPE

4 lbs Gravel or 3 lbs. Slag per sq.ft. embedded in pouring

3 or 4 alternate moppings and plies of felt

Base used only when roofing is asphalt and asphalt felt.

Concrete, poured gypsum, book tile or insulation.

GRAVEL OR SLAG SURFACING

Total Plies	Bond Years	Base	Type of Roofing	Roof Slope	Wt. Per Sq. Ft.†
3	15	Pd.Con.*	Pitch and Tarred Felt	0" to 1"	6.20
4	20	Pd.Con.*	Pitch and Tarred Felt	0" to 1"	6.60
3	15	Pd.Con.*	Asphalt and As. Rag Felt..........	1" to 4"	5.80
4	20	Pd.Con.*	Asphalt and As. Rag Felt..........	1" to 4"	6.20

Pd.Con. refers to poured concrete, poured gypsum, precast concrete, book tile, approved rigid insulation. † Weight given is in pounds per square foot with gravel surfacing; for slag surfacing the weight is 1# less. All roofs in table are Underwriters Class A.

CHOICE OF ROOFING. The selection of the proper specification will depend upon the slope of the deck, the number of plies desired, and whether a pitch or asphalt type is preferred as shown in the table. See page 447.

POURED CONCRETE AND POURED GYPSUM DECKS. This type of deck must not be either wet or frozen. Sharp or abrupt ridges or depressions must be made smooth by filling with mortar or hammering down the high spots. The deck must be swept clean of all loose material. *If the deck is to be made of poured gypsum, felts must be nailed, as over boards.*

PRECAST CONCRETE DECKS. The blocks must be dry. If pronounced ridges or depressions occur, they must be leveled off before the roofing is applied.

BOOK TILE DECKS. These must be covered with a brush coat of cement mortar, which is allowed to set and dry, so that a smooth, even surface is obtained to receive the roof.

Two Plies #15 Rag Felt Nailed Dry to Roof Deck. Apply Roof as over Concrete.

Board Deck - Gravel Roof

One Ply of #15 or #30 Rag Felt Mopped to Roof Deck. Insulation Laid in Hot Mopping. Follow Specifications as over Concrete.

Concrete Deck - Gravel Roof

INSULATION. Insulation must be of a type appropriate for roofing applications and compatible with the roofing system being applied. It must be thoroughly dry. Insulation thicknesses vary, usually from ½" to 5½". The illustrations at left show the application of the insulation to different types of decks. Manufacturers' application instructions must be followed.

INSULATED STEEL DECK. Roofing may be applied to insulation over a steel deck. For slopes of less than 2" per foot, the insulation is held by mopping; over 3" per foot, by screws or clips.

457

ROOFS OVER CONCRETE, GYPSUM, BOOK TILE, OR INSULATION, 1″ TO 6″ SLOPE

2 or 3 plies of asphalt saturated asbestos felt embedded in asphalt (or) 2 plies of double-coverage mineral surfaced roofing embedded in asphalt.

1 or 2 plies of asphalt or asbestos base felt embedded in asphalt.

Concrete primer.

Concrete roof, etc.

Asphalt finish except where double-coverage roofing is used

SMOOTH SURFACING

Total Plies	3	3	4	3	5
Bond Years	10	10	15	20	20
Roof Slope, in ins. per ft	1″-6″	3″-6″	1″-6″	1″-6″	1″-6″
Wt. in lbs. per sq. ft	1.70	2.18	2.00	2.10	2.15
Mopped Plies Asphalt Felt	1	1	2	1	2
Mopped Plies Asbestos Felt	2	—	2	2	3
Mopped Plies Double-Coverage	—	2	—	—	—
Total Moppings	4	2	5	4	6
Cold Coating Surface Finish	x	—	x	x	x
Surface Fin. Double-Coverage	—	x	—	—	—

CHOICE OF ROOFING. The selection of the proper specification will depend upon the number of plies desired, and whether an asphalt surface finish or a mineral surface roll roofing (double-coverage) is preferred. See page 447.

POURED CONCRETE AND POURED GYPSUM DECKS. This type of deck must not be either wet or frozen. Sharp or abrupt ridges or depressions must be made smooth by filling with mortar or hammering down the high spots. The deck must be swept clean of all loose material.

PRECAST CONCRETE AND PRECAST GYPSUM DECKS. The blocks must be dry. Pronounced ridges or depressions must be leveled off before the roofing is applied.

BOOK TILE DECKS. These must be covered with a brush coat of cement mortar, which is allowed to set and dry, so that a smooth, even surface is obtained to receive the roof.

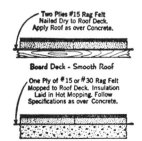

Two Plies #15 Rag Felt Nailed Dry to Roof Deck. Apply Roof as over Concrete.

Board Deck - Smooth Roof

One Ply of #15 or #30 Rag Felt Mopped to Roof Deck. Insulation Laid in Hot Mopping. Follow Specifications as over Concrete.

Concrete Deck - Smooth Roof

INSULATION. Insulation must be of a type appropriate for roofing applications and compatible with the roofing system and materials being applied. It must be thoroughly dry. Insulation thicknesses vary, usually from ½″ to 5½″. The illustrations at left show two methods of application of the insulation to different types of decks. Manufacturers' application instructions must be followed.

INSULATED STEEL DECK. Roofing may be applied to insulation over a steel deck. For slopes of less than 2″ per foot, the insulation is held by mopping; over 2″ per foot, by screws or clips.

ROOFS OVER INSULATION SLOPES 1″ TO 6″ PER FT., UNDER PROMENADE DECKS

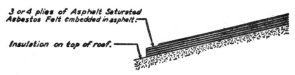

3 or 4 plies of Asphalt Saturated Asbestos Felt embedded in asphalt.

Insulation on top of roof.

SMOOTH SURFACING

3-PLY 15-YEAR SPECIFICATION. This is a roof especially developed for use only over insulation, for slopes of 1″ to 6″ per foot. It consists of 3 plies of asphalt-impregnated asbestos felt and three moppings of asphalt. The felt itself constitutes the surface finish, no final mopping being applied. The insulation to which the roofing is applied may be over any type of deck—wood, steel, concrete, gypsum, book tile, etc.—but must be held securely to its base material.

4-PLY 20-YEAR SPECIFICATION. This is a roofing especially developed for use only over insulation, for slopes of 1″ to 6″ per foot. It consists of 4 plies of asphalt-impregnated asbestos felt and five moppings of asphalt. The last mopping of asphalt constitutes the surface finish. The insulation to which the roofing is applied may be over any type of deck—wood, steel, concrete gypsum, book tile, etc.—but must be securely held to its base material.

INSULATION. See other *Data Sheets* in this section.

PROMENADE DECKS

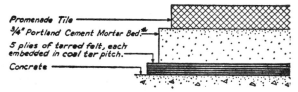

Promenade Tile
*¾″ Portland Cement Mortar Bed.**
5 plies of tarred felt, each embedded in coal tar pitch.
Concrete

*Or Pitch Base Plastic Cement.

5-PLY 20-YEAR SPECIFICATION. This is a special type of roofing for use under promenade tile, having slopes from 0″ to 1″ per foot. It consists of 5 plies of tarred felt and 6 moppings of coal tar pitch. This roof has Underwriters Classification A. If rigid insulation is placed on the concrete slab, the roofing may be applied to the insulation in the same manner as it would be if directly on concrete.

SLOPED ROOFS
OVER GYPSUM PLANK

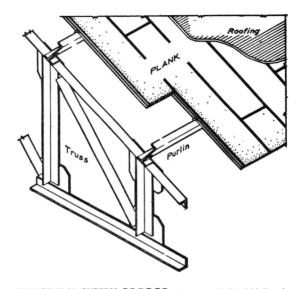

STRUCTURAL SYSTEM OF ROOF. Gypsum plank is laid directly over the purlins of a truss roof, over beams acting as rafters on either sloping or flat decks. When the supporting members run horizontally, the courses of plank run up the slope. When the supporting members run with the slope, the plank is coursed horizontally.

PITCHED ROOFS. On steep roofs, provision must be made to prevent the sliding action of the construction. A stop angle should be used at the eave. It may also be necessary to bolt the plank through rafters or purlins, as the pitch of the roof requires. Clips are used at every intersection of the plank with a support.

EAVES AND RAKES. Gypsum plank may overhang up to 18″ beyond support at the eave and up to 6 inches at the rake, where the courses of plank run up the slope (as shown above). If the plank runs horizontally, these overhangs will be reversed. All openings, except those for small pipes, vents, or downspouts, should be framed out.

VALLEYS, GUSSETS, AND COVES. These are readily formed from gypsum, poured and screeded to the desired contours.

BUILT-UP ROOFING. Application is same as over wood decks—nailing the first layer of paper or felt and spot-mopping if desired, but not mopping the entire area.

SHINGLE OR SLATE. Roofing of this type may be nailed directly to the plank. Nails should be square cut, preferably hard copper, penetrating 1½″.

DETERMINATION OF
ROOF LEADERS

$$\text{Required area in } \square'' \text{ of leaders} = \frac{\text{Area of roof surface*}}{\text{Constant shown in table**}}$$

Arizona 500	Maine 400	Oklahoma 264
Alabama 68	Maryland 345	Ohio 370
Arkansas 208	Massachusetts . 416	Oregon 400
California 400	Michigan 238	Pennsylvania .. 263
Colorado 303	Minnesota 294	Rhode Island .. 435
Connecticut ... 370	Mississippi 345	South Carolina. 294
Delaware 330	Missouri 56	South Dakota . 275
Florida 244	Montana 285	Tennessee 181
Georgia 222	Nebraska 200	Texas 322
Idaho 285	Nevada 285	Utah 1000
Illinois 345	New Hampshire 500	Vermont 345
Indiana 303	New York 312	Virginia 275
Iowa 135	North Carolina. 232	Washington ... 400
Kansas 345	North Dakota . 312	West Virginia . 384
Kentucky 454	New Mexico .. 200	Wisconsin 244
Louisiana 208	New Jersey ... 222	Wyoming 910

Type of Leader	Leader Size	Area in square ins.
Plain Round Leader	3″ 4″ 5″ 6″	7.07 12.57 19.63 28.27
Corrugated Round Leader	3″ 4″ 5″ 6″	5.94 11.04 17.72 25.97
Square Corrugated Leader	1¾″ x 2¼″ (2″) 2⅜″ x 3¼″ (3″) 2¾″ x 4¼″ (4″) 3¾″ x 5″ (5″)	3.80 7.73 11.70 18.75
Plain Rectangular Leader	1¾″ x 2¼″ 2″ x 3″ 2″ x 4″ 3″ x 4″ 4″ x 5″ 4″ x 6″	3.94 6.00 8.00 12.00 20.00 24.00

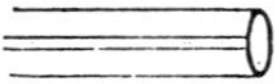

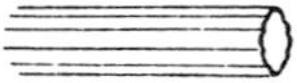

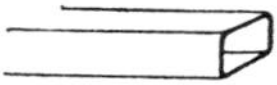

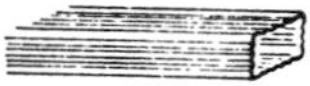

Seventy-five feet is the maximum spacing for leaders.
All outlets should be provided with screens or strainers.
Scuppers should be provided for all roofs with encircling parapets.
Round leaders should not be less than 3″ in diameter.
Rectangular leaders should not be less than 1¾″ x 2¼″.

---

*This is square feet of actual roof surface, not horizontal proj'n.
**Table calculated on the basis of 1□″ of leader per 1″ of rainfall per 1200 □ feet per hour for maximum rate in different localities as shown by report of Chief of Weather Bureau.

GOOSENECKS, LEADER
HEADS AND STRAPS

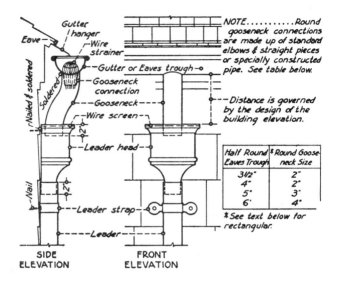

NOTE..........Round
gooseneck connections
are made up of standard
elbows & straight pieces
or specially constructed
pipe. See table below.

Gutter
hanger
Eave
Wire
strainer
Gutter or Eaves trough
Gooseneck
connection
Gooseneck
Wire screen
Leader head
Leader strap
Leader

Distance is governed
by the design of the
building elevation.

Nailed & soldered
Soldered
Nail

SIDE
ELEVATION

FRONT
ELEVATION

Half Round Eaves Trough	*Round Goose-neck Size
3½"	2"
4"	2"
5"	3"
6"	4"

*See text below for
rectangular.

RECTANGULAR GOOSENECK CONNECTIONS. The rectangular gooseneck is much more efficient than the standard round type, in handling the water flowing through it. In section, it should be as long as the gutter width; the gooseneck width should be ⅔ of the gutter width.

LEADER HEADS. Leader heads are primarily ornamental. They also effect transitions between goosenecks and leaders of different cross-sectional shapes, as well as providing a "magazine" space for the collection of water. Because of a limited range of standard leader heads, many architects use special designs. The dimensions of a leader head are entirely at the discretion of the designer—no rules of hydraulics enter the problem other than that a smooth path should be provided for the water.

LEADER STRAPS. Leader straps are available in many stock designs but vary from locality to locality. The architect would do well either to make the designs or to require the successful bidder on the sheet metal work to submit samples of available styles.

STANDARD SHEET METAL
GUTTERS AND EAVES TROUGHS

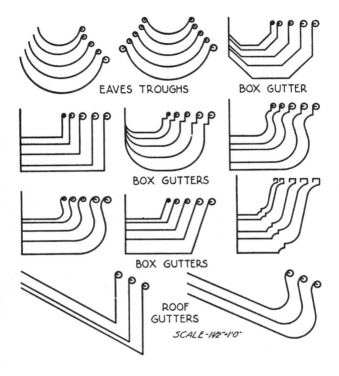

EAVES TROUGHS   BOX GUTTER

BOX GUTTERS

BOX GUTTERS

ROOF
GUTTERS

SCALE -1½"=1'0"

GUTTER SIZES.   Gutters are made in overall widths of 3½", 4", 5", 6", 7", and 8". Exceptions are the gutter shown at lower left, which is made in depths of 4½", 5¾", and 6¼" overall, and the gutter shown at lower right, which is made in depths of 4", 5", and 6" overall. However, numerous variations in design, shape, and size appear from time to time from various manufacturers, and countless differences occur in locally custom-made guttering systems.

GUTTER DESIGN.   Gutters smaller than 4" should not ordinarily be used unless demanded by architectural design. The size of the gutters is determined by the leaders and their spacing. For leader spacings up to 50', use a gutter not less than the equivalent circular diameter of the leader. For leader spacings from 50' to 70', use a gutter 1" wider than the equivalent circular diameter of the leader. For leader spacings from 70' to 90', use a gutter 2" wider than the equivalent circular diameter of the leader.

DEAD-LEVEL GUTTERS.   Accurately installed dead-level gutters drain readily and are usually more desirable than sloping gutters from an architectural design standpoint.

GUTTER HANGERS

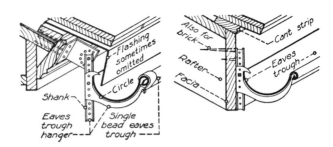

Flashing sometimes omitted — Circle — Shank — Eaves trough hanger — Single bead eaves trough

Also for brick — Cant strip — Rafter — Facia — Eaves trough

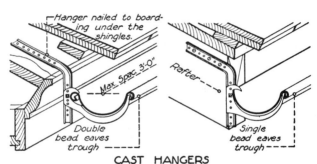

Hanger nailed to boarding under the shingles.

Max. Spac. 3'-0"

Double bead eaves trough

Rafter

Single bead eaves trough

CAST HANGERS

For hanging gutters, the half round eaves trough in single or double bead is used. Other gutter types are for box construction.

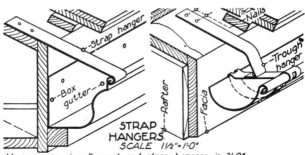

Strap hanger — Box gutter — Nails — Trough hanger — Rafter — Facia

STRAP HANGERS
SCALE 1½"=1'-0"

*Maximum spacing for cast and strap hangers is 3'-0".
Good practice is 2'-6".*

SNOW GUARDS, LEADER HOOKS, STRAINERS

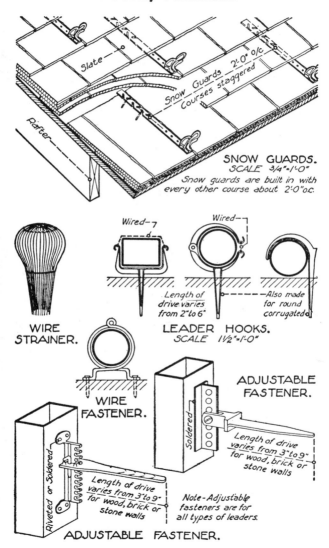

Slate

Snow Guards 2'0" o/c
Courses staggered

Rafter

SNOW GUARDS.
SCALE 3/4"=1'-0"
Snow guards are built in with
every other course about 2'-0"o.c.

Wired

Wired

Also made
for round
corrugated

**WIRE
STRAINER.**

Length of
drive varies
from 2" to 6"

LEADER HOOKS.
SCALE 1½"=1'-0"

**WIRE
FASTENER.**

**ADJUSTABLE
FASTENER.**

Soldered

Length of drive
varies from 3"to 9"
for wood, brick or
stone walls

Riveted or Soldered

Length of drive
varies from 3"to 9"
for wood, brick or
stone walls

Note - Adjustable
fasteners are for
all types of leaders.

ADJUSTABLE FASTENER.

465

OPEN AND CLOSED
VALLEYS

OPEN VALLEYS. Where the adjacent roof surfaces are of different areas or different slopes, a baffle rib prevents the larger or faster descending volume of water from forcing its way up under the roofing on the opposite side, as shown at A and D. If the slopes and areas are the same, a smooth valley may be used, as shown at B. Separate sheets lapped 2″ provide for expansion and contraction. It is preferred by many to the usual locked and soldered cross seams in the valleys.

CLOSED VALLEYS. The closed valley may be used for slopes of 45° or steeper where adjacent roof surfaces are of similar slopes and areas. One of several methods is shown at C where the sheets are laid in long pieces directly on the paper or felt which covers the roof sheathing. They may be of any length and should lap 4″. The center crimp stiffens the valley flashing and forms a straight line to which the slates or shingles are set.

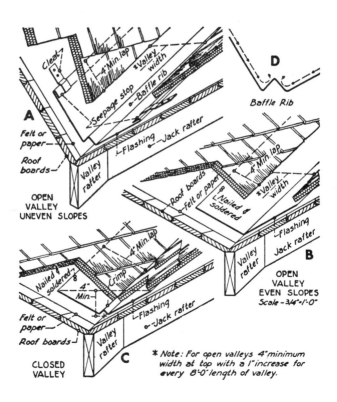

OPEN VALLEY
UNEVEN SLOPES

Baffle Rib

OPEN
VALLEY
EVEN SLOPES
Scale - 3/4″=1′-0″

CLOSED
VALLEY

*Note: For open valleys 4″minimum width at top with a 1″increase for every 8′-0″length of valley.

SLOPE OF
ROOF VALLEYS

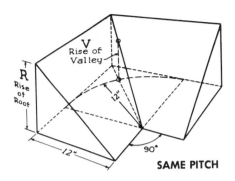

SAME PITCH

It is frequently necessary to find the slope of a gutter between two
roofs of the same pitch intersecting at right angles. The chart below
shows that two intersecting roofs having 6″ rise per foot will have
a gutter whose rise is about 4¼″ per foot.

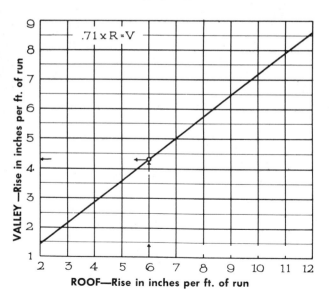

SLOPE OF
ROOF VALLEYS

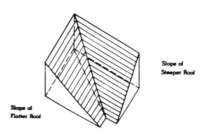

Slope of
Steeper Roof

Slope of
Flatter Roof

The slope of a valley (or hip) formed by roofs intersecting at right angles, but having different slopes, can be roughly determined from the chart below. EXAMPLE: A roof having a slope of 8″ rise per foot of run intersects at right angles with a roof having a slope of 12½″ per foot of run; the chart shows that the ridge or hip will slope at about 6¾″ per foot of run.

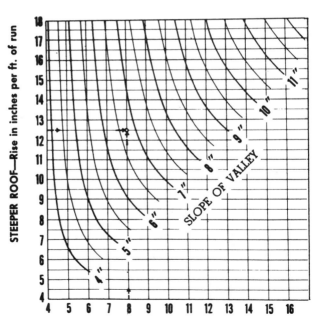

SHEET METAL RIDGES
AND HIPS

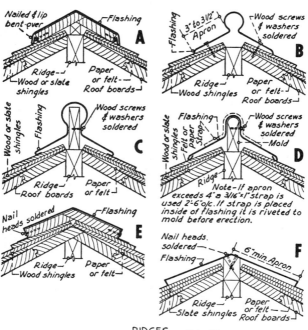

A Nailed & lip bent over — Flashing — Ridge — Wood or slate shingles — Paper or felt — Roof boards

B Flashing — 3" to 3½" Apron — Wood screws & washers soldered — Ridge — Wood shingles — Paper or felt — Roof boards

C Wood or slate shingles — Flashing — Wood screws & washers soldered — Ridge — Roof boards — Paper or felt

D Wood or slate shingles — Flashing — Felt or paper strap — Wood screws & washers soldered — Mold — Ridge — Note – If apron exceeds 4" a 3/16"x1" strap is used 2'-6" o/c. If strap is placed inside of flashing it is riveted to mold before erection.

E Nail heads soldered — Flashing — Ridge — Wood shingles — Paper or felt

F Nail heads soldered — Flashing — 6" min. Apron — Ridge — Slate shingles — Paper or felt — Roof boards

RIDGES
Scale-1½"=1'-0"   Note: The above ridges are sometimes used as hips.

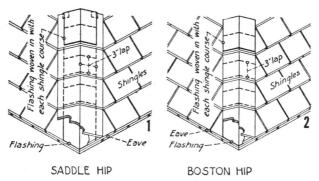

Flashing woven in with each shingle course — 3" lap — Shingles — Flashing — Eave

SADDLE HIP

Flashing woven in with each shingle course — 3" lap — Shingles — Eave — Flashing

BOSTON HIP

CHANGE OF ROOF SLOPE

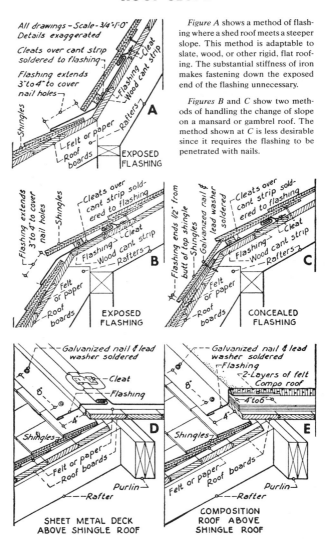

All drawings – Scale – 3/4"=1:0"
Details exaggerated

Cleats over cant strip soldered to flashing

Flashing extends 3" to 4" to cover nail holes

Flashing
Cleat
Wood cant strip

Shingles
Felt or paper
Roof boards
Rafters

A

EXPOSED FLASHING

Figure A shows a method of flashing where a shed roof meets a steeper slope. This method is adaptable to slate, wood, or other rigid, flat roofing. The substantial stiffness of iron makes fastening down the exposed end of the flashing unnecessary.

Figures B and *C* show two methods of handling the change of slope on a mansard or gambrel roof. The method shown at *C* is less desirable since it requires the flashing to be penetrated with nails.

Flashing extends 3" to 4" to cover nail holes

Cleats over cant strip soldered to flashing

Shingles

Flashing
Cleat
Wood cant strip
Rafters

Felt or paper
Roof boards

B

EXPOSED FLASHING

Cleats over cant strip soldered to flashing

Flashing ends 1/2" from butt of top shingle

Galvanized nail & lead washer soldered

Shingles

Flashing
Cleat
Wood cant strip
Rafters

Felt or paper
Roof boards

C

CONCEALED FLASHING

Galvanized nail & lead washer soldered

Cleat
Flashing

8"
4"

Shingles

Felt or paper
Roof boards

Purlin
Rafter

D

SHEET METAL DECK ABOVE SHINGLE ROOF

Galvanized nail & lead washer soldered
Flashing
2-Layers of felt Compo roof

8"
4" to 6"
4"

Shingles

Felt or paper
Roof boards

Purlin
Rafter

E

COMPOSITION ROOF ABOVE SHINGLE ROOF

DORMER
FLASHING

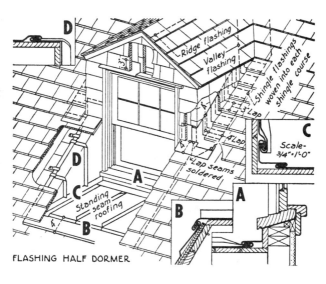

D

Ridge flashing

Valley flashing

Shingle flashings woven into each shingle course

3" Lap

C

Scale—3/4"=1'-0"

4" Lap

D

C

Standing seam roofing

A

Lap seams soldered

B

B

A

FLASHING HALF DORMER

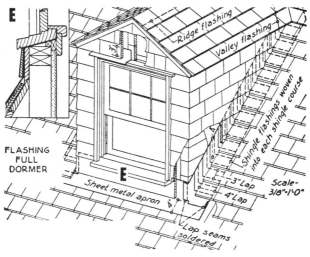

E

Ridge flashing

Valley flashing

Shingle flashings woven into each shingle course

FLASHING FULL DORMER

E

Sheet metal apron

3" Lap

4" Lap

Scale—3/8"=1'-0"

Lap seams soldered

471

CHIMNEY FLASHING

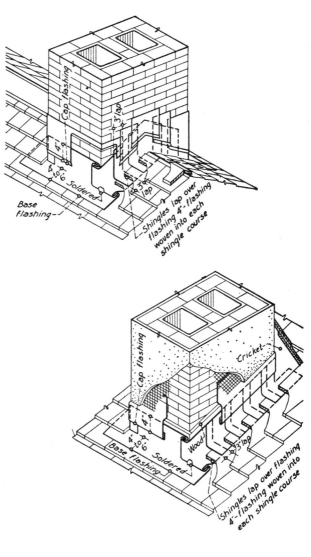

Cap flashing
3" lap
4"
30/6
Soldered
Base flashing
1" lap
Shingles lap over flashing 4"- flashing woven into each shingle course

Cap flashing
Cricket
4"
30/6
Soldered
Base flashing
Wood
3" lap
Shingles lap over flashing 4"- flashing woven into each shingle course

FLASHING ROOF
PENETRATIONS

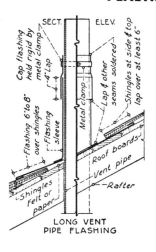

LONG VENT
PIPE FLASHING

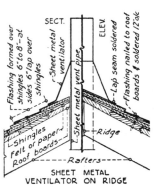

SHEET METAL
VENTILATOR ON RIDGE

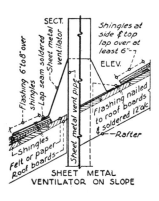

SHEET METAL
VENTILATOR ON SLOPE

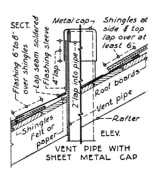

VENT PIPE WITH
SHEET METAL CAP

FLASHING OF
STONE JOINTS

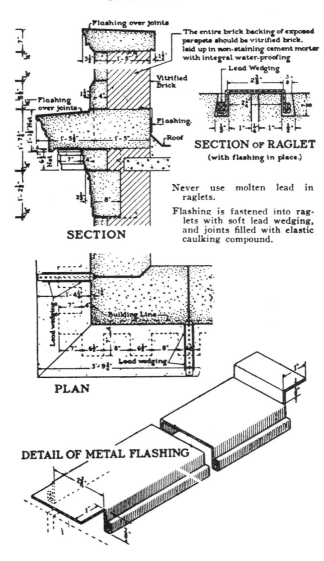

Flashing over joints

The entire brick backing of exposed parapets should be vitrified brick, laid up in non-staining cement mortar with integral water-proofing

Vitrified Brick

Flashing over joints

Flashing.

Roof

SECTION

Lead Wedging

SECTION OF RAGLET
(with flashing in place.)

Never use molten lead in raglets.

Flashing is fastened into raglets with soft lead wedging, and joints filled with elastic caulking compound.

Building Line

Lead wedging

Lead wedging

PLAN

DETAIL OF METAL FLASHING

STANDING SEAM
ROOFING

STANDING SEAM. This type of roofing is available in several variations of the seaming method. Because it makes the most watertight sheet metal roofing, it should be used on roof slopes of less than 4" in 12" and is effective on slopes as slight as 2" in 12". With proper application procedures, it can be used on slopes down to ¼" in 12". Water would have to take the course shown at x and y to get through the tightly swaged seams and end locks. Various materials are available, such as stainless steel, copper, T-C-Z alloy, terne, and copper-bearing steel. Widths vary from 14" to 25", and lengths run from 4' to 60', depending upon specific manufacturer and material. There is also a variety of thicknesses, again depending upon the material. Seams may be locked, double-locked, soldered, or welded, and the panels are held in place by cleats of various sorts. Consult manufacturers for specific details and dimensions.

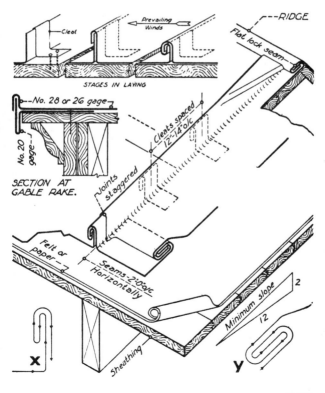

RIBBED SEAM
ROOFING

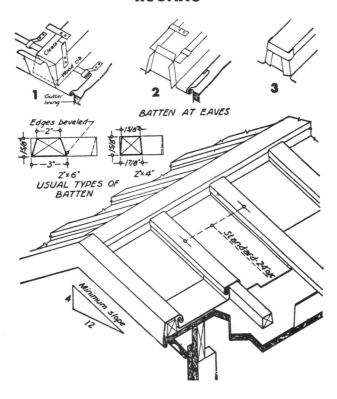

BATTEN AT EAVES

USUAL TYPES OF
BATTEN

RIBBED SEAM ROOFING. This style roofing is formed by the sheet metal contractor, over wood battens placed by the carpenter. Adaptable to roofs of steep slope, Ribbed Seam Roofing should not be used on slopes flatter than 4″ in 12″ and preferably 6″ in 12″. This roofing is formed from flat sheet metal, therefore the battens may be spaced any distance apart up to the limits of a 48″ wide sheet. The size and shape of the battens used are at the designer's option, triangular, semicircular and other sections sometimes being employed.

CONSTRUCTION. Roofing sheets are secured to the wood battens by cleats as shown at 1. spaced 12″ to 14″ apart and alternating on top and side of batten.

V-BEAM AND V-CRIMP ROOFING

5 V-CRIMP ROOFING. This roofing may be used unsealed on roof slopes of 3″ or more, and on slopes down to 1½″ if the laps are sealed. It is available in several colors, and in metal or fiberglass. Roofings with more or fewer Vs may be obtained. Heavier weights may be applied without the wood nailing strips shown in the drawing.

V-BEAM ROOFING. This style has largely superseded crimp roofing. The crimps, instead of being in the shape of a perfect inverted V, are squared off at the top to form slant-sided beams. Several colors are available in steel and aluminum, in widths from about 35″ to 42″ and in lengths of 30′, from 18-gauge to 24-gauge. Slopes are the same as for crimp roofing. The material is frequently applied without the need for cleats.

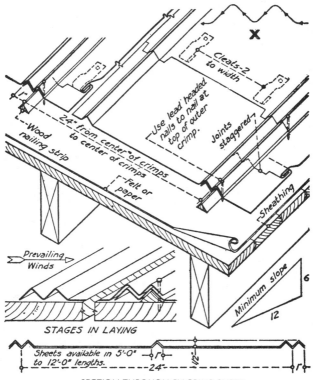

SECTION THROUGH 5V-CRIMP SHEET

CORRUGATED IRON
FOR ROOFS AND WALLS

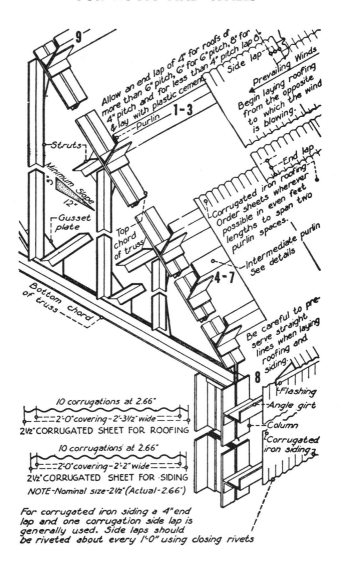

Allow an end lap of 4" for roofs of more than 6" pitch, 6" for 6" pitch, 8" for 4" pitch and for less than 4" pitch 6" & lay with plastic cement.

Side lap

Prevailing Winds

Begin laying roofing from the opposite to which the wind is blowing.

Purlin 1-3

Struts

Minimum Slope 3" 12"

Gusset plate

Top chord of truss

End lap

Corrugated iron roofing. Order sheets wherever possible in even feet lengths to span two purlin spaces.

4-7

Intermediate purlin See details

Bottom chord of truss

Be careful to preserve straight lines when laying roofing and siding.

8

Flashing

Angle girt

Column

Corrugated iron siding

10 corrugations at 2.66"
2'-0" covering - 2'-3½" wide
2½" CORRUGATED SHEET FOR ROOFING

10 corrugations at 2.66"
2'-0" covering - 2'-2" wide
2½" CORRUGATED SHEET FOR SIDING
NOTE-Nominal size-2½" (Actual-2.66")

For corrugated iron siding a 4" end lap and one corrugation side lap is generally used. Side laps should be riveted about every 1'-0" using closing rivets

478

CORRUGATED IRON
CONNECTION DETAILS

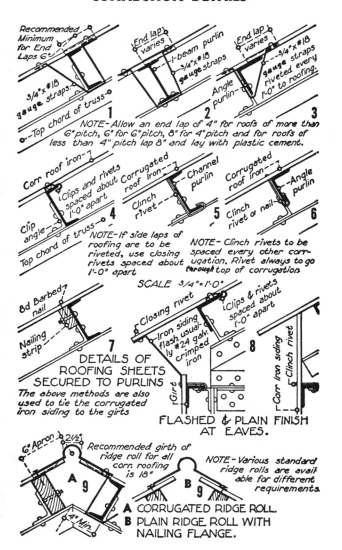

Recommended Minimum for End Laps 6"

3/4" × #18 gauge straps

Top chord of truss

End lap varies

I-beam purlin

3/4" × #18 gauge straps

End lap varies

3/4" × #18 gauge straps riveted every 1'-0" to roofing.

Angle purlin

1   **2**   **3**

NOTE- Allow an end lap of 4" for roofs of more than 6" pitch, 6" for 6" pitch, 8" for 4" pitch and for roofs of less than 4" pitch lap 8" and lay with plastic cement.

Corr roof iron

Clips and rivets spaced about 1'-0" apart

Clip angle

Top chord of truss

4

Corrugated roof iron

Channel purlin

Clinch rivet

5

Corrugated roof iron

Angle purlin

Clinch rivet or nail

6

NOTE- If side laps of roofing are to be riveted, use closing rivets spaced about 1'-0" apart

NOTE- Clinch rivets to be spaced every other corrugation. Rivet always to go through top of corrugation

SCALE 3/4" = 1'-0"

8d Barbed nail

Nailing strip

7

Closing rivet

Iron siding flash. usually #24 galv. crimped iron

Girt

Clips & rivets spaced about 1'-0" apart

Corr iron siding

Clinch rivet

8

DETAILS OF
ROOFING SHEETS
SECURED TO PURLINS

The above methods are also used to tie the corrugated iron siding to the girts

FLASHED & PLAIN FINISH
AT EAVES.

6" Apron   2½"

Recommended girth of ridge roll for all corr. roofing is 18"

A   **9**

4" Min.

B   **9**

NOTE- Various standard ridge rolls are available for different requirements.

A CORRUGATED RIDGE ROLL.
B PLAIN RIDGE ROLL WITH
NAILING FLANGE.

ATTIC
VENTILATION

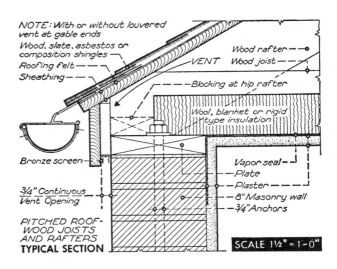

NOTE: With or without louvered vent at gable ends

Wood, slate, asbestos or composition shingles

Roofing felt

Sheathing

Wood rafter

VENT Wood joist

Blocking at hip rafter

Wool, blanket or rigid type insulation

Bronze screen

Vapor seal

Plate

Plaster

8" Masonry wall

¾" Anchors

¾" Continuous Vent Opening

PITCHED ROOF-
WOOD JOISTS
AND RAFTERS
TYPICAL SECTION

SCALE 1½" = 1'-0"

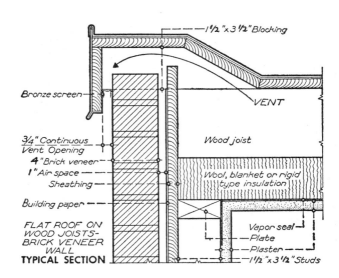

1½"x 3½" Blocking

Bronze screen

VENT

¾" Continuous Vent Opening

4" Brick veneer

1" Air space

Sheathing

Wood joist

Wool, blanket or rigid type insulation

Building paper

FLAT ROOF ON
WOOD JOISTS-
BRICK VENEER
WALL
TYPICAL SECTION

Vapor seal

Plate

Plaster

1½"x 3½" Studs

480

ATTIC
VENTILATION

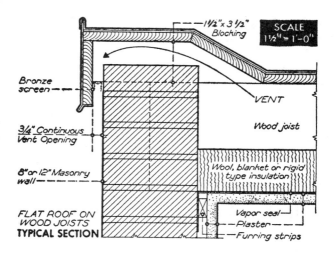

SCALE
1½" = 1'-0"

1½" x 3½" Blocking

Bronze screen

VENT

¾" Continuous Vent Opening

Wood joist

8" or 12" Masonry wall

Wool, blanket or rigid type insulation

FLAT ROOF ON WOOD JOISTS
TYPICAL SECTION

Vapor seal
Plaster
Furring strips

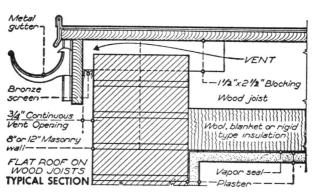

Metal gutter

VENT

Bronze screen

1½" x 2½" Blocking

Wood joist

¾" Continuous Vent Opening

Wool, blanket or rigid type insulation

8" or 12" Masonry wall

FLAT ROOF ON WOOD JOISTS
TYPICAL SECTION

Vapor seal
Plaster

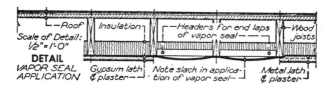

Roof

Insulation

Headers for end laps of vapor seal

Wood joists

Scale of Detail:
½" = 1'-0"
DETAIL
VAPOR SEAL APPLICATION

Gypsum lath & plaster

Note slack in application of vapor seal

Metal lath & plaster

481

SYMBOLS FOR WINDOWS
AND DOORS

WINDOWS

(Viewed in elevation from outside)

Horizontally Pivoted

Top Pivoted

Bottom Pivoted

Vertically Pivoted

Hinged at Left (Casement)

Hinged at Right (Casement)

Top Hinged to project out at bottom
Bottom Hinged to project in at top

Counterbalanced Window (Double Hung)

Horizontally Rolling toward left

Horizontally Rolling toward right

DOORS

Hinged Side

Left Hand to open in

Left Hand to open out

Right Hand to open in

Right Hand to open out

Horizontally Rolling toward right

Horizontally Rolling toward left

The symbols shown are widely used, following both conventional practice and various official and semiofficial recommendations. They are applicable to all types of windows and vents, irrespective of the frame/sash materials involved. Some minor variations may be encountered, especially as to notations.

METAL WINDOW IN STUCCO
OR BRICK VENEER WALL

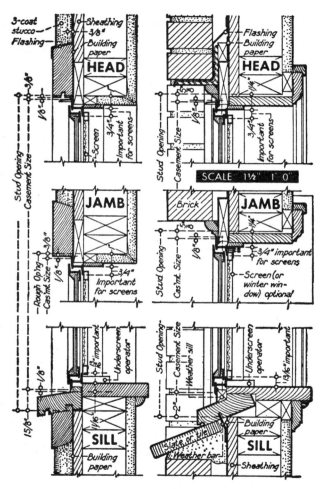

Caulking, mastic, wood strips, moldings, flashings, trimwork, structural steel, and similar installation materials are not supplied by the window manufacturer; head drips and head flashing sometimes are. Consult with the window manufacturer to determine what additional installation materials are needed.

METAL WINDOW IN SHINGLE OR CLAPBOARD WALL

WOOD SURROUNDS. Various forms of specially milled shapes provide a time- and trouble-saving shortcut to better installation of frameless window sash. Any durable close-grained wood may be used; redwood makes an ideal material. The surrounds must be milled to exact size and should be provided with interlocking joints mitered at the upper corners and dovetailed or tenoned into the sill. Assembly is recommended at the mill or shop where frames should be accurately squared, glued with waterproof glue, and shipped or transported with temporary diagonal braces to ensure their squareness in transit. Windows are best mounted in the frames on the flat, and then the frame and window together put into place. This procedure avoids racking of the metal sash, which can be one of the principal causes of air infiltration in this type of fenestration.

Wood surrounds may be used with brick veneer and stucco to completely frame the units.

Mullions may be cut from standard 4" nominal width stock, (actual width 3½").

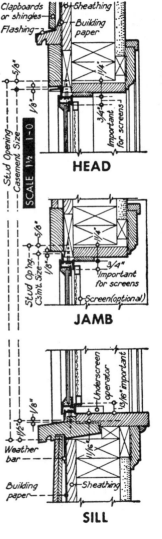

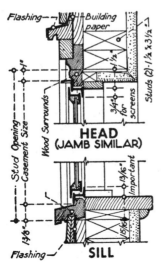

OUTSWINGING WOOD CASEMENT
IN BRICK VENEER

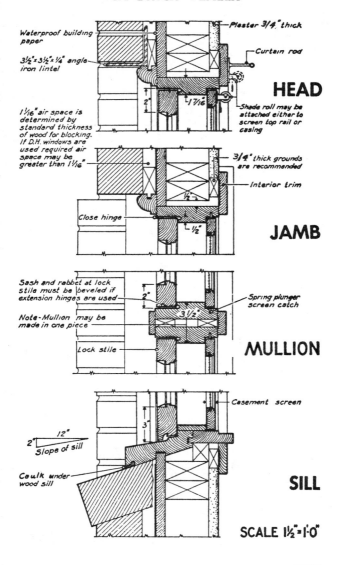

Waterproof building paper

3½"x3½"x¼" angle iron lintel

1¹⁄₁₆" air space is determined by standard thickness of wood for blocking. If D.H. windows are used required air space may be greater than 1¹⁄₁₆"

Plaster 3/4" thick

Curtain rod

HEAD

2"

1⁷⁄₁₆"

Shade roll may be attached either to screen top rail or casing

3/4" thick grounds are recommended

Interior trim

Close hinge

½"

½"

JAMB

Sash and rabbet at lock stile must be beveled if extension hinges are used

Note-Mullion may be made in one piece

Lock stile

2"

3½"

Spring plunger screen catch

MULLION

Casement screen

3"

12"
2"
Slope of sill

Caulk under wood sill

SILL

SCALE 1½"=1'0"

485

OUTSWINGING WOOD CASEMENT
IN STUCCO FRAME WALL

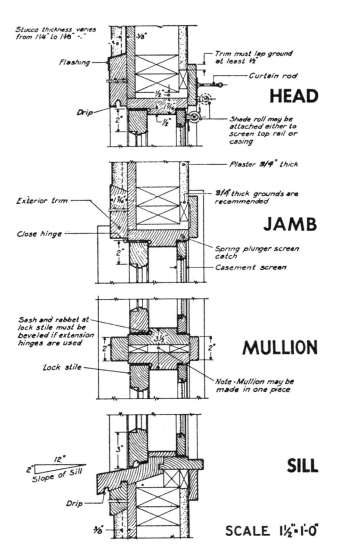

Stucco thickness varies from 1¼" to 1⅜"

⅛"

Flashing

Trim must lap ground at least ½"

Curtain rod

½"

1¾₆"

HEAD

Drip

2"

½"

Shade roll may be attached either to screen top rail or casing

Plaster ¾" thick

1¾₆"

Exterior trim

¾" thick grounds are recommended

JAMB

Close hinge

2"

Spring plunger screen catch

Casement screen

Sash and rabbet at lock stile must be beveled if extension hinges are used

2"

3½"

2"

MULLION

Lock stile

Note - Mullion may be made in one piece

12"

2"

Slope of Sill

3"

SILL

Drip

⅜"

SCALE 1½"·1·0"

486

LENGTH OF SIDES
OF 30°—60° BAYS

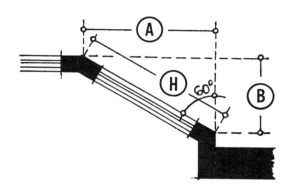

A	B	H	A	B	H
2'- 7 3/16"	1'- 6"	3'- 0"	4'-10 7/8"	2'-10"	5'- 8"
2'- 8 15/16"	1'- 7"	3'- 2"	5'- 0 5/8"	2'-11"	5'-10"
2'-10 5/8"	1'- 8"	3'- 4"	5'- 2 3/8"	3'- 0"	6'- 0"
3'- 0 3/8"	1'- 9"	3'- 6"	5'- 4 1/16"	3'- 1"	6'- 2"
3'- 2 1/8"	1'-10"	3'- 8"	5'- 5 13/16"	3'- 2"	6'- 4"
3'- 3 13/16"	1'-11"	3'-10"	5'- 7 9/16"	3'- 3"	6'- 6"
3'- 5 9/16"	2'- 0"	4'- 0"	5'- 9 15/16"	3'- 4"	6'- 8"
3'- 7 5/16"	2'- 1"	4'- 2"	5'-11"	3'- 5"	6'-10"
3'- 9 1/16"	2'- 2"	4'- 4"	6'- 0 3/4"	3'- 6"	7'- 0"
3'-10 3/4"	2'- 3"	4'- 6"	6'- 2 1/2"	3'- 7"	7'- 2"
4'- 0 1/2"	2'- 4"	4'- 8"	6'- 4 3/16"	3'- 8"	7'- 4"
4'- 2 1/4"	2'- 5"	4'-10"	6'- 5 5/16"	3'- 9"	7'- 6"
4'- 3 15/16"	2'- 6"	5'- 0"	6'- 7 11/16"	3'-10"	7'- 8"
4'- 5 11/16"	2'- 7"	5'- 2"	6'- 9 7/16"	3'-11"	7'-10"
4'- 7 7/16"	2'- 8"	5'- 4"	6'-11 1/8"	4'- 0"	8'- 0"
4'- 9 3/16"	2'- 9"	5'- 6"	7'- 0 13/16"	4'- 1"	8'- 2"

LENGTH OF SIDES
OF 45° ANGLE BAYS

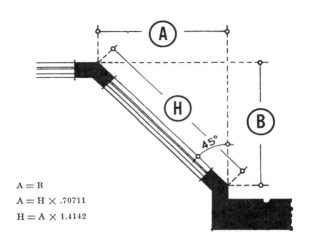

$A = B$

$A = H \times .70711$

$H = A \times 1.4142$

A	B	H	A	B	H
1'- 6"	1'- 6"	2'- 1 7/16"	2'-10"	2'-10"	4'- 0 1/16"
1'- 7"	1'- 7"	2'- 2 7/8"	2'-11"	2'-11"	4'- 1 1/2"
1'- 8"	1'- 8"	2'- 4 1/4"	3'- 0"	3'- 0"	4'- 2 15/16"
1'- 9"	1'- 9"	2'- 5 11/16"	3'- 1"	3'- 1"	4'- 4 5/16"
1'-10"	1'-10"	2'- 7 1/8"	3'- 2"	3'- 2"	4'- 5 3/4"
1'-11"	1'-11"	2'- 8 1/2"	3'- 3"	3'- 3"	4'- 7 1/8"
2'- 0"	2'- 0"	2'- 9 15/16"	3'- 4"	3'- 4"	4'- 8 9/16"
2'- 1"	2'- 1"	2'-11 3/8"	3'- 5"	3'- 5"	4'-10"
2'- 2"	2'- 2"	3'- 0 3/4"	3'- 6"	3'- 6"	4'-11 3/8"
2'- 3"	2'- 3"	3'- 2 3/16"	3'- 7"	3'- 7"	5'- 1 5/16"
2'- 4"	2'- 4"	3'- 3 5/8"	3'- 8"	3'- 8"	5'- 2 1/4"
2'- 5"	2'- 5"	3'- 5"	3'- 9"	3'- 9"	5'- 3 5/8"
2'- 6"	2'- 6"	3'- 6 7/16"	3'-10"	3'-10"	5'- 5 1/16"
2'- 7"	2'- 7"	3'- 7 7/8"	3'-11"	3'-11"	5'- 6 1/2"
2'- 8"	2'- 8"	3'- 9 1/4"	4'- 0"	4'- 0"	5'- 7 7/8"
2'- 9"	2'- 9"	3'-10 11/16"			

MINIMUM GLASS AREAS FOR ROOMS

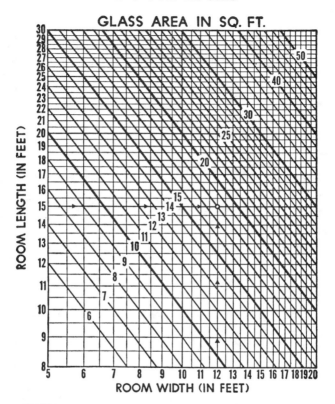

GLASS AREA IN SQ. FT.

ROOM LENGTH (IN FEET)

ROOM WIDTH (IN FEET)

GLASS AREA. The Uniform Building Code is typical of most model, as well as local, building codes in its window glass area requirements. All habitable rooms in a dwelling must have exterior glazed openings with a total minimum area of 10% of the floor area of each respective room, and each such room must have at least 10 square feet of glass area, regardless of room size. Bathrooms, laundry rooms, and similar rooms must be fitted with *openable* glazing amounting to no less than 5% of the floor area of each room, with a minimum of 1½ square feet of glass area.

PROPORTION OF OPENABLE SASH. For ventilating purposes, the glazing in each habitable dwelling room must be openable to an extent of 5% of the floor area of each respective room, with a minimum openable glazing area of 5 square feet. All bathroom, laundry room, and similar room glazing must be openable.

ALTERNATIVE. An alternative to openable glazing for ventilation is usually available. In habitable dwelling rooms, a mechanical ventilating system may be installed that will provide at least two air changes per hour. In bathrooms, laundry rooms, and similar rooms, a mechanical ventilating system can be installed to provide at least five air changes per hour.

HOW TO USE THE CHART. A room 15' long by 12' wide is shown by the chart to require 18 square feet of glass area.

LIFE OF NON-FERROUS
INSECT-SCREEN CLOTH

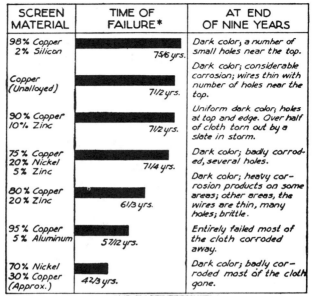

SCREEN MATERIAL	TIME OF FAILURE*	AT END OF NINE YEARS
98% Copper 2% Silicon	7 5/6 yrs.	Dark color; a number of small holes near the top.
Copper (Unalloyed)	7 1/2 yrs.	Dark color; considerable corrosion; wires thin with number of holes near the top.
90% Copper 10% Zinc	7 1/2 yrs.	Uniform dark color; holes at top and edge. Over half of cloth torn out by a slate in storm.
75% Copper 20% Nickel 5% Zinc	7 1/4 yrs.	Dark color; badly corroded, several holes.
80% Copper 20% Zinc	6 1/3 yrs.	Dark color; heavy corrosion products on some areas; other areas, the wires are thin, many holes; brittle.
95% Copper 5% Aluminum	5 7/12 yrs.	Entirely failed most of the cloth corroded away.
70% Nickel 30% Zinc (Approx.)	4 2/3 yrs.	Dark color; badly corroded most of the cloth gone.

*Failure was deemed to have occurred when there was a break in the wire in at least 1 place, as a result of corrosion.

ATMOSPHERIC-EXPOSURE TESTS. Research Paper RP803 of the National Bureau of Standards records the results of atmospheric exposure tests on 7 compositions of non-ferrous screen wire cloth, made by the National Bureau of Standards in cooperation with the A.S.T.M. over a period of about 9 years. The specimens were exposed at Pittsburgh, Pa., a heavy-industrial atmosphere; at Portsmouth, Va., and Cristobal, Canal Zone, a temperate and tropical sea-coast atmosphere, respectively, with some industrial contamination; and at Washington, D. C., a normal inland atmosphere. The bar chart above gives results of the tests at Pittsburgh.

MATERIALS. Seven non-ferrous materials in the form of 16-mesh insect-screen cloth woven from wire 0.0113″ in diameter were used. Of the 7 compositions, unalloyed copper and the 90-copper 10-zinc alloy were commercially available at the time the program was started and have continued to be since. The other alloys were not on the market at the time.

LABORATORY TESTS. Accelerated-corrosion tests were also made to determine the relative corrodibility of the different materials. The accelerated-corrosion tests consisted of salt spray and intermittent-immersion tests in salt solutions and dilute acid. The results were not consistent with the results of the exposure tests in any of the 4 locations and could not have been used to predict the behavior of the screen material in actual service.

DETAILS OF
WINDOW BOXES

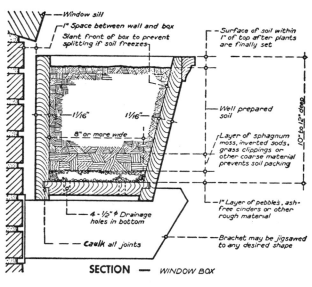

- Window sill
- 1" Space between wall and box
- Slant front of box to prevent splitting if soil freezes
- Surface of soil within 1" of top after plants are finally set
- Well prepared soil
- Layer of sphagnum moss, inverted sods, grass clippings or other coarse material prevents soil packing
- 1" Layer of pebbles, ash-free cinders or other rough material
- 10" to 12" deep
- 1 1/16"   1 1/16"
- 8" or more wide
- 4 - 1/2" ∮ Drainage holes in bottom
- Bracket may be jigsawed to any desired shape
- Caulk all joints
- Bracket may be jigsawed to any desired shape

SECTION — WINDOW BOX

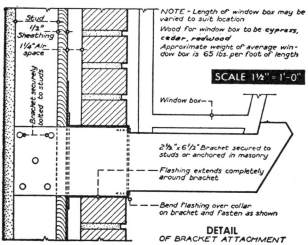

- Stud
- 1/2" Sheathing
- 1 1/4" Air space
- Bracket securely bolted to studs

NOTE - Length of window box may be varied to suit location

Wood for window box to be **cypress, cedar, redwood**

Approximate weight of average window box is 65 lbs. per foot of length

SCALE 1 1/2" = 1'-0"

- Window box
- 2 1/2" x 6 1/2" Bracket secured to studs or anchored in masonry
- Flashing extends completely around bracket
- Bend flashing over collar on bracket and fasten as shown

DETAIL
OF BRACKET ATTACHMENT

PLANTING FOR
WINDOW BOXES

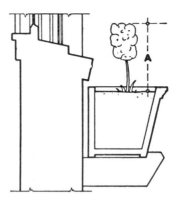

Ageratum grows 12″ or less high, compact, with white, blue, or purple flowers.

Chinese pink grows about 12″ high, with single or double flowers of white or shades of red.

Sweet alyssum is a spreading plant with white sweet-scented flowers, varying in height from 4″ to 8″. It blooms continually, covering the surface of the box and trailing over its edge.

Candytuft attains a height of 12″ and more, with upright stalks of white or purplish flowers. It is not a continual bloomer.

Lobelia grows from 6″ to 12″ high, with flowers that are white or shades of blue. It is upright and compact with good foliage. When given plenty of water in hot weather, it blooms continually during a long season.

Mignonette grows to a height of 15″ and more. It is chiefly valuable for its sweet fragrance. Its greenish-yellow to brownish flowers are attractive, though not showy.

Dwarf nasturtiums grow about 12″ high, with large, showy yellow, orange, or red flowers. Manure should not be added to the soil for these plants.

Petunias will grow about 12″ high without support, although the branches will grow several feet long and, if permitted to droop over the edges of the box, make a beautiful showing. They grow best in a warm, sunny situation. There are many varieties, from white to a rich royal purple.

Verbenas grow less than 12″ high, but the long stems will droop gracefully over the edges of the box. There are white, scarlet, and purple varieties, which thrive in full sunshine and bloom freely for a long season.

Calliopsis, snapdragon, and *helichrysum* or *strawflower* are upright, easily grown annuals that attain a height of 18″.

Vines or trailing plants adapted to use in window boxes are *kenisworth, ivy, wandering Jew, Vinca major, climbing nasturtiums, Ageratum rostrata, Asparagus sprengeri, Ficus pumila,* and *English ivy.*

Porch and outdoor window boxes planted with evergreens may be used effectively. More permanent appearing and dignified summer effects may often be obtained by evergreens, especially in connection with more formal buildings. They are the only plants that can be widely used for winter effects.

492

CAULKING OF JOINTS
AND OPENINGS

CAULKING COMPOUNDS. These are any of several plastic compounds composed of various (chiefly synthetic) materials. They are proof in various degrees to heat, cold, moisture, or acid fumes. When set, they form a tough skin on the surface but remain relatively pliable and elastic underneath. They adhere well to wood, stone, terra cotta, concrete, steel, glass, and most other building materials. Specific compounds should be selected upon the basis of the requirements of the job at hand.

PREVENTING INFILTRATION, ETC. Caulking compound is used as a plastic filler for spaces between exterior window and door frames and the surrounding structure to prevent the leakage of water, air, and dust into the building, and the escape by exfiltration of heated air from the building. Caulking is also used for the same purpose at drip caps, siding corners, meeting-points of sills and foundations, masonry/siding joints, penetration points of pipes and cables into the structure, and so on.

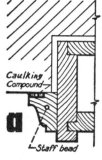

Caulking Compound

a

Staff bead

WINDOW AND DOOR FRAME CAULKING. In frame construction, caulking is liberally applied to drip caps, sills, and all finish trim to seal off any gaps through which air or moisture might enter.

In masonry construction, staff beads are sometimes detailed to receive caulking without removal, as shown in Figure a. The caulking rabbet should be $3/16''$ to $1/4''$ wide by $1/2''$ to $3/4''$ deep.

If the staff beads are removable, as shown at Figure b, the mortar of masonry joints behind frames should be raked out to a depth of $1/2''$ to $3/4''$. The joint between frame and masonry should be filled with plumber's oakum, caulking yarn, or fiberglass (check for compatibility with caulk). Fill the space with caulking compound and form a fillet corner in the angle. Then replace the staff bead, putty the nail holes, and paint.

Caulking cotton or oakum

Caulking Compound

b

Staff bead

CAULKING MASONRY. The joints in masonry provide a point of attack for the entrance of moisture, particularly in copings, corners, gutters, belt courses, base courses, and other projecting members. Caulking compound can be used for pointing and forms a permanently sealed masonry joint. Though it must be carefully applied, it will not stain the masonry, will not run or melt, and will not shrink or crack from temperature cycling or minor movement of the construction. Joints to be caulked should be kept back, or raked out, to a depth of about $1''$ to receive the caulking compound.

CAULKING WOOD, ETC. The caulking of wood and other building materials should be done by forcing the material into cracks, and/or by applying continuous beads to flat surfaces of members that will be compressed together during construction. In most cases, surfaces must be clean and dry for best results.

Wood window sill

c

Caulking Compound

APPLICATION. Both small and large hand caulking guns are most useful for small jobs. Care must be taken to force the compound into cracks or joints; a putty knife, small trowel, or caulking iron is useful for this process. For large jobs, a power caulking gun that forces the compound into place under pressure is most efficient. For laying beads on flat surfaces, however, no pressurizing is needed.

TYPES OF CAULKING. Several varieties of caulking are available, such as oil- or resin-based types, elastomerics such as polyurethanes and silicones, and latex-, polyvinyl-, and butyl-based compounds. They are packaged in small and large tubes for hand caulking guns, and in bulk forms, and can be obtained in several colors.

493

HAND AND BEVEL
OF DOORS

Bevel of Door—The free edge of doors over 1⅜" thick is bevelled ⅛" in 2" to clear the rabbet. If a mortise lock is used its front must be bevelled to correspond.

Bevel of Lock—Term used to describe the direction in which the latch bolt is inclined, for either mortise or rim locks, and corresponds to the door bevel always.

"Outside"—If the key functions from one side only, that is the "outside." It is usually the exterior side of an entrance door, the hall side of a room door, and the room side of a closet door.

Hand and Bevel of Doors and Locks

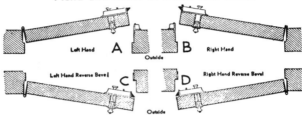

Hand of a Lock—With either mortise or rim locks if the key functions from one side only, stand on that side and if the butts are on the right it is a right hand lock. If on the left it is a left hand lock. If it is a mortise lock having the key function the same on both sides, determine the hand from the side from which the butts are not seen, as at A and B.

Bevel of a Lock—Standing as for determining hand, if the door opens toward you it requires a reverse bevel. If it opens away from you it requires a regular bevel. If no bevel is designated it is understood as regular bevel.

Hand of Door Itself—Determined by the side that is hinged, standing on the side from which the butts are not seen. A and D are left hand doors, and B and C are right hand doors.

Book Case or Cupboard

Book Case or Cabinet Locks are made with reverse bevel bolts as such doors regularly open outwards. Designated as "right hand" or "left hand" only.

Casement Windows

Casements and French Doors—Hand taken from the *inside*, which is the side on which casement fasteners or cremone bolts are applied.

LOCATION OF
HARDWARE

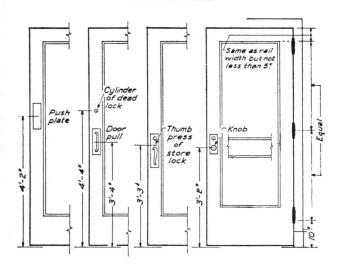

Heights are from finished floor.

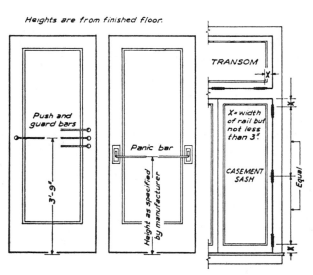

STOCK WOOD STILE-AND-RAIL DOOR PATTERNS

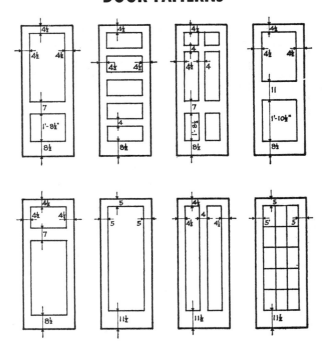

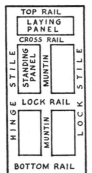

Dimensions shown are exclusive of moldings and vary with each manufacturer. The moldings on stock doors are generally ½" or less in width, but can vary. Numerous other stock door patterns are also available, with and without lights.

One-panel doors should be not less than 1¾" thick, with the panel not less than ½" thick. However, some interior one-panel doors are 1⅜" thick.

Doors over 3'-0" wide or over 7'-0" high should be not less than 1¾" thick. However, some 3'-6" wide doors are made in 1⅜" thickness.

Flush doors should be at least 1⅜" thick, whether hollow-core or solid-core.

Standard sizes: exterior doors 1¾" thick, interior doors 1⅜" thick; both types available 1'-6" to 3'-6" wide, 6'-6" to 7'-0" high.

BUTTS AND HINGES

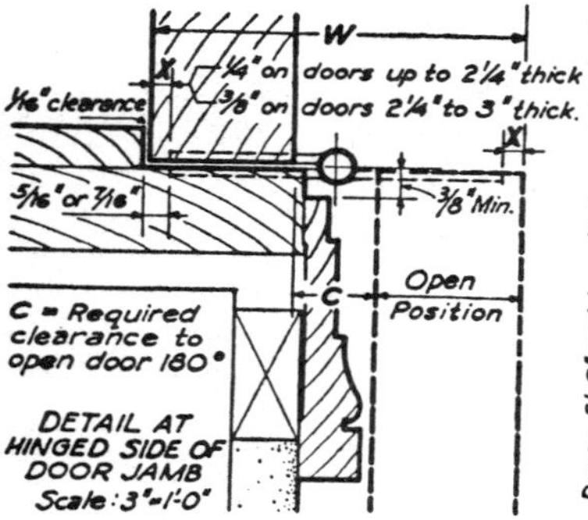

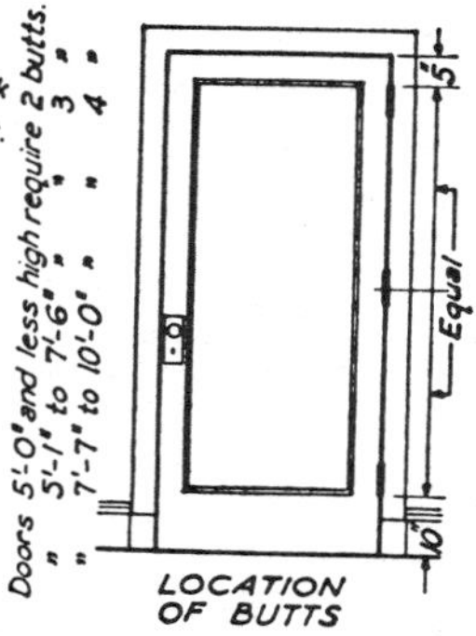

WIDTH OF BUTTS = W — 2X. Butts come in multiples of ½" widths, fractional sizes resulting from formula take next higher width.

Type of Door	Door Thickness	Door Width	Height of Butt
Cupboard Doors..........	¾" to ⅞"	up to 2'-0"	2½"
Screen Doors............	⅞" to 1⅛"	up to 3'-0"	3"
Wood Doors............	1⅛"	up to 3'-0"	3½"
Steel Doors............	1⅛"	up to 3'-0"	4½"
Wood Doors............	1¼" to 1⅜"	up to 2'-8"	3½"
Wood Doors............	1¼" to 1⅜"	2'-9" to 3'-1"	4"
Steel Doors............	1¼" to 1⅜"	up to 2'-8"	4½"
Steel Doors............	1¼" to 1⅜"	2'-9" to 3'-1"	5"
Steel or Wood Doors......	1 9/16" to 1⅞"	up to 2'-8"	4½"
Steel or Wood Doors......	1 9/16" to 1⅞"	2'-9" to 3'-1"	5"
Steel or Wood Doors......	1 9/16" to 1⅞"	3'-2" to 3'-7"	5"*
Steel or Wood Doors......	1 9/16" to 1⅞"	3'-8" to 4'-2"	6"*
Steel or Wood Doors......	2" to 2½"	up to 3'-7"	5"*
Steel or Wood Doors......	2" to 2½"	3'-8" to 4'-2"	6"*

*** = Extra Heavy.**

HINGES AND BUTTS. A *hinge* is a device that allows a door to swing. That type of hinge in which the leaves close together when the door is closed, is called a *butt hinge* or *butt*.

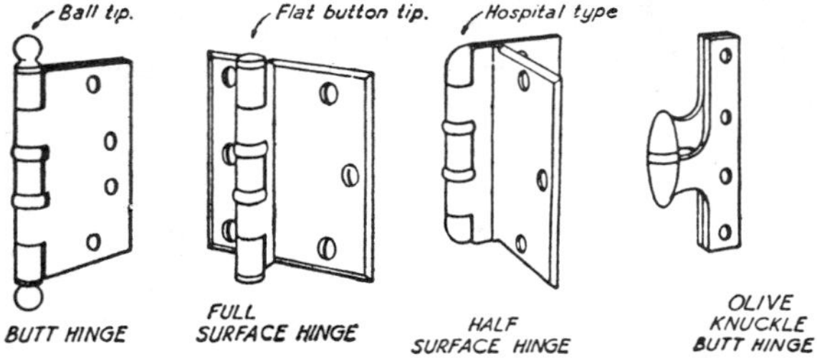

DETAILS OF SECTIONAL
GARAGE DOORS

Doors roll up vertical tracks and rest on horizontal tracks when fully open. The illustration below shows a typical sectional door installation. Special tracks are made for full vertical lift and for high lift and low headroom situations. Doors are made of three or more horizontal hinged sections, typically at least four. The hardware, with the exception of the handles and lock cylinder face, is inside and protected from the weather. The proper size and type of spring is furnished at the factory. The operation of the door allows the garage to be practically the same depth inside as the car is long, thus saving plan space. Doors are securely locked by heavy lock rods engaging in the tracks. The opening and closing of the door is easy, smooth, and quiet.

Doors are furnished in the following forms: in wood with hardboard panels, either sanded and plain or factory primed for painting; in steel primed or painted; in factory-finished aluminum; and in fiberglass with integral coloring. Most styles are available with sandwiched insulation, generally polyurethane. Numerous panel/glazing patterns are offered as stock units, and special designs can be custom-built.

Brackets and hangers for mounting the track are supplied. In planning, avoid posts, beams, pipes, angle walls, truss braces, gables, or other obstructions that might cut down side or head room or interfere with the door installation.

Service doors can be had in many door styles; they are uncommon in residential garage doors but standard options in many commercial types.

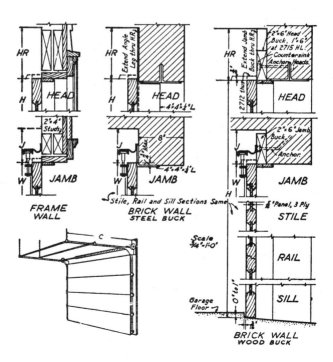

ACCORDION
DOORS

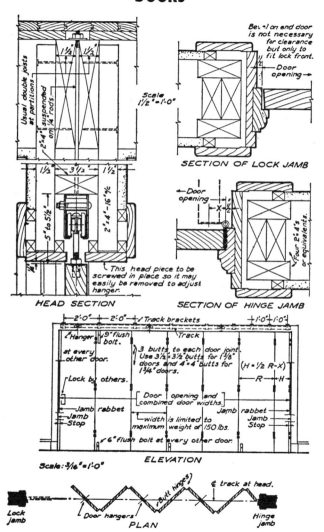

Bevel on end door
is not necessary
for clearance
but only to
fit lock front.

Door
opening →

SECTION OF LOCK JAMB

Usual double joists at partitions

2"x4" suspended on ¼" rods.

Scale
1½" = 1'-0"

1½   3½   1½

5" to 5½"   2"x4" - 16 ⅝"

This head piece to be screwed in place so it may easily be removed to adjust hanger.

HEAD SECTION

← Door opening →

X — ½

Four 2"x4's or equivalents.

SECTION OF HINGE JAMB

2'-0"   2'-0"   Track brackets   1'-0" 1'-0"

Hanger   9" flush bolt.   Track

at every other door.

3 butts to each door joint. Use 3½ x 3½ butts for 1⅜" doors and 4"x 4" butts for 1¾" doors.

Lock by others.

(H = ½ R - X)

Jamb rabbet   Jamb   Stop

Door opening and combined door widths.

width is limited to maximum weight of 150 lbs.

R — H

Jamb rabbet   Jamb   Stop

6" flush bolt at every other door.

Scale: ⅜" = 1'-0"

ELEVATION

Lock jamb

Door hangers   (Butt hinges)   ₵ track at head.

PLAN

Hinge jamb

499

ACCORDION
DOORS

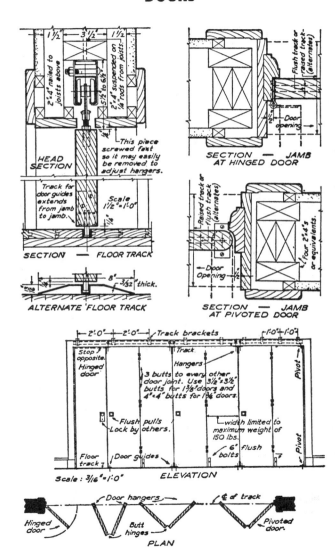

HEAD
SECTION

2"x4" nailed to joists above

2"x4" suspended on 1/4" rods from joists.

This piece screwed fast so it may easily be removed to adjust hangers.

SECTION — JAMB
AT HINGED DOOR

Flush track or raised track (alternates)

Door opening

Track for door guides extends from jamb to jamb.

Scale 1 1/2" = 1'0"

SECTION — FLOOR TRACK

ALTERNATE FLOOR TRACK

8"   3/32" thick.

Raised track or flush track (alternates)

Four 2"x4's or equivalents

Door Opening

SECTION — JAMB
AT PIVOTED DOOR

Track brackets

Stop opposite. Hinged door

Track

Hangers

3 butts to every other door joint. Use 3 1/2"x3 1/2" butts for 1 3/8" doors and 4"x4" butts for 1 3/4" doors.

Flush pulls
Lock by others.

width limited to maximum weight of 150 lbs.

Pivot

Pivot

Floor track

Door guides

6" flush bolts

Scale: 3/16" = 1'0"   ELEVATION

Hinged door

Door hangers

Butt hinges

¢ of track

Pivoted door.

PLAN

500

SIZES AND CAPACITIES
REVOLVING DOORS

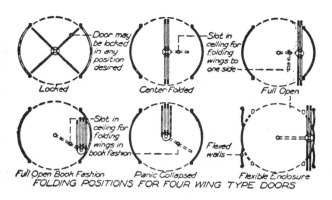

Door may be locked in any position desired

Locked

Slot in ceiling for folding wings to one side

Center Folded

Full Open

Slot in ceiling for folding wings in book fashion

Full Open Book Fashion

Panic Collapsed

Flexed walls

Flexible Enclosure

FOLDING POSITIONS FOR FOUR WING TYPE DOORS

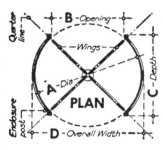

B - Opening
Wings
A - Dia
Quarter line
PLAN
C - Depth
Enclosure post
D - Overall Width

A 6'-6" Dia door is recommended for average use. At entrances accommodating people with luggage use 7'-0" Dia. to 8'-0" Dia doors.

Based on 10 to 12 revolutions per minute the capacity per hour both in and out of a revolving door with four wings is about 2600 persons.

Dimensions of doors of different manufacturers may vary 1" from those given in table.

All manufacturers may not regard all diameters given as standard.

A - DIAMETER	B	C	D
6'-6"	4'-5"	5'-1"	6'-9"
6'-8"	4'-6 1/2"	5'-2 1/2"	6'-11"
6'-10"	4'-8"	5'-4"	7'-1"
7'-0"	4'-9 1/2"	5'-5 1/2"	7'-3"
7'-2"	4'-10 3/4"	5'-7"	7'-5"
7'-4"	5'-0"	5'-8"	7'-7"
7'-6"	5'-1 1/2"	5'-9 1/2"	7'-9"
8'-0"	5'-6"	6'-11 1/2"	8'-3"

TABLE OF DIMENSIONS

Revolving doors are made in numerous styles, including special security types, and may be manual or motorized. Though "standard" sizes are listed here, there are others as well. Most are constructed of aluminum, bronze, or stainless steel; the use of plastics is uncommon, and wood has fallen into disfavor. Glazing is done with several types of glass, including laminates. Consult manufacturers for detailed specifications.

VARIOUS PLANS FOR
REVOLVING DOORS

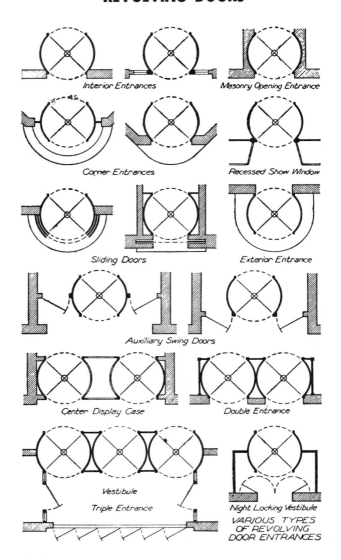

Interior Entrances

Masonry Opening Entrance

Corner Entrances

Recessed Show Window

Sliding Doors

Exterior Entrance

Auxiliary Swing Doors

Center Display Case

Double Entrance

Vestibule

Triple Entrance

Night Locking Vestibule

VARIOUS TYPES
OF REVOLVING
DOOR ENTRANCES

SECTION AND DETAILS
REVOLVING DOORS

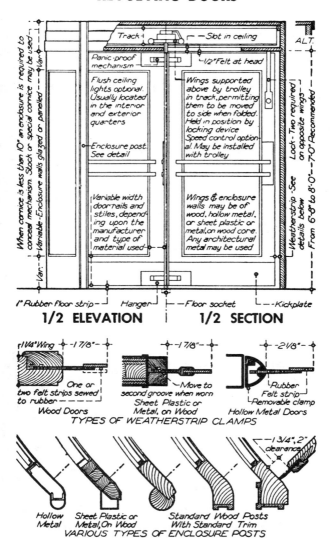

Track — Slot in ceiling

ALT.

Panic-proof mechanism — 1/2" Felt at head

Flush ceiling lights optional. Usually located in the interior and exterior quarters

Wings supported above by trolley in track, permitting them to be moved to side when folded. Held in position by locking device. Speed control optional. May be installed with trolley

Enclosure post. See detail

Variable width door-rails and stiles, depending upon the manufacturer and type of material used

Wings & enclosure walls may be of wood, hollow metal, or sheet plastic or metal, on wood core. Any architectural metal may be used

When cornice is less than 10" an enclosure is required to conceal mechanism. Stock or special cornice may be used

Var. — Variable-Enclosure walls glazed or paneled

Var.

1" Rubber floor strip — Hanger — Floor socket — Kickplate

Lock - Two required on opposite wings — Weatherstrip - See details below — From 6'-8" to 8'-0"--7'-0" Recommended

1/2 ELEVATION   1/2 SECTION

1 1/4" Wing — 1 7/8"

One or two felt strips sewed to rubber

Wood Doors

1 7/8"

Move to second groove when worn

Sheet Plastic or Metal, on Wood

2 1/8"

Rubber
Felt strip
Removable clamp

Hollow Metal Doors

TYPES OF WEATHERSTRIP CLAMPS

1 3/4", 2" clearance

Hollow Metal   Sheet Plastic or Metal, On Wood   Standard Wood Posts With Standard Trim

VARIOUS TYPES OF ENCLOSURE POSTS

503

CORNER POST DETAILS
REVOLVING DOORS

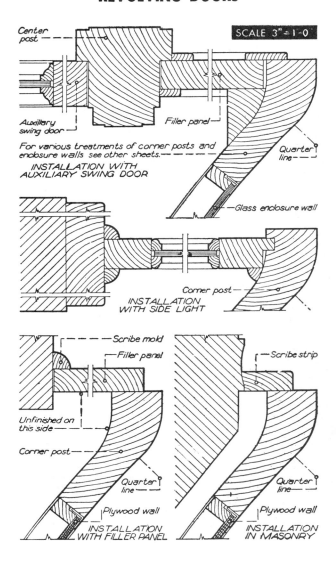

SCALE 3"=1'-0"

Center post

Auxiliary swing door

Filler panel

Quarter line

For various treatments of corner posts and enclosure walls see other sheets.

INSTALLATION WITH
AUXILIARY SWING DOOR

Glass enclosure wall

Corner post

INSTALLATION
WITH SIDE LIGHT

Scribe mold

Filler panel

Unfinished on this side

Corner post

Quarter line

Plywood wall

INSTALLATION
WITH FILLER PANEL

Scribe strip

Quarter line

Plywood wall

INSTALLATION
IN MASONRY

HOW TO SELECT FIRE DOORS

DEFINITION OF FIRE DOOR. A fire door is a special door, specifically constructed in any of a variety of types and configurations and intended for proper installation in a suitable wall as a component part of a fire door *assembly*. The complete fire door assembly consists of a combination of door or doors, hardware, frame, and accessory equipment installed to resist the passage of heat, smoke, and flame for a specified length of time under specified conditions. The fire door must be a labeled item, and must be installed together with the appropriate labeled frame, hardware, and so forth, in order to constitute a valid fire door installation. If any component parts of the fire door assembly are not labeled and/or are not in compliance with the appropriate standards and codes, or if a fire door is installed with ordinary hardware or in an ordinary frame, the opening protection is not considered equal to the labeled protection of any labeled components; or the opening may be considered entirely unprotected. In short, a complete fire door assembly of appropriate type and rating must be installed for the protection it provides to be effective.

DESIGN PROCEDURE. Since codes vary and the local inspector is given considerable latitude in making determinations of exactly what constitutes a proper installation in any given application, the architect will find it advantageous to consult an expert, and the local inspector as well, during planning and specification stages. A manufacturer's representative can be of great service in such a capacity, since that person will, without obligation, freely give the architect advice based on an intimate knowledge not only of the specific products, but also of general and local requirements, interpretations of those requirements, costs, installation procedures, and the like. Selection of the fire door assembly will depend upon the desired appearance, desired or required amount of protection to be provided, type of openings to be protected and their use, location of the openings with respect to structural components, building areas, potential fire hazard, and numerous other factors.

DOOR CONSTRUCTION. There are several different constructions commonly used in fire doors, each having various advantages and disadvantages, different applications, and varying fire protection ratings depending upon their configuration and use. Though terminology may differ somewhat throughout the industry, the following terms are widely used to identify fire door constructions.

Composite. Composite doors are built up around a solid core material, by bonding steel, plastic, or wood sheeting directly to the core. They may be either flush or panel type.

Metal-clad. Also called Kalamein doors, these consist of a wood frame completely covered with tight-fitting galvanized sheet steel. If made in flush style, the wood frame may be a solid core; if of the panel type, wood rails and stiles are used with insulated panels. The steel covering is generally 24-gauge, with 26-gauge sometimes used.

Hollow Metal. Flush-style hollow metal fire doors are made up of minimum 20-gauge steel sheet supported by internal metal bracing and/or honeycomb coring. Panel style doors contain stiles and rails to support the metal, with the panels being insulated. Insulation may also be used in the flush doors.

Sheet Metal. These doors are composed of 22-gauge or lighter sheets of either flat or corrugated steel, and may be either panel or flush style.

Tin-clad. One of the oldest constructions, tin-clad doors consist of a wood core covered with multiple sheets of terneplate in a maximum size of 14″ x 20″. (Terneplate — or "terne" — is a composite material composed of steel sheet coated with a tin/lead alloy — with a standard 20-pound weight in this application, equivalent to 30-gauge steel sheet.) The sheets are laid with ½″ seams, nailed underneath. Alternatively, 30-gauge galvanized sheet steel in maximum 14″ x 20″ pieces, or 24-gauge galvanized sheet steel pieces no more than 48″ wide, may be used. The wood core must be of either two or three plies. In all cases the face sheeting must be vented.

Wood Core. The frame or core of this construction may be of wood or particle board, to which wood, plastic, or hardboard sheets are bonded. The edges of the door remain open and untreated.

Rolling Steel. This type of door is made up of a series of narrow panels or slats that interlock. The panels are attached to an overhead drum, and the door operates much like a window shade. The panels are of minimum 22-gauge steel. This type of door is a complete assembly in itself, including guides, operating mechanism, counterbalances, and enclosure.

Curtain. This is made up of either interlocking steel slats or blades, or continuous spring steel, contained in a frame.

Sectional. This type is made up of steel panels hinged together, much like a garage door.

HOW TO SELECT FIRE DOORS

DOOR CONFIGURATIONS. Fire door assemblies may be obtained in a number of different configurations, but not all constructions are available in all configurations; some are restricted to certain styles. The configurations are:

> Swinging—single.
> Swinging—pair.
> Horizontal level sliding—single.
> Horizontal sliding—pair, center parting.
> Horizontal inclined sliding—single.
> Vertical sliding—single.
> Horizontal level sliding—one speed.
> Horizontal level sliding—two speed.
> Rolling.

In addition, there are other arrangements for special applications, such as for dumbwaiters, access ports, freight elevators, and chutes.

FIRE DOOR CLASSIFICATION. Fire door assemblies are classified according to certain parameters. Any of three designating systems may be used, and all commonly occur.

One classification system is expressed in terms of hours. The time period noted is the length of time that a representative assembly of that particular type was able to withstand certain standard fire test exposure successfully, as measured and recorded by an accredited testing laboratory under specified conditions. This time period is called the fire protection rating. In the case of fire doors, there are six designations: $\frac{1}{3}$-hour (or 20-minute), $\frac{1}{2}$-hour (or 30-minute), $\frac{3}{4}$-hour (or 45-minute), 1-hour, $1\frac{1}{2}$-hour, and 3-hour.

Another classification system uses the capital letter A, B, C, D, or E. Each designation is assigned to certain types of doorway openings that are to be protected; then a suitable door is chosen, and the designation is then used for the door as well. The classes are as follows:

Class A. Openings in fire walls and in other walls that divide the interior of a building into separate fire areas.

Class B. Openings in 2-hour rated partition walls that provide fire separations and in vertical enclosures such as shaftways that communicate from floor to floor of a building.

Class C. Openings in partitions and walls between rooms, or between rooms and corridors, where the wall fire rating is 1-hour or less.

Class D. Openings in exterior walls where severe fire exposure from the outside is a potential hazard.

Class E. Openings in exterior walls where light or moderate fire exposure from the outside is a potential hazard.

The third method of classification is simply to use both of the above systems together. Thus, a fire door assembly might be classified as a 3-hour Class A, or a $\frac{3}{4}$-hour Class C.

DOOR FRAMES. Door frames for use with fire doors must be labeled and approved for the purpose. They are obtainable in pressed or channel steel, aluminum, and wood. Steel frames are suitable for the highest ratings. The maximum rating for aluminum is 45 minutes; for wood, 20 minutes.

DOOR HARDWARE. Only labeled and approved hardware may be used in fire door assemblies. There are two classifications: Fire Door Hardware and Builder's Hardware. The latter classification includes a subclassification, Fire Exit Hardware. Requirements specify what hardware can be used in the various kinds of fire door assemblies.

DOOR SIZES. Numerous sizes and thicknesses of doors are available. Certain size restrictions are placed upon all types of fire doors and vary depending upon the construction and the classification. These maximum sizes, which are subject to change periodically, should be verified before plans are made final.

EXPOSED GLASS AREA. In no case is any exposed glass area allowable in any Class A, 3-hour fire door. Some constructions, such as rolling or curtain doors, cannot have glass installed in them; in certain other specialty doors, such as access or chute doors, glass is not permitted. When installed in Class B openings, exposed glass area is restricted to a maximum of 100 square inches per door. The same restriction is imposed on a few Class C installations, but in most cases a maximum of 1,296 square inches of glass per light is permitted. No glass is permitted in Class D installations. The maximum allowable exposed glass in Class E installations is 1,296 square inches per light, with one exception: swinging, single, wood-core fire doors with wood faces are restricted to 324 square inches of glass per door.

BATTEN DOOR CONSTRUCTION

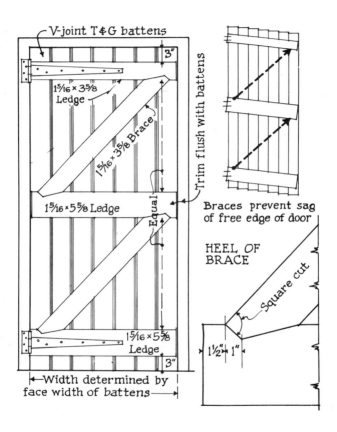

V-joint T&G battens

3"

1⁵⁄₁₆ × 3⅝ Ledge

1⁵⁄₁₆ × 3⅝ Brace

Trim flush with battens

1⁵⁄₁₆ × 5⅝ Ledge

Equal

1⁵⁄₁₆ × 5⅝ Ledge

3"

Width determined by face width of battens

Braces prevent sag of free edge of door

HEEL OF BRACE

Square cut

1½" 1"

Ledged and battened doors will sag out of square unless braced with braces slanting upwards from the hinged edge. For best appearance, the slope of the braces should be the same, requiring the center ledge to be midway between the top and bottom ledges. Properly constructed, these doors may be used for openings of almost any width, so long as the height is slightly greater than twice the width. Strap hinges are generally used with doors of this type.

GENERAL INFORMATION ON AUTOMATIC DOOR OPERATION

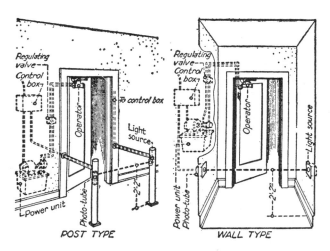

POST TYPE   WALL TYPE

PURPOSE. Automatic operation is a simple, safe, and reliable means of opening and closing doors in all sorts of applications, including residences, restaurants, hospitals, hotels, shops, factories, office buildings, and shipping rooms. The equipment may be incorporated in new installations, or retrofitted to existing doors, and may be either surface-mounted or concealed. Most kinds of doors can be automated, including single or double swinging, sliding, revolving, and overhead doors of various styles; even fire doors may be so equipped. First cost is moderate, and power consumption of the control and actuating system is low. Automatic door operation increases efficiency, promotes good will with employees and the public, and provides both added convenience and an excellent egress safety factor. Air conditioning and humidity control are made more efficient and economical when doors are automatically controlled.

OPERATION. Actual operation of the doors is accomplished by one of many different designs of devices mounted nearby and connected to the door. These door operators are of three general types: electromechanical, consisting of an electric motor and a mechanical drive system; pneumatic, requiring a compressed air supply; and electrohydraulic. When the operator is actuated, the door is opened under mechanical pressure derived from the power unit and remains open for a predetermined length of time, usually adjustable from 0-30 seconds. At the end of the time period, the door is released and closes. In the event of electric power failure to the unit, or electrical malfunction in the circuitry, the door is not blocked—it will operate manually. Should the door touch an individual passing through, a slight manual pressure will open the door without danger of personal injury or injury to the apparatus.

The power unit and operator can be actuated by any of several devices, all of which act as simple switches to turn the power to the unit on. A commonly used arrangement over past years (one that is still available)—photoelectric actuating devices, either post-mounted or flush-mounted in entryway walls—is shown above. One of the most widely used devices today is the motion sensor, which is mounted above the door and detects motion within a certain preset range below it; sensitivity to motion is adjustable. Another established device is the carpet or mat switch, which actuates when pressure is applied to it. Simple switches, such as pushbutton, toggle, or pull types, are often used. In addition, wired or wireless touch switches, delayed-action switches, and encoded radio-controlled switches are available. Control systems to suit any possible kind of application can be readily developed and installed.

SUITABILITY OF WOOD FOR FRAMING AND SHEATHING

FRAMING (HOUSE)

Usual Requirements: High stiffness, good bending strength, good nail-holding power, hardness, freedom from pronounced warp or shrinkage. For this use, dryness and size are more important factors than inherent properties of the different woods.

Highly Suitable: Widely used—Douglas-fir, western larch, southern pines. Sometimes used, but harder to obtain and to work—ashes, beech, birches, maples, oaks. Often used for special purposes—redwood.

Good Suitability: Widely used—hemlocks, balsam fir, lodgepole pine, spruces. Less often used because of adaptability to more exacting uses—white pines, ponderosa pine, sugar pine, western redcedar. (Low strength may be compensated for by use of larger members.) Seldom used—yellow poplar.

Grades Used: Select Structural and Dense Select Structural are the top grades, usually used for special purposes or exceptionally fine construction; they are not found in all species and grading systems. No. 1 grade is usually employed for all framing members in both high- and medium-class construction. No. 2 grade renders satisfactory service but is not as straight or as easily fabricated. No. 3 grade is serviceable for economical construction of modest demands. In some locales and species, Stud grade stock is available, and is suitable for that purpose. Hardwoods are no longer used in commercial construction and only very infrequently in owner-built structures.

ROOF SHEATHING (HOUSE)

Usual Requirements: Strength, rigidity, good nail holding, little tendency to warp, ease of working.

Highly Suitable: Used almost exclusively—plywood, generally Douglas-fir. Exterior grade should be employed around the perimeter of the roof; interior grade with exterior glue is satisfactory within the perimeter. Sometimes used (board sheathing)—Douglas-fir, western larch, southern pines. Rarely used (board sheathing)—baldcypress, ashes, beech, birches, elms, maples, oaks, tupelo.

Good Suitability: Occasionally used (board sheathing)—hemlocks, ponderosa pine, white fir, spruces, white pines, sugar pine, yellow poplar.

Grades Used: For plywood, Engineered grades C-D INT-APA and C-C EXT-APA, and where extra strength is required, Structural I in the same designations. For board sheathing, No. 2 and No. 3 grades are used extensively; the latter is serviceable but not as tight as No. 2. No. 4 and No. 5 grades are usable but less satisfactory and involve waste and extra cutting. Hardwoods are not used in commercial construction.

WALL SHEATHING (HOUSE)

Usual Requirements: Easy working, easy nailing, moderate shrinkage, good nail holding for some types of finish siding. All woods can be used for board sheathing with satisfactory results, although some are less time-consuming to work with than others.

Highly Suitable: Widely used—plywood, generally Douglas-fir but other species as well. Seldom used because of cost and labor factors—cedars, baldcypress, hemlocks, white pines, sugar pine, ponderosa pine, spruces, white fir, basswood, yellow poplar.

Good Suitability: Seldom used because of cost and labor factors—Douglas-fir, western larch, southern pines, cottonwoods. (Vegetable-fiber structural insulating board and rigid thermal insulation board also have good suitability and are in widespread use.)

Grades Used: For plywood, usually Engineered grade C-D INT-APA with exterior glue (often called CDX). For board sheathing, No. 3 grade of softwoods is serviceable when covered with a good grade of building paper. No. 4 grade is usable, but entails extra cost, labor, and waste. Hardwoods are not used for sheathing in commercial building applications.

BALLOON FRAMING

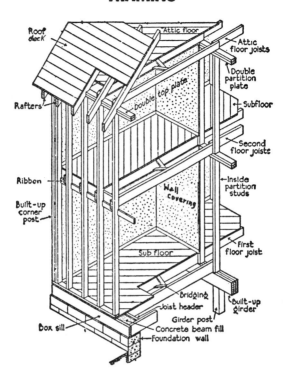

This traditional type of frame was first developed about 1850, but now it has generally been superseded by other methods. The one-piece studs, extending the full height of the wall and tied together by the ribbon at the second floor line, reduce to a minimum the shrinkage factor. The frame is strong and rigid. For these reasons, it is particularly to be preferred for two-story brick or stone veneer or stucco construction, although the same advantages are important to any type of wall covering. Balloon framing is more difficult and expensive than platform framing.

This frame requires careful fire-stopping. It is more efficient when the interior studding is set directly on top of girders or bearing partitions.

Studs, joists, and rafters are spaced 16″ on centers for proper nailing. If the stud spacing is changed, for example to 12″ o/c as it sometimes is for back-plastered stucco, then the joists and rafters must correspond. Subflooring boards laid diagonally give added strength, but when laid at right angles they provide economy of materials. Diagonal board sheathing aids in tying the superstructure to the sill. However, plywood subflooring and roof/wall sheathing afford greatest strength and rigidity and can be applied considerably faster.

PLATFORM
FRAMING

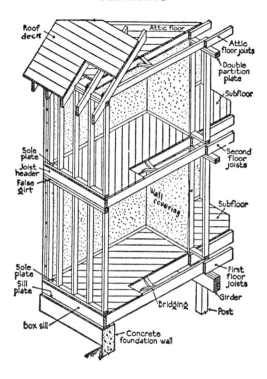

The platform frame is sometimes known as the western frame. This type is distinguished by floor platforms independently framed. The second and third floors are supported by studs that are one story in height. In this type, the studs and floor joists need not be spaced the same distance apart. Studding may be either 2″ x 4″ or 2″ x 6″, with spacing 16″ or 24″ on centers.

The chief merit of this type of framing is its ease and economy of construction; it is by far the most widely used framing system. Shrinkage is fairly equal, though likely to be greater than with other types because of the boxed sill construction at each floor line.

The platform frame is best used with all-wood construction, although it is also widely employed with brick, stone, and stucco veneer, especially in single-story houses. Fire-stopping is well taken care of with no further additions; however, this frame system is weaker than other types. Plywood sheathing is best used throughout, since this gives greatest strength and rigidity and furnishes almost the only tie from one story to another in two- and three-story structures.

SILLS IN
FRAME CONSTRUCTION

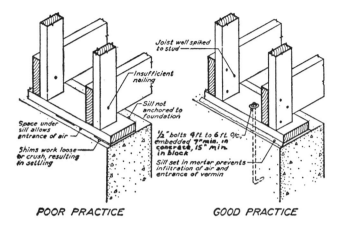

POOR PRACTICE

Joist well spiked to stud

Insufficient nailing

Sill not anchored to foundation

Space under sill allows entrance of air

Shims work loose or crush, resulting in settling

½" bolts 4 ft. to 6 ft. O/c, embedded 7" min. in concrete, 15" min. in block

Sill set in mortar prevents infiltration of air and entrance of vermin

GOOD PRACTICE

SIZE OF SILL. For small buildings of light frame construction, a 2″ x 6″ sill is large enough under most conditions, but 2″ x 8″ is also used. For two-story structures and in localities subject to earthquakes or high winds, a sill 4″ in (nominal) thickness, combined with other strengthening measures, is desirable. This affords more nailing surface for the sheathing and permits a much more satisfactory lap splice.

ANCHORING. Where high winds are at all probable, it is important that the building be strongly anchored to the foundation. In fact, solid anchoring is desirable and a good practice in all localities. It is best accomplished by embedding ½″ minimum diameter anchor bolts, 4′-6′ o/c to a minimum depth of 7″ in poured concrete or 15″ in masonry. There must be at least two anchor bolts per sill piece, and one within 12″ of each corner. The bolts should project sufficiently through the sill to receive a good-sized washer and allow the nut to be fully threaded on. Severe wind conditions may require anchoring as shown at left, or even sturdier methods; check local codes.

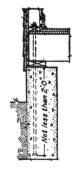

SHEATHING. So that the full advantages of anchoring may be realized, especially where windstorms occur, it is essential to use board wall sheathing applied diagonally or (preferably) plywood sheathing nailed securely to the sills *and* wall plates. This provides a tie between the sill and the structure above.

SPLICING THE SILL. Where a 2″ x 6″ is used as a sill, the satisfactory and usual practice is to butt the ends if the pieces are properly anchored. Where the sill is built of doubled 2″ x 6″ pieces, the joints in the two courses should be staggered. A solid sill 4″ or more thick may be butted where properly anchored but is better lap-spliced and anchored.

TYPE OF WOOD. Wood sills in contact with concrete or masonry foundations should be of treated wood or redwood. Those not in contact with concrete or masonry, such as wood posts or all-weather wood foundation systems, may be of any kind.

DORMERS IN WOOD FRAMING

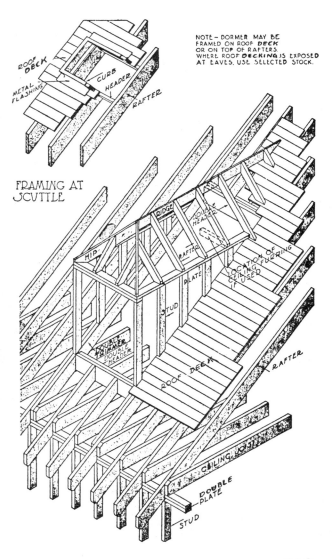

NOTE – DORMER MAY BE FRAMED ON ROOF **DECK** OR ON TOP OF RAFTERS WHERE ROOF **DECKING** IS EXPOSED AT EAVES, USE SELECTED STOCK.

ROOF **DECK**

METAL FLASHING

CURB

HEADER

RAFTER

FRAMING AT SCUTTLE

RIDGE

DOUBLE HEADER

RAFTER

LOCATION OF FURRING IF CEILING IS USED

HIP

PLATE

STUD

DOUBLE TRIMMER

DOUBLE HEADER

ROOF DECK

RAFTER

CEILING JOIST

DOUBLE PLATE

STUD

513

FRAMING AROUND CHIMNEYS

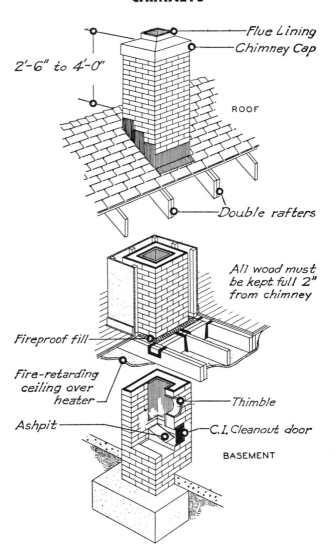

Flue Lining
Chimney Cap
2'-6" to 4'-0"
ROOF
Double rafters
All wood must be kept full 2" from chimney
Fireproof fill
Fire-retarding ceiling over heater
Thimble
Ashpit
C.I. Cleanout door
BASEMENT

FRAMING FOR
CANTILEVER PLATFORMS

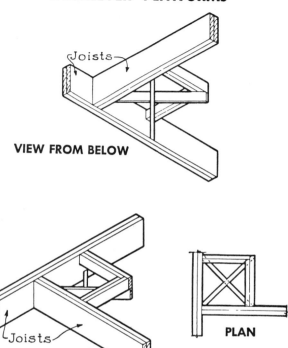

VIEW FROM BELOW

VIEW FROM ABOVE

PLAN

There are many locations in framing where a cantilever platform is required. These places most commonly occur in fireplace hearths and at turning of stairways. The same principle of support may also be applied to shelves that must support relatively heavy weights, occurring in inside corners.

The trimmers that form the nailing for the subflooring are doubled, as shown in the drawing, but they do not need to be as deep as the two supporting members beneath, since they do not actually carry the load themselves. The lower diagonal member must be a sound piece of wood because it carries a concentrated load at its middle.

Resting on the lowest member is the cantilever piece, which carries the corner of the platform. The end of this piece should be nailed in the corner with nails sloping upward, since this end will tend to move up. The size of all the members will depend upon the depth of the joists and the size of the platform to be supported.

515

FRAMING OVER
3 FT. MAX. OPENINGS

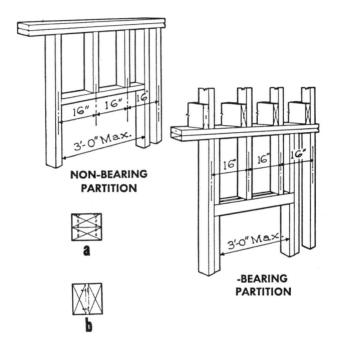

NON-BEARING PARTITION

16" 16" 16"

3'-0" Max.

a

b

-**BEARING PARTITION**

16" 16" 16"

3'-0" Max.

The framing of an opening depends on the width of the opening and whether the partition in which it occurs is bearing or nonbearing. The illustrations show the proper method of framing openings 3'-0" or less in width.

In a nonbearing partition, a single 2" x 4" is satisfactory as a header. It is sufficiently strong and lessens the likelihood of plaster cracks due to movement of a double member. However, it often happens that the use of wide trim requires doubled lengths of 2" x 4" to provide nailing.

In load-bearing partitions or walls, the header should be doubled and should rest on doubled studs, as shown. Lengths of 2" x 4" placed horizontally, as at a, do not provide the strength of studs laid vertically, as at b. The strength of double horizontal studs depends upon very secure nailing together. Vertical studs require spacers of plywood, hardboard, or other appropriate material to bring their total thickness to 3½", to match that of the surrounding framing.

STAIR CARRIAGE FRAMING

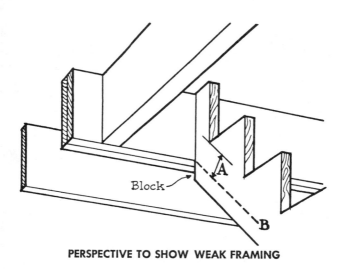

PERSPECTIVE TO SHOW WEAK FRAMING

**SECTION TO SHOW
CORRECT FRAMING**

The weakness lies in the small effective depth marked *A* on the drawing together with the low resistance of wood to splitting along the dotted line *B*. This error could be corrected in some measure by nailing a substantial block under the double joists at the heel of the stringer so that the vertical face of the stringer would have bearing.

The only correct way to frame the stairs, however, is shown in the small section. The double joists should be placed far enough from the face of the top riser so that the line of the underside of the stringer intersects the lowest corner of the double joists.

FIBER-REINFORCED PAPER
OVER SHEATHING

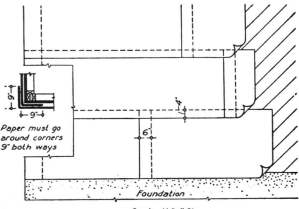

Paper must go around corners 9" both ways

Foundation

Figure-1   Scale-1/4"=1'-0"

OVER SHEATHING. To prevent drafts and expensive heating/cooling losses through a structure's outer walls, air infiltration must be controlled. When sheet sheathing such as plywood, rigid insulation, or structural fiberboard is applied, no further precautions need generally be taken, especially if the sheet edges are sealed with caulk or glue. However, if horizontal or diagonal board sheathing is applied an air-permeable barrier such as rosin paper or 15-pound roofing felt should be applied over the sheathing and beneath the exterior siding. It is important that this barrier be air-permeable and not impervious to moisture passage, particularly if an impervious vapor barrier such as polyethylene sheeting is installed (as it should be) on the inner side of the wall.

The paper should be installed so that it is smooth but not taut, shingle-fashion, and fastened with staples or roofing barbs. Edge laps should be at least 4", and end laps at least 6". Since the corners of a building offer the easiest point of attack for wind, the paper should be turned around both inside and outside corners for at least 9" in both directions, as shown in Figure 1.

FLASHING FRAMES. When the frames for door, window, or other openings are in place at the time the paper is applied, the paper should be carried over the frame as shown in Figure 2—creating an effective "windbreak."

When the frames are set after the application of the paper, a strip of sufficient width should be stapled around the frame, as shown in Figure 3, to cover the joint beween the frame and the wall construction. Vinyl-clad windows with vinyl mounting flanges, however, do not need such added protection if the flanges are sealed down with a bead of caulk.

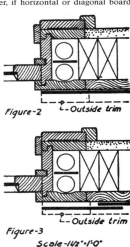

Figure-2   └─Outside trim

Figure-3   └─Outside trim

Scale-1/2"=1'-0"

518

MASONRY WALL CONSTRUCTION

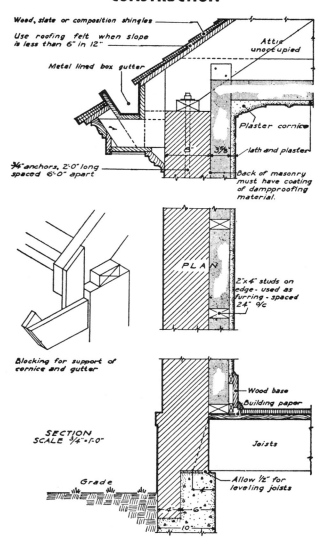

Wood, slate or composition shingles

Use roofing felt when slope is less than 6" in 12"

Metal lined box gutter

Attic unoccupied

Plaster cornice

lath and plaster

3⅝

Back of masonry must have coating of dampproofing material.

¾" anchors, 2'-0" long spaced 6'-0" apart

PLAN

2"x4" studs on edge - used as furring - spaced 24" o/c

Blocking for support of cornice and gutter

SECTION
SCALE ¾" = 1'-0"

Wood base
Building paper

Joists

Grade

Allow ½" for leveling joists

4" 6"

10"

519

WOOD SHINGLE SIDEWALLS

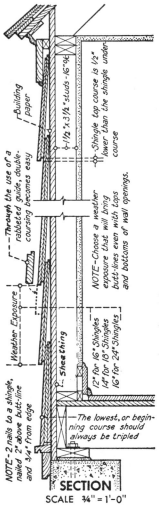

Building paper

1½" x 3½" studs—16" o.c.

Shingle top course is ½" lower than the shingle under-course

—Through the use of a rabbeted guide, double-coursing becomes easy

NOTE—Choose a weather exposure that will bring butt-lines even with tops and bottoms of wall openings.

Weather Exposure

Sheathing

12" for 16" Shingles
14" for 18" Shingles
16" for 24" Shingles

NOTE—2 nails to a shingle, nailed 2" above butt-line and ¾" from edge

The lowest, or beginning course should always be tripled

SECTION
SCALE ¾" = 1'-0"

DOUBLE COURSES. Sidewalls covered with shingles that are given a very wide exposure create a strikingly attractive appearance. It is particularly adaptable to modern interpretation of the Colonial styles, as well as lending individuality to the designer's treatment of other architectural periods. The wide exposure requires deep butt shadows to be effective—both being readily obtainable with standard shingles at surprisingly low cost.

APPLICATION. The weather exposure of shingles in single courses should not exceed half the shingle length minus ½". When double coursing is employed with "butt-nailing," much longer exposures become possible, greatly reducing the cost of application. Use 5d small-head hot-dipped zinc coated nails, two nails per shingle, placed near the edges of the shingles and not more than 3" above the butts.

The following table shows that double coursing is economical because it allows greater exposure of the shingles.

Length of Shingles (in inches)	Exposure of Shingles (in inches)	
	Single Course	Double Course*
16"	6" to 7½"	8" to 12"
18"	6" to 8½"	8" to 14"
24"	8" to 11½"	12" to 16"

*Assuming exposed course is face or butt-nailed.

GRADES. The exposed shingles in each course should be No. 1 Blue Label, which comprises all clear edge-grain and 100% heartwood shingles. However, in the interest of economy, No. 2 Red Label shingles may be substituted; this grade is a reasonably good one for most applications. The under-coursing may be No. 3 Black Label or No. 4 Undercoursing grades. When stained shingles are applied, unstained shingles may be used successfully in the concealed courses.

SHINGLE, SIDING, OR CLAPBOARD WALL CONSTRUCTION

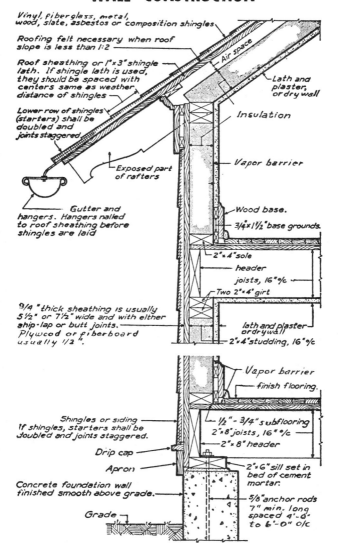

Vinyl, fiberglass, metal, wood, slate, asbestos or composition shingles.

Roofing felt necessary when roof slope is less than 1:2

Roof sheathing or 1"×3" shingle lath. If shingle lath is used, they should be spaced with centers same as weather distance of shingles

Lower row of shingles (starters) shall be doubled and joints staggered.

Exposed part of rafters

Gutter and hangers. Hangers nailed to roof sheathing before shingles are laid

Air space

Lath and plaster, or dry wall

Insulation

Vapor barrier

Wood base.

3/4"×1½" base grounds.

2"×4" sole

header

joists, 16" °/c

Two 2"×4" girt

lath and plaster or dry wall

2"×4" studding, 16" °/c

9/4" thick sheathing is usually 5½" or 7½" wide and with either ship-lap or butt joints. Plywood or fiberboard usually 1/2".

Vapor barrier

finish flooring.

Shingles or siding. If shingles, starters shall be doubled and joints staggered.

Drip cap

Apron

Concrete foundation wall finished smooth above grade.

Grade

1/2" - 3/4" subflooring
2"×8" joists, 16" °/c
2"×8" header

2"×6" sill set in bed of cement mortar.

5/8" anchor rods 7" min. long spaced 4'-0" to 6'-0" °/c

521

SHEATHING PAPER AND
FLASHING IN BRICK VENEER

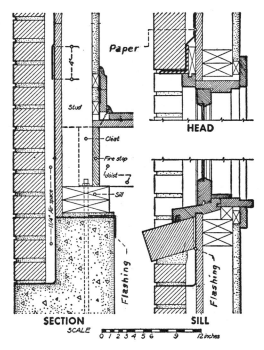

PROTECTING FOUNDATION SILLS. When workmanship is careless, mortar droppings can fill the dead air space between the sheathing and veneer to a substantial depth. This should be avoided, since moisture is readily conducted from the brick veneer by the mortar. Rotting of sheathing and sills can take place as a result, with serious damage to the building. A wide strip of metal flashing installed (as shown above) all around the building will obviate this hazard. All field and corner laps should be sealed with mastic and nailed 2″ o/c with copper roofing nails. In addition, air-permeable sheathing paper should be applied to the sheathing and lapped down over the base flashing.

SILL MATERIALS. The sills should uniformly contact the foundation top. This can be accomplished by setting the sills in a bed of mortar, or by installing a sill seal material between the sill and the foundation. Sills should be made of redwood or an approved pressure-treated wood. Where sill pieces are doubled, the upper course may be of any wood.

WINDOW FLASHING. Moisture and air infiltration around the periphery of a window and at the sill can cause rotting and warping of the frame. This can be obviated by the use of metal flashing, as shown in the detail above. Heavy polyethylene film is an alternative; this material is light and flexible, making it easy and economical to install. Polyethylene sheeting applied to the inner side of the wall as a vapor barrier can be lapped around the window construction to serve as an added barrier against air infiltration.

BRICK VENEER
CONSTRUCTION

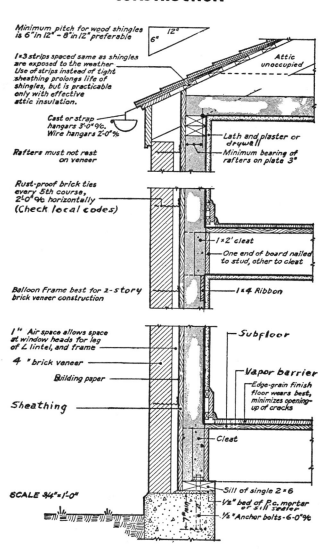

Minimum pitch for wood shingles is 6" in 12" - 8" in 12" preferable

12"

6"

1×3 strips spaced same as shingles are exposed to the weather. Use of strips instead of tight sheathing prolongs life of shingles, but is practicable only with effective attic insulation.

Attic unoccupied

Cast or strap hangers 3'-0" o.c. Wire hangers 2'-0" o.c.

Lath and plaster or drywall

Minimum bearing of rafters on plate 3"

Rafters must not rest on veneer

Rust-proof brick ties every 5th course, 2'-0" o.c. horizontally (Check local codes)

1×2' cleat

One end of board nailed to stud, other to cleat

1×4 Ribbon

Balloon Frame best for 2-story brick veneer construction

1" Air space allows space at window heads for leg of ∟ lintel, and frame

Subfloor

4 " brick veneer

Vapor barrier

Building paper

Edge-grain finish floor wears best, minimizes opening-up of cracks

Sheathing

Cleat

SCALE 3/4" = 1'-0"

Sill of single 2×6

½" bed of p.c. mortar or sill sealer

½" Anchor bolts - 6-0" o.c.

1-7"

CORRECT STUCCO
CONSTRUCTION

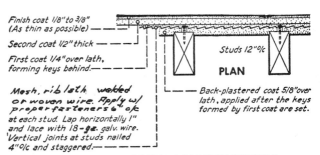

Finish coat 1/8" to 3/8"
(As thin as possible) ————

Second coat 1/2" thick ————

First coat 1/4" over lath,
forming keys behind.————

Studs 12" o/c

PLAN

Mesh, rib lath welded
or woven wire. Apply w/
proper fasteners 6" o/c
at each stud. Lap horizontally 1"
and lace with 18-ga. galv. wire.
Vertical joints at studs nailed
4" o/c and staggered.————

Back-plastered coat 5/8" over
lath, applied after the keys
formed by first coat are set.

BACK-PLASTERED CONSTRUCTION

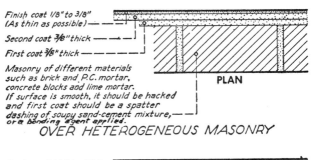

Finish coat 1/8" to 3/8"
(As thin as possible) ————

Second coat 3/8" thick ————

First coat 3/8" thick ————

Masonry of different materials
such as brick and P.C. mortar,
concrete blocks and lime mortar.
If surface is smooth, it should be hacked
and first coat should be a spatter
dashing of soupy sand-cement mixture,——
or a bonding agent applied.

PLAN

OVER HETEROGENEOUS MASONRY

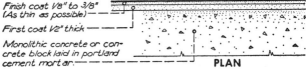

Finish coat 1/8" to 3/8"
(As thin as possible)————

First coat 1/2" thick ————

Monolithic concrete or concrete block laid in portland
cement mortar.————

PLAN

Three coat work should always be used except on masonry walls of
homogeneous materials, which present a true, even, clean, sound surface, where the second or leveling coat may be omitted.

Where the wall surface is too smooth to provide proper bond it
should be hacked and dashed with a coat of soupy sand-cement
mixture, or a bonding agent applied.

Due to its rough texture and the similarity of its composition to
that of stucco, concrete masonry is well adapted as a stucco base.

OVER HOMOGENEOUS MASONRY

CORRECT STUCCO CONSTRUCTION

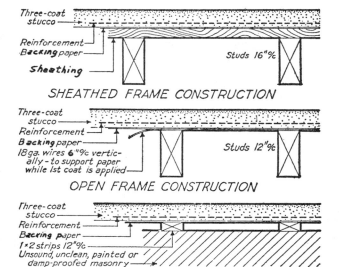

Three-coat stucco
Reinforcement
Backing paper
Sheathing
Studs 16"⁰⁄ₒ

SHEATHED FRAME CONSTRUCTION

Three-coat stucco
Reinforcement
Backing paper
18 ga. wires 6"⁰⁄ₒ vertically - to support paper while 1st coat is applied
Studs 12"⁰⁄ₒ

OPEN FRAME CONSTRUCTION

Three-coat stucco
Reinforcement
Backing paper
1×2 strips 12"⁰⁄ₒ
Unsound, unclean, painted or damp-proofed masonry

OVER UNSUITABLE OLD MASONRY

FINISH COAT. This should be at least ⅛" thick and sufficient to bring the overall stucco thickness to at least ⅞", but no thicker than necessary to provide the desired finish texture.

REINFORCEMENT. Metal mesh should be applied over sheathed or unsheathed wood frame construction, steel frame construction, poorly bonding concrete or masonry, chimneys, and flashings. Diamond mesh, stucco mesh, flat rib lath, ⅜" lath, and woven-and-welded wire are all suitable.

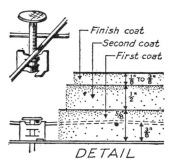

Finish coat
Second coat
First coat

⅛ TO ⅜

½

⅝

⅜

DETAIL

FASTENERS. They can be 1½" 11-gauge barbs with ⁷⁄₁₆" head, or ⅞" 16-gauge staples with ¾" crown; ⅜" rib lath, however, requires 1¼" 16-gauge staples. Fasteners are spaced 6" o/c maximum, except that ⅜" rib lath requires staples 4½" o/c. Woven-and-welded wire can be attached with 1½" 12-gauge furring nails with ⅜" head.

BACKING PAPER. Use quality building paper, #15 tar paper or roofing felt.

FRAMING. Balloon type, heavily braced and bridged, is best, but platform type is commonly used.

525

STUCCO OVER OLD CLAPBOARD WALLS

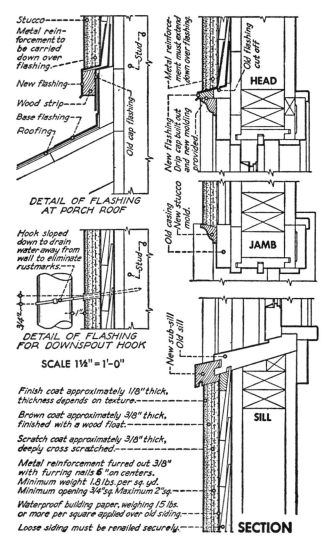

Stucco

Metal reinforcement to be carried down over flashing.

New flashing

Wood strip

Base flashing

Roofing

ℓ-Stud

Old cap flashing

DETAIL OF FLASHING AT PORCH ROOF

Hook sloped down to drain water away from wall to eliminate rustmarks.

¾"

1"

ℓ-Stud

DETAIL OF FLASHING FOR DOWNSPOUT HOOK

SCALE 1½" = 1'-0"

Finish coat approximately 1/8" thick, thickness depends on texture.

Brown coat approximately 3/8" thick, finished with a wood float.

Scratch coat approximately 3/8" thick, deeply cross scratched.

Metal reinforcement furred out 3/8" with furring nails 6" on centers. Minimum weight 1.8 lbs. per sq. yd. Minimum opening ¾" sq. Maximum 2" sq.

Waterproof building paper, weighing 15 lbs. or more per square applied over old siding.

Loose siding must be renailed securely.

Metal reinforcement must extend down over flashing.

Old flashing cut off

HEAD

New flashing. Drip cap built out and new molding provided.

Old casing

New stucco mold.

JAMB

New sub-sill

Old sill

SILL

SECTION

OPEN FRAME
STUCCO CONSTRUCTION

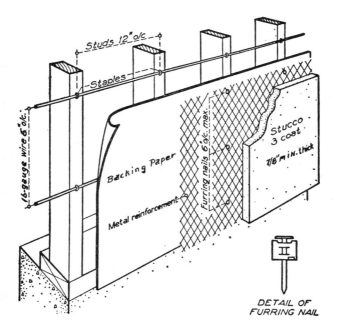

Studs 12" o/c

Staples

16-gauge wire 6" o/c

Backing Paper

Metal reinforcement

Furring nails 6" o/c max.

Stucco
3 coat

7/8" min. thick

DETAIL OF
FURRING NAIL

OPEN FRAME CONSTRUCTION.   So-called "open frame" construction has been widely used in various parts of the country because of its economy. With wood lath and plaster interior finish and stud spaces filled with rock wool, the heat transmission coefficient is 0.10.

BRACING REQUIREMENTS.   The corners of exerior walls should be braced diagonally with 1" x 6" pieces let into the studs on the interior face. The studding should be bridged with 2" x 4" braces at least once in every story height.

BACKING PAPER.   The dryness of the frame and the air leakage potential of the open frame stucco wall depend in considerable measure upon the quality of the backing paper used. The backing paper should be applied in shingle-fashion directly over the studs. Horizontal joints should be lapped 4" minimum, and vertical joints should be lapped over studs. The paper is held temporarily with tacks or staples. The fasteners used in the subsequent application of the metal reinforcing hold it permanently. When the stucco is applied, a quality backing paper will not belly back and require excessive amounts of stucco.

METAL REINFORCING.   Use expanded metal lath or wire fabric with relatively large openings (¾" to 2") so that the stucco will completely embed the metal, forming a reinforced concrete slab. No form of metal or wood strips or rods should be used since they reduce the thickness of the stucco section with resulting cracks and discoloration. Horizontal and vertical joints of the metal reinforcing must be lapped one full mesh. Horizontal joints should be tied in each stud space with No. 18 annealed tie wire.

527

LOG CABIN
WALLS

SUITABLE SPECIES.   Northern white cedar is perhaps the best for log house construction. Western redcedar, baldcypress, Douglas-fir, and lodgepole pine follow closely. Eastern redcedar, white pines, Engelmann spruce, hemlocks, red and southern pines, tamarack, and redwood are also very suitable. Cottonwoods, willows, aspens, birches, and basswood are not so suitable. Other species can be used, but it is advisable to choose the more durable woods.

SIZE OF LOGS.   Logs from 4″ to 12″ in diameter are usually employed. Logs should be straight and have little taper, insofar as is possible. They should also be longer than the length of the room to allow for joining, but those longer than 20′ are heavy and difficult to handle. Wall logs can be joined end to end with half-lap, squared, or tongue splices; all such joints should be widely staggered and used as infrequently as possible. Short lengths can be set between wall openings.

TIME OF CUTTING.   If the logs are to be peeled, winter is the best felling time for the trees. If felled in spring while the sap is running, logs deteriorate through the development of stain and decay organisms. Bark will adhere to the logs if the trees are cut in late summer. To avoid damage by insects, cutting should be postponed until after the first frost. To increase adhesion of bark, a narrow strip or score can be cut off on two sides of the entire length. The logs should then be seasoned by properly stacking them in shade to allow thorough circulation of air around them until at least the following spring. Scores, ends, and notches should be coated with creosote or other preservative a few days after felling and again just before the logs are used. Peeled logs are less subject to rotting and can be better protected against insect attack.

FINISHING.   Logs should not be painted, varnished, or urethaned. Staining is not recommended but can be done. Periodic coatings of wood preservative, wood sealer, or linseed oil are all that is necessary.

INSECT PROTECTION.   For termite protection, see pages 530–532. For protection against other insects, such as carpenter ants or wood borers, periodic treatments with a preservative such as pentachlorophenol is sufficient.

CHINKING.   Flatted logs should be sealed at the joints with continuous strips of foam weatherstripping or caulk beads, and perhaps a spline as well. The exterior cracks along the joints can be further sealed with caulk, after a year or two has passed. The cracks or spaces between round logs must be sealed, but this should be delayed as long as possible to allow shrinking and settling to take place. Narrow cracks can be filled with caulking yarn or fiberglass, and then sealed over with caulk. Wide spaces between logs can be closed with short, narrow slab pieces or quartered thin logs, bedded in cement mortar and nailed in place. The mortar bed should be keyed in place with a series of partly-driven nails. Moderately wide openings between logs can be stuffed with fiberglass. This is held in place with strips of metal lath or hardware cloth stapled in place. The whole is then covered with cement mortar or stucco mix, troweled in place and faired off to suitable contours. Clay or wattle chinking can be applied in similar fashion, but it is not as long-lived except in arid areas.

POLE AND SIDING LOG CABIN WALLS

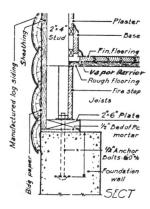

LOG SIDING CONSTRUCTION

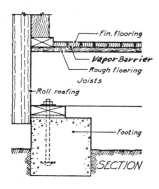

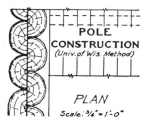

Manufactured shiplapped log siding may be applied horizontally over sheathing as shown, in the same manner as any other exterior siding. This material may also be used to simulate palisade construction, if a proper waterproof building paper is placed between the logs and the sheathing.

Buildings with walls consisting of logs placed on end are referred to as *palisade* or *stockade* types. Such structures are easier to build than horizontal-log buildings because the short, often smaller-diameter logs can be easily handled by one person. Moreover, corner joining is not required, and fitting is less problematical.

The logs should be not less than 4″ to 5″ in diameter. If full round logs are to be used for palisade construction, they should be hewn, milled, or sawn on the sides and matched to fit tightly together, making a weathertight wall. Joints can be sealed and/or chinked as with horizontal construction. A good foundation should be provided, to which the sill logs are bolted. In some locales, pressure-treated timbers may be required. Sills should be equal in diameter or cut dimensions to the wall logs. The top surface of the sill must be square and level to provide proper bearing for the squared ends of the wall logs. Similar logs or timbers must be used as plates atop the vertical wall logs. Sills and plates should be half-lapped at the corners. The corner vertical logs are set up on the sills first; then the top plates are secured to the corner logs. The remaining wall logs are then cut, fitted, and spiked in place.

Split logs and slabs are sometimes placed as shown, in two layers with staggered joints. Proper edge-matching is important, and an exterior-grade plywood core or waterproof building paper can be used between the inner and outer layers.

529

LOG CABIN
CORNER JOINTS

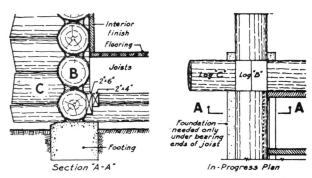

Section "A-A"

In-Progress Plan

METHOD OF FRAMING SILL AND FLOOR JOISTS
SCALE ½"=1'-0"

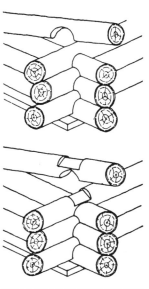

**TWO METHODS OF NOTCHING
LOGS FOR CORNERS**

The simplest footing for temporary log structures consists of large, flat stones laid on firm ground. Where permanence is required, a full masonry wall is advisable under the bearing ends of the joists. If there is no cellar under the structure, ventilation may be ensured by omitting the footing or foundation of the nonbearing sides of the building, as indicated above. If a continuous footing is used, completely enclosing the space under the building, a number of screened holes should be provided to ventilate the underside of the floor to reduce the danger of rotting.

The heaviest and best logs should be used for sills or bottom logs. Various methods are used for joining the logs at the corners. Two types of notches are shown.

CASEMENT IN
LOG CABIN

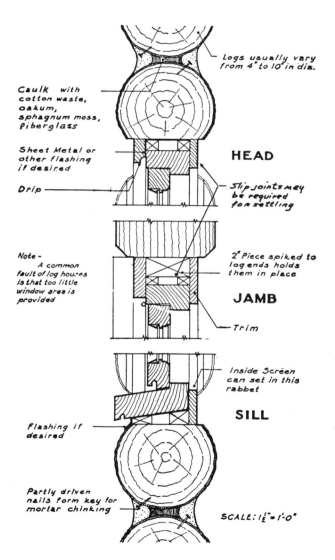

Logs usually vary from 4" to 10" in dia.

Caulk with cotton waste, oakum, sphagnum moss, fiberglass

Sheet Metal or other flashing if desired

Drip

HEAD

Slip joints may be required for settling

Note -
A common fault of log houses is that too little window area is provided

2" Piece spiked to log ends holds them in place

JAMB

Trim

Inside Screen can set in this rabbet

SILL

Flashing if desired

Partly driven nails form key for mortar chinking

SCALE: 1½" = 1'-0"

531

FACTS ABOUT
TERMITES

EXTENT OF TERMITE DAMAGE.   Damage to buildings by termites remains a consistently serious problem because preventive measures are not always or completely taken and because periodic preventive maintenance is not carried out; this holds true for both new and existing buildings. Destruction by termites in terms of dollar value is impossible to assess accurately, but it surely runs to many millions of dollars annually in replacement costs. Wood treated under pressure with creosote or certain salt preservatives is termite-proof. Its liberal use, together with correct application of poisons and proper construction methods, will definitely prevent termite attack.

It should be noted that while damage to buildings is serious, decay of wood in structures causes even greater losses. Untreated wood that is subjected to moisture tends to decay; treated wood does not. As often as not, destruction by rot and by termites goes hand in hand.

EFFECTIVE PROTECTIVE MEASURES.   The most effective protective measures that may be taken against possible termite attack in new construction include complete site clearing, soil poisoning, the use of pressure-treated wood, tight foundations, and observance of proper construction techniques. Remedies for termite infestation usually involve renovation and rebuilding (often substantial) and institution of the same preventive measures used in new construction—particularly soil poisoning. Architects, builders, and homeowners are cautioned not to accept any new and easy control methods, "secret" formulas or materials, or other miracle cures; only established measures should be considered, since this is fertile ground for various scams and con games. Advice can be obtained from local authorities and from the Forest Products Laboratory, U.S. Department of Agriculture, the National Pest Control Association, and the Bureau of Entomology.

TYPES AND RANGE.   Termites superficially resemble ants in size, general appearance, and the trait of living in colonies. Hence, they are frequently called "white ants." They are not true ants, however, but are more closely related to cockroaches. Termites are of two main classes: (1) the ground-inhabiting or subterranean termites, and (2) the drywood termites. A third type, called the Formosan termite, has recently made its appearance in this country. Subterranean termites are found in every state except Alaska and are responsible for most termite damage to wood structures. Drywood termites are found along the southern rim of the country and in Hawaii. The Formosan termite is at present found only in parts of Louisiana and Florida. The map on the following page shows the approximate ranges within which significant damage is inflicted by subterranean and drywood termites in the United States.

HABITS OF LIFE.   Subterranean termites develop their colonies in the ground, live in the dark and shun light, and require warmth and moisture to survive. The requirements of warmth and moisture mean that they must usually maintain contact with the ground and that colder climates do not favor colony development; termite activity is consequently lessened in these climates. The principal staple of their diet is wood cellulose, which they seek through foraging tunnels. These tunnels may be in the ground, or may be earthen tubes on exterior surfaces of any kind. Upon reaching wood, they tunnel within and digest the cellulose in the process. However, the destruction thus caused takes place over very long periods of time, since even a well-established colony of 60,000 termites eats only about $\frac{1}{5}$ ounce of wood per day. Surface tubes can extend as high as 14' to 15', while unsupported tubes

FACTS ABOUT TERMITES

or columns may reach a height of 1'. If the soil in which they live is poisoned or the moisture is removed or the connection between the earth and their wood food supply is destroyed, the termites will die or attempt to move elsewhere. In early spring or fall, winged, fertile female termites go into a short swarming flight, and if they reach a suitable home they start new colonies. These swarms, or piles of wings that have been shed, are often the only indication of the presence of a colony.

Drywood termites do not multiply as rapidly as the subterranean type. They require no moisture or ground contact, and they can live indefinitely in dry wood. They establish new colonies by swarming flights, and also by simply being carried to new locations in lumber, cordwood, or even in infested furniture. Preventive measures include site-clearing, complete screening of building openings, periodic inspection of premises, and checking of all in-coming goods and materials that might harbor them. Eradication is accomplished by spraying or, if necessary in the case of large infestations, by fumigation. As happens with subterranean termites, the destruction of wood proceeds very slowly.

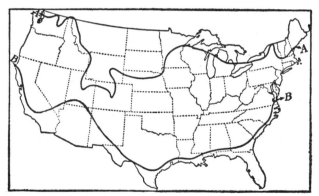

FIGURE 1.—*Map showing approximately (line A-A) the northern limit of damage done by subterranean termites in the United States and (line B-B) the northern limit of damage done by dry-wood or non-subterranean termites.*

Less is known about the Formosan termite, but what is known is rather unsavory. The range of this termite is being increased primarily by mechanical transportation of infected materials. Cold weather and climate greatly inhibit breeding, but in warm climates the spread can be relatively rapid. Unlike its cousins, the Formosan termite is unfortunately able to gnaw through mortar, plaster, and even wood preservatives to reach its food supply. As yet, little concerted effort has been made to control or even to conduct studies of the creature and of effective preventive measures against it.

TERMITE CONTROL METHODS

Food and moisture are the factors that must be controlled to prevent damage by termites. Protective measures are designed: (1) to render the soil surrounding the structure uninhabitable; (2) to prevent termites from reaching wood that they need as their food supply; (3) to render wood that may become subject to attack inedible; and (4) to force termites to build visible shelter tubes so that their presence can be discovered, the tubes destroyed, and immediate control measures taken. All of the following precautions should be taken to provide buildings maximum protection against damage by subterranean termites.

1. WOOD DEBRIS.   Remove all stumps, roots, wood debris, and paper litter from a wide area around the building. Remove all wood forms that have been used in concrete work, to a point at least 18″ above finished grade. Make sure that no wood form spreaders or grade/leveling stakes used in concrete foundation work are left in place. Take particular care that no wood scraps or paper trash become buried in nearby fill or backfill. Insofar as is possible, keep the area beneath the building dry during and after construction.

2. FOUNDATIONS.   Construct foundation walls and piers using poured concrete. Reinforce foundation walls with not less than two #3 reinforcing rods placed not more than 4″ below the top of the wall. Reinforcing should be continuous throughout the length of every wall and around corners. Lap rods not less than forty rod diameters. Interior foundation walls abutting exterior foundation walls must be joined with special care, as shown in drawings A and B. The tops of foundation walls should be not less than 6″ above finished grade (check local codes). Masonry foundation walls should be heavily reinforced and capped with a layer of poured concrete no less than 6″ thick, also reinforced.

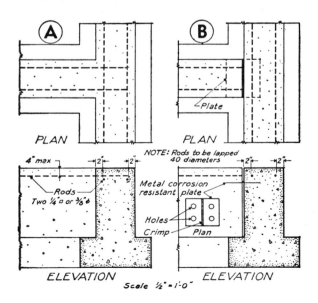

TERMITE CONTROL
METHODS

3. VENTILATION. For buildings without basements, openings must be provided through foundation walls for (a) cross ventilation, (b) access for inspection, and (c) lighting of the area. See drawing C.

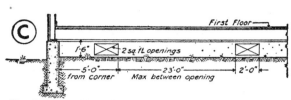

FOUNDATION OPENINGS FOR BUILDINGS WITHOUT BASEMENT
Scale ⅛" = 1'-0"

4. BASEMENT FLOORS. The basement floor should be of poured concrete, a minimum of 3½" thick, and of 1:3:6 mixture. While wooden columns, partitions, door casings, sleepers, wood flooring, and so forth may generally be safely installed on a good concrete floor, the best practice is not to depend upon perfection in the concrete. Pressure treated wood should be used. Place partitions on 6"-high curbs, which should be poured monolithically with the floor slab. See drawing D. Columns and stair carriages should rest on similar 6" high bases.

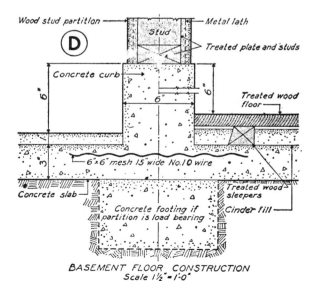

BASEMENT FLOOR CONSTRUCTION
Scale 1½" = 1'-0"

TERMITE CONTROL
METHODS

5. WINDOWS AND DOORS. Window and door frames, with their trim, installed in basements should be made of pressure-treated wood, vinyl-clad wood, steel, or aluminum.

6. TERMITE BARRIERS. Termite barriers or "shields" of noncorroding metal may be constructed so as to cut off all access of termites from the ground to untreated wood, as shown in drawing E. Attack in more than 90% of cases of termite damage can be traced to contact of wood with the ground. It has also been found through experience that termite shields, because of improper installation, inferior or unsuitable materials, and deterioration of both shields and buildings with time, generally cause more problems than they solve. As a result, they have fallen into disfavor and are little used.

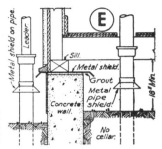

7. DRAINAGE. Provide adequate drainage of soil beneath and around the structure.

8. JOINT RECESSES. Do not seal the ends of first-floor wood members entering masonry or concrete. Provide recesses which allow an air space of not less than 1" at each side of the member.

9. PRESSURE-TREATED LUMBER. Use pressure-treated wood for all lumber up to and including the first subfloor for protection against rot as well as against termites.

10. PORCHES. Patios, porches, and steps should not be higher than the top of the foundation wall. Failure to observe this precaution has been the largest single cause of termite damage. Where patio, porch, or steps are not resting on earth fill, provision must be made for the removal of concrete form work from underneath them and ventilating openings should be left—even under reinforced concrete slabs. See drawing F.

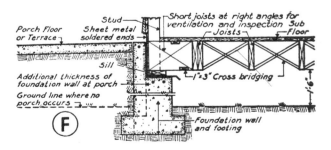

11. EXTERIOR FINISH. Wood siding should not be closer than 6" to ground level. Stucco should be stopped at least 3" above the ground line.

12. SOIL POISONING. This most important protective measure must be carried out at regular intervals by licensed, trained professionals, using the latest approved pesticides and techniques. Frequent reinspection is essential.

PRESSURE-TREATED WOOD
FOR TERMITE PROTECTION

PRESSURE-TREATED WOOD. Proper use of pressure-treated wood below the first floor in residences and other buildings offers protection against termites, is practical, and costs relatively little. The primary characteristics of a good wood preservative are toxicity, permanence, and freedom from damaging effects on the wood. Other special characteristics often sought in treated wood are the ability to accept and retain paint, low electrical conductivity, low flammability, the ability to be worked with edged tools, harmlessness to the health of workers using normal care, and reasonably low weight increase over untreated wood. Above all, the treated wood must give service that will ultimately save the consumer money.

EFFECTIVE PREPARATIONS. Creosote and creosote solutions are effective in preventing damage by subterranean termites, drywood termites, and decay. The odor and color resulting from the treatment of wood with creosote may be objectionable, however, and creosote-treated wood will neither accept nor retain paint. Oil-borne preservatives containing pentachlorophenol are equally effective and overcome the creosote difficulties. Water-borne preservatives such as chromated copper arsenate, ammoniacal copper arsenate, acid copper chromate, fluor chrome arsenate phenol, and chromated zinc chloride, are also fully effective and in widespread use. These solutions are relatively odorless and colorless, are paintable, and will not leach out in the continued presence of moisture.

METHODS OF APPLICATION. Several methods of applying protective treatments are in use. *Brush and spray treatments* result in only a thin protective coating on the wood. The protection is also of short duration. Nailholes, checks, etc., provide passages for the entrance of termites to the untreated wood beneath the surface. *The open tank method of treatment,* when carefully and properly applied, will usually show a full penetration of the sapwood. *Pressure treatments* are the most effective method of securing maximum impregnation of the timber by the preservative. Pressure treatments are applicable only to full-length treatment of timbers. All wood should be sized, framed, and bored before treatment. If further shaping of the treated wood on the job is unavoidable, the freshly cut surface should be treated with several generous coats of the preservative.

FOR BUILDINGS, PORCHES, AND EXTENSIONS WITHOUT BASEMENTS. Pressure creosote treatment of: foundation timbers in contact with ground; supporting posts, pillars, and footings in contact with ground.

Pressure salt treatment of: siding up to 18″ above ground, lattices, first floor joists, and first sub-floor, sleepers, leaders, and plates embedded in or laid on concrete or concrete-masonry foundations or walls; all other structural timbers within 18″ of the ground.

FOR BUILDINGS WITH BASEMENTS HAVING CONCRETE FOUNDATIONS. Pressure salt treatment of: all wood used in basement, stairs, door and window casings, partitions, studding, lath sleepers, leaders, plates and joists embedded in or laid on concrete or masonry, and sills; all structural timbers within 18″ of the ground.

RODENT-PROOFING
FRAME BUILDINGS

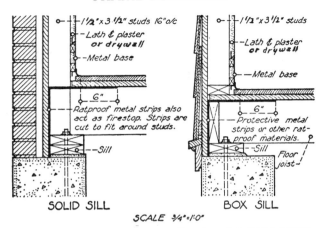

1½"x3½" studs 16"o/c
Lath & plaster or drywall
Metal base
6"
Ratproof metal strips also act as firestop. Strips are cut to fit around studs.
Sill

SOLID SILL

1½"x3½" studs
Lath & plaster or drywall
Metal base
6"
Protective metal strips or other rat-proof materials.
Sill
Floor joist

BOX SILL

SCALE ¾"=1'0"

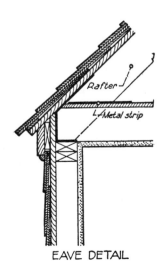

Rafter
Metal strip

EAVE DETAIL

Normal practices of good construction will usually render a frame building proof against birds, rodents, reptiles, and other inquisitive small animals. However, in areas where there is danger from Norway rats, where bushy-tailed woodrats are prevalent, or where snakes, scorpions, field mice, skunks, porcupines and other wild creatures abound, and in buildings such as summer camps that are left unattended and closed up for long periods, extra precautions may be in order.

Hollow walls above the cellar must be protected from invasion from the cellar, intermediate floors, or attic. (See drawings.) Pipelines can be fitted with metal collars. Decorative grilles should have openings no larger than ¼". All windows, vents, and doors must be carefully screened. Masonry or metal appliance chimneys must be fitted with caps or fine-mesh screening, preferably of stainless steel for durability, within or atop the flue. All holes through which piping or wiring enter the building must be carefully sealed. Cellar floor drains must be fitted with caps. Extensive fire-stopping in partitions and between ceilings and second floors and attics, as well as along eaves, affords good protection from intruding creatures as well as from rapid spread of fire. Above all, high quality building materials and careful construction techniques should be used throughout the structure. In the event that small creatures do manage to encroach upon the premises, an electronic device that emits certain high-pitched sound frequencies can be installed that will drive many of them away.

FOUNDATION FOOTING DRAINS

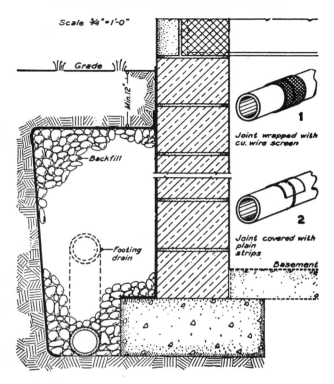

Scale ¾"=1'-0"

Grade

Min. 12'

Backfill

Footing drain

Basement

1 Joint wrapped with cu. wire screen

2 Joint covered with plain strips

FOOTING DRAIN. Every well-constructed basement should have a footing drain. This drain should lead to an outlet such as a dry well, open watercourse, or storm sewer, to dispose of the collected water. The 4″ draintile, as shown in the figure above, is laid either dead level or with a very slight slope along the footing. The joints should be kept open and a strip of copper screening or 30-pound 6″ x 9″ roofing felt should be placed over the top of each joint. Perforated plastic drain pipe should be laid at a slight slope with the perforations down and the lengths coupled together. Slotted flexible plastic drain tubing can be laid in a continuous run, slightly down-sloped; care should be taken to avoid dips and low spots in the line. In all cases the drain line should rest upon a layer of gravel about 4″ deep.

BACKFILL. The drain line is then covered with a layer of gravel to a depth of at least 2″ above the top of the footing. The gravel can then be covered, if desired, with a 2″ layer of straw or a strip of heavy polyethylene sheeting; this prevents loose dirt from filtering down into the gravel bed and clogging it. Backfill material, free of rocks any larger than fist-size, is next hand-placed to a depth of about 1′. Then the trench can be filled to grade level by mechanical means with whatever fill is available.

WALL DAMPPROOFING. Added protection against dampness can be provided by mopping bituminous material over the footing top and foundation walls. This coating can be shielded from injury when coarse backfill is placed on it by sticking a layer of 30-pound roofing felt to the mopping and by performing further mopping as necessary to keep it in place.

539

DAMPPROOFING
RESIDENCE FOUNDATIONS

DAMPPROOFING VS. WATERPROOFING. Dampness appearing on the inside of basement surfaces may be from either of two causes: (1) condensation or (2) capillarity. In neither case is there a static head to cause water to enter under pressure, which is a condition that requires waterproofing.

CONDENSATION. If walls, floors, or ceiling are below the dewpoint for the relative humidity of the air, droplets will be condensed on these surfaces. The cure is either adequate air movement, absorption of the air vapor with chemicals, or insulation of the surfaces to raise their surface temperature above the dewpoint. Condensation is most prevalent during relatively warm and humid times of the year.

CAPILLARITY. Water will climb by capillarity in coarse sands from 2' to 3', and in fine sands, silts, loams, and clays, from 5' to 8'. In doubtful soils, borings should be made to ensure that the permanent ground water level is a safe distance below the basement floor. If there is the slightest question, the precautions shown in the drawing should be taken because, after the building is up, corrective measures are prohibitively expensive.

WATERPROOFING. Where the permanent (or intermittent, due to rains or spring snowmelt) ground water level is above the foundations, complete membrane *waterproofing* is needed, as shown on the next page.

EXCAVATIONS. Footings should always rest on undisturbed earth. If a condition of dampness is foreseen, special care should be exercised in mixing the concrete for the footings so that they will be as impermeable as possible. A 4" to 6" cushion of crushed rock, coarse gravel, or cinders should be tamped in place below the floor slab. The voids in this material will break capillary action and reduce water pressure.

FLOOR SLAB. Heavy polyethylene sheeting should be laid over the cushion. A two-layer slab, as shown below, provides good protection against dampness. The alternative, a 4" gravel concrete single slab poured in the same location as the 2" finish slab shown below, is more common and has proved satisfactory. In either case, the placement of reinforcing mesh in the slab is advisable.

DAMPPROOF COAT. A double coating of bituminous dampproofing is usually sufficient to protect a poured concrete foundation wall; if desired, a layer of 15-pound or 30-pound roofing felt can also be mopped in place for added protection. If the foundation walls are of masonry, such as concrete block, it is advisable to parge the exterior surface with cement plaster before applying the bituminous coatings.

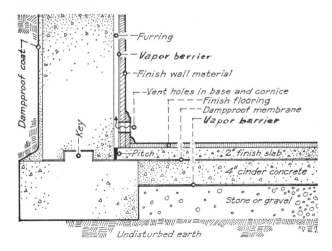

WATERPROOFING
RESIDENCE FOUNDATIONS

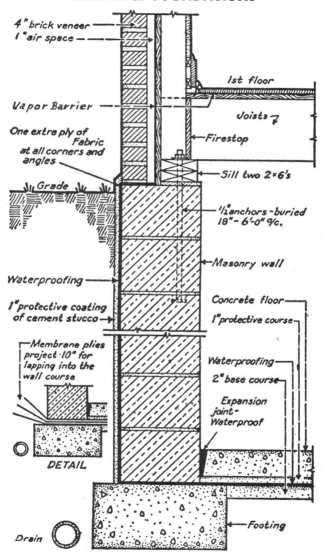

4" brick veneer

1" air space

Vapor Barrier

One extra ply of Fabric at all corners and angles

Grade

Waterproofing

1" protective coating of cement stucco

Membrane plies project 10" for lapping into the wall course

DETAIL

Drain

1st floor

Joists

Firestop

Sill two 2×6's

½" anchors - buried 18" - 6'-0" o/c.

Masonry wall

Concrete floor

1" protective course

Waterproofing

2" base course

Expansion joint - Waterproof

Footing

MEMBRANE WATERPROOFING

MEMBRANE WATERPROOFING. Foundations and similar constructions can be waterproofed by any of several methods: applying bituthene, a rubberized asphalt coated with polyethylene; applying polyethylene sheeting embedded in mastic; applying a membrane of butyl rubber, neoprene, or similar material; spraying binderized bentonite; attaching of bentonite panels; applying any of several liquid polymer membranes; or applying a traditional built-up membrane using coal-tar pitch.

None of these methods is absolutely foolproof, and all have potential flaws. The built-up membrane is the best known and perhaps most reliable of them; it is still the form of protection recommended by many experts. The other methods show promise and have been used successfully in some applications but have not fared well in others. As yet, they have not proved themselves over time.

BUILT-UP MEMBRANE. This waterproofing is constructed in place; the procedure is to build up a strong, waterproof, and impermeable blanket with overlapping plies of tar-coated glass fiber fabric, tarred felt, or tar-saturated fabric, the first of these materials now being preferred. The plies are coated and cemented together with hot coal-tar pitch. The number of applications of pitch is always one greater than the number of plies, except when it is necessary to lay a dry sheet on a wet surface in order to start work. If properly and carefully constructed, this type of waterproofing will prevent the entrance of water regardless of hydrostatic head, capillary attraction, concrete cracks, and expansion joints.

FABRIC. Applications of organic fabrics of relatively open weave, such as cotton, that have been well saturated with coal-tar pitch have a long record of successful use. The creosote contained in pitch does an excellent job of preserving the fabric fibers. However, since eventual rot is likely, especially in the constant presence of water, the preferred fabric material is a glass fiber coated with coal-tar pitch. Care must be taken that the fabric becomes fully filled with pitch — a difficult job that requires close attention on vertical surfaces.

FELT. Tarred felt is less costly than fabric and has been used in the membrane-waterproofing of many important structures. It may be used alone or with alternate plies of fabric. However, it is considerably more prone to eventual damage or decay than is glass fabric.

PITCH. Waterproofing pitch contains only coat-tar products and is the same material as is applied to roofs. It is unaffected by prolonged submersion in water, is resistant to attack by termites, and possesses reasonably good ductility. Unfortunately, its self-healing, self-sealing capabilities, which make it excellent for built-up roofing, are largely negated in an underground environment because temperatures there are not high enough. Nonetheless, coal-tar pitch's general overall effectiveness as a waterproofing is well established.

Recommended Number of Thicknesses of Membrane Waterproofiing Materials for Different Water Pressures

Head of Water (Feet)	Felt and Pitch or Fabric and Pitch	
	Plies of Tarred Felt or Tar Saturated Fabric	Mopping of Waterproofing Pitch
1-3	2	3
3-6	3	4
6-9	4	5
9-12	5	6
12-18	6	7
18-25	7	8
25-35	10	11
35-50	11	12
50-75	13	14
75-100	14	15

WATERPROOFING OF
SIDEWALK VAULT

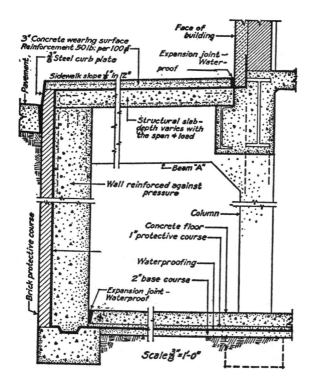

3" Concrete wearing surface
Reinforcement 50 lb. per 100 ft.

⅜" Steel curb plate

Sidewalk slope ½" in 12"

Face of building

Expansion joint — Waterproof

Pavement

Structural slab depth varies with the span & load

Beam "A"

Wall reinforced against pressure

Column

Concrete floor

1" protective course

Waterproofing

2" base course

Expansion joint — Waterproof

Brick protective course

Scale ⅜"=1'-0"

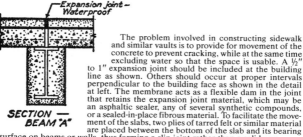

Expansion joint — Waterproof

SECTION — BEAM "A"

The problem involved in constructing sidewalk and similar vaults is to provide for movement of the concrete to prevent cracking, while at the same time excluding water so that the space is usable. A ½" to 1" expansion joint should be included at the building line as shown. Others should occur at proper intervals perpendicular to the building face as shown in the detail at left. The membrane acts as a flexible dam in the joint that retains the expansion joint material, which may be an asphaltic sealer, any of several synthetic compounds, or a sealed-in-place fibrous material. To facilitate the movement of the slabs, two plies of tarred felt or similar material are placed between the bottom of the slab and its bearing surface on beams or walls, thus forming a slip joint rather than a solid one.

543

WATERPROOFING OF
DEEP FOUNDATIONS

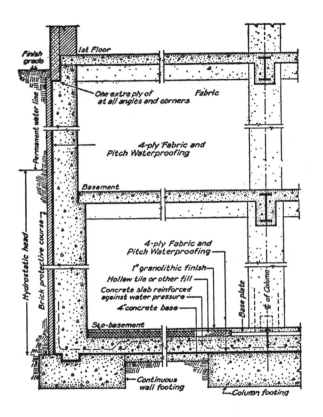

Finish grade

1st Floor

Permanent water line

One extra ply of
at all angles and corners

Fabric

4-ply Fabric and
Pitch Waterproofing

Basement

Hydrostatic head

Brick protective course

4-ply Fabric and
Pitch Waterproofing

1" granolithic finish

Hollow tile or other fill

Concrete slab reinforced
against water pressure

4" concrete base

Sub-basement

Base plate

C of Column

Continuous
wall footing

Column footing

Shown here is a traditional (and effective) waterproofing method used in the construction of deep basements that have unfavorable water conditions. A four-ply fabric and pitch membrane will withstand a hydrostatic head of up to 9'. For heads from 9' to 12', five plies are used. A sheet of 20-ounce soft rolled copper should be placed between the plies beneath all columns, posts, or walls under which the pressure exceeds 400 pounds per square inch. The fabric may be any of several felt compositions or may be fiberglass. As an alternative, flexible membranes such as neoprene or butyl applied with special sealers may be used.

The subbasement floor is reinforced as a flat slab to resist the upward pressure of the water. Inverted beams may be used, but these necessitate more fill to make the construction level so that it can properly receive the wearing surface. If plies are used, the footings shown above become the pile cappings. No dowels need be run through the membrane.

544

SUBTERRANEAN TUNNEL
TO CONNECT BUILDINGS

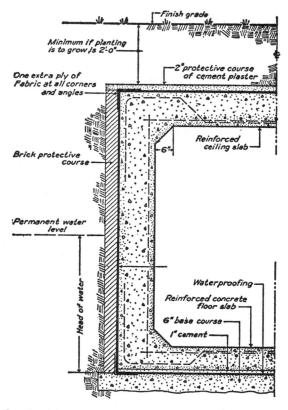

Sometimes it is necessary to construct subterranean tunnels from one building to another, or to construct underground rooms or vaults for storage or other purposes. If the soil will retain groundwater or if the construction extends below the permanent water level, effective waterproofing is necessary; dampproofing will not suffice. An unbroken envelope through which water cannot penetrate is required. A typical installation is shown above. One of several waterproofing membranes (for example, a built-up membrane of fabric and asphalt, a combination of polyethylene sheet and adhesive, sprayed or paneled bentonite, or a flexible membrane of butyl, neoprene, or similar material) may be selected, depending upon conditions of the job.

The thickness and reinforcing of the floor slab depend on the span and the head of water to be resisted. The thickness and reinforcing of the walls depend on their height and the water head. The detail of the roof or ceiling slab depends on the span and the weight of earth to be sustained.

DEPTHS FOR
FOUNDATIONS

The foundation depths conventionally considered safe in various regions are given in the table. However, these figures should be checked with local builders, county agricultural agents, or (particularly) local building office authorities for building code compliance and/or recommendations. Building code regulations may require a foundation depth different than shown here. In addition, safe depth varies to a great extent, depending usually upon the depth to which frost penetrates and upon the effect frost has in the soil. Dry soils ordinarily do not heave when freezing, but damp clay may heave enough to cause serious damage to the building unless its footings are below frost depth.

The depths given in the table are past recommendations made by state agricultural colleges and may require amending in the event of new data. The depths are considered sufficient to prevent damage by frost but are not the total depths to which frost penetrates. Notice the soil conditions at these depths; if soil is not firm or if it is subject to change in volume due to alternate wetting and drying, footings must be made wider, given additional reinforcement, or carried deeper than indicated in the table.

In regions having little frost, footings should be set below the topsoil on firm ground; if they are placed too close to the surface, animals can burrow under them, and over time wind, rain, or floods may erode the soil from beneath, causing the building to settle. In some localities the firm soil is a relatively thin layer overlaying soft ground. If the firm soil is cut through, a secure bearing is almost impossible. Under such conditions, shallow footings may be protected from erosion by banking soil against the foundations. This fill requires sodding and providing rain gutters, downspouts, and a proper foundation drainage system as protection from erosion caused by drip from the roof. Typically, model building codes require minimum footing depths of 12″ for a single-story building, 18″ for a two-story building, and 24″ for a three-story building. They also require that all footings for any type of building be below frost level; however, this regulation is frequently modified to allow footings to be placed at shallower levels than the extreme frost depth. This is particularly true in the case of small outbuildings, such as detached garages, or in situations in which the potential frost depth far exceeds practical footing depth.

All footings of buildings are best set upon the same type of soil. They must be level but need not be at the same elevation. Where the ground slopes or where there is a basement under only a portion of the building, step the footing down gradually to avoid undermining the higher portion.

DEPTHS FOR FOUNDATIONS

The ratio at which the stepping can be done safely varies with the type of soil, but for average conditions a vertical rise of not more than 2'-0" in a horizontal distance of 4'-0" is generally satisfactory. In the case of sloping ground surfaces, building codes typically require that the foundation top and bottom be level and continuous or level and stepped.

When one part of the foundation rests on rock and another on soil, the footing of the portion on soil should be made at least twice as wide as is called for by the normal soil bearing area. Under such circumstances, some building codes require the rock surface to be cut so that a 6" layer of sand can be placed on top of the rock. Occasionally a relatively thin rock stratum overlays soft clay or loose sand; such a bed is unsafe for heavy buildings or concentrated pier loads. Care must be taken to see that the rock is not merely a large boulder that might be loosened by the weight of the building.

When the rock stratum slopes, the surface may be cut to form level steps to prevent the footing from sliding. Sometimes slight slopes are merely heavily chipped; at times the surfaces are doweled. Where outcroppings of rock strata have been exposed to long-term weathering and the surfaces are likely to be rotten or loose, cut the rotten layers away to reach solid material.

In the case of expansive soil conditions, special designs and precautions may have to be undertaken in order to devise a footing and foundation system that will withstand the forces of the moving soil and prevent structural damage to the building.

State	Mild Areas	Colder Areas	Local Consideration
Alabama	1'-6"	1'-6"	Reinforce footings and floor, and use piles in Blackbelt area
Arizona	1'-6"	3'-0"	Closeness of irrigation a factor
Arkansas	1'-4"	1'-4"	Continuous foundations preferred
California	0'-6"—1'-0"	1'-6"—2'-0"	————————
Connecticut	2'-0"—3'-6"	2'-6"—4'-0"	————————
Florida	surface	0'-6"—1'-0"	Wide footings near surface; sandy soil
Georgia	———	———	Conditions variable; seek local advice

DEPTHS FOR FOUNDATIONS

State	Mild Areas	Colder Areas	Local Consideration
Idaho	2'-0"	3'-0"	
Illinois	3'-6"	5'-6"	Reinforcement advised
Indiana	2'-0"—3'-0"	2'-0"—3'-0"	
Iowa	3'-0"	3'-6"	
Kansas	5'-0"	5'-0"	Reinforce; heavy footings needed on swelling and shrinkage soils
Kentucky	1'-6"—2'-0"	2'-6"	
Louisiana	0'-2"—1'-0"	0'-2"—1'-0"	Wide footings on alluvial soils
Maine	4'-0"—5'-0"	5'-0"—6'-0"	Use batter on outside face of wall or use a footing; less depth is required in gravelly soils
Maryland	—	—	Conditions variable; seek local advice
Massachusetts	2'-0"—4'-0"	2'-0"—4'-0"	Soil conditions fairly uniform
Michigan	3'-0"	3'-0"	
Minnesota	5'-0"	5'-0"	
Mississippi	Depth to uniform soil	Depth to uniform soil	
Missouri	1'-6"	2'-0"	
Montana	—	—	Seek local advice
Nebraska	1'-6"	2'-0"	Guard against roof water and rooting animals
Nevada	0'-0"—0'-6"	1'-6"	
New Hampshire	6'-0"—8'-0"	6'-0"—8'-0"	
New Jersey	1'-4"	3'-0"	
New Mexico	0'-9"—1'-0"	1'-0"—1'-3"	
New York	4'-0"	4'-0"	
North Carolina	2'-0"	2'-6"—3'-0"	
North Dakota	5'-0"—8'-0"	5'-0"—8'-0"	Reinforce
Ohio	1'-6"	1'-6"	Reinforce
Oklahoma	1'-6"	2'-0"	
Oregon	1'-0"	2'-0"	
Pennsylvania	4'-0"—6'-0"	4'-0"—6'-0"	
South Carolina	1'-2"	1'-6"	
South Dakota	4'-6"—5'-0"	4'-6"—5'-0"	Use continuous foundations
Tennessee	2'-0"	2'-0"	Guard against termites
Texas	1'-0"	2'-6"	Guard against erosion
Utah	1'-8"	2'-6"	
Vermont	5'-0"	5'-0"	Conditions vary widely; carry to firm soil
Virginia	2'-0"	2'-0"	
Washington	—	—	Conditions variable; seek local advice
West Virginia	1'-6"—2'-0"	2'-0"—2'-6"	
Wisconsin	3'-0"	4'-0"	

HOW MANY RISERS?

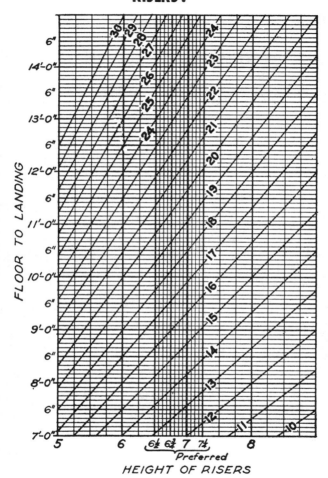

HEIGHT OF RISERS

Read across on the proper horizontal line to whichever intersection with a slanting line showing the number of risers gives a desirable riser height. For example a 10'-0" story height requires 17 risers at approximately 7 1/16". Twelve risers of this same height would give 7'-1" from floor to an intermediate landing, for determining head room, or thirteen would give 7'-8".

PROPORTIONING
RISERS AND TREADS

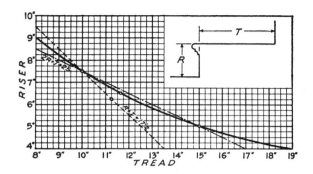

The graph shows a comparison between the two best known stair "laws" and a curve based on actual stairs that have been tested for comfort.

Because it is the easiest to use, the rule that the sum of a riser and a tread should equal 17½ has been generally adopted. Actual testing, however, demonstrates that this rule gives steps that are too large for steep stairs and steps that are too small for gradual stairs.

The rule that the sum of two risers and one tread should equal 25 results in very good stairs when the risers are less than 7½". But for steeper stairs than this, the steps become too small for comfort.

The curve shown by the heavy solid line follows very closely the old rule that the product of a riser and a tread should equal 75. This curve gives a proportion of riser and tread that is most comfortable for the resulting angle of climb. By reference to the chart it will be seen that a 6½-11⅝ proportion is better than a 6¼-11¼ proportion, although both occupy approximately the same space in plan and section.

The following table gives the correct proportion of riser and tread, as determined from the recommended curve:

Riser	Tread	Riser	Tread	Riser	Tread
4	19	5	15	6	12⅝
4⅛	18¼	5⅛	14⅝	6⅛	12⅜
4¼	17⅝	5¼	14⅜	6¼	12⅛
4⅜	17	5⅜	14	6⅜	11⅞
4½	16⅝	5½	13¾	6½	11⅝
4⅝	16⅛	5⅝	13⅜	6⅝	11½
4¾	15¾	5¾	13⅛	6¾	11¼
4⅞	15⅜	5⅞	12⅞	6⅞	11
7	10¾	8	9¼	9	8
7⅛	10⅝	8⅛	9⅛		
7¼	10⅜	8¼	9		
7⅜	10¼	8⅜	8¾		
7½	10	8½	8⅝		
7⅝	9⅞	8⅝	8½		
7¾	9⅝	8¾	8⅜		
7⅞	9½	8⅞	8⅛		

550

STAIRWAY
HEADROOM

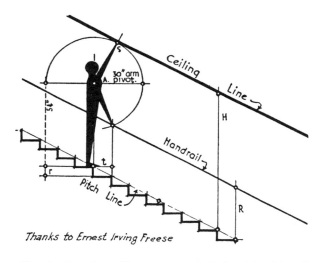

Thanks to Ernest Irving Freese

The drawing above illustrates a method for determining the headroom and handrail heights on stairways, as published in the British *Architect's Journal,* and reprinted here by permission. The height of the arm pivot is taken as 4'-6" for an average person. Using this as a center, a circle of 2'-6" radius is described. The ceiling or line of head clearance should be not lower than the tangent (parallel to the stair slope) of this circle.

The handrail is found by erecting a perpendicular from the nosing of the next lower tread, the tread representing the length of stride of a person ascending or descending the stairs. Where this perpendicular intersects the arm circle, a point is established determining the handrail height.

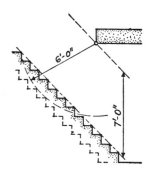

At the lower left is a composite drawing showing two other methods of arriving at the headroom. Using a 7'-0" height for a line parallel to the stair slope does not give an entirely satisfactory clearance on steep stairs.

Another rule is to use a circle of 6'-0" radius from the lowest point to be cleared, with the stair slope tangent to the circle. This provides considerably more headroom than the 7'-0" parallel scheme, on steep stairs, as shown by the drawing. On very gradual stairs it would be unsatisfactory.

551

HEADROOM AND HANDRAIL
HEIGHTS FOR STEPLADDERS

For stepladders (with or without handrails) to trap doors, fire escapes, fly galleries, belfries, attics, etc. If risers are not open, handrails should be provided on both sides for safety. Maximum ladder width 2'-0", minimum 1'-9". Without handrails risers must be left open for hand holds. Building codes should be consulted.

The Workmen's Compensation Bureau considers ladders to be steps or rungs inclining from 50° to 90° with the horizontal, the preferred angle being from 75° to 90° with the horizontal. Within this range, step- or ship's ladders are preferred for angles from 50° to 75°, and rung ladders for steeper angles.

The handrail heights given in the table are determined by the intersection of the arm circle with a vertical from the stair nosing as shown. The headroom heights are determined by a tangent to the arm circle, which is parallel to the stair slope.

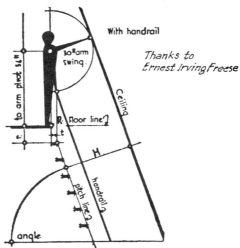

With handrail

*Thanks to
Ernest Irving Freese*

Riser r in Inches	Tread t in Inches	Angle in Degrees and Minutes	Headroom H in Inches	Handrail R in Inches
9⅜	7½	51.21	64	34½
9¾	7	54.19	62	34½
10⅛	6½	57.18	59	35
10½	6	60.16	57	35
10⅞	5½	63.10	54	35½
11¼	5	66.2	52	35½
11⅝	4½	68.50	50	36
12	4	71.34	47	36
12⅜	3½	74.12	45	36½
12¾	3	76.46	42	37

WOOD STAIR
CONSTRUCTION

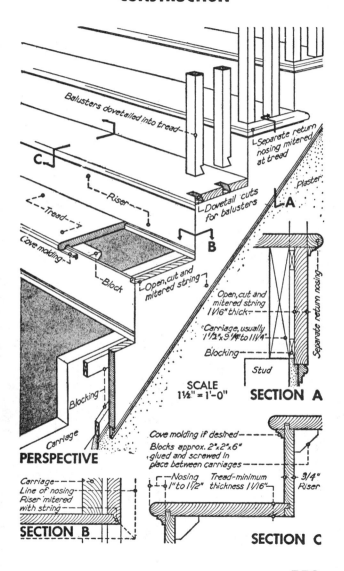

Balusters dovetailed into tread

Separate return nosing mitered at tread

Plaster

C

Riser

Dovetail cuts for balusters

A

Tread

B

Cove molding

Block

Open, cut and mitered string

Open, cut and mitered string 1¹/₁₆" thick

Carriage, usually 1½"x9¼"to 11¼"

Blocking

Separate return nosing

Stud

SCALE
1½" = 1'-0"

SECTION A

Cove molding if desired

Blocks approx. 2"x2"x6" glued and screwed in place between carriages

Nosing 1" to 1½"   Tread-minimum thickness 1¹/₁₆"

3/4" Riser

Blocking

Carriage

PERSPECTIVE

Carriage
Line of nosing
Riser mitered with string

SECTION B

SECTION C

CLOSED STRING
STAIRS

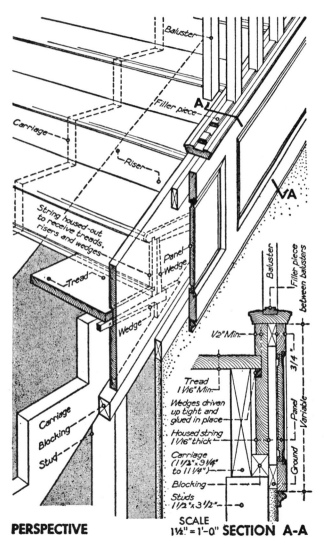

Baluster

Filler piece

A

Carriage

Riser

String housed-out
to receive treads,
risers and wedges.

Tread

Panel

Wedge

Wedge

A

Carriage

Blocking

Stud

PERSPECTIVE

Baluster

Filler piece
between balusters

1/2" Min.

Tread
1 1/16" Min.

Wedges driven
up tight and
glued in place

Housed string
1 1/16" thick

Carriage
(1 1/2" x 9 1/4"
to 11 1/4")

Blocking

Studs
1 1/2" x 3 1/2"

3/4

Panel — Variable

Ground

SCALE
1 1/2" = 1'-0" **SECTION A-A**

CONCRETE STAIR
CONSTRUCTION

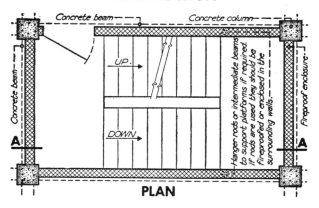

Concrete beam — Concrete column —

Concrete beam —

UP.

DOWN

Hanger rods or intermediate beams to support platforms if required. If rods are used they should be fireproofed or enclosed in the surrounding walls.

Fireproof enclosure —

A   A

PLAN

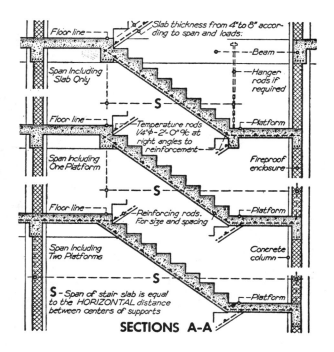

Floor line — — —

Slab thickness from 4" to 8" according to span and loads.

— Beam

Span Including Slab Only

— Hanger rods if required

S

Floor line — — —

Temperature rods ¼"φ–2'-0" %c at right angles to reinforcement—

— Platform

Span Including One Platform

Fireproof enclosure

S

Floor line — — —

Reinforcing rods. For size and spacing

— Platform

Span Including Two Platforms

Concrete column —

S

S - Span of stair slab is equal to the HORIZONTAL distance between centers of supports

— Platform

SECTIONS A-A

FINISHES FOR
CONCRETE STAIRS

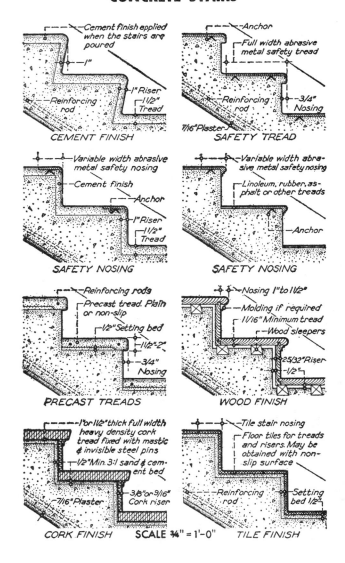

Cement finish applied when the stairs are poured

1"

1" Riser

1½" Tread

Reinforcing rod

CEMENT FINISH

Anchor

Full width abrasive metal safety tread

Reinforcing rod

¾" Nosing

7/16" Plaster

SAFETY TREAD

Variable width abrasive metal safety nosing

Cement finish

Anchor

1" Riser

1½" Tread

SAFETY NOSING

Variable width abrasive metal safety nosing

Linoleum, rubber, asphalt or other treads

Anchor

SAFETY NOSING

Reinforcing rods

Precast tread. Plain or non-slip

½" Setting bed

1½"–2"

¾" Nosing

PRECAST TREADS

Nosing 1" to 1½"

Molding if required

11/16" Minimum tread

Wood sleepers

25/32" Riser

½"

WOOD FINISH

1" or 1½" thick full width heavy density cork tread fixed with mastic & invisible steel pins

½" Min. 3:1 sand & cement bed

7/16" Plaster

3/8" or 9/16" Cork riser

CORK FINISH

Tile stair nosing

Floor tiles for treads and risers. May be obtained with non-slip surface

Reinforcing rod

Setting bed ½"

TILE FINISH

SCALE ¾" = 1'-0"

MARBLE STAIR
OVER STEEL

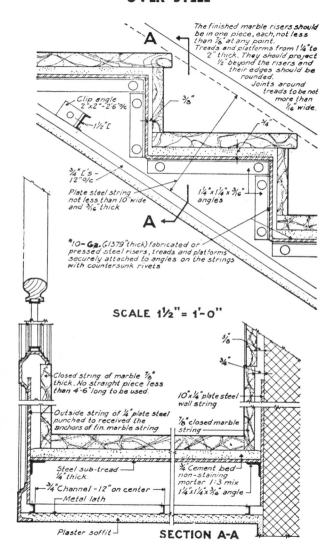

The finished marble risers should be in one piece, each, not less than 7/8" at any point.
Treads and platforms from 1¼" to 2" thick. They should project ½" beyond the risers and their edges should be rounded.
Joints around treads to be not more than 1/16" wide.

A

Clip angle 2"x2"-2'-6"%

1½" [

3/8"

3/4"

¾" ['s - 12"%c

Plate steel string not less than 10" wide and 3/16" thick

1¼"x1¼"x3/16" angles

A

*10-Ga. (.1379"thick) fabricated or pressed steel risers, treads and platforms securely attached to angles on the strings with countersunk rivets

SCALE 1½"= 1'-0"

Closed string of marble 7/8" thick. No straight piece less than 4'-6" long to be used.

Outside string of ¼" plate steel punched to received the anchors of fin. marble string

Steel sub-tread ¼" thick.

¾" Channel - 12" on center

Metal lath

Plaster soffit

5/8"

3/4"

10"x¼" plate steel wall string

7/8" closed marble string

¾" Cement bed non-staining mortar 1:3 mix

1¼"x1¼"x3/16" angle

SECTION A-A

MARBLE STAIR
OVER CONCRETE

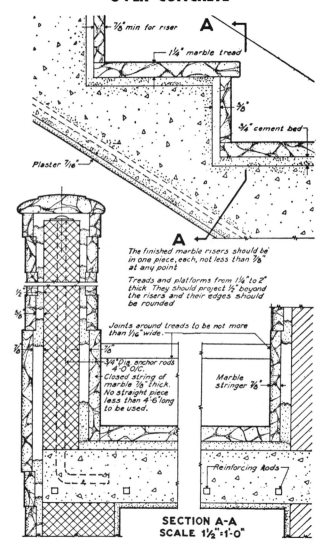

7/8" min for riser

1¼" marble tread

5/8"

¾" cement bed

Plaster 7/16"

A

The finished marble risers should be in one piece, each, not less than 7/8" at any point

Treads and platforms from 1¼" to 2" thick. They should project ½" beyond the risers and their edges should be rounded

Joints around treads to be not more than 1/16" wide.

½

5/8

7/8

7/8"

¾" Dia. anchor rods 4'-0" O/C.
Closed string of marble 7/8" thick.
No straight piece less than 4'-6" long to be used.

Marble stringer 7/8"

Reinforcing Rods

SECTION A-A
SCALE 1½":1'-0"

TYPICAL ALL-STEEL STAIR SECTIONS
WITH BRICK ENCLOSING WALLS

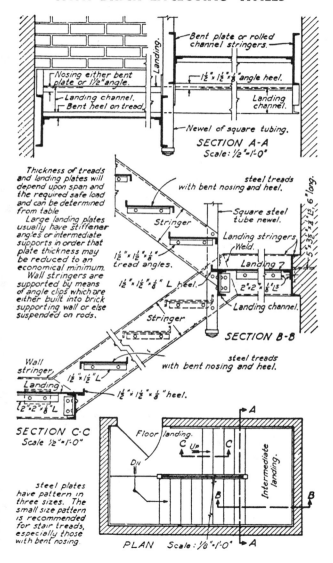

Nosing either bent plate or 1/2" angle.

Landing channel.

Bent heel on tread.

Bent plate or rolled channel stringers.

1 1/2" × 1 1/2" × 3/8" angle heel.

Landing channel.

Newel of square tubing.

SECTION A-A
Scale: 1/2"=1'-0"

Thickness of treads and landing plates will depend upon span and the required safe load and can be determined from table

Large landing plates usually have stiffener angles or intermediate supports in order that plate thickness may be reduced to an economical minimum.

Wall stringers are supported by means of angle clips which are either built into brick supporting wall or else suspended on rods.

steel treads with bent nosing and heel.

Stringer

1 1/2" × 1 1/2" × 3/8" tread angles.

1 1/2" × 1 1/2" × 3/8" L heel.

Stringer

Square steel tube newel.

Landing stringers.

Weld.

Landing

2" × 2" × 1/8" Ls

Landing channel.

5" × 3 3/4" × 3/8", 6" long.

SECTION B-B

Wall stringer.

Landing

1/2" × 1 1/2" L

1 1/2" × 1 1/2" × 3/8" heel.

1 1/2" × 2" × 3/8" L

steel treads with bent nosing and heel.

SECTION C-C
Scale 1/2"=1'-0"

steel plates have pattern in three sizes. The small size pattern is recommended for stair treads, especially those with bent nosing.

Floor landing.

C Up C

Dn

B

Intermediate landing.

B

A

PLAN   Scale: 1/8"=1'-0"   A

559

STAIR TREADS OF
STEEL FLOOR PLATES

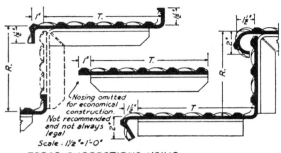

TREAD SUGGESTIONS USING
STEEL PLATE WITH BENT NOSING AND HEEL

The use of checkered steel plate treads and landings in stair construction is both safe and economical because the slip-proof projections extend in two directions, at right angles to each other, and are so arranged that the plates may be easily cleaned and drained. Plates are cut to size and bent to form nosing and heel by the stair contractor.

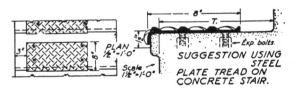

SUGGESTION USING STEEL PLATE TREAD ON CONCRETE STAIR.

Treads with turned down nosing make an ideal wearing and slip-proof surface for concrete stairs. They are easily secured, either by means of expansion bolts or anchors set in concrete when poured.

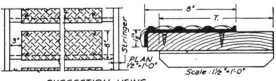

SUGGESTION USING
STEEL PLATE TREAD ON WOOD STAIR.

Wood treads can be made to last for a long time when protected with the hard wearing surface of steel plate. Treads which have been worn down should be blocked out to a true level before steel plate is applied, otherwise the wood screws may work loose.

FINISHES FOR
STEEL STAIRS

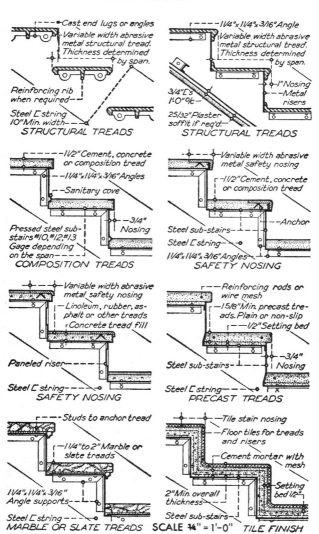

Cast end lugs or angles
Variable width abrasive metal structural tread. Thickness determined by span.
Reinforcing rib when required
Steel ⊏ string 10" Min. width
STRUCTURAL TREADS

1 1/4"x 1 1/4"x 3/16" Angle
Variable width abrasive metal structural tread. Thickness determined by span.
1" Nosing
Metal risers
3/4" ⊏'s 1'-0" o/c
25/32" Plaster soffit if req'd
STRUCTURAL TREADS

1 1/2" Cement, concrete or composition tread
1 1/4"x 1 1/4"x 3/16" Angles
Sanitary cove
3/4" Nosing
Pressed steel sub-stairs #10,#12,#13 Gage depending on the span
COMPOSITION TREADS

Variable width abrasive metal safety nosing
1 1/2" Cement, concrete or composition tread
Anchor
Steel sub-stairs
Steel ⊏ string
1 1/4"x 1 1/4"x 3/16" Angles
SAFETY NOSING

Variable width abrasive metal safety nosing
Linoleum, rubber, asphalt or other treads
Concrete tread fill
Paneled riser
Steel ⊏ string
SAFETY NOSING

Reinforcing rods or wire mesh
1 5/8" Min. precast treads. Plain or non-slip
1/2" Setting bed
3/4" Nosing
Steel sub-stairs
Steel ⊏ string
PRECAST TREADS

Studs to anchor tread
1 1/4" to 2" Marble or slate treads
1 1/4"x 1 1/4"x 3/16" Angle supports
Steel ⊏ string
MARBLE OR SLATE TREADS

Tile stair nosing
Floor tiles for treads and risers
Cement mortar with mesh
Setting bed 1/2"
2" Min. overall thickness
Steel sub-stairs
SCALE 3/4" = 1'-0"   **TILE FINISH**

561

PIPE
RAILINGS

Fittings for pipe railing systems are made from galvanized malleable iron, brass, chrome-plated steel, aluminum, stainless steel, or aluminum/magnesium, for use with appropriate piping. Both slip-on and threaded fittings are produced in variable conformations. Drawings and dimensions given here are for a typical galvanized iron railing system; consult specific manufacturers for details and specifications.

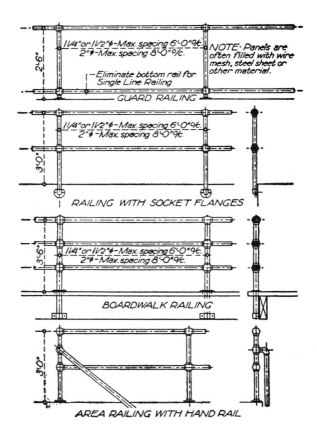

NOTE: Panels are often filled with wire mesh, steel sheet or other material.

GUARD RAILING

RAILING WITH SOCKET FLANGES

BOARDWALK RAILING

AREA RAILING WITH HAND RAIL

PIPE
RAILINGS

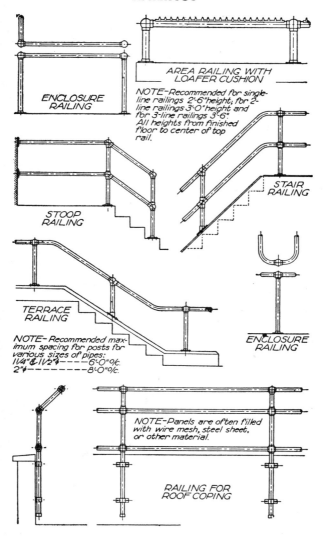

ENCLOSURE
RAILING

AREA RAILING WITH
LOAFER CUSHION

NOTE—Recommended for single-line railings 2'-6" height; for 2-line railings 3'-0" height; and for 3-line railings 3'-6". All heights from finished floor to center of top rail.

STOOP
RAILING

STAIR
RAILING

TERRACE
RAILING

NOTE—Recommended maximum spacing for posts for various sizes of pipes:
1¼"&1½"⌀————6'-0"%c.
2"⌀————————8'-0"%c.

ENCLOSURE
RAILING

NOTE—Panels are often filled with wire mesh, steel sheet, or other material.

RAILING FOR
ROOF COPING

PIPE RAIL FITTINGS

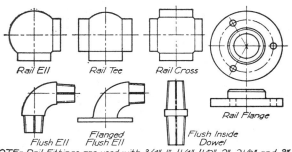

Rail Ell   Rail Tee   Rail Cross   Rail Flange

Flush Ell   Flanged Flush Ell   Flush Inside Dowel

NOTE- Rail Fittings are used with 3/4", 1", 1 1/4", 1 1/2", 2", 2 1/2", and 3" pipes. Flush Fittings are used with 1 1/4" and 1 1/2" pipes.

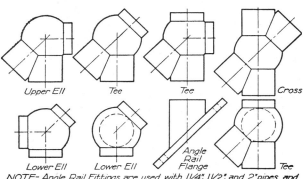

Upper Ell   Tee   Tee   Cross

Lower Ell   Lower Ell   Angle Rail Flange   Tee

NOTE- Angle Rail Fittings are used with 1 1/4", 1 1/2", and 2" pipes and for all angles from 27 1/2° to 46 1/2°

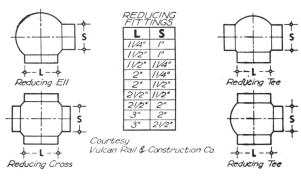

Reducing Ell   Reducing Tee

Reducing Cross   Reducing Tee

REDUCING FITTINGS

L	S
1 1/4"	1"
1 1/2"	1"
1 1/2"	1 1/4"
2"	1 1/4"
2"	1 1/2"
2 1/2"	1 1/2"
2 1/2"	2"
3"	2"
3"	2 1/2"

Courtesy
Vulcan Rail & Construction Co.

HOW PORCELAIN ENAMEL
PANELS ARE MADE

MANUFACTURING PROCEDURE. Numerous steps are required to produce a finished porcelain enamel installation; they are somewhat variable, depending upon the nature of the product and the demands of the specific application. Some manufacturers perform only one or two of the steps, while others undertake many or all of them.

1. Manufacture of the frit.
2. Formulation of the ground coat material.
3. Formulation of the cover coat material.
4. Manufacture of the metal enameling sheets.
5. Manufacture of the core material.
6. Manufacture of the backing material.
7. Production of shop drawings, as required, for various stock and custom-designed products.
8. Forming of the panels, pieces, or stock bulk-material coils.
9. Cleaning of the panels or sheet stock.
10. Ground-coating of the panels, pieces, or sheets.
11. Cover-coating of the panels, pieces, or sheets.
12. Coring/backing of the panels, pieces, or sheets, as required.
13. Additional processing/manufacturing, as required.
14. Job installation of the finished pieces.

MANUFACTURE OF THE FRIT. Two types of frit coatings are required. Both consist of an opaque glass that is composed entirely from minerals and has no organic ingredients.

Frit for the *ground coat* contains *adherence oxides* that are so compounded as to have an affinity for the metal and will actually fuse with it. The mineral ingredients for the ground coat are melted together at temperatures of up to 2,500°F. When suddenly immersed in water, the material is shattered by thermal action into light, airy particles called *frit*.

Frit for the cover coat is produced in a similar manner, but the composition is different. Various combinations of minerals, such as feldspar (aluminum silicate), cryolite (soidum aluminum fluoride), and fluorspar (calcium fluoride) may be used. Feldspar has been widely used in making artificial teeth and opalescent glass. Cryolite produces opacity in glass.

GRINDING OF THE FRIT. The frit is ground with plain water and clays to make a liquid that passes a 200-mesh screen, comparable in fineness to flour or face powder. The mixture for cover coats may contain color pigments in the form of mineral oxides. When prepared for use, the aqueous mixture is about the consistency of thin cream.

ENAMELING SHEETS. Sheets of special composition have all the metallurgical properties necessary for fine porcelain enamel work; they are generally made to particular manufacturers' specifications in order to be best suited for their particular products. These sheets are very different from the usual sheet iron or sheet steel, being uniform in composition, absolutely flat, and possessed of the ductility required for forming, stamping, drawing, and so on. The composition is such as to provide a tenacious bond between the porcelain enamel and the metal when the porcelain is fired. For unbacked panels or other items, 16- or 18-gauge sheets are used. The standard gauge for backed and cored panels is 28, but 20- and 24-gauge sheets are optionally available.

SHOP DRAWINGS. There are many methods of attachment for both standard veneer and insulated panels, as well as for plain and custom-designed panels, whether for securing to facades, storefronts, interior walls, curtain walls, or window walls, or as add-ons to other materials. Some systems are patented and others are not. Some require certain forming of panel edges. Attempting to detail the attachment of every panel before taking figures may automatically and unnecessarily restrict the architect's options.

For work that requires competitive bidding, it is usual for the architect to indicate on the relevant drawings only the design, dimensions, desired joining, colors, and similar general specifications, including provisions for metal or wood furring strips if required. Each bidder should be required to submit a sample showing the proposed attachment system. The specifications should also require that the successful contractor furnish complete shop drawings showings the contractor's recommended jointing and attachment methods for the architect's approval.

HOW PORCELAIN ENAMEL
PANELS ARE MADE

FORMING THE PANELS. If bulk coils of enameled metal are being made up for future remanufacture, no preliminary fabrication is needed. Standard veneer or insulated panels (or other products and custom designs) are sheared, bent, drawn, punched, formed, or otherwise processed as required before being enameled. Fluting, reeding, louvers, or other special forming may be needed. Flanges, clips, and pieces for installation connections are welded in place and filed smooth. Holes or cutouts are best made at this time, though special hole-saw blades chucked in a portable electric drill can be used if necessary at the job site; raw edges should be painted for protection.

The fabricated sheets are chemically washed, rinsed, and dried to remove all traces of dirt, grease, scale, and so forth. The chemical action prepares the surface of the sheet for proper adhesion and fusion of the ground coat of enamel.

ENAMELING THE PANELS. The metal is dipped or sprayed uniformly on all surfaces with the liquid mixture of frit. The sheets are then fired at about 1,500°F, which fuses the frit. A glasslike ground coating is created, which is bonded intimately with the metal.

For monochrome panels, a single cover coat is standard, though two thin cover coats may be applied to achieve a superior finish. The cover coat is applied in the same manner as the ground coat, but the firing temperature is a bit lower in this and any subsequent coatings (such as for a polychrome design) so that the ground coat will not be loosened from the metal and so that any other previous coatings will likewise not be disturbed.

In panels of porcelain enamel done in polychrome, the lightest color in the design generally must be applied over the entire piece. After firing, the next lightest color is dipped or sprayed on and dried. A stencil is then imposed over the piece, and the unwanted areas are brushed off. The piece is fired and the process repeated until all the colors of the design have been applied.

BACKING OF THE SHEETS. Backed sheets may be made in a number of ways. The usual arrangement with veneer panels is to core the porcelainized metal face with a thin ($^3/_{16}$", ¼", or $^5/_{16}$") layer, called a *stabilizer*, of oil-tempered hardboard or cement-asbestos board. This is then covered with a back, which typically is paintable 25-gauge steel, 0.015" prime-painted aluminum, porcelain-enameled steel the same as the face, or random porcelain-enameled steel. Insulated panels are made differently. The porcelainized face is bonded to a face stabilizer of hardboard or cement-asbestos board. A suitably thick layer of rigid insulation, generally extruded polystyrene, perlite, fiberglass, polyurethane, or polyisocyanurate, is bonded to the face stabilizer. Then a back stabilizer of hardboard or cement-asbestos board may be bonded to the insulation, and a back applied to complete the sandwich panel.

Panels or pieces of other than standard manufacture may be similarly backed with hardboard, plywood, rigid insulation, or various combinations of these and other materials. Panels may also be sprayed with various materials for sound-deadening purposes.

JOB INSTALLATION. Labor experienced in the installation of porcelainized panels should be used when available. In lieu of that, skilled and experienced mechanics should be employed—never unskilled or even semiskilled labor. The craft from which the labor should be selected will vary with the nature of the job, the attachment method of the panels, and local building codes and labor regulations.

Porcelain enamel panels are often erected by a subcontractor under a general contract, from the enameler's shop drawings. Sometimes installation is done by the enameler or the enamel jobber.

Particular attention should be paid to the installation of the first or lower course of panels, if this course will affect all that follow. Where caulking is required, it should be forced into the joints to the full depth of the caulking recess, without gaps or air bubbles. Smears of caulk can be removed with an appropriate solvent (none of which affect the porcelain enamel), and soap and water is usually sufficient for general cleaning up.

CONTOUR OF SURFACES
FOR PORCELAIN ENAMEL

FABRICATING COMPLEX PARTS. Any shape which can be made in sheet metal by rolling, stamping, braking, spinning, or cutting and welding, can be porcelain enameled. If there is a repetition of special parts sufficient to justify the making of dies, the price of a complicated piece may be brought to an economical level.

Some manufacturers have stock dies for certain contours. The designer would do well to consult the manufacturer with whom he is working, to determine the dimensions of such pieces. Die stamped pieces are easier to enamel satisfactorily than welded pieces. It often happens that a manufacturer can suggest a small change in dimensions to make possible a more economical method of fabrication (and erection) than if the original design were insisted upon. It should be noted, however, that since the industry is not standardized, one manufacturer may be at variance with another as a result of differences in their equipment for fabrication. It will often be best to take figures on the original design and then call in selected bidders for their advice after figures are in.

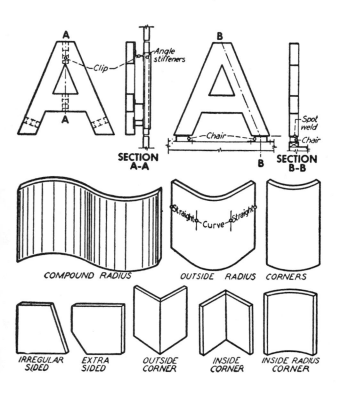

CONTOUR OF SURFACES
FOR PORCELAIN ENAMEL

BULLNOSE PANEL

OUTSIDE CORNER BULLNOSE

RETURN BULLNOSE

RADIUS CORNER BULLNOSE

FLUTED PANELS AND PILASTERS

  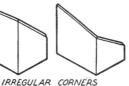

IRREGULAR CORNERS

FLUTED COLUMNS

IRREGULAR WITH CUT-OUT

COPING PANEL

SINGLE OFFSET PANEL

SPECIAL ORNAMENTS

LOUVERED PANEL

EMBOSSED PANEL

METHODS OF ATTACHING
PORCELAIN SHEETS

ATTACHMENT SYSTEMS IN GENERAL. There are numerous methods and systems available for attaching porcelain enamel panels to create a finished interior or exterior wall surface. The system used depends upon the nature of the wall structure (facade, storefront, interior, curtain, or window, for example), the type of panel or sheet (unbacked, veneer, or insulated), and the recommendations of the manufacturer of the panels. Some veneer panels are designed to be mounted in "standard" extruded glazing moldings, and various methods are used to mount insulating panels. Construction details of various representative attachment systems are shown on pages 566-575, primarily for unbacked and veneer sheets.

PROVISION FOR FASTENING. Porcelain enamel panels can be installed over any type of rough wall or wall frame, be it wood frame, steel skeleton with or without curtain walls, or masonry of any type, either in new construction or as a retrofit. It is necessary to provide some means of receiving the screws from the selected attachment device and holding the screws firmly. If furring strips are to be used over masonry, wood strips should be built into the joints so that furring strips can be securely nailed to the rough wall. The wood furring strip is a widely used and effective method of receiving the attachment screws. Usually, a furring strip must be provided beneath each joint; additional strips should be introduced and so spaced that the material will be adequately supported from behind, as recommended. In some cases both horizontal and vertical furring strips will be required.

FLAT SHEETS. In Figure 1A is shown the meeting of two plain flat sheets. The use of flat sheets is generally confined to interior surfaces, and is a seldom-used system because the sheets do not have the rigidity of flanged plain sheets, or of veneer or insulated panels. A successful plain flat sheet installation depends upon the installer's observing one or more of three possible precautions:

1. Use sheet of relatively smaller area.
2. Use metal of relatively heavier gauge.
3. Install backing of rigid insulation or plywood.

SHEETS WITH FORMED EDGES. A group of typical methods of flanging or forming sheet edges is shown in the figures below. The turning or bending of the edges adds rigidity to the panel, as well as creating a contour that lends itself to a particular attachment method. The joints created by most of these systems allow and/or require pointing or caulking. The caulked joints make the resulting wall weatherproof and provide for the expansion and contraction of the porcelain enamel panels under the natural changes of temperature to be expected in exterior construction. Backing the panels or using prebacked panels is possible with most of the attachment systems.

1A

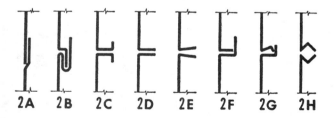

2A   **2B**   **2C**   **2D**   **2E**   **2F**   **2G**   **2H**

MASONRY BACKED SHEETS. A third classification of porcelain enamel sheets for finished wall surfaces consists of masonry-backed sheets. The porcelain enamel sheets and the backing material form an integral unit obtainable in total thicknesses of 1" up to 8" or more. Due to the nature of the masonry backing and the rigidity of the units, the material is installed in exactly the same manner as comparable thicknesses of slate, marble, cut stone, or other masonry would be.

ATTACHMENT OF SHEET AND VENEER PANELS

ATTACHMENT OF SHEETS AND VENEER PANELS. Flat sheets are generally used in interior work, under dry conditions, for decorative purposes; veneer panels may be used for any purpose. For sound deadening or for heat-transfer prevention, these attachment methods allow the use of a rigid insulation if desired. Some are patented, as noted.

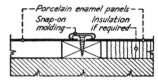

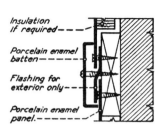

SNAP-ON MOLDINGS. The panels can be removed. Molding covered joints may run horizontally, vertically, or both, depending upon shape and size of panels and their arrangement. Usually, it is only necessary to hold two opposite edges of any flat porcelain enamel sheet panel. The other two edges can be butted and caulked. By caulking under the snap-on molding track and using the snap-on molding over vertical joints, a sanitary interior wall surface can be created, which may be washed down. Not patented.

FLASHING AND BATTENS. This is a flat sheet panel method that can be used for exterior work. The horizontal joints have a small bent piece that acts as flashing. Panels and flashing are screwed into wood furring strips, as shown in the drawing above. The joint is then covered with a batten of porcelain enamel or other metal. The vertical joints are assembled from the back with blind fastenings, so that an entire horizontal course is placed at one time. Not patented.

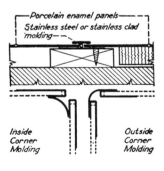

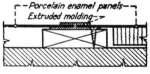

ROLLED MOLDINGS. A patented series of stainless steel moldings is shown above. These moldings allow for the expansion and contraction of the porcelain enamel sheets. An edge trim molding (not shown) completes the system.

EXTRUDED MOLDINGS. A variety of extruded moldings, chiefly of aluminum, can be used to attach flat sheet or veneer panels. The panels are installed progressively with the moldings, as shown. Horizontal or vertical courses can be assembled with blind fastenings so that the moldings only occur in one direction. Some are patented, others not.

ATTACHMENT OF SHEET AND VENEER PANELS

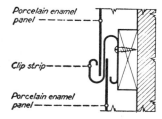

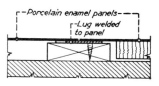

CLIP STRIP. This device is made of stainless steel. Clip strips are screwed to the wall or furring in a horizontal direction, one above the other, the proper distance apart. The top edge of a porcelain enamel sheet is inserted in the deep rear groove of the clip strip. Then, with a downward motion, the bottom edge of the same sheet is slipped into the outer groove of the next lower strip. The ends of the porcelain enamel sheets are overlapped. Holes at the proper intervals in the middle leg of the strip allow the introduction of a screwdriver for attaching the strips to the furring. This is a proprietary product.

WELDED LUG. This is the same system as is used in the lug and pan attachment, as applied to formed edge panels. This method of attachment is not patented. It is an interlocking system, the panels being erected in sequence from the lower left of the facade upward and to the right. Furring strips are placed horizontally and vertically behind the joints. No intermediate strips are needed. This system is adaptable to interior use. With careful installation, the joints can be made very fine.

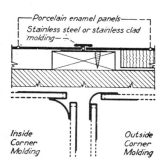

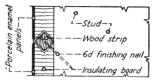

ROLLED MOLDINGS. A patented series of stainless steel moldings is shown above. These moldings allow for the expansion and contraction of the porcelain enamel sheets. An edge trim molding (not shown) completes the system.

VEE CLAMP. Vee clamps may be used on interior or exterior joints. A rigid insulation is cemented as a backing for flat panels. The edge of the insulation is grooved. The lower edge of the panels is fitted over the top edge of a continuous, square wooden strip. At the top of the course another strip of wood is installed and nailed into the studding or furring, and the process is repeated. The joints are subsequently caulked. The vertical joints may be handled in the same way as horizontal joints, thus holding the panels on all four sides. Vertical joints may also be caulked or may be covered with molding. Proprietary method.

PAN-AND-LUG
ATTACHMENT

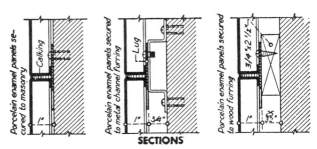

SECTIONS

One of the common methods of attachment for sheets having formed edges is known as the *pan and lug system*. This method of attachment is not patented. The depths of the pans may vary, generally from ½" to 2", with 1" being common. This is an interlocking method, as shown at the lower right—the panels proceeding from the lower left of the facade upward and to the right. If a masonry wall is extremely smooth and plumb, plugs may be used to receive the attachment screws. Metal or wood furring strips make a satisfactory base, and allow shimming to arrive at a planer wall. In *X* and *Y* are shown two variations of the formed edge. The method shown in *X* forms a key for the caulking so that it cannot work out of the joint.

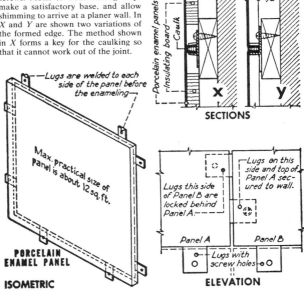

NOTE— On exterior work the joints are filled with caulk. On interior work the joints may be made so narrow that no filler is required.

DETAILS OF PAN-AND-LUG ATTACHMENT

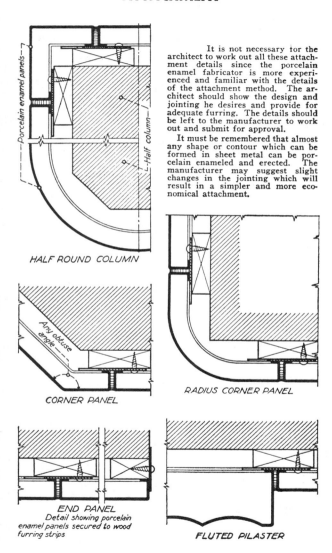

It is not necessary for the architect to work out all these attachment details since the porcelain enamel fabricator is more experienced and familiar with the details of the attachment method. The architect should show the design and jointing he desires and provide for adequate furring. The details should be left to the manufacturer to work out and submit for approval.

It must be remembered that almost any shape or contour which can be formed in sheet metal can be porcelain enameled and erected. The manufacturer may suggest slight changes in the jointing which will result in a simpler and more economical attachment.

HALF ROUND COLUMN

RADIUS CORNER PANEL

CORNER PANEL

END PANEL
Detail showing porcelain enamel panels secured to wood furring strips

FLUTED PILASTER

DETAILS OF PAN-AND-LUG ATTACHMENT

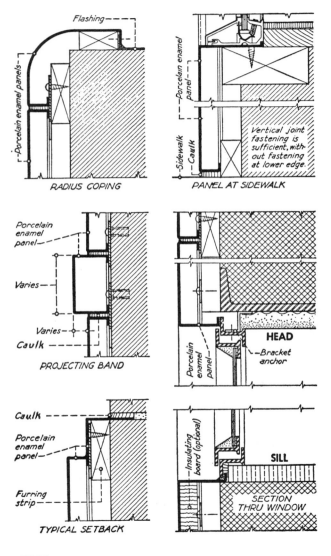

Flashing

Porcelain enamel panels

RADIUS COPING

Porcelain enamel panel

Sidewalk

Caulk

Vertical joint fastening is sufficient, without fastening at lower edge.

PANEL AT SIDEWALK

Porcelain enamel panel

Varies

Varies

Caulk

PROJECTING BAND

Porcelain enamel panel

Bracket anchor

HEAD

Caulk

Porcelain enamel panel

Furring strip

TYPICAL SETBACK

Insulating board (optional)

SILL

SECTION THRU WINDOW

ATTACHMENT METHODS FOR FORMED EDGE PANELS

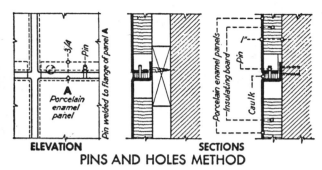

ELEVATION　　　　**SECTIONS**

PINS AND HOLES METHOD

PINS AND HOLES METHOD. The bottom and two sides of the panel have plain flanges. The top edges have turned-up flanges that flash the horizontal joints. This flashing can be made continuous by adding a small piece of light-gauge metal at each vertical joint. The top edge should be screwed 18″ o/c to wood or metal furring strips, or directly to the masonry. Pins in the top flange engage holes in the bottom flange of the next course above. Installation is begun at the bottom and proceeds in vertical rows. Patent rights in doubt.

LAP JOINT. For interior work with exposed fastenings. Panels should be backed if placed where people may lean or push on the panels. Two adjacent edges are formed, the other two being flat. Not patented.

LOCK CLIP. A slot in the edge of the porcelain enamel panel is engaged by the flanges of a clip, as shown below. The panels are placed, the clips are then fastened to masonry plugs or furring strips, and the next panel is placed to engage the free clip leg. Proprietary product.

EXTRUDED MOLDING. Numerous extruded moldings are suitable for various thicknesses of panels, for both interior and exterior use. Panels may be held by the moldings in horizontal joints, vertical joints, or both. Panels can be made to touch where joints have no molding. Pan-shaped panels may be used with or without backing. Some are patented, others not.

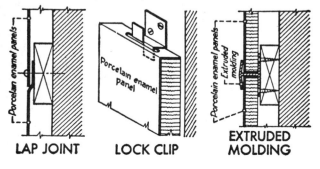

LAP JOINT　　　**LOCK CLIP**　　　**EXTRUDED MOLDING**

ATTACHMENT METHODS FOR FORMED EDGE PANELS

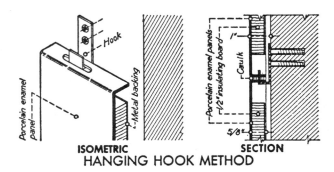

ISOMETRIC
SECTION
HANGING HOOK METHOD

HANGING HOOK.   The flanges of the pan-shaped panels have slotted holes. The hooks, attached to the wall or to furring strips, allow the independent removal of any panel.

VEE CLAMP.   The lower edges of the panels in a horizontal course are fitted over the top edge of the continuous horizontal angle. At the top of the course, another angle is installed and the process is repeated. Angles, strips, or caulking can be used on the vertical joints as desired. Proprietary product.

MASONRY-BACKED PANELS.   Panels are anchored to the lightweight concrete backing. Larger units have steel lifting hooks to facilitate handling. The units are load-bearing and are set on a mortar bed similar to other masonry. Proprietary product.

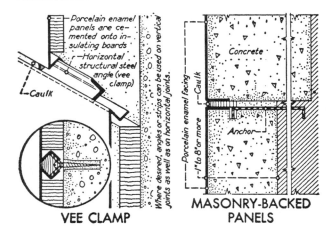

VEE CLAMP
**MASONRY-BACKED
PANELS**

ATTACHMENT METHODS FOR FORMED EDGE PANELS

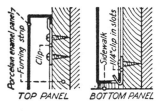

SLOT CLIP METHOD. This method allows removal of any individual panel. The bottom row of panels is attached with a simple hook engaging the slots on the bottom edge. Work proceeds from bottom upward and from left to right or from right to left. Once one panel is in place, adjoining panels are slipped into place so that the clips engage the slots. Clips are then placed in the slots on the exposed edges of the last panel, and the process is repeated. Proprietary product.

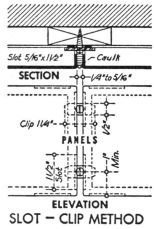

SLOT — CLIP METHOD

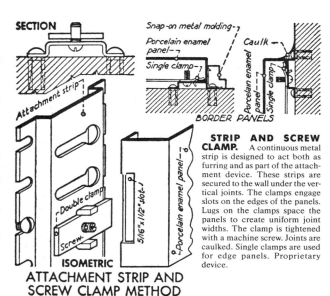

ATTACHMENT STRIP AND SCREW CLAMP METHOD

STRIP AND SCREW CLAMP. A continuous metal strip is designed to act both as furring and as part of the attachment device. These strips are secured to the wall under the vertical joints. The clamps engage slots on the edges of the panels. Lugs on the clamps space the panels to create uniform joint widths. The clamp is tightened with a machine screw. Joints are caulked. Single clamps are used for edge panels. Proprietary device.

577

ATTACHMENT METHODS FOR
FORMED EDGE PANELS

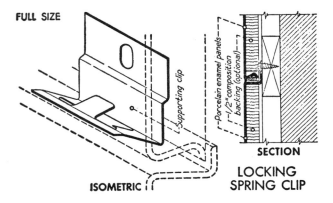

FULL SIZE

ISOMETRIC

SECTION

LOCKING SPRING CLIP

LOCKING SPRING CLIP. Steel furring strips are recommended, although wood furring strips may be used. Bottom clips are applied first; then the panel is placed in position, and the top clips are applied. Clips are applied to metal furring with self-tapping screws and to wood furring with wood screws. The ⅛″ joints are caulked after panels are in place. Caulking is keyed by the shape of the flanges. Each panel is individually suspended, so that any panel may be removed without disturbing the adjacent panels. Proprietary product.

HANGING CLIP. In this method, channels (or wood furring strips) are applied vertically, two to each panel. The upstanding flange on the top edge of a horizontal course of panels is attached to the steel channels with self-tapping screws. The bottom and left-hand edges of the next row of panels have a hanging clip welded to the flange. The panel is dropped over the upstanding flange, engaging the hanging clips. Then the top edge of this panel is again fastened, and the process is repeated. The joint is self-flashing and is caulked. Proprietary device.

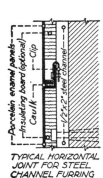

TYPICAL HORIZONTAL
JOINT FOR STEEL
CHANNEL FURRING

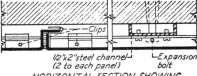

½″x2″steel channel
(2 to each panel)

Expansion
bolt

HORIZONTAL SECTION SHOWING
TYPICAL VERTICAL JOINT

SECTIONS

HANGING CLIP METHOD

578

ATTACHMENT METHODS FOR FORMED EDGE PANELS

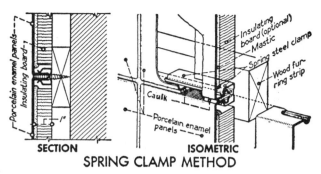

SECTION ISOMETRIC

SPRING CLAMP METHOD

SPRING CLAMP. The "clips" or clamps are attached to the wall or to furring strips, and the flanges of the panels have hollows that engage the spring clamps. Each panel can be independently removed and does not rest or depend on adjacent panels. For interior work, the joints can be made extremely narrow and left without caulking. For exterior work, the joints are caulked. Proprietary method.

SPRING CLIP. Furring is placed on the building at the centerlines of vertical joints and on horizontal joints where needed (around windows and at the edges of walls). Work may start at any point. The clips at the top of each panel act as hangers to support the weight of the panel and also serve as a holding device. Proprietary device.

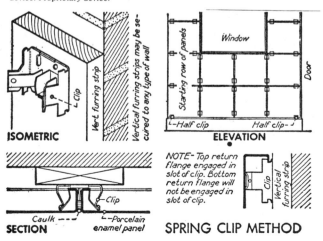

ISOMETRIC ELEVATION

SECTION

NOTE- Top return flange engaged in slot of clip. Bottom return flange will not be engaged in slot of clip.

SPRING CLIP METHOD

PORCELAIN ENAMEL
CHARACTERISTICS AND FINISHES

CHARACTERISTICS. Porcelain enamel is a completely versatile material that can be applied to supporting framework of wood, steel, or any form of masonry. Porcelain enamel sheets are durable and have unusual resistance to abrasion. The material has almost unlimited capacity for shaping into desired surface contours. Being a vitreous material, it is nonporous, nonabsorbent, and as easily cleaned as an enameled kitchen pot. It is light in weight, the finished product usually weighing less than 3 pounds per square foot. Porcelain enamel requires practically no maintenance. The panels are not generally damaged by fire or violent changes in temperature.

USES. Porcelain enamel is particularly well suited to uses where rigid sanitation is a requisite, where the appearance of absolute cleanness is a commercial asset, and where the character of the building demands the high attention value contributed by the color and brilliance of porcelain.

COLOR AND DECORATIVE PATTERNS. Porcelain enamel is essentially an opaque glass. It should not be confused with either brushed or baked organic enamels, which belong in the category of painter's materials. Porcelain enamel is composed entirely from minerals having no organic ingredients. History records no permanent colors except in the field of glass and ceramics. Porcelain enamel offers a complete range of lasting colors, of any value or intensity. The complexity of polychrome designs that can be achieved is limited only by the designer's ingenuity and the building budget available. Stippled effects can be produced readily. Designs of diversified character can be printed by a screen process and fired.

COVER COAT ENAMELS. Cover coats, if subjected to unusually harsh conditions, should be selected on the basis of the specific application involved. Enamel surfaces should withstand the Porcelain Enamel Institute's standard tests for specified properties and classification, and should meet or exceed the Institute's general standards. Where corrosive conditions are encountered, the cover coat must be sufficiently thick and nonporous to repel the attack of acids to which the material will be subjected.

ENAMEL FINISHES. Two general types of finishes may be specified. The first is variously termed *glossy, glaze, lustrous,* or *semi-matte.* This finish is used in applications demanding a brilliant surface and/or a high attention value. The second type of finish is known as *matte.* A dead matte surface is not practical in porcelain enamel because such a surface would readily collect dirt and would not have good weathering properties. Therefore, even so-called matte enamel produces a fair image of reflection in a flat area.

SURFACE TEXTURES. The breaking up of flat surfaces by means of corrugations and other embossed overall patterns presents many interesting decorative possibilities. Narrow corrugations or reeding about $3/16''$ o/c produces a surface not unlike tooled stone. The corrugations create a dull or matte effect and correct the tendency of slight waves in the panels to be accented. Great stiffness is added to the metal by this treatment, making it possible to use lighter gauges of metal.

580

2″ SOLID
PLASTER PARTITION

COST. 2″ solid metal lath and plaster partitions have been widely used in many hospitals, offices, hotels, and apartments, and in many other types of buildings. This construction offers relatively low initial cost, structural soundness, and minimum repair and replacement costs during a long period of amortization. Simplicity of erection allows, in some sections of the country, a cost quite favorable with that of wood stud, lath, and plaster walls. Several different proprietary systems exist.

USEFUL FLOOR AREA. Space economy is secured by the exceptionally small amount of floor area required, as compared with other thicker partitions, resulting in up to 7% more usable space. If the number and size of rooms are to remain constant, construction costs can often be somewhat reduced through a diminished gross building area.

SOUND INSULATION. 2″ solid partitions with a noise reduction factor of 37.7 decibels are effective as a noise insulator; scientific tests have shown them to be satisfactory for use in apartments, schools, offices, hotels, and similar buildings. Generally the reduction in sound depends on the comparative weights of partitions. The solid structure of this system is superior to 4″ or 5″ wood stud walls of the same weight.

The reason for the exceptional sound insulation properties of the 2″ solid partition lies in its great density. Its steel core provides unusual resistance to the transmission of sound, much as comparatively thin plate glass does in telephone booths, recording studios, and the like.

CRACK AND IMPACT RESISTANCE. The 2″ solid partition is a system built from steel and gypsum to form a monolithic unit, rigidly anchored to floor and ceiling. The base for its plaster body is a two-way reinforcement of metal lath securely attached over sturdy metal channel studs. The final 2″ slab is resistant alike to shear, tension, impact, and vibration. This resistance to shocks and cracks (and the resulting absence of repairs) accounts for the preference for two-inch solid partitions over past decades by federal housing authorities.

FIRE PROTECTION. Composed entirely of solid metal and gypsum, the 2″ solid partition is an excellent fire barrier, making possible the heading-off of a fire with resulting safety to life and property. In official tests, a 2″ partition system was subjected to intense heat and flame for four hours without failure. At the end of these tests the temperature had reached 2,000°F—hot enough to melt glass and destroy certain types of masonry—but the solid partition continued to stand up. Securely attached and two-way reinforced from floor to ceiling, it provides a continuous fire barrier.

WEIGHT REDUCTION. 2″ solid partitions weigh only about 17.5 pounds per square foot, as compared with 27.5 pounds per square foot for 3″ clay tile and plaster walls and 45.5 pounds per square foot for plain 4″ hollow concrete partition block walls. Worthwhile savings can thus be effected in the steel or concrete framework of modern buildings. The increased weight of heavier materials affects the cost of steel columns as the square of the height.

ADAPTABILITY. 2″ solid plaster partitions are admirably suited for use as non-load-bearing partitions or enclosures such as office partitions that may be frequently altered to meet tenant requirements, partitions in apartments between tenants and around corridors, enclosures around elevators and stairways; and separation planes in schools, factories, and homes.

2" SOLID
PLASTER PARTITION

A

Prong Ceiling Runner applied to underside of floor slab

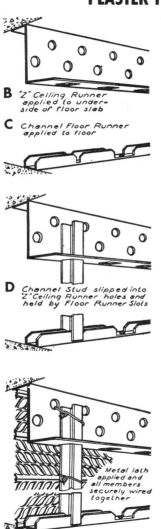

B *"Z" Ceiling Runner applied to underside of floor slab*

C *Channel Floor Runner applied to floor*

D *Channel Stud slipped into "Z" Ceiling Runner holes and held by Floor Runner Slots*

Metal lath applied and all members securely wired together

A. PRONG CEILING RUNNERS.
These allow utmost flexibility and speed of erection. The nailing surface is flat for rapid and accurate attachment to concrete ceilings with stub nails. The vertical prong is rigid, and long enough to "take up" variations in ceiling height. The top runner is fabricated from 20-gauge black steel sheets. Vertical prongs are spaced at the factory for any job.

B. "Z" CEILING RUNNERS.
These are attached with concrete stub nails or rawl drives. The lower horizontal surface is perforated for steel channel studs. Variations up to 2" in ceiling height are automatically taken care of. No cutting or reshaping is necessary on the job.

C. CHANNEL STUDS.
These are ¾" cold-rolled channel studs fabricated from best quality open-hearth 16-gauge steel. They are available in either 16'-0" or 20'-0" lengths.

D. CHANNEL FLOOR RUNNERS.
Side flanges are punched every 2" to receive vertical studs. The flat section contains holes 12" o/c for direct attachment to the floor. The studs drop securely in place; no wiring is necessary. Channel floor runners may be used at door and window frames.

WEIGHTS OF METAL LATH

Type of Lath	Weight Lbs. per Sq. Yard	Spacing of Supports
Diamond Mesh Lath	2.5	16"
	3.4	16"
Flat Rib Lath	2.75	16"
	3.4	24"
⅜" Rib Lath	3.4	24"
	4.0	24"

BASE
DETAILS

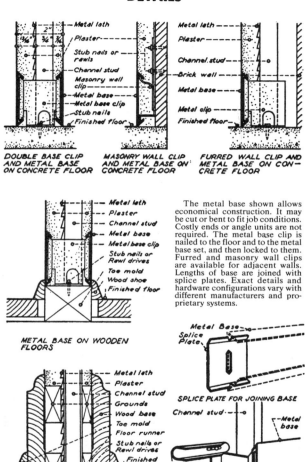

DOUBLE BASE CLIP AND METAL BASE ON CONCRETE FLOOR

Labels: Metal lath / Plaster / Stub nails or rawls / Channel stud / Masonry wall clip / Metal base / Metal base clip / Stub nails / Finished floor

MASONRY WALL CLIP AND METAL BASE ON CONCRETE FLOOR

Labels: Metal lath / Plaster / Stub nails or rawls / Channel stud / Masonry wall clip / Metal base / Metal base clip / Stub nails / Finished floor

FURRED WALL CLIP AND METAL BASE ON CONCRETE FLOOR

Labels: Metal lath / Plaster / Channel stud / Brick well / Metal base / Metal clip / Finished floor

METAL BASE ON WOODEN FLOORS

Labels: Metal lath / Plaster / Channel stud / Metal base / Metal base clip / Stub nails or Rawl drives / Toe mold / Wood shoe / Finished floor

WOOD BASE
Showing channel floor runner applied to wood sub-floor

Labels: Metal lath / Plaster / Channel stud / Grounds / Wood base / Toe mold / Floor runner / Stub nails or Rawl drives / Finished floor

SPLICE PLATE FOR JOINING BASE

Labels: Metal Base / Splice Plate

DOUBLE BASE CLIP

Labels: Channel stud / Metal base / Stub nails

SCALE
3" = 1'-0"

The metal base shown allows economical construction. It may be cut or bent to fit job conditions. Costly ends or angle units are not required. The metal base clip is nailed to the floor and to the metal base set, and then locked to them. Furred and masonry wall clips are available for adjacent walls. Lengths of base are joined with splice plates. Exact details and hardware configurations vary with different manufacturers and proprietary systems.

583

TREATMENT OF OPENINGS
IN 2" SOLID PARTITION

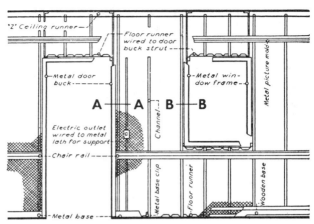

NOTE Erect steel channel with recommended spacing which depends on type and weight of metal lath

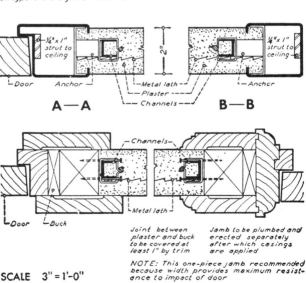

A—A

B—B

Joint between plaster and buck to be covered at least 1" by trim

Jamb to be plumbed and erected separately after which casings are applied

NOTE: This one-piece jamb recommended because width provides maximum resistance to impact of door

SCALE   3" = 1'-0"

INSTALLATION OF TRIM
AND ELECTRICAL WORK

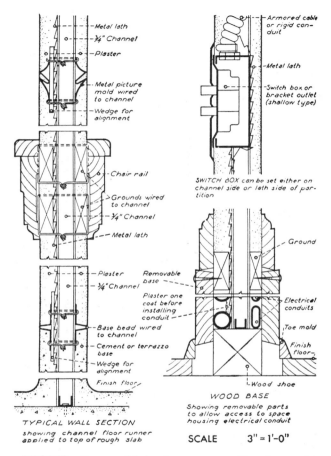

Metal lath
¾" Channel
Plaster
Metal picture mold wired to channel
Wedge for alignment

Chair rail
Grounds wired to channel
¾" Channel
Metal lath

Plaster
¾" Channel
Base bead wired to channel
Cement or terrazzo base
Wedge for alignment
Finish floor

Removable base
Plaster one coat before installing conduit

TYPICAL WALL SECTION
showing channel floor runner applied to top of rough slab

Armored cable or rigid conduit
Metal lath
Switch box or bracket outlet (shallow type)

SWITCH BOX can be set either on channel side or lath side of partition

Ground
Electrical conduits
Toe mold
Finish floor
Wood shoe

WOOD BASE
Showing removable parts to allow access to space housing electrical conduit

SCALE   3" = 1'-0"

ELECTRIC. The types of devices, boxes, raceways, cable, and accessories used with the 2″ solid partition vary with the exact details of construction, wall location, and so forth. In all cases wiring must conform to the latest requirements of the National Electrical Code.

PLUMBING. It is recommended that no pipe with a nominal diameter exceeding 1″ be installed in a 2″ solid partition. Run any work other than short, simple runs in pipe chases of 2″ solid partition construction, fit with access doors. Approved hangers on substantial construction must be used for any fixtures and must be installed before the final plaster coat is applied.

WOOD FINISHED
INTERIOR WALLS

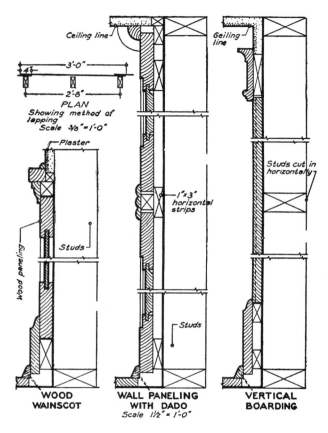

PLAN
Showing method of lapping
Scale 3/8" = 1'-0"

3'-0"

2'-8"

4"

Plaster

Wood paneling

Studs

**WOOD
WAINSCOT**

Ceiling line

1"x3" horizontal strips

Studs

**WALL PANELING
WITH DADO**
Scale 1 1/2" = 1'-0"

Ceiling line

Studs cut in horizontally

**VERTICAL
BOARDING**

Plaster or drywall may safely be omitted from behind nominal 1"-thick (generally about ¾"-thick actual) wainscots, paneling, or vertical planking if the construction shown in the drawings above is followed. The use of 6-mil polyethylene sheet film behind the wood effectively stops air movement and protects the wood from moisture, which causes the joints to open and the paneling to warp. A further advantage is that wood grounds or furring strips for nailing may be eliminated entirely or kept to a minimum. Where the horizontal members are cut between the studs, the room dimensions are increased by the thickness ordinarily occupied by the plaster or drywall. In the case of wood wainscots, the plane of the finish wood projects relatively little from the plane of the plaster or drywall above, minimizing the width of the cap molding.

Notice, however, that plaster or plasterboard should be installed as a backing behind thin (usually about ¼") wall paneling or stripping, to provide adequate support for the wallcovering and to afford better room acoustics.

INTERIOR WALLS
OF SOFT PINE

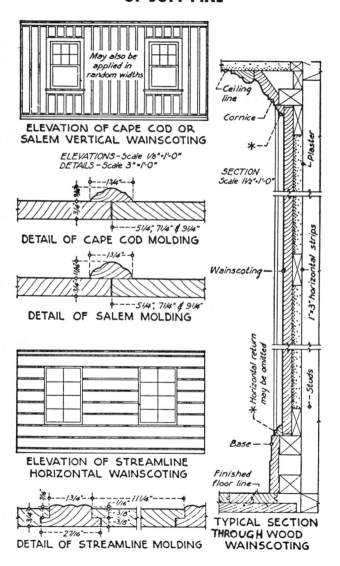

ELEVATION OF CAPE COD OR SALEM VERTICAL WAINSCOTING

ELEVATIONS – Scale 1/8" = 1'-0"
DETAILS – Scale 3" = 1'-0"

May also be applied in random widths

DETAIL OF CAPE COD MOLDING

1 3/4" 3/4" 3/4"
5 1/4", 7 1/4" & 9 1/4"

DETAIL OF SALEM MOLDING

1 3/4" 3/4" 11/16"
5 1/4", 7 1/4" & 9 1/4"

ELEVATION OF STREAMLINE HORIZONTAL WAINSCOTING

DETAIL OF STREAMLINE MOLDING

1/16" 1 3/4" 1 1/4"
6/16" 3/8" 3/4" 3/8"
2 7/16"

Ceiling line
Cornice
*
Plaster

SECTION
Scale 1 1/2" = 1'-0"

Wainscoting

1"x 3" horizontal strips

Horizontal return may be omitted

6 – Studs

Base

Finished floor line

TYPICAL SECTION THROUGH WOOD WAINSCOTING

CABINETWORK
ESSENTIALS

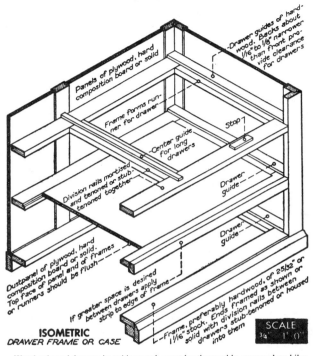

Drawer guides of hardwood. Backs about 1/16 to 1/8 narrower than front provide clearance for drawers

Panels of plywood, hard composition board or solid

Frame forms runner for drawer

Center guide for long drawers

Stop

Drawer guide

Division rails mortised and tenoned or stubtenoned together

Drawer guide

Dustpanel of plywood, hard composition board or solid. Top face of panel end of frames or runners should be flush

If greater space is desired between drawers apply strip to edge of frame

Frame, preferably hardwood, or 25/32" or 1 1/16" stock. Ends framed as shown or solid with division rails between drawers stub-tenoned or housed into them

ISOMETRIC
DRAWER FRAME OR CASE

SCALE
3/4" 1' 0"

Wood selected for use in cabinetwork must be thoroughly seasoned or kiln-dried to 12%-15% m.c.; it should be without defects in any exposed parts. The use of dry material reduces warping, shrinking, or swelling to a minimum. Installation of the cabinetwork should be left until all the moisture within the building has evaporated and plaster or masonry has fully cured.

The case forms the containing framework of all typical cabinet construction, whether housing drawers or serving as a cupboard. Successful operation of drawers and doors depends upon proper construction of case, doors, and drawers.

The drawing above shows one common construction of a drawer case; the case here should be made so that it has only sufficient contact with the drawers to support and guide them. The case is mortised and tenoned together, with division rails added between drawers as required. In better work, dust panels are installed in the frame between drawers. Notice that if drawers are to operate properly, without sticking, the side guides must be narrower at the back than at the front to provide the necessary clearance. A popular alternative, usually less problematical, is to install steel slide mechanisms with ball-bearing nylon rollers.

The drawer itself consists of a box constructed with sturdy joints. Dovetails, fairly small and with little taper, are preferable front and back for high-quality work; however, lock, dovetail dado, dado overlap, flush rabbet, and beveled rabbet joints can also be used. All joinery should fit snugly and be closely spaced to form perfectly secure glued joints.

CABINET DRAWER
DETAILS

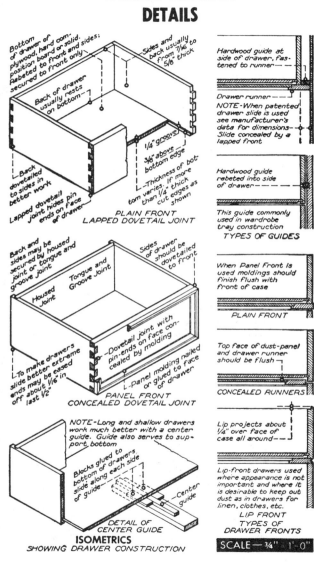

Bottom of drawer of plywood, hard composition board or solid. Rebeted to front and sides; secured to front only

Sides and back usually from 7/16 to 5/8 thick

Back of drawer usually rests on bottom

Back dovetailed to sides in better work

Lapped dovetail joint hides pin ends on face of drawer

1/4" groove

3/8" above bottom edge

Thickness of bottom varies. If more than 1/4" thick cut edges as shown

**PLAIN FRONT
LAPPED DOVETAIL JOINT**

Back and sides may be secured by housed joint or tongue and groove joint

Housed Joint

Tongue and Groove Joint

Sides of drawer should be dovetailed to front

Dovetail joint with pin-ends on face concealed by molding

To make drawers slide better extreme ends may be eased off about 1/16 in.

Panel molding nailed or glued to face of drawer

**PANEL FRONT
CONCEALED DOVETAIL JOINT**

NOTE—Long and shallow drawers work much better with a center guide. Guide also serves to support bottom

Blocks glued to bottom of drawers slide along each side of guide

Center guide

DETAIL OF CENTER GUIDE

ISOMETRICS
SHOWING DRAWER CONSTRUCTION

Hardwood guide at side of drawer, fastened to runner

Drawer runner

NOTE-When patented drawer slide is used see manufacturer's data for dimensions- Slide concealed by a lapped front

Hardwood guide rebeted into side of drawer

This guide commonly used in wardrobe tray construction

TYPES OF GUIDES

When Panel Front is used moldings should finish flush with front of case

PLAIN FRONT

Top face of dust-panel and drawer runner should be flush

CONCEALED RUNNERS

Lip projects about 1/4" over face of case all around

Lip-front drawers used where appearance is not important and where it is desirable to keep out dust as in drawers for linen, clothes, etc.

**LIP FRONT
TYPES OF
DRAWER FRONTS**

SCALE—3/4" = 1'-0"

589

CABINET DOOR
DETAILS

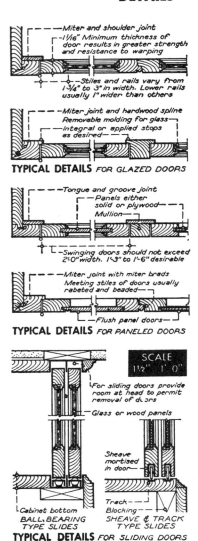

Miter and shoulder joint

1-1/16" Minimum thickness of door results in greater strength and resistance to warping

Stiles and rails vary from 1-3/4" to 3" in width. Lower rails usually 1" wider than others

Miter joint and hardwood spline

Removable molding for glass

Integral or applied stops as desired

TYPICAL DETAILS *FOR GLAZED DOORS*

Tongue and groove joint

Panels either solid or plywood

Mullion

Swinging doors should not exceed 2'-0" width. 1'-3" to 1'-6" desirable

Miter joint with miter brads

Meeting stiles of doors usually rabeted and beaded

Flush panel doors

TYPICAL DETAILS *FOR PANELED DOORS*

SCALE 1½" 1'-0"

For sliding doors provide room at head to permit removal of doors

Glass or wood panels

Sheave mortised in door

Cabinet bottom

BALL BEARING TYPE SLIDES

Track

Blocking

SHEAVE & TRACK TYPE SLIDES

TYPICAL DETAILS *FOR SLIDING DOORS*

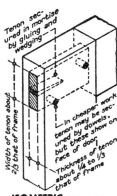

Tenon secured in mortise by gluing and wedging

Width of tenon about 2/3 that of frame

In cheaper work tenon may be secured by dowels but these show on face of door

Thickness of tenon about 1/4 to 1/3 that of frame

ISOMETRIC *SHOWING MORTISE AND TENON*

This joint should be used in constructing cabinet doors

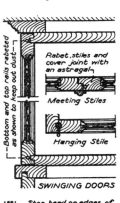

Bottom and top rails rabeted as shown to keep out dust

Rabet stiles and cover joint with an astragal

Meeting Stiles

Hanging Stile

SWINGING DOORS

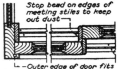

Stop bead on edges of meeting stiles to keep out dust

Outer edge of door fits into rabet to keep out dust

SLIDING DOORS

DETAILS *DUST-PROOF CABINET DOORS*

MARBLE
WAINSCOTS

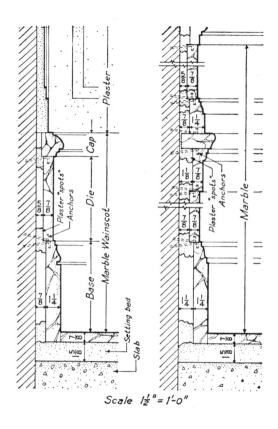

Scale $1\frac{1}{2}'' = 1'\text{-}0''$

The very least space that should be figured from rough wall face to finish marble face for wainscot or ashlar is $1\frac{1}{2}''$. Some marbles containing a large number of natural faults require reinforcing liners and need a minimum of $2\frac{1}{2}''$. Concealed anchors are used for fastening the marble to the rough wall. These are usually of 9-gauge copper, brass, or aluminum wire. The number of anchors to be used should be left to the discretion of the marble contractor. The space behind the marble should never be filled solid with plaster of paris; filling in spots only allows for contraction and expansion and thus prevents cracks.

SHEET
LATH

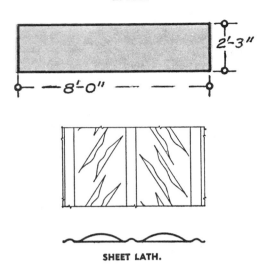

SHEET LATH.

SHEET LATH. Sheet lath is a metal lath formed by stamping a pattern of perforations into full-size sheets of steel. It is an exceptionally rigid and sturdy type of metal lath.

USES. Sheet lath is used as a combination keying and formwork for concrete floor and roof constructions, as centering for concrete slabs, and (especially) as a base for pneumatically applied concrete. It may be used for solid plaster partitions and for ceiling work; because of its rigidity, it is particularly appropriate as a backing for ceramic clay tile floors or walls.

STANDARDS OF THE INDUSTRY. The individual members of the industry have generally accepted the weight of sheet lath as 4.5 pounds per square yard. Sheet lath has a painted finish. Heavier weights than the standard one are available upon request.

FLAT RIB EXPANDED
METAL LATH

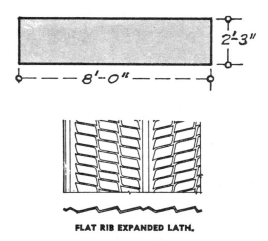

FLAT RIB EXPANDED LATH.

FLAT RIB EXPANDED LATH. This is coil steel continuously slit with roller cutters in a certain pattern, after which the metal is stretched to expand the slits into characteristic parallelogramlike openings in a herringbone pattern, bordered by slope-sided steel strands. During manufacture, ribs are impressed into the material to a depth of no more than $\frac{1}{8}''$.

USES. Flat rib lath is quite rigid in the direction of the ribs, and therefore is particularly suitable for furred and suspended ceilings and plain wall areas; the weight to be used depends upon the spacing of the supports. It is widely used for all sorts of plastering applications and is especially good as nail-on lath. No flat rib lath is recommended for cornice work, special detail work, or any contour type of lathing work.

STANDARDS OF THE INDUSTRY. The individual members of the industry have generally accepted 2.75 and 3.4 pounds per square yard as standard weights for flat rib lath.

DIAMOND MESH METAL LATH

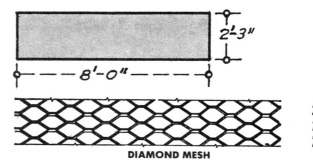

DIAMOND MESH

DIAMOND MESH METAL LATH.   This is the term used to indicate a metal lath that is fabricated from coil steel by slitting and expanding so that a uniform diamond mesh is formed.

USES.   Diamond mesh is a multipurpose base that can be used with either wood or steel framing. It is relatively nonrigid and can be easily formed and contoured into curved surfaces. It can be employed for all general purposes, including ornamental plastering, furred walls, stucco applications, and structural fireproofing of steel members. It is available in a paper-backed form on special order.

SELF-FURRING DIAMOND MESH.   This is similar to diamond mesh, but contains a continuous series of indentations that hold the mesh about ¼″ away from its mounting surface. This eliminates the need for additional furring and allows full keying of the plaster.

USES.   This mesh may be applied to most substrates, including masonry or concrete, sheathing, etc. It is particularly useful in replastering applications over old surfaces. It is also used for fireproofing columns, and as a cement plaster base. It is available with a paper backing on special order.

STANDARDS OF THE INDUSTRY.   The individual members of the industry have generally accepted 2.5 and 3.4 pounds per square yard as standard weights for both diamond mesh and self-furring diamond mesh. Both types are available either painted or galvanized.

3/8" RIB
METAL LATH

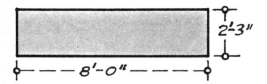

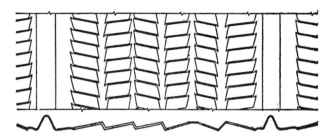

3/8" RIB EXPANDED LATH

⅜" RIB LATH. This metal lath is manufactured in a herringbone pattern similar to flat rib lath, but with ⅜"-deep longitudinal ribs for good rigidity and stability. Edges are terminated in ⅛"-deep inverted ribs. It is available in paper-backed form on special order.

USES. ⅜" rib lath is a very rigid form. It is widely used as centering for concrete floor and roof slabs, as a form for concrete when attached to joist tops, and as a plastering base when attached to joist bottoms. It is used in studless solid partitions and, because of its rigidity, can be attached directly to steel joists or other horizontal members up to a spacing of 24" (up to 27" under concrete joists). It may also be used for partitions and furring.

STANDARDS OF THE INDUSTRY. The individual members of the industry have generally accepted 3.4 and 4.0 pounds per square yard as standard weights for painted ⅜" rib lath and 3.4 pounds per square yard as the weight for the galvanized variety.

595

3/4" RIB
METAL LATH

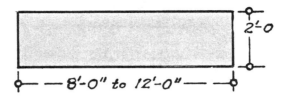

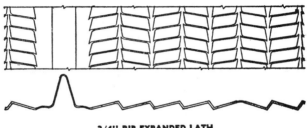

3/4" RIB EXPANDED LATH

¾" RIB LATH. This metal lath is similar to ⅜" rib lath, except the rib depth is ¾". The result is a sturdier, more rigid lath.

USES. This lath is designed primarily as reinforcement for concrete floors and roofs, serving, in addition, as a form upon which wet concrete may be poured. Laid over wood joists, it is ideal as reinforcing for the concrete base for tile, terazzo, or composition flooring. Solid plaster partitions may be constructed with ¾" lath, requiring no studs since the rigidity of the ribs allows the lath to span from floor to ceiling.

STANDARDS OF THE INDUSTRY. The individual members of the industry have generally accepted 0.60 and 0.75 pounds per square yard as standard weights for both painted and galvanized ¾" rib lath.

PLASTER GROUNDS
FOR BASEBOARDS

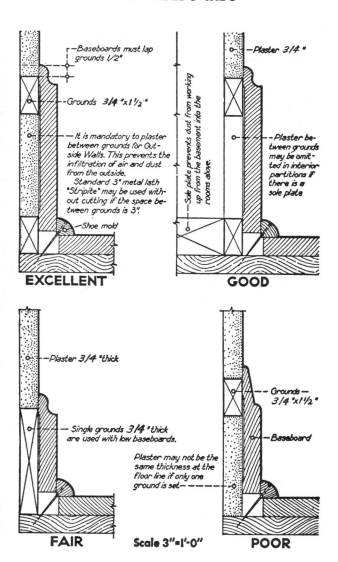

Baseboards must lap grounds 1/2"

Grounds 3/4" x 1 1/2"

It is mandatory to plaster between grounds for Outside Walls. This prevents the infiltration of air and dust from the outside.
Standard 3" metal lath "Stripite" may be used without cutting if the space between grounds is 3".

Shoe mold

EXCELLENT

Plaster 3/4"

Sole plate prevents dust from working up from the basement into the rooms above.

Plaster between grounds may be omitted in interior partitions if there is a sole plate

GOOD

Plaster 3/4" thick

Single grounds 3/4" thick are used with low baseboards.

Plaster may not be the same thickness at the floor line if only one ground is set

FAIR

Scale 3"=1'-0"

Grounds 3/4" x 1 1/2"

Baseboard

POOR

597

PLASTER GROUNDS
FOR TRIM

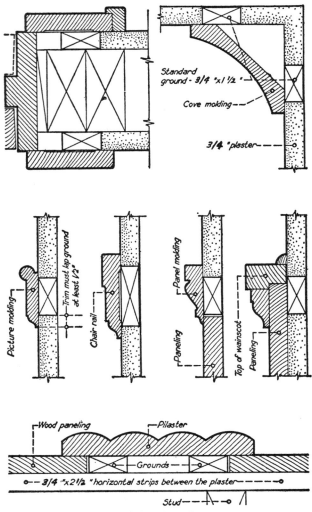

Standard ground - 3/4 "x 1 1/2 "

Cove molding

3/4 " plaster

Picture molding

Trim must lap ground at least 1/2"

Chair rail

Panel molding

Paneling

Top of wainscot

Paneling

Wood paneling — Pilaster

Grounds

3/4 "x 2 1/2 " horizontal strips between the plaster

Stud

Scale 3"=1'-0"

SIX ACOUSTIC DEFECTS

USUAL ACOUSTICAL DEFECTS. There are six defects normally to be considered in acoustic design:

1. Echo
2. Sound Foci
3. Insufficient Loudness
4. Reverberation Time
5. Excessive Loudness
6. Sound Transference

In the customary sense of the term, *Echo* results from the reflection of sound in such a way as to cause a definite or articulate repetition of the sound after an interval at least equal to the total duration of the original sound. *Sound Foci* result from a concentration or convergence of sound rays reflected from an extended concave surface, in exactly the manner that a headlight reflector concentrates light rays. Echo and sound foci are both defects wherever encountered, although they are of relatively infrequent occurrence in buildings. *Insufficient Loudness* is a problem of there being insufficient sound energy transfer for speech (and music, less often) to be intelligible. *Reverberation* is a confused or inarticulate prolongation of the original sound and up to a certain point is desirable. *Excessive Loudness* occurs when the generated sound reaches high intensities and requires noise quieting. *Sound Transference* is the audibility of sounds generated in some other room of the building, through walls, ducts, floors, or ceilings.

ECHO. Echo arises by regular reflection of sound from smooth walls, ceilings, or other surfaces, just as a mirror reflects a beam of light without either focusing or scattering it. Echo is generally produced by the reflection of sound waves from plane surfaces. If, however, the surface of a mirror is roughened, the reflected light will be diffused in all directions. Similarly, if the walls and ceiling of a room are irregular (on a sufficiently large scale), reflected sound will be scattered and broken up and its articulate character destroyed. In this case, echo becomes reverberation.

The lapse of time before an echo is heard is due to the longer path traveled by the reflected sound, compared to the path traveled by the sound that comes directly from the source. The longer this path difference is, the greater the time lapse grows. The shortest path difference at which an echo is audible is about 65'.

Echoes occur mostly in large rooms, although they can also occur in small ones, especially those that are relatively empty and of highly sound-reflective construction. In larger rooms, echoes are caused primarily by high ceilings and great distance to rear walls; the problem is compounded by concave surfaces. Echoes seldom cause serious difficulty in hearing but are regarded as a distinct annoyance. They may be reduced or eliminated by either of two methods:

(1) Introduce large irregularities on the surface that is causing the echo in order to scatter or diffuse the reflected sound and prevent regular reflection. This is frequently done by coffering in the case of ceilings. The dimensions which should be assigned to such coffering are not a matter of taste or accident. If the wave length of the incident sound is very large, in comparison to the size of the irregularities it encounters, there will be little dispersive effect; if very small, the smooth surface inside the coffering may act as regular reflectors. Depressions of about 4' square containing a succession of steps totaling a depth of about 8" or 10" provide the proper treatment for an average wave length between the male and female voice.

(2) Treat the surface with a highly sound absorptive material (one that absorbs at least 75% of the sound incident upon it).

SOUND
FOCI

SOUND FOCI. A sound focus is caused by reflection from a curved surface, which concentrates the sound rays in the same manner as a headlight reflector concentrates light rays. Depending on the curvature of the surface and the relative positions of the sound source and listener, focussing action may be heard as an abnormally loud echo, or as sound apparently coming from a source quite remote from its true source. These effects, when noticeable, are at least distracting and disturbing, and may sometimes cause serious difficulty in hearing. In a few extreme cases auditoriums have been rendered *totally useless* by this one defect.

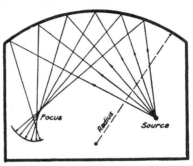

Trouble of this kind is usually caused when barreled or domed ceilings are laid out with the center of curvature near the floor line, or when the center of curvature of a rear wall is near the front of the stage. An empirical rule for curved surfaces is; *The radius of curvature of ceiling surfaces should be less than half or more than twice the perpendicular distance to the source of sound: the radius of curvature for walls should be very small, as a coved corner, or more than twice the distance to the source of sound.*

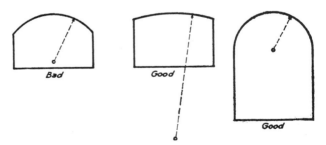

The best cure for focussing action is to change the curvature of the offending surface in accord with the above rule. In extreme cases this is the only possible means to a complete elimination of the difficulty Occasionally sound foci may be overcome by breaking up the offending surface by means of coffering, or by the use of a highly efficient sound-absorptive material, as described for the reduction of echoes.

INSUFFICIENT LOUDNESS

Insufficient loudness is more serious for speech than music, since adequate loudness throughout an auditorium is necessary for speech to be understood. The larger an auditorium is, the louder a speaker must talk to be heard. Since the average speaker's voice power is limited, loudspeakers and an efficient public address system are necessary in auditoriums larger than about 500,000 cubic feet. In rooms this large, electronic amplification is required even though all other acoustic conditions are perfect. On the other hand, loudspeakers are of little or no help in any auditorium, and indeed may make matters worse, unless other acoustic conditions are satisfactory.

For legitimate theater productions, the distance from the curtain line to the last row of seats will be limited so that delicate voice shading is audible. One authority has established this distance as 75'-0" for a theater with a balcony, and 100'-0" for a theater without a balcony.

The volume of an auditorium also has a bearing on the number of instruments that are suitable for musical renditions. In Circular No. 380 of the Bureau of Standards an empirical rule is given as follows:

Volume of Room	Number of Instruments
50,000	10
100,000	20
200,000	30
500,000	60
800,000	90

Loudness may be increased somewhat by locating the speaker or musicians near hard, sound-reflecting surfaces that reinforce the direct sound. A stage should be furnished with veneer "flats" or similar surfaces rather than heavy, sound-absorbing curtains. Musicians particularly prefer a sound-reflecting stage.

Loudness is sometimes insufficient because of an excessively wide seating area. Listeners in the front corners do not receive the full loudness because the speaker's voice is directed away from them at a wide angle. If the seats are arranged within the proper angle for correct vision, they will generally be satisfactory for hearing, too.

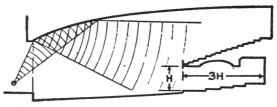

Loudness is sometimes inadequate in excessively deep under-balcony spaces. The depth of such spaces should not be more than three times the height of the opening, as shown in the illustration. In the average auditorium loudness is usually adequate in the front and center of the seating area, but insufficient at the sides and rear. The use of a fan-shaped floor plan and a ceiling sloping up from the stage will help to overcome this defect. Sound from the stage is reflected by the walls and ceiling to the sides and rear, where it increases the loudness by reinforcing the direct sound. If such a design is impractical, a proscenium having soffit and sides at a 45° angle is of benefit.

TIME OF
REVERBERATION

A sound produced in a room is reflected back and forth from the walls, floors, and ceilings, losing part of its energy by absorption at each reflection. These reflections continue after the sound source is stopped, and are heard as a prolongation of the original sound which gradually dies out to inaudibility. This effect is called *"reverberation,"* and the length of time required for a sound of standard intensity to die out to inaudibility is called *"reverberation time."*

Excessive reverberation causes overlapping and confusion of spoken syllables and musical tones that render hearing unsatisfactory. A working rule may be stated as follows:

Reverberation Time	Hearing Conditions
Over 3 seconds	Poor
2 to 3 seconds	Fair
1 to 2 seconds	Good

The most desirable reverberation time for a given room depends on its size and purpose. A chart of optimum values is given at the bottom of this page.

An effective formula for computing the reverberation time of a room is:

$$T = 0.049 \ V/a$$

in which T = reverberation time in seconds, V = volume of room in cubic feet, and a = total absorption in the room (which is the sum of the number of units absorbed by the walls, floor, ceiling, furnishings, audience, and so forth) in sabins.

The number of sabins absorbed by a given wall or other surface is the product of the surface's area in square feet and its absorption coefficient. The *absorption coefficient* of a material is the percentage of sound absorbed by the surface when sound strikes it. If an open window may be said to "absorb" all the sound that falls upon it, its coefficient of absorption is unity, or 1 unit per square foot (1 sabin). A material that absorbs half the sound that falls upon its surface would have an absorption coefficient of 0.50 and would absorb 0.5 sabin in every square foot of area. A surface having an area of 100 square feet and an absorption coefficient of 0.50 would absorb 100 x 0.50 or 50 sabins.

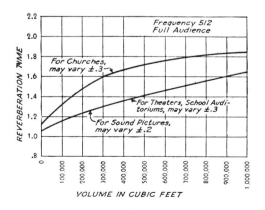

NOISE
REDUCTION

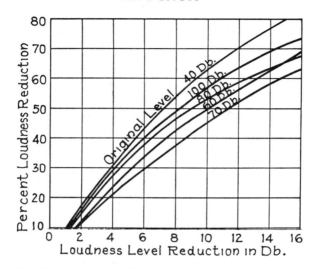

NOISE REDUCTION CALCULATIONS. The ear does not judge loudness in direct proportion to the physical intensity expressed in decibels. Bell Telephone Laboratories identified a relation between apparent loudness and actual intensity level, which is shown in the graph above.

The reduction of sound intensity expressed in decibels can be easily determined by the following formula, after which a reference to the graph above will indicate the reduction in loudness as judged by the human ear:

$$\frac{\text{Reduction}}{\text{in decibels}} = 10 \log \frac{\text{treated room absorption}}{\text{untreated room absorption}}$$

In determining the absorption needed to quiet noisy offices, restaurants, and so forth, the numerical averages of absorption at 125, 250, 500, 1,000, 2,000, and 4,000 Hz are used. The resulting figure for each material so tested is called the material's "noise reduction coefficient" (NRC).

DESIGN PROCEDURE. It is good practice in noise reduction to treat enough area to produce reduction of at least 6 decibels. The graph above indicates that this means reducing apparent loudness from 30% to about 44% — depending upon the original noise level. Because the total absorption of the room is calculated before treatment, the desirable absorption after treatment can be found from the formula.

APPLYING THE FORMULA. Assume that the calculation indicates 4,000 absorption units are required in a room to reduce the noise to a desired level. A further calculation reveals that the room without treatment, using plastered walls and ceiling, has a total absorption of 1,900 units. Therefore, the difference, 2,100 units, must be supplied by replacing plaster with a sound-absorbing material.

SOUND INTENSITIES

Decibels	Relative Energy	
		Threshold of painful feeling
		Thunder
		Artillery firing
110	100,000,000,000	Unmuffled airplane engine
		Large steam whistle
		Boiler factory
		Structural steel riveter at 15 ft.
100	10,000,000,000	In subway car
		Pneumatic Jackhammer drill 10 ft. away
		Newspaper press room
		Noise in untreated airplane cabin
'90	1,000,000,000	Elevated trains from street
		Automobile horn at 23 ft.
		Noisiest street corner, New York
		Fire siren at 75 ft.
		Large public address system
80	100,000,000	Police whistle at 15 ft.
		Average machine shop
		Interior of electric interurban train
		Snow shoveling on cement walk
		Motor truck without muffler
70	10,000,000	Noise in a stenographic room
		Average factory
		Busy street traffic
		Full volume of modern home radio
		Noisy ventilating system, grille 3 ft. away
60	1,000,000	Average busy street
		Congested department stores
		Average public building
		Church bells at 1200 ft.
		Average store
50	100,000	Moderate restaurant clatter
		Noisy residence
		Average office
		Quiet automobile
		Satisfactory high school ventilating system
40	10,000	Ordinary school class room
		Public library
		Average residence
		Quiet office
		Silent-movie theatre
30	1,000	Quiet residence
		Legitimate theatre
		Private office acoustically treated
		Planetarium
		Rustling paper
20	100	Average whisper
		Quiet church
		Underground vault
		Broadcasting studio
		Sound-film studio
10	10	Breathing through nose
		Very quiet studio for making sound pictures
0	1	Threshold of audibility

Deafening Noise

Distracting Noise

Range of Conversation

Extreme Quiet

Sound Proof Chambers

SOUND
TRANSFERENCE

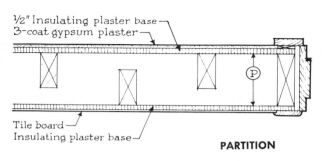

½" Insulating plaster base
3-coat gypsum plaster

Tile board
Insulating plaster base

PARTITION

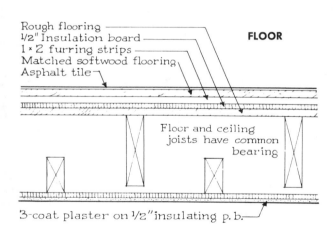

Rough flooring
½" Insulation board
1 × 2 furring strips
Matched softwood flooring
Asphalt tile

FLOOR

Floor and ceiling
joists have common
bearing

3-coat plaster on ½" insulating p.b.

The problem of sound transference can be analyzed in advance of construction or as a corrective measure in existing buildings. Publications dealing with noise control and acoustics in general construction, as well as the specifications provided by manufacturers of materials, should be consulted.

Numerous tests indicate the effectiveness of staggered studs and floor joists, among other construction methods, in reducing the transmission of room noises. The location of bathrooms and noise-producing areas (workshops, playrooms, and so forth) will determine when walls and/or floors require this treatment. (A bathroom located over the kitchen, for example, would not necessarily need the floor treatment, but one over the dining or living room would). Soil stacks should be wrapped with hair felt or other insulation, meaning that the dimension "P" for a partition with a 4″ standard-weight pipe must be at least 7½″.

COEFFICIENTS (NRC)
OF ORDINARY MATERIALS

Material	NRC
Auditorium chair, wood veneer	0.25
Brick wall, painted	0.018
Brick wall, unpainted	0.04
Carpet, on concrete	0.30
Carpet, on foam pad	0.55
Concrete, painted	0.05
Concrete block, painted	0.076
Concrete block, rough	0.35
Fabrics, hung straight	
Heavy, draped, 18 oz. per sq. yd.	0.50
Light, 10 oz. per sq. yd.	0.11
Medium, 14 oz. per sq. yd.	0.13
Floors	
Concete or terrazzo	0.015
Sheet vinyl, asphalt, rubber, or cork tile on concrete	0.03-0.09
Wood	0.09
Glass	0.168
Marble or glazed tile	0.00
Metal or wood chairs (each)	0.17
Openings	
Deep balcony, upholstered seats	0.50-1.00
Grilles, ventilating	0.15-0.50
Stage, depending on furnishings	0.25-0.75
Pew cushions	1.45-1.90
Plaster, rough finish on lath	0.035
Plaster, smooth finish on tile or brick	0.028
Plasterboard, on stud wall	0.10
Plywood	0.15
Seated audience	0.80
Theater chairs, Leatherette	1.6
Wood paneling over fiberglass insulation	0.65

Complete tables of coefficients, which are occasionally updated, of the various materials that normally constitute the interior finish of rooms may be found in various books on architectural and general acoustics, and further information can be obtained from manufacturers of acoustic and general building materials. This short list will be useful in making simple calculations of the reverberation in rooms.

GRADING BETWEEN
HOUSE AND SIDEWALK

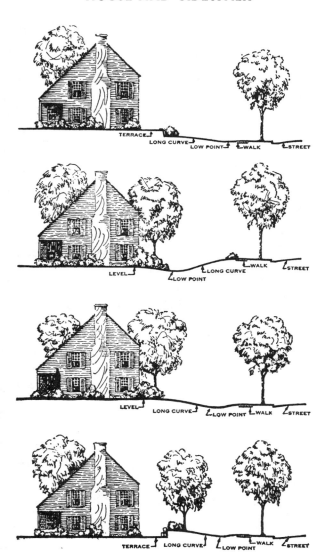

WATER LILY
GARDEN POOL

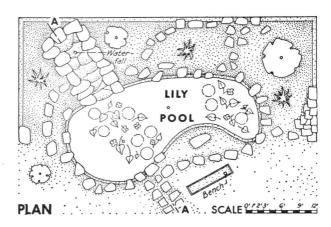

PLAN . . . . A . . SCALE 0' 1' 2' 3'   6'   9'   12'

CHARACTER OF POOL.   The general style of the garden—whether it is formal or informal—will suggest the pool treatment and its size. Colorful fish and aquatic plants make the garden pool a main point of interest for the "outdoor living room."

SIZE AND CONSTRUCTION.   Small species of water lilies require a pool 3'-0" or more across at its least dimension. Larger varieties require 6'-0". The minimum height from waterline to top of soil in the tubs should be 22". Pools with vertical sides can be built with wood or 20-gauge sheet metal forms. Pools with sloping or curved sides can be made in firm soil by plastering a stiff mixture against the earth, placing the reinforcement, and then completing the slab by further plastering. Floor and walls, in any case, should be placed in one operation to avoid joints.

WATERFALL.   It may be advisable to set the stones for the waterfall in concrete so that the surrounding earth does not become soggy and form stagnant puddles where mosquitoes might breed.

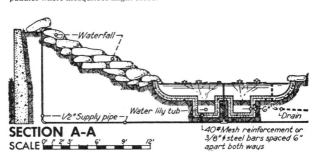

SECTION A-A
SCALE 0' 1' 2' 3'   6'   9'   12'

40# Mesh reinforcement or 3/8"∅ steel bars spaced 6" apart both ways

SLAB-TYPE CONCRETE DRIVES

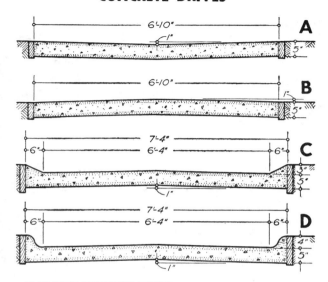

SLAB-TYPE CONCRETE DRIVES. The slab-type drive is less problematical to an unskilled driver than the ribbon type. When this type is built with curbs, a careless driver is less likely to run over adjoining plantings. It is more expensive than the ribbon type but is the only choice for drives that curve or require turnarounds. Combinations of colored concrete, brick, and textured surfaces can also be used to bring the driveway into greater harmony with the surroundings. Widths shown are minimum; a width of 8' to 10' is recommended.

SUBGRADE. The area upon which the slab is to be laid should be brought to subgrade and well compacted before pouring. All soft and yielding material and all loose rocks or boulders must be removed or broken off several inches below subgrade and the holes refilled with tamped material. Settlement of the subgrade is likely to cause cracking and heaving. Construction on fresh fill should be delayed at least a year, and the material should be compacted several times meanwhile. If the soil is gravelly and porous no subbase or cushion is needed. However, if the soil is clayey or if ground moisture is present, a 4" to 6" layer of gravel, crushed stone, or cinders should first be placed.

FORMS. Plywood strips or 2" x 4", 2" x 6", or 2" x 8" lumber can be used for forms. In ground likely to be infested with termites, care should be taken to remove all lumber and stakes after the concrete has cured; however, treated lumber or redwood can be left in place as edgings, if desired.

EXPANSION JOINTS. No expansion joints are needed for drives less than 40' long. On longer drives, an expansion joint should be set every 20' to 30'. Control joints should be cut about 1" deep into the surface every 10'.

THICKNESS OF SLAB. Residential drives need be only 4" thick—or 5" thick in areas of severe winter weather. If the drive will be used by heavy trucks, its thickness should be increased to 6". Reinforcing mesh should be laid at midthickness in any case.

RIBBON-TYPE CONCRETE DRIVES

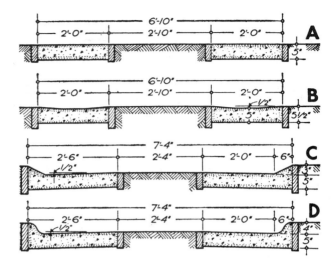

RIBBON-TYPE DRIVES. For straight drives, the ribbon type is often considered more in keeping with the landscape treatment because of the area of turf that breaks up the driveway area. It is also more economical than solid full-width pavements. Ribbon drives without curbs should not be used on curves, no matter how slight they are. The dimensions given in the drawings above may be taken as adequate, but they might well be checked against the track width of the vehicles operating on the drive. Ribbons as narrow as 1'-6", with 3'-4" between them, represent a minimum for straight drives.

SUBGRADE. The areas upon which the slabs are to be laid should be brought to subgrade and well compacted before pouring. All soft and yielding material and all loose rocks and boulders must be removed or broken out to a depth several inches below subgrade and the holes refilled with tamped materials. Settlement of the subgrade is particularly likely to cause cracking and heaving with ribbon-type drives. Construction on fresh fill should be delayed at least a year, and the material should be compacted several times meanwhile. If the soil is gravelly and porous, no subbase or cushion is required. However, if the soil is clayey or if ground moisture is present, a 4" to 6" course of gravel, crushed stone, or cinders should first be placed.

FORMS. Plywood strips or 2" x 4", 2" x 6", or 2" x 8" lumber can be used for forms. In ground likely to be infested with termites, care should be taken to remove all lumber and stakes after the concrete has cured; however, treated lumber or redwood can be left in place as edgings, if desired.

EXPANSION JOINTS. No expansion joints are needed for drives less than 40' long. On longer drives, an expansion joint should be set every 20' to 30'. Control joints should be cut about 1" deep into the surface every 10'.

THICKNESS OF SLAB. Residential drives need be only 4" thick—or 5" thick in areas of severe winter weather. If the drive will be used by heavy trucks, the thickness should be increased to 6". Reinforcing mesh should be laid at midthickness in any case.

CONSTRUCTION OF
DRIVEWAY CURBS

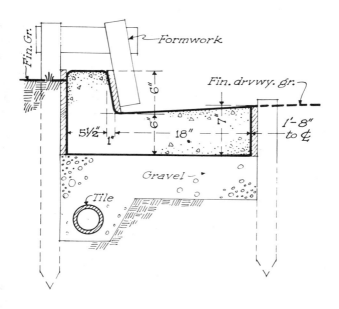

QUANTITIES REQUIRED FOR 100 FT.
7.8 barrels of cement (use *white* for night visibility)
2.6 cubic yards of sand
3.5 cubic yards of stone (1½″ max.)

PROCEDURE—If soil requires a sub-base, gravel or cinders to a thickness of 6″ should be used. If the nature of either the soil or the slope makes it necessary, provide open 4″ clay drain tile, as shown. On curves, the distance to the center line (CL) of the roadway should be increased to 3′-9″. Provide expansion joints of asphaltic felt at least every 50 linear feet, which separate the sections from top to bottom. A good finish can be obtained by removing the forms as soon as possible and troweling and rubbing the surface.

NIGHT-SAFE
DRIVEWAY CURB

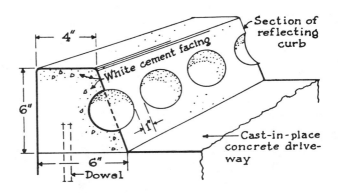

Section of reflecting curb

White cement facing

4"

6"

6"

Dowel

1"

Cast-in-place concrete driveway

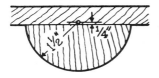

FORM FOR HOLLOWS

Any wooden sphere, such as a croquet ball, which is 3" to 4" in diameter, should be cut to ¼" less than half to make the reflecting hollows in the curb.

1½"

¼"

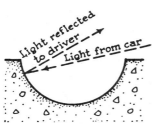

Light reflected to driver

Light from car

SECTION

Note that light from the headlamps of the car will be reflected to the driver's eyes from <u>some</u> part of the spherical surface, no matter from what angle it strikes either horizontally or vertically.

This is a simple adaptation for private driveways of the reflective highway curb used with great success in New Jersey. This curb is visible at night because (1) it is a good reflector of light and (2) it is designed to reflect light to the driver's eyes.

The curb may be precast or cast in place. White concrete should be used for the top and reflecting side in a 1 : 2 : 3½ mix with white quartz sand as a fine aggregate.

In rainy weather, when ordinary curbs are difficult or impossible to see, the reflective curb becomes a better reflector than when dry and its visibility is increased.

CONCRETE
SIDEWALKS

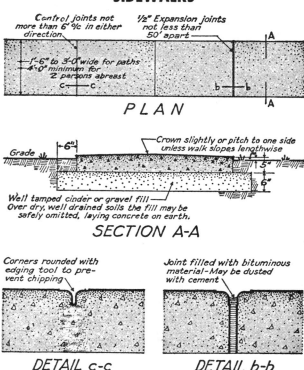

Control joints not more than 6' %c in either direction.

½" Expansion joints not less than 50' apart

1'-6" to 3'-0" wide for paths
4'-0" minimum for
2 persons abreast

PLAN

Grade

6"

Crown slightly or pitch to one side
unless walk slopes lengthwise

5"

6"

Well tamped cinder or gravel fill
Over dry, well drained soils the fill may be
safely omitted, laying concrete on earth.

SECTION A-A

Corners rounded with
edging tool to pre-
vent chipping

Joint filled with bituminous
material-May be dusted
with cement

DETAIL c-c

DETAIL b-b

NOTES

When walks are built around trees, provision must be made for the growth of the tree, to avoid its raising or cracking the walk.

Plain concrete sidewalk slabs are not designed to act as bridges. Therefore, the subgrade must be of uniform load-bearing capacity. If the slab is to be laid directly on the subgrade, all soft spots and vegetation must be dug out and filled with solid material; exceptionally well-compacted spots must be loosened and tamped.

Control joints are made after the concrete has taken its initial set and is being finished. Edge-rounded cuts are made in the green concrete with a grooving tool, to a depth of about 1", providing a weakened plane that controls cracking. A total slab thickness of 5" is more than adequate for localities with severe winter climates, while 4" is sufficient elsewhere.

The expansion joints allow movement of the sections of the walk, and provide cushions to absorb those movements and ease stresses on the control joints.

613

CONCRETE
STEPS

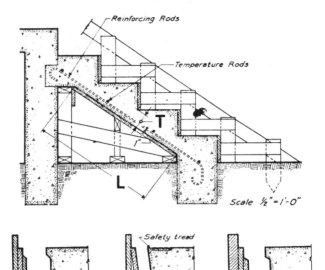

Scale ½" = 1'-0"

TYPES OF RISER FORMS

Scale ¾" = 1'-0"

Reinforced concrete steps that are independent of the ground beneath can be depended upon not to crack if properly constructed. The entire slab should be concreted at one time. The longitudinal reinforcement should be placed before the forms for the risers are attached. The mixture used should be 1:2:4. The side and riser forms can be removed 24 hours after concreting, but the forms and shoring supporting the stair slab should be left in place at least 4 weeks unless high-early-strength concrete is used.

L	T	Reinforcing rods		Temperature rods	
Length of Slab	Thickness	Dia.	Spacing	Dia.	Spacing
2 to 3 feet	4″	¼″	10″	¼″	12″ to 18″
3 to 4 feet	4″	¼″	7″	¼″	12″ to 18″
4 to 5 feet	5″	¼″	6″	¼″	18″ to 24″
5 to 6 feet	5″	¼″	5½″	¼″	18″ to 24″
6 to 7 feet	6″	¼″	5″	¼″	18″ to 24″
7 to 8 feet	6″	⅜″	4″	¼″	18″ to 24″
8 to 9 feet	7″	½″	5″	¼″	18″ to 24″

GARDEN STEPS AND
WALKS OF BRICK

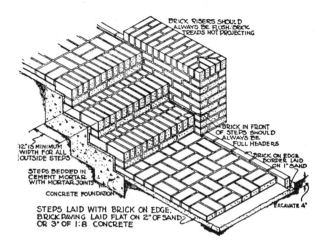

BRICK RISERS SHOULD ALWAYS BE FLUSH. BRICK TREADS NOT PROJECTING

BRICK IN FRONT OF STEPS SHOULD ALWAYS BE FULL HEADERS

2 BRICK ON EDGE BORDER LAID ON 1" SAND

12" IS MINIMUM WIDTH FOR ALL OUTSIDE STEPS

STEPS BEDDED IN CEMENT MORTAR WITH MORTAR JOINTS

CONCRETE FOUNDATION

EXCAVATE 4"

STEPS LAID WITH BRICK ON EDGE. BRICK PAVING LAID FLAT ON 2" OF SAND OR 3" OF 1:8 CONCRETE

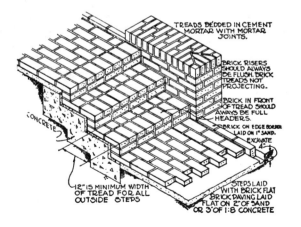

TREADS BEDDED IN CEMENT MORTAR WITH MORTAR JOINTS.

BRICK RISERS SHOULD ALWAYS BE FLUSH. BRICK TREADS NOT PROJECTING.

BRICK IN FRONT OF TREAD SHOULD ALWAYS BE FULL HEADERS.

BRICK ON EDGE BORDER LAID ON 1" SAND.

CONCRETE

EXCAVATE 4"

12" IS MINIMUM WIDTH OF TREAD FOR ALL OUTSIDE STEPS

STEPS LAID WITH BRICK FLAT. BRICK PAVING LAID FLAT ON 2" OF SAND OR 3" OF 1:8 CONCRETE

The undersurface of the concrete base should not slope, or it will tend to slide out of place. It should be stepped and reinforced, unless placed on undisturbed earth.

DETAILS OF
BRICK WALKS

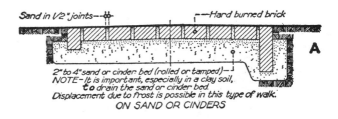

Sand in 1/2" joints-- | --Hard burned brick

2" to 4" sand or cinder bed (rolled or tamped)--
NOTE-It is important, especially in a clay soil,
to drain the sand or cinder bed.
Displacement due to frost is possible in this type of walk.
ON SAND OR CINDERS

A

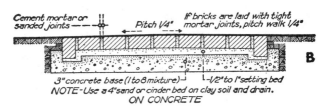

Cement mortar or | Pitch 1/4" | If bricks are laid with tight
sanded joints-- | | mortar joints, pitch walk 1/4"

3" concrete base (1 to 8 mixture)-- | --1/2" to 1" setting bed
NOTE-Use a 4" sand or cinder bed on clay soil and drain.
ON CONCRETE

B

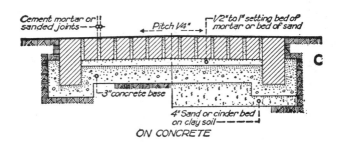

Cement mortar or | Pitch 1/4" | 1/2" to 1" setting bed of
sanded joints-- | | mortar or bed of sand

--3" concrete base

4" Sand or cinder bed
on clay soil -----

ON CONCRETE

C

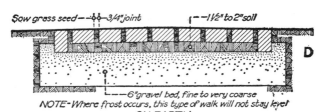

Sow grass seed-- 3/4" joint | --1 1/2" to 2" soil

--6" gravel bed, fine to very coarse
NOTE-Where frost occurs, this type of walk will not stay level
ON EARTH

D

SECTIONS   SCALE 3/4" = 1'-0"

616

GARDEN
WALKS

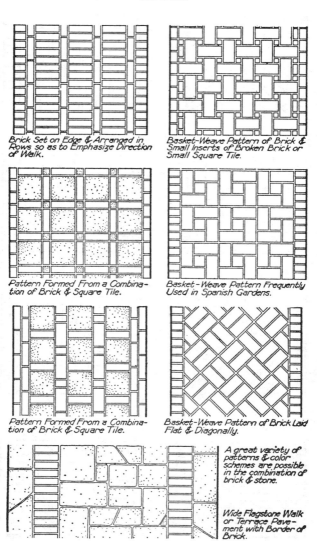

Brick Set on Edge & Arranged in
Rows so as to Emphasize Direction
of Walk.

Basket-Weave Pattern of Brick &
Small Inserts of Broken Brick or
Small Square Tile.

Pattern Formed From a Combination of Brick & Square Tile.

Basket-Weave Pattern Frequently
Used in Spanish Gardens.

Pattern Formed From a Combination of Brick & Square Tile.

Basket-Weave Pattern of Brick Laid
Flat & Diagonally.

A great variety of
patterns & color
schemes are possible
in the combination of
brick & stone.

Wide Flagstone Walk
or Terrace Pave-
ment with Border of
Brick.

BRICK PATTERNS
FOR WALKS

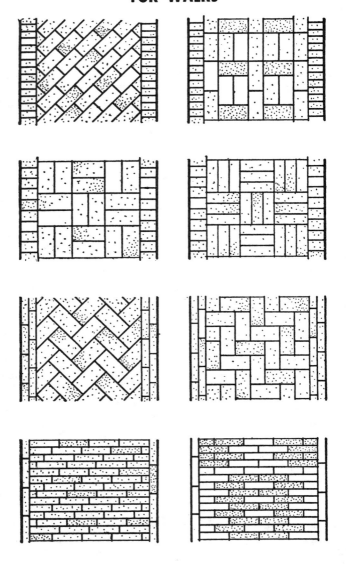

FLAGSTONE PAVING

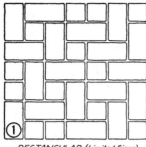

① RECTANGULAR (Limited Sizes)

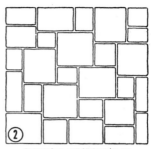

② RANDOM RECTANGULAR

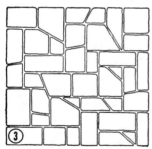

③ RANDOM SEMI-IRREGULAR

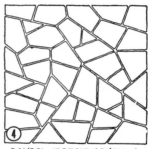

④ RANDOM IRREGULAR (Fitted)

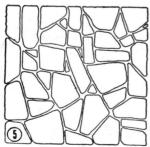

⑤ RANDOM IRREGULAR (Not Fitted)

The paving of walks and terraces with flagstones furnishes a desirable transition from the manmade geometrical formality of the building to the freedom and naturalness of the lawn and garden. Bluestone, limestone, stratified natural stones from the vicinity, cast stone and slate are commonly used materials. For terraces, it is important that the stones have level surfaces and that they be laid on concrete if furniture is to be used — see *Detail A* on following page. The method shown in *Detail C* on the following page may eventually result in tipping and movement of the stones out of level.

FLAGSTONE
PAVING

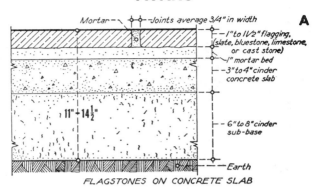

A

Mortar ⌐ ◇◇ Joints average 3/4" in width

— 1"to 1 1/2" flagging,
(slate, bluestone, limestone,
or cast stone)
— 1" mortar bed
— 3"to 4" cinder
concrete slab

— 6"to 8" cinder
sub-base

11" ‒ 14 1/2"

— Earth

FLAGSTONES ON CONCRETE SLAB

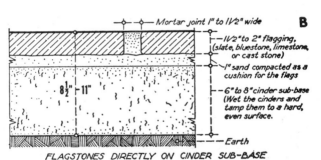

B

◇◇ Mortar joint 1" to 1 1/2" wide

— 1 1/2"to 2" flagging,
(slate, bluestone, limestone,
or cast stone)
— 1" sand compacted as a
cushion for the flags
— 6"to 8" cinder sub-base
(Wet the cinders and
tamp them to a hard,
even surface.

8 1/2" ‒ 11"

— Earth

FLAGSTONES DIRECTLY ON CINDER SUB-BASE

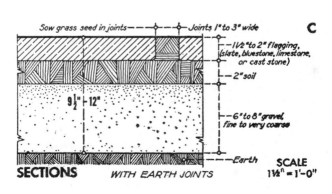

C

Sow grass seed in joints ‒ ◇ ◇ Joints 1" to 3" wide

— 1 1/2"to 2" flagging,
(slate, bluestone, limestone,
or cast stone)
— 2" soil

— 6"to 8" gravel,
fine to very coarse

9 1/2" ‒ 12"

— Earth

SECTIONS   WITH EARTH JOINTS

SCALE
1 1/2" = 1'-0"

CONCRETE
FLAGSTONES

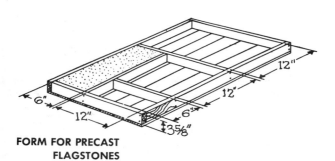

FORM FOR PRECAST FLAGSTONES

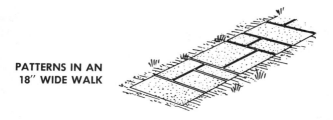

PATTERNS IN AN 18″ WIDE WALK

The sizes of flagstones obtained by using the simple forms shown, may be placed in a number of interesting designs. The forms should be made so they may be easily taken down for removal of the cast pieces and re-use. The wood should be well oiled each time before concreting.

Mineral pigments are often introduced in several shades to produce flagstones of different shades so that they vary not only in pattern but in color as well. The stones may be laid on a concrete base or on cinders with earth joints so that grass may grow between them. A number of different textures can be given to the stones by brooming, troweling, patting with a wire brush, etc. A 1:2¼:3 mix with maximum aggregate 1½″ will produce a good quality concrete. Using very wet sand and pebbles, about 4¼ gallons of water should be used in the mix to each 1-sack batch. Using damp sand and pebbles, 5½ gallons should give a workable mixture.

SIDEWALK
AREA GRATINGS

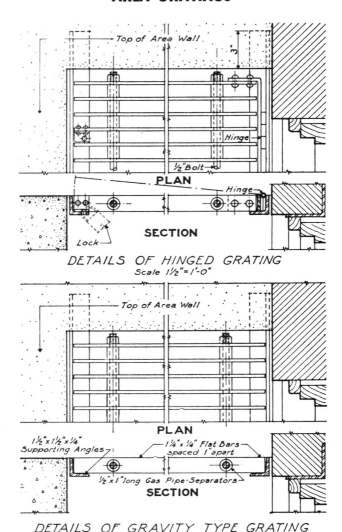

PLAN

SECTION

DETAILS OF HINGED GRATING
Scale 1½"=1'-0"

PLAN

SECTION

DETAILS OF GRAVITY TYPE GRATING

SIDEWALK COVERS, HATCHWAY COVERS, AND SUMP PIT COVERS

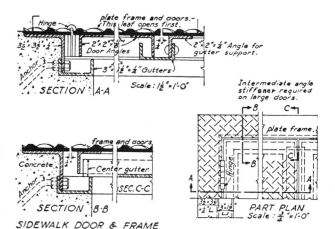

SECTION A-A

Scale: 1½"=1'-0"

PART PLAN
Scale: ¾"=1'-0"

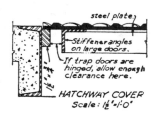

SIDEWALK DOOR & FRAME
WITH GUTTER CONSTRUCTED
OF COMMON STEEL SHAPES.

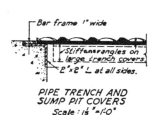

HATCHWAY COVER
Scale: 1½"=1'-0"

PIPE TRENCH AND
SUMP PIT COVERS
Scale: 1½"=1'-0"

The need of safety precautions is as important outside a building as it is within. Steel plate with slip-proof projections extending in two directions is ideal for sidewalk doors. The top three illustrations above show an economical method of constructing an all-steel sidewalk door using plates and common-size structural steel shapes.

Hatchway, manhole and sump pit covers, etc., are easily constructed of slip-proof plates and standard-size steel bars and structural shapes. The plates can be scribed with a hacksaw to fit irregular surfaces or openings.

Plates of large area should have stiffener angles in order to reduce plate thickness to an economical size.

Special slip-proof steel plates are well-adapted for use in industrial buildings as a slip-proof floor surface, especially at furnaces, machines, etc., and also at trucking areas and loading platforms.

623

HORIZONTAL
SUNDIAL

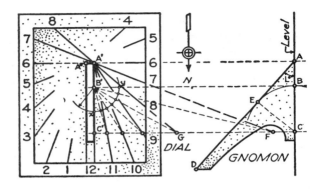

Draw a line *AC* of any convenient length. Draw *AD*, making the included angle *L°* equal to the latitude of the place. These two lines form the outline of the gnomon, stile, or rod, as it is variously termed. *AC* represents the plane of the dial plate and *AD* is the plane of the two edges of the gnomon, which casts the time-telling shadow. The gnomon may be cut away underneath to any desired shape as long as these two planes are not violated.

About point *C* describe an arc with radius *CE* equal to the perpendicular distance from *C* to line *AD*, cutting *AC* at *B*. Draw lines through *A*, *B*, and *C* at right angles to *AC*. These will interesect a convenient line parallel to *AC* at points *A'*, *B'*, and *C'*. Line *A'A* becomes the 6 o'clock mark on the dial.

From point *B'* describe a quadrant of convenient radius as *xy*. Divide the quadrant into six equal parts of 15° each. From *B'* draw lines through the five division marks until they intersect *C'C*. Lines connecting the points on *C'C* thus found to point *A'* are the 7, 8, 9, 10, and 11 o'clock dial marks.

If it is desired to show half- or quarter-hour marks, the quadrant is divided into 12 or 24 equal parts, the procedure being the same. Five- or one-minute divisions on the dial face can then be made accurately enough by eye.

The interesection of the planes of the sides of the gnomon with the dial plate becomes the two 12 o'clock marks. Continuation of the lines *FA'* and *GA'* become the 7 and 8 o'clock P.M. marks.

Since the dial is symmetrical, the 4 and 5 A.M. and the 1 to 5 P.M. marks that converge at *A"* are easily found. The sundial must be set with the gnomon in a true north and south direction, with the plate absolutely level.

SOUTH VERTICAL
SUNDIAL

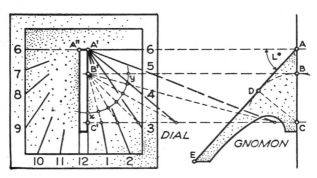

Draw a line *AC* of any convenient length. Draw *AE*, making the included angle *CAE* equal to the complement of the angle of latitude (90° − lat.) of the place. These two lines form the outline of the gnomon. *AC* represents the plane of the dial plate, and *AE* is the plane of the two edges of the gnomon, which casts the time-telling shadow. The gnomon may be cut away underneath to any desired design as long as these two planes are not violated.

About point *C* describe an arc with radius *CD* equal to the perpendicular distance from point *C* to line *AE*, cutting *AC* at *B*. Draw lines through *A*, *B*, and *C* at right angles to *AC*. These will intersect a convenient line parallel to *AC* at points *A'*, *B'*, and *C'*. Line *A'A* becomes the 6 o'clock mark on the dial.

From point *B'* describe a quadrant of convenient radius as *xy*. Divide the quadrant into six equal parts of 15° each. From *B'* draw lines through the five division marks until they intersect *C'C*. Lines connecting the points on *C'C* thus found to point *A'* are the 1 to 5 o'clock marks on the dial.

If it is desired to show half- or quarter-hour marks, the quadrant is divided into 12 or 24 equal parts, the procedure then being the same. Five- or one-minute divisions on the dial face can then be made accurately enough by eye.

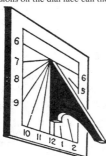

The intersections of the planes of the sides of the gnomon with the dial plate become the two 12 o'clock marks. Hours after 6 P.M. or before 6 A.M. cannot be shown on a south vertical dial.

Since the dial is symmetrical, the 7 to 11 A.M. marks that converge at *A"* are easily found. The sundial must be set exactly vertical and must face true (not magnetic) south.

NORTH VERTICAL
SUNDIAL

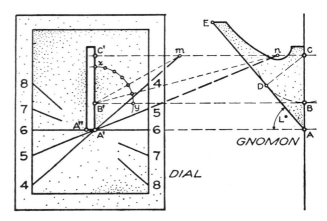

Draw a line *AC* of any convenient length. Draw *AE*, making the included angle *CAE* equal to the complement of the angle of latitude of the place (90° − latitude°). *AC* represents the plane of the dial plate, and *AE* is the plane of the two edges of the gnomon, which casts the time-telling shadow. The gnomon may be cut away as shown to any desired shape so long as these two planes are not violated.

About point *C* describe an arc with radius *CD* equal to the perpendicular distance from *C* to line *AE*, cutting *AC* at *B*. Draw lines through *A*, *B*, and *C* at right angles to *AC*. These will intersect a convenient line parallel to *AC* at points *A'*, *B'*, and *C'*. Line *AA'* becomes the 6 o'clock mark on the dial.

From point *B'* describe a quadrant of convenient radius as *xy*. Divide the quadrant into six equal parts of 15° each. From *B'* draw lines through the two lower division marks until they intersect *CC'* at *m* and *n*. Lines from *m* and *n* continued through *A'* become the 4 and 5 o'clock dial marks.

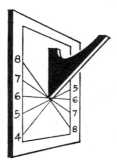

If it is desired to show half- or quarter-hour marks, the quadrant is divided into 12 or 24 equal parts, the procedure then being the same. Five- or one-minute divisions on the dial face can then be made accurately enough by eye.

Since the dial is symmetrical, the 7 and 8 o'clock marks that converge at *A"* are easily found. The sundial must be set with the plate vertically plumb and the gnomon in a true (not magnetic) north direction.

626

WEST VERTICAL SUNDIAL

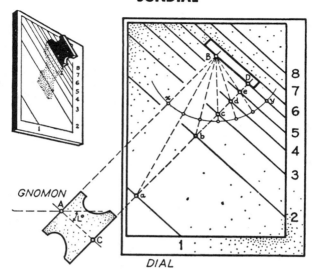

GNOMON

DIAL

Through a convenient point *A* draw the line *AC* making an angle *L°* with the horizontal which is equal to the latitude of the place. The distance *AC* is the height of the gnomon, which may be of any desired design so long as the top and bottom edges remain as parallel planes.

Draw parallel lines *AB* and *CD* at right angles to *AC*, making *BD* parallel to *AC*. Draw a line parallel to *BD* at a distance from it equal to the thickness of the gnomon. These two lines are the six o'clock marks on the dial, and locate the position of the gnomon.

From point *B* describe a quadrant of convenient radius as *xy*. Divide the quadrant into six equal parts of 15° each. From point *B* draw lines to the five division marks until they intersect line *CD*. Lines drawn parallel to *BD* through points *a, b, c, d,* and *e,* thus found, on *CD,* are the 1 to 5 o'clock dial marks.

If it is desirable to show half or quarter hour marks the quadrant is divided into 12 or 24 equal parts, these intermediate dial marks then being found in the same manner as the hours. Five- or one-minute divisions can then be made accurately enough on the dial face by eye.

The 8 and 7 o'clock marks are symmetrical about the gnomon with the 4 and 5 o'clock marks. The dial plate must be set exactly vertical and facing true west.

EAST VERTICAL
SUNDIAL

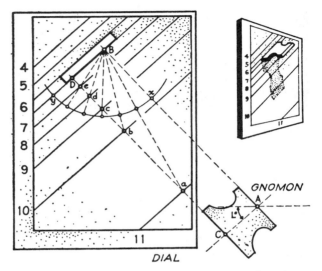

DIAL

Through a convenient point *A* draw the line *AC* making an angle *L°* with the horizontal which is equal to the latitude of the place. The distance *AC* is the height of the gnomon, which may be of any desired design so long as the top and bottom edges remain as parallel planes.

Draw parallel lines *AB* and *CD* at right angles to *AC*, making *BD* parallel to *AC*. Draw a line parallel to *BD* at a distance from it equal to the thickness of the gnomon. These two lines are the six o'clock marks on the dial, and locate the position of the gnomon.

From point *B* describe a quadrant of convenient radius as *xy*. Divide the quadrant into six equal parts of 15° each. From point *B* draw lines to the five division marks until they intersect line *CD*. Lines drawn parallel to *BD* through points *a*, *b*, *c*, *d*, and *e*, thus found on *CD*, are the 7 to 11 o'clock dial marks.

If it is desirable to show half or quarter hour marks the quadrant is divided into 12 or 24 equal parts, these intermediate dial marks then being found in the same manner as the hours. Five- or one-minute divisions can then be made accurately enough on the dial face by eye.

The 5 and 4 o'clock marks are symmetrical about the gnomon with the 7 and 8 o'clock marks. The dial plate must be set exactly vertical and facing due east.

OUTDOOR GRILL

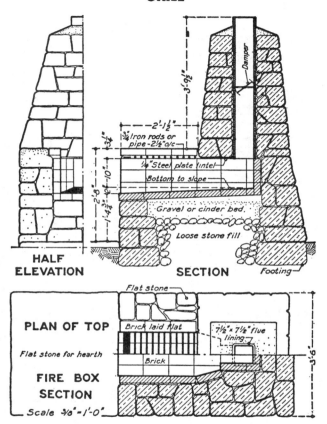

HALF ELEVATION

SECTION

2'-1½"

1¾" Iron rods or pipe - 2½"o/c

¼" Steel plate lintel

Bottom to slope

Gravel or cinder bed.

Loose stone fill

Footing

Damper

3'-9½"

3"

2'-8"

10"

1'-4½"

PLAN OF TOP

Flat stone

Brick laid flat

7½" x 7½" flue lining

Flat stone for hearth

Brick

FIRE BOX SECTION

Scale ¾" = 1'-0"

3'-6"

A simple type of outdoor fireplace is shown above. The footings should be carried below the frost line to prevent heaving. Rubble, ashlar, brick, or other masonry are equally suitable building materials. The firebox should be lined with well-burned brick — it is not necessary to use firebrick. The cooking grill can be formed out of round or square iron rods, steel pipe, or sidewalk grating, set into the mortar joint. An angle or plate lintel should be used to carry the chimney wall over the firebox opening. The masonry should be laid up in portland cement mortar made of one part portland cement, one part putty or hydrated lime, and six parts sand. Pleasing effects may be obtained by using colored mortar joints. The flue should be lined; only two lengths of standard 7½" x 7½" square flue lining are required.

629

OUTDOOR GRILL

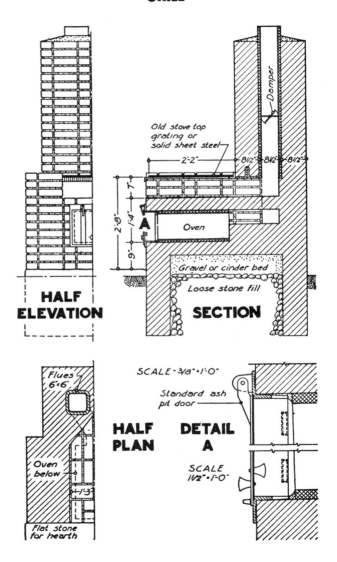

Old stove top grating or solid sheet steel

Damper

2'-2"   8½"   8½"   8½"

7"

2'-8"   1'-4"

9"

A   Oven

Gravel or cinder bed

Loose stone fill

HALF ELEVATION

SECTION

Flues 6×6"

Oven below

1'-3"

Flat stone for hearth

HALF PLAN

SCALE - ⅜" = 1'-0"

Standard ash pit door

DETAIL A

SCALE 1½" = 1'-0"

ORIGIN OF FIRES
IN RESIDENCES

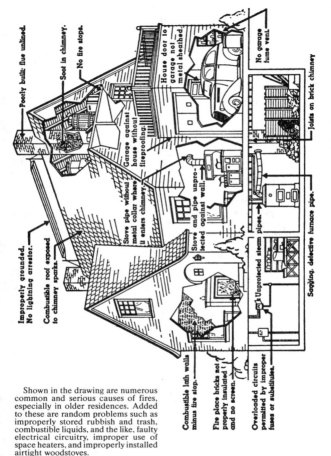

Shown in the drawing are numerous common and serious causes of fires, especially in older residences. Added to these are random problems such as improperly stored rubbish and trash, combustible liquids, and the like, faulty electrical circuitry, improper use of space heaters, and improperly installed airtight woodstoves.

In 1982, the percentages of multiple-death fires at different points of origin in residences were as follows:

Living room	50.3%
Bedrooms	20.8%
Kitchen	11.7%
Egress means	7.6%
Structural areas	5.1%
Heating equipment areas	1.5%
Other	3.0%

FIRE RESISTIVE
VAULTS

FUNDAMENTAL REQUIREMENTS.   In the design of a fire resistive vault, a number of requirements must not be overlooked if the structure is to withstand the effects of a severe fire, successfully protecting the records it contains.

1. Wall, floor, and roof construction must be of materials having sufficient fire resistance to resist the action of the most severe fire and having sufficient heat insulating resistance to prevent destruction of contents from high temperatures due to heat transmitted to the interior of the vault. Floors must be not less than 6″ thick, and greater if necessary to support the full load; if exposed to fire from outside the vault, floors must meet the same standards required for walls. Roofs must be at least 6″ thick, and greater if subject to unusual impact; if exposed to fire from outside the vault, roofs must meet the same standards required for walls.
2. Foundations and other supporting members should be designed and constructed to carry the weight of the vault and its contents safely when these supports are subjected to fire.
3. Provision must be made against the impact of falling building members and building contents such as machinery and other heavy objects.
4. Independence of the vault structure from the building members, at least to the extent that failure of the building will not cause failure of the vault, must be ensured.
5. Proper protection of door openings must be provided.
6. Vault must be ventilated only through door openings. Walls, floors, and roofs must not be pierced. There may be not more than two door openings.

VAULT CLASSIFICATION.   Vaults are classified in two groups according to type of support—ground supported vaults and structure supported vaults. Each has a subdivision based upon the resistance periods to fire—6-hour, 4-hour, and 2-hour vaults.

GROUND SUPPORTED VAULTS.   Ground supported vaults are supported directly on the ground and independent of the building in which they are located. They afford full protection to their contents, even when the building is completely destroyed.

Foundations must be of reinforced concrete. Structural members supporting vaults must have steelwork protected by at least 4″ of fire-proofing.

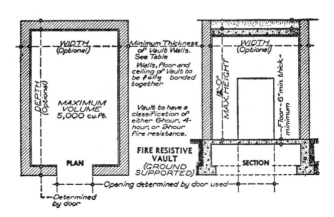

FIRE RESISTIVE
VAULTS

STRUCTURE SUPPORTED VAULTS. Structure supported vaults are supported by the framework of buildings of fire resistive construction. These vaults may be located individually on any floor and are designed to afford full protection to their contents, assuming the integrity of the supporting structure.

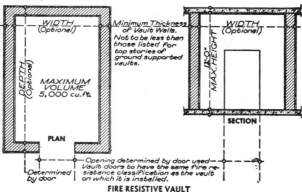

FIRE RESISTIVE VAULT
(STRUCTURE SUPPORTED)

Structure supporting vault shall be of adequate strength to carry full building load as well as entire weight of vault structure and contents. Structural members which support vault shall have steel protected by at least 4 inches of fire-proofing.

SUGGESTED MINIMUM THICKNESS OF WALLS FOR GROUND-SUPPORTED VAULTS

The following table suggests minimum thicknesses to take care of ordinary structural conditions and ordinary vault loads. The line for the "Top" floor may be considered as minimum wall thicknesses for Structural Supported Vaults.

	Thickness of Wall						Hollow Concrete Masonry
Floor No.	Reinforced Concrete			Brick			
	6 hr.	4 hr.	2 hr.	6 hr.	4 hr.	2 hr.	2 hr.
Top.............	10″	8″	6″	12″	12″	8″	8″
2nd from top....	10″	8″*	8″	12″	12″	12″	12″
3rd from top....	10″	10″	10″	12″	12″	12″	12″
4th from top....	12″*	10″	10″	16″†	16″†	16″†	16″†
5th from top....	12″	12″	12″	16″	16″	16″	16″
6th from top....	12″	12″	12″	16″	16″	16″	16″
7th from top....	12″‡	12″‡	12″‡	16″‡	16″‡	16″‡	16″‡
8th from top....	12″	12″	12″	16″	16″	16″	16″
9th from top....	12″	12″	12″	16″	16″	16″	16″
10th from top....	14″	12″	12″	16″	16″	16″	16″

* Thickness in panel construction may be 2″ less.
† Thickness in panel construction may be 4″ less.
‡ These thicknesses apply to panel construction.

FILE STORAGE
VAULT

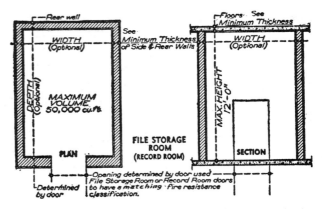

A file room is an enclosure that is fire resistive but provides less protection than a vault. It is intended for record storage where the volume of material is too large and of insufficient importance to justify from an economic standpoint the provision of standard record vaults or safes, but where values do warrant a certain amount of special protection.

File rooms may be located within buildings whose construction is either fire resistive or not fire resistive. In the latter case, all building structural members that support the file room must have a fire resistance equal to that of the file room, and the file room roof construction must be independent of the building construction. File rooms must not be constructed below ground level. The protection specified for file rooms and file room doors must be equal and must have a rating of 1, 2, 4, or 6 hours, as required by proper calculation. Practical structural requirements may necessitate structural thicknesses that have a higher fire-resistance classification than the minimum in some cases.

MINIMUM THICKNESSES. *Side and rear walls:* reinforced concrete, 6"; brick, solid or hollow, 8"; hollow concrete masonry units, plastered ½" on each side, 8". *Floor and roof:* not less than 6" thick, and greater if necessary to support the full load or resist unusual impact; if exposed to fire from outside the room, floor and roof must meet thickness standards required for wall.

OPENINGS. The openings in all walls must be restricted to doorways and to required lighting, sprinkler, and heat piping; the latter openings must be fully sealed. Roofs or floors above grade must not be pierced for any reason. Doorways must be protected with approved file-room-type doors. Doors may not open into any kind of shaftway.

VENTILATION. Ventilation of file storage rooms must be through door openings. Walls, floors, or ceilings must not be pierced.

HEATING. Heating must be by hot water or steam. If by steam, radiator units must be placed so that records cannot come into contact with them. Piping must be overhead. Pipe holes through walls must be lined with noncombustible sleeving and completely sealed with an approved material after pipe installation. No other type of heating, including auxiliary heaters of any sort, must be used. Floors (except on-grade type) and roofs must not be pierced for heat piping.

LIGHTING. Fixed lamps wired in conduit must be used to provide adequate lighting; there must be no pendant lamps, extension cords, or other electrical devices. Lamps must be vapor- or explosion-proof type, controlled by two-pole switches with pilot lights outside the file room. Installation of all electrical equipment must be in accordance with the latest provisions of the National Electrical Code.

REFERENCE. See "Protection of Records," NFPA publication #232, by the National Fire Protection Association, Quincy, Massachusetts.

MERCHANDISE VAULT

Merchandise vaults are for the storage of soft goods and other merchandise. They do not include vaults for the storage of film, pyroxylin plastics, or other highly inflammable materials. Floors and ceilings must be equivalent in strength and fire resistance to the walls. Doors must have a 4-hour or longer fire-resistance classification.

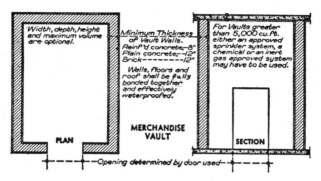

PLAN / SECTION — MERCHANDISE VAULT

Width, depth, height and maximum volume are optional.

Minimum Thickness of Vault Walls.
Reinf'd concrete—8"
Plain concrete—12"
Brick——————12"

Walls, floors and roof shall be fully bonded together and effectively waterproofed.

For Vaults greater than 5,000 cu. Ft. either an approved sprinkler system, a chemical or an inert gas approved system may have to be used.

Opening determined by door used

SUPPORTS. Vaults should be supported from the ground up by a properly protected steel or reinforced concrete framework having a minimum 4-hour fire-resistance classification. The supporting walls or framework should be of adequate strength to carry the weight of the vault structure and contents, together with any building loads they will be called upon to bear.

Vaults should be structurally independent of non-fireproof buildings, and any connection must be designed not to endanger the stability and fire-resistive qualities of the vault, in event of collapse of the building.

VENTILATION. Some means of ventilation may be required by the inspection department; it must be designed in such a manner as to prevent fire passing through the opening.

LIGHTING. Vaults must be adequately lighted by electricity. Wiring must be installed in accordance with the National Electrical Code; all exposed wiring must be in conduit. Pendant or extension cords must not be used inside the vault.

REFRIGERATION. Refrigeration systems, if used, must conform to applicable building, electrical, and mechanical code requirements and recommendations, or to those of the local inspecting authority.

FIRE EXTINGUISHING EQUIPMENT. If the vault contains valuables that are subject to water damage, a system using an inert gas is recommended. If practical, vaults protected by automatic sprinklers should be provided with suitable floor drains.

REFERENCE. For further information on general inside storage requirements, see NFPA 30, "Inside General Storage," by the National Fire Protection Association, Quincy, Massachusetts.

635

AUTOMATIC SPRINKLER LOCATION

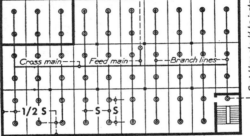

Sprinklers in fire section of small area may be fed from riser in another section if warranted. Holes in partition walls allowing sprinklers to distribute water to either side are not effectual

Each fire section should have one or more separate risers. Each riser to be of sufficient size to supply all sprinklers on riser on any one floor

Sprinklers should be installed under stairs. If stair tower has independent riser consider as one fire section. Non-combustible shafts require sprinklers only at top and bottom. However, if two or more separate fire sections are served by shafts, sprinklers are required at each landing

Cross main — Feed main — Branch lines →

1/2 S — S — S

S — *Maximum distance between lines and between sprinklers on lines*

A — *Maximum square foot protection area allotted per sprinkler*

1/2 S — *Maximum distance from wall or partition to first sprinkler is 1/2 allowable distance between sprinklers in the same direction.*

SPRINKLER SPACING. The maximum distances (S) between the sprinkler-branch lines and the sprinklers on the branch lines are as follows:

Light hazard	15'
Ordinary hazard	15'
With high-piled storage	12'
Extra hazard	12'

SPRINKLER PROTECTION AREA. The maximum allowable sprinkler protection areas (A) per individual sprinkler are as follows:

Light hazard -	
Smooth ceiling	200 sq. ft.
Beam & girder	200 sq. ft.
Hydraulically designed	225 sq. ft.
Open wood joist	130 sq. ft.
All other	168 sq. ft.
Ordinary hazard -	
All constructions	130 sq. ft.
All high-piled storage	100 sq. ft.
Hydraulically designed	130 sq. ft.
Extra hazard -	
All constructions	90 sq. ft.
Hydraulically designed	100 sq. ft.

AUTOMATIC SPRINKLER
INSTALLATION

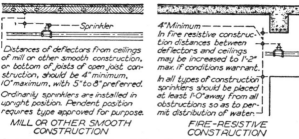

Distances of deflectors from ceilings of mill or other smooth construction, or bottom of joists of open joist construction, should be 4" minimum, 10" maximum, with 5" to 8" preferred.

Ordinarily sprinklers are installed in upright position. Pendent position requires type approved for purpose.

MILL OR OTHER SMOOTH CONSTRUCTION

4" Minimum
In fire resistive construction distances between deflectors and ceilings may be increased to 1'-0" max. if conditions warrant.

In all types of construction sprinklers should be placed at least 1'-0" away from all obstructions so as to permit distribution of water.

FIRE-RESISTIVE CONSTRUCTION

Sprinkler deflectors installed parallel to ceilings, roofs, or incline of stairs. When in peak of pitched roof sprinklers should be horizontal.

Dimensions given above should be adhered to so as to ensure proper distribution of water on ceilings; proper protection of area below sprinklers, and quick response to early heat waves from a fire.

INSTALLATION OF SPRINKLERS

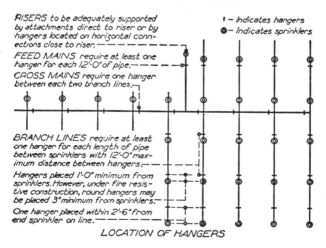

RISERS to be adequately supported by attachments direct to riser or by hangers located on horizontal connections close to riser.

FEED MAINS require at least one hanger for each 12'-0" of pipe.

CROSS MAINS require one hanger between each two branch lines.

| – Indicates hangers
● – Indicates sprinklers

BRANCH LINES require at least one hanger for each length of pipe between sprinklers with 12'-0" maximum distance between hangers.

Hangers placed 1'-0" minimum from sprinklers. However, under fire resistive construction, round hangers may be placed 3" minimum from sprinklers.

One hanger placed within 2'-6" from end sprinkler on line.

LOCATION OF HANGERS

Hangers should be of round wrought iron U-type or approved adjustable type. Cast iron hangers or parts of hangers should be malleable.

Pipes should be supported by hangers attached directly to structural members, by means of floor plates and bolts, or by approved inserts set in concrete when the suitability of the concrete has been definitely determined.

Where pipes run through concrete beams proper sleeves should be provided. Such sleeves should not be used for support of pipes.

AUTOMATIC SPRINKLER
WATER SUPPLIES

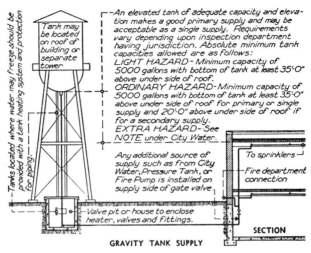

Tank may be located on roof of building or separate tower

Tanks located where water may freeze should be provided with a tank heating system and protection for piping.

—An elevated tank of adequate capacity and elevation makes a good primary supply and may be acceptable as a single supply. Requirements vary depending upon inspection department having jurisdiction. Absolute minimum tank capacities allowed are as follows:
LIGHT HAZARD - Minimum capacity of 5000 gallons with bottom of tank at least 35'-0" above under side of roof.
ORDINARY HAZARD - Minimum capacity of 5000 gallons with bottom of tank at least 35'-0" above under side of roof for primary or single supply and 20'-0" above under side of roof if for a secondary supply.
EXTRA HAZARD - See NOTE under City Water.

Any additional source of supply such as from City Water, Pressure Tank, or Fire Pump is installed on supply side of gate valve.

Valve pit or house to enclose heater, valves and fittings.

To sprinklers

Fire department connection

SECTION

GRAVITY TANK SUPPLY

A pressure tank makes a good primary supply and in some cases may be acceptable as a single supply. Minimum requirements are as follows:
LIGHT HAZARD - 2000 gallons minimum amount of available water.
ORDINARY HAZARD - 3000 gallons minimum amount of available water.
EXTRA HAZARD - See NOTE under City Water.
Pressure tank ordinarily kept 2/3 full of water and an air pressure of 75 lbs. maintained.
Pressure tank should not be used to supply other than sprinklers or hand hose attached to sprinkler piping.

Pressure tank preferably located above top level of sprinklers, but if conditions warrant tank may be located in basement or elsewhere:—

Sprinkler supply piping

Fire department connection

Sprinklers

Any additional source of supply such as from Gravity Tank or Fire Pump is installed on tank side of valve.

SECTION

PRESSURE TANK SUPPLY

FIRE PUMP SUPPLY
A properly located fire pump of adequate capacity and reliability makes a good secondary supply. An electrically driven and automatically controlled fire pump, taking water from an adequate source, may be acceptable as a single supply.
LIGHT HAZARD - 250 G.P.M. min. pump capacity.   ORDINARY HAZARD - 500 G.P.M. min. pump capacity. EXTRA HAZARD - See NOTE under City Water.

AUTOMATIC SPRINKLER
WATER SUPPLIES

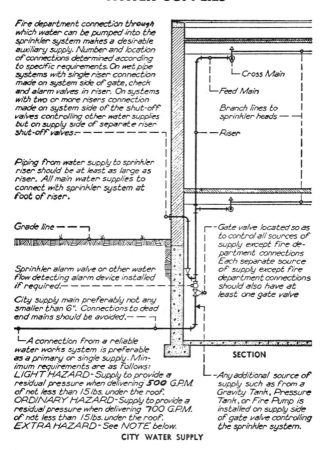

Fire department connection through which water can be pumped into the sprinkler system makes a desirable auxiliary supply. Number and location of connections determined according to specific requirements. On wet pipe systems with single riser connection made on system side of gate, check and alarm valves in riser. On systems with two or more risers connection made on system side of the shut-off valves controlling other water supplies but on supply side of separate riser shut-off valves. — — — — —

Piping from water supply to sprinkler riser should be at least as large as riser. All main water supplies to connect with sprinkler system at foot of riser.

Grade line — — —

Sprinkler alarm valve or other water flow detecting alarm device installed if required. — — — — — — —

City supply main preferably not any smaller than 6". Connections to dead end mains should be avoided. — —

A connection from a reliable water works system is preferable as a primary or single supply. Minimum requirements are as follows:
LIGHT HAZARD - Supply to provide a residual pressure when delivering **500** G.P.M. of not less than 15 lbs. under the roof.
ORDINARY HAZARD - Supply to provide a residual pressure when delivering **700** G.P.M. of not less than 15 lbs. under the roof.
EXTRA HAZARD - See NOTE below.

CITY WATER SUPPLY

Cross Main

Feed Main

Branch lines to sprinkler heads — —

Riser

Gate valve located so as to control all sources of supply except fire department connections Each separate source of supply except fire department connections should also have at least one gate valve

SECTION

Any additional source of supply such as from a Gravity Tank, Pressure Tank, or Fire Pump is installed on supply side of gate valve controlling the sprinkler system.

NOTE Supply needed for EXTRA HAZARD and other various occupancies must be determined by a study of the conditions in each case. Consideration should be given to number of sprinklers that may operate.

Every automatic sprinkler system should have at least one automatic water supply of adequate pressure, capacity and reliability. The necessity for a second independent supply, which is desirable, should be determined by a study of conditions in each case and consultation with the inspection department having jurisdiction.

SIDEWALK ELEVATORS

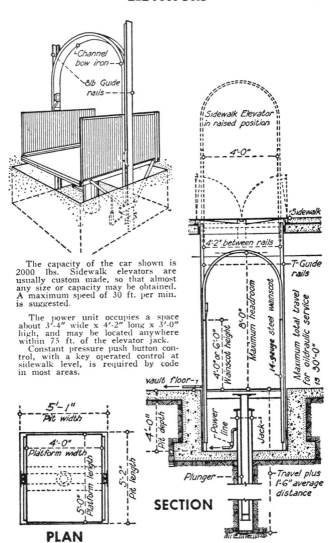

Channel bow iron

8lb Guide rails

Sidewalk Elevator in raised position

4'-0"

Sidewalk

4'-2" between rails

T-Guide rails

8'-0" Maximum headroom

14-gauge steel wainscot

4'-0" or 6'-0" Wainscot height

Maximum total travel for oildraulic service is 30'-0"

vault floor

Power line

Jack

Plunger

Travel plus 1'-6" average distance

SECTION

The capacity of the car shown is 2000 lbs. Sidewalk elevators are usually custom made, so that almost any size or capacity may be obtained. A maximum speed of 30 ft. per min. is suggested.

The power unit occupies a space about 3'-4" wide x 4'-2" long x 3'-0" high, and may be located anywhere within 75 ft. of the elevator jack.

Constant pressure push button control, with a key operated control at sidewalk level, is required by code in most areas.

5'-1" Pit width

4'-0" Platform width

5'-0" Platform length

4'-0" Pit depth

5'-2" Pit length

PLAN

FACTORS IN
ELEVATOR DESIGN

SELECTION OF PROPER EQUIPMENT. Selecting the proper number of elevators, capacity, speed, type of control, size of cars, type of doors, etc., depends on a number of factors requiring careful analysis of each building.

In general, at least two passenger elevators are needed in the ordinary building that requires elevators for its proper functioning.

In larger and higher buildings, the number of passenger elevators, their arrangement, and their control can be determined only after careful study and analysis of all the factors involved.

Selection of freight elevator capacities, speed, size of platforms, and proper controls must be determined by the type of goods or materials to be handled and the proposed flow through the building. Increased demands on freight elevators can be anticipated by specifying larger capacities than those required immediately.

The preliminary selection of passenger elevators involves first determining the characteristics of the building in which the installation will be made—the number of floors, the floor-to-floor heights, elevator travel, optimum location, and the characteristics, travel patterns, and tempo of the building population. This is followed by an assessment of potential round-trip times, passenger waiting intervals, and passenger-carrying capacity of the system, as well as of individual cars.

For final information needed to draw up detailed specifications for elevator equipment, consult elevator manufacturers. Their engineers are fully trained, familiar with architectural problems, and competent to deal with them. They are happy to honor architects' requests for expert engineering assistance.

FREQUENTLY USED DESIGN SPEEDS AND LOADS. The speeds, loads, and platform sizes given in the table below are subject to variation but represent values frequently employed.

Type Building	Speed	Load	Car Size
Office			
Up to 5 stories..........	100–150	2000	6'-4" x 4'-6"
Up to 15 stories	150–600	2500	7'-0" x 5'-0"
Small Apartments	100–150	1500	5'-0" x 4'-6"
Small Hotels	100–150	2000	6'-4" x 4'-6"
Apartment Houses	150–450	2500	7'-0" x 5'-0"
Hotels			
Up to 15 stories.........	150–450	2000	6'-4" x 4'-6"
15 Stories and Up........	450–800	2500	7'-0" x 5'-0"
Hospitals	100—450	4000	5'-8" x 8'4"
Department Stores			
Small..................	100–150	2500	7'-0" x 5'-0"
Large..................	150–300	4000	8'-0" x 6'-0"
Freight	50–150	3000	60 sq. ft.

SECTION DIMENSIONS
ELEVATOR HATCHWAYS

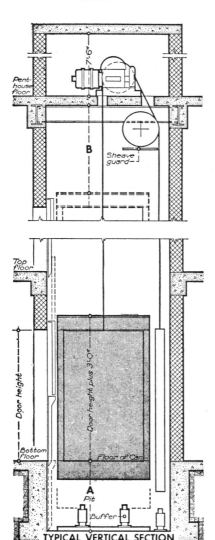

TYPICAL VERTICAL SECTION

The section shown here illustrates the essential parts of any passenger or freight elevator installation, which influence the space which must be allowed vertically: the car, the elevator machinery and the pit.

Feet per min.	A	B
to 200	4'-0"	4'-0"
to 300	6'-9"	5'-0"
to 400	7'-6"	5'-6"
to 550	8'-2"	6'-3"
to 700	12'-6"	6'-9"
to 800	13'-10"	8'-8"

The *Safety Code for Elevators* requires that freight elevators having a travel of more than 2 floors above the main street floor and all passenger elevators shall be installed in fire-resistant hatchways.

Local laws or ordinances should be consulted before proceeding with the erection of hoistways, to determine legally acceptable construction which satisfies local fire-resistance requirement.

PLAN CLEARANCES
ELEVATOR HATCHWAYS

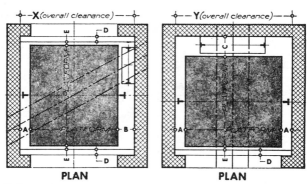

PLAN	**PLAN**

D –Depends on doors.
E –Depends on elevators. "E" is 1" for Passenger Elev. and 1¼" for Freight Elev.

Speed of car NOT over 200 ft. per min.

	Weight of Car Tee in lbs.	A	*B	*C
Up to 10,000 lbs. capacity and 100 sq. ft. platform area......	15	7″	1′–1″	1′–0½″
Up to 20,000 lbs. capacity and 150 sq. ft. platform area......	22	8″	1′–1″	1′–0½″
Over 20,000 lbs. or over 150 sq. ft. of platform area, or both...	30	9″	1′–1″	1′–0½″

For speed over 200′ per min., B or C must be 1′-2″.

Size of guide rails that must be used determines the clearances in the horizontal plane. The *Safety Code for Elevators* indicates the guide rails that are required for cars of different weights and capacities.

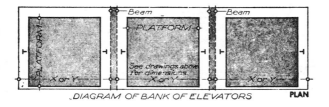

DIAGRAM OF BANK OF ELEVATORS   **PLAN**

PASSENGER ELEVATOR
DOOR TYPES

PASSENGER ELEVATOR DOORS. The nature of the traffic, space available, and architectural effect will govern the selection of the doors. It should be observed that the type of door selected will govern the detail of the door sill.

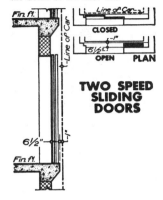

TWO SPEED SLIDING DOORS

Two-Speed Sliding Doors. These are used where a wider opening into the car is desirable. They allow an opening equal to about ⅔ the width of the car. These are probably the most commonly used type, especially where doors are power-operated.

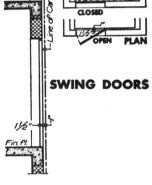

SWING DOORS

Single Swing Doors. Swing doors are specified for passenger elevators for apartments, hospitals, small office buildings, etc., because they are easy to operate, quiet, and have very little equipment to be maintained or get out of order. They are also the least expensive.

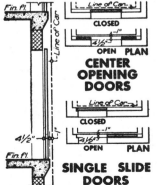

CENTER OPENING DOORS

SINGLE SLIDE DOORS

Center Opening Doors. These doors usually allow an opening about ½ the width of the car but more readily permit a successful architectural treatment since both doors are in the same plane. Center opening doors are sometimes advantageous because they operate faster than the other types.

Single Slide Doors. These are used where a door opening about ½ the width of the car is acceptable.

FREIGHT ELEVATOR
DOOR TYPES

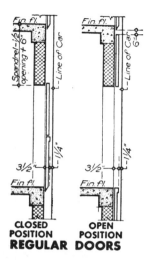

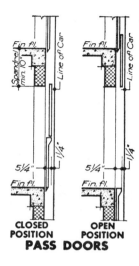

CLOSED POSITION   **OPEN POSITION**

REGULAR DOORS

CLOSED POSITION   **OPEN POSITION**

PASS DOORS

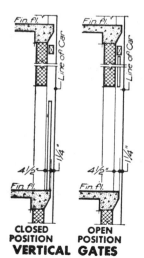

CLOSED POSITION   **OPEN POSITION**

VERTICAL GATES

REGULAR DOORS. This type is for general use in freight elevator openings in commercial or industrial buildings. Examination of the sections at the upper left show that the spandrel height must be equal to half the door opening height plus 6".

PASS DOORS. Where the spandrel height is not sufficient to accommodate regular type doors, pass type doors may be utilized. Pass doors may be installed where the spandrel height is as little as 10". This type is suitable for general use in freight elevator openings in commercial, industrial, and other types of buildings.

VERTICAL GATES. This type of elevator opening protection is used where local laws do not demand fireproof door protection, because it is least expensive.

645

GARAGE DOORS –
SWINGING OR SLIDING

DESCRIPTION.   Though most garage doors installed today are of the roll-up variety, there are still occasions when swinging or sliding doors are desired. Fir is the preferred material for swinging or sliding doors intended for paint finish. Some stock types are available in various sizes, and doors of any style or size can be custom-made by a door and sash mill.

These doors are usually hung in pairs on hinges—generally to swing outward— or on rolling hangers that allow the doors to meet at center and slide back to either side. Tracks may be mounted either inside or outside of the building. An alternative method suitable for large doors is a bypassing track arrangement. Sets of three doors may be hung on a track to slide around a corner, or one leaf may be hinged with the remaining two on a track hanger to fold in accordion fashion. Multiple doors may be hung with combination arrangements of tracks and hinges. It should be obvious that this type of door is not adaptable to overhead hardware.

THICKNESS.   Though these doors, especially in the smaller sizes, may be made in 1⅜″ thickness, the use of this size thickness is not recommended because of its susceptibility to warping and bowing and consequent problems of an ill-fitting and poorly operating installation. The minimum thickness employed should be 1¾″, allowing much greater strength and stability. Larger size doors, of the barn-door variety, may exceed 2″ in thickness.

SIZES.   Doors may be custom-made in any size desired, to fit special applications. Generally, however, framing rough openings to fit stock door sizes or modular sizes even when the doors are custom-made, is less expensive. Widths starting at 2'-0″ and increasing in increments of 6″ and heights starting at 7'-0″ and increasing in increments of 6″ prove the most satisfactory.

DESIGN.   The designs shown on this and the following two pages suggest the use of these doors where appearance is not of special importance and the design used need not follow other existing designs or where they would comple- ment existing architectural niceties of the building. Some of the designs shown are old, standard ones that have not changed for decades. However, they remain useful as they stand, and their patterns or dimensions may also be helpful in designing special doors.

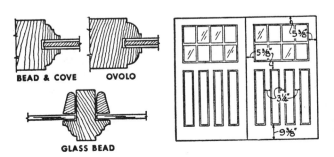

BEAD & COVE   OVOLO

GLASS BEAD

GARAGE DOORS —
SWINGING OR SLIDING

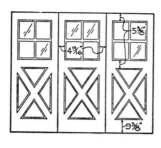

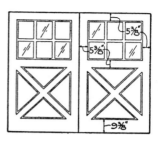

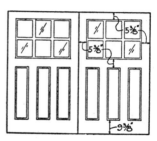

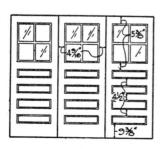

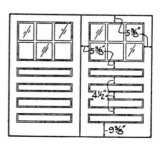

GARAGE DOORS —
SWINGING OR SLIDING

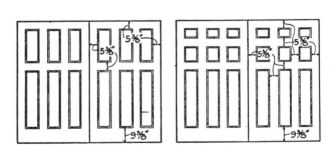

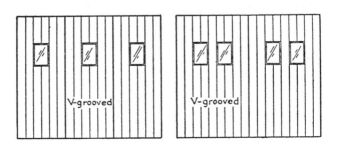

PRESERVATIVES
FOR WOOD

PRESSURE TREATMENTS. These have been established by experience over many years as a definitely superior means of obtaining maximum service against decay and insect attack. Pressure treatments generally begin with placing the wood that is to be treated into a large steel cylinder, which is then sealed and filled with an appropriate preservative. Various temperature and pressure combinations are used to produce the desired retention and penetration levels, after which the wood is removed and readied for use. Two general processes are used; full-cell, and empty-cell.

The full-cell process is used when high retention of creosote preservatives is required and when water-borne preservatives are involved. The cell walls of the wood absorb the preservative, and the cells themselves are also left full.

The empty-cell process is always used with oil-borne preservatives when the desired retention can be achieved by that method. In this process, a variable amount of the air naturally in the wood cells is trapped there, or the wood is subjected to additional air pressure, so that the cells cannot fill with preservative. A proper balancing of pressures and temperatures results in the desired preservative retention. The treated wood is drier and has quite deep, uniform preservative penetration.

NONPRESSURE TREATMENTS. These should not be used for maximum service when it is practicable to use pressure treatments, since the protection afforded is far less; they should never be used for wood that will remain in contact with the ground unless they are intended for only temporary protection.

The most successful of the nonpressure treatments are the vacuum processes, whereby air is exhausted from the pores of the wood under a vacuum, and the preservative replaces the air under atmospheric pressure. The "brief dip" process is less effective than a pressure treatment but can be useful at times; the wood is simply submerged in a tank of preservative for at least 3 minutes.

Since penetration of the wood is slight with these processes, they are primarily useful for millwork products such as window, door, and trim parts that are exposed to moisture only occasionally. Commonly used solutions are petroleum solvents containing zinc naphthenate or pentachlorophenol. Wood that is cut or bored after treatment should always be re-treated by being liberally brushed at the cuts with a preservative prior to assembly.

BRUSH AND SPRAYING TREATMENTS. These should be applied in at least two coats. Penetrations obtained will usually be less than $\frac{1}{16}''$. Brush and spray treatments are used on surfaces cut after treatment. However, it is more practical than is commonly supposed to design wood structures so that all cutting, framing, and boring of holes may be done before treatment.

RECOMMENDED PRACTICE. For full protection against decay and attack by termites and other pests, only pressure-treated wood stock should be considered. Sufficient impregnation of the wood by the preservative cannot be obtained in any other way, and the greater the penetration is, the longer the useful life span of the wood will be.

The specific preservative treatments recommended vary with the location of the wood (not in contact with the ground, in contact with the ground, buried in the ground, marine environment, and so forth) and with the use of the wood (wood foundation, sills, plates, sheathing, siding, millwork, fencing, posts, and so forth). Different preservatives—and there are many—are used for different purposes; the specifications for the treated wood stock are established in terms of the particular preservative, the wood upon which it is used, the moisture content of the wood when treated, and (perhaps most importantly) the minimum retention of the preservative in the wood, in terms of pounds of preservative per cubic foot of wood.

Current recommendations and standards issued by the American Wood Preservers Association and the American Wood Preservers Bureau should be followed for pressure-treated wood products. The National Woodwork Manufacturers Association certifies non-pressure-treated commercially manufactured millwork under its industry standard.

PRESERVATIVES
FOR WOOD

DESCRIPTION.   The life of wood placed under conditions favorable to decay, attack of insects, or marine borers can be considerably extended by treatment with suitable preservatives. The penetrability of preservatives varies considerably depending upon tree species. Sapwood is more easily penetrated than heartwood. Treatment does not appreciably affect the ultimate strength of woods. Some few woods, such as the heartwood of baldcypress and the butt section of redwood, have a natural decay resistance without preservative treatment.

Some preservatives are more effective than others. All possess certain disadvantages that limit their use, as well as advantages that make them especially suitable for specific purposes. They fall into three general classes:

1. Those commonly called "preservative oils," which are relatively insoluble in water.
2. Salts injected into the wood in the form of water solutions.
3. Toxic material combined with a solvent, usually volatile, other than water.

1. PRESERVATIVE OILS.   Preservative oils are ordinarily applied to posts, poles, ties, and any material that will be in water, in contact with the soil, or in any other situation where high-moisture conditions prevail. If oil-treated material is to be used in buildings or where bleeding is especially undesirable, only straight creosote should be used.

Coal-tar creosote is the most important and most generally useful wood preservative. Coal-tar creosote is a black or brownish oil made by distilling coal-tar. The character of the various coal-tar creosotes available may vary considerably but satisfactory results may be expected from any good grade.

The advantages of coal-tar creosote are: (1) its toxicity to wood-destroying fungi and insects; (2) its relative insolubility in water and its low volatility, which impart to it a great degree of permanence under the most varied conditions; (3) its ease of application; (4) the ease with which its depth of penetration can be determined; and (5) its general availability and low cost.

For some purposes, coal-tar creosote has disadvantageous properties. Freshly creosoted timber can be ignited and will burn, producing a dense smoke. After seasoning, however, the creosoted wood usually is little, if at all, easier to ignite than sound untreated wood and is less flammable than untreated but decayed wood.

Creosoted wood can be used in sills and foundation timbers, in floor sleepers embedded in or resting on concrete, and even in subflooring—though unless it is sealed off from the interior, occupants are likely to be aware of its odor for some while (especially in the case of freshly treated stock). Foodstuffs that are easily affected by odors should not be stored near creosoted wood; the vapors are harmful to growing plants.

Many persons object to and may be physically upset by the odor of creosote. Workers often object to handling creosoted wood because it soils clothes and can cause skin burns similar to sunburn. Gloves furnish protection against creosote burn, and hand creams or lotions can be used to soothe it.

For the most part, it is impossible to paint over creosoted wood; however, certain types of aluminum paint may be used alone or as a primer over some kinds of creosote-treated wood—especially after the wood has weathered for a year or so, allowing reasonable paint receptivity if finish coating of a different color is desired.

Creosote solutions consist of solutions in various proportions of coal-tar creosote to either coal-tar or petroleum oils. These solutions have been used for many years and have a good performance record, through they are not as effective as 100% creosote. Principal applications are for structural and bridge timbers, piles, and railroad crossties and switchties. Several different classes are in common use, containing 80%, 70%, 65%, 60%, and 50% creosote by volume.

Chlorinated phenol solutions consist mainly of pentachlorophenol in heavy petroleum oils or in liquid petroleum gas. These solutions are in widespread use for preserving poles and timbers, but they are not recommended for marine applications. The chlorinated phenols have a high degree of toxicity and must be handled carefully.

Other preservative oils include wood-tar creosote, water-gas-tar creosote, oil-tar creosote, coal tars, and various toxic compounds in oils. These have not found wide use in building construction because of cost, lack of effectiveness, lack of service test record, or other reasons.

PRESERVATIVES
FOR WOOD

2. WATER SOLUBLE SALTS.   Inorganic salts and similar materials used in water solutions as wood preservatives include chromated zinc chloride, acid copper chromate, chromated copper arsenate in three varieties (called Types I, II, and III), ammoniacal cooper arsenite, and fluor chrome arsenate phenol. Preservative salts are ordinarily used in buildings where oil-treated wood is unacceptable because of odor, color, oily surface, or unsatisfactory receptivity to paint. All of these preservatives afford a high degree of protection for woods not in direct contact with the ground and not in permanent contact with water. Chromated copper arsenate and ammoniacal copper arsenite processes are suitable for such ground-contact and fresh-water-contact purposes as wood foundations, pilings, and poles, since they have a high resistance to leaching and have a very good performance record in such uses. The same two preservatives, applied with very high retentions, are suitable for salt-water-contact uses. However, the most effective protection against marine borers is achieved by means of high retention treatment with a copper-containing preservative followed by coal-tar creosote treatment.

All salt-treated lumber for use in buildings or other places where high moisture content or shrinkage after installation would be a disadvantage should be properly seasoned after treatment and before use. Preferably, the level of drying after treatment should approximate the moisture content level to which the wood will be subject during service.

Chromated zinc chloride consists either of approximately 80% zinc oxide and 20% chromium trioxide, or of approximately 80% zinc chloride and 20% sodium dichromate, under Federal Specification TT-W-551 and AWPA Standard P5. This "old reliable" preservative is not especially effective in ground-contact or permanently wet applications, but it is excellent for relatively dry uses and is low in cost. It also has fire retardant capabilities and, as *chromated zinc chloride (FR)* consisting of 80% chromated zinc chloride, 10% ammonium sulfate, and 10% boric acid and used at a retention level of 1½ to 3 pounds per cubic foot, is employed for that purpose.

Chromated copper arsenate is a mixture of three chemicals: chromium trioxide, copper oxide, and arsenic pentoxide. These may be mixed in three different combinations, under Federal Specification TT-W-550, to produce Type I (Erdalith, Tanalith), Type II (Boliden), and Type III (Wolman). Chemical substitutes are allowable: arsenic acid or sodium arsenate can be used for arsenic pentoxide; potassium or sodium dichromate for chromium trioxide; and basic copper carbonate, copper sulfate, or copper hydroxide for copper oxide. All three types have excellent service records.

Ammoniacal copper arsenate is covered by Federal Specification TT-W-549. It consists of approximately 50% copper hydroxide and 50% arsenic trioxide or arsenic pentoxide, plus a bit of acetic acid. This preservative affords excellent protection against termites and rot and, used at high retention levels, can be employed in certain marine applications.

Acid copper chromate is covered by Federal Specification TT-W-546; it is composed of 31.8% copper oxide, copper sulfate, sodium dichromate, or potassium dichromate, and 68.2% chromic acid. This preservative has a good record and affords good protection against decay and termites, but it is not especially effective against marine borers.

Fluor chrome arsenate phenol is covered by Federal Specification TT-W-535; it consists of a particular mix of four chemicals: sodium or potassium fluoride, chromium trioxide, arsenic pentoxide, and dinitrophenol. Certain substitutions of chemicals may be made, such as sodium pentachlorophenate for dinitrophenol, in order to avoid objectionable staining of the wood. This preservative, sometimes known simply as FCAP, is widely used to treat wood used in above-ground structures such as decks and porches, as well as for some ground-contact purposes. One of its best-known names in the field is Wolman salts.

3. TOXIC MATERIAL IN NONAQUEOUS VOLATILE SOLVENTS.
This group of preservatives meets the need for a clean treatment that does not swell wood but leaves it odorless and paintable. Many such preservatives are sold under trade names.

Some preservatives of this type are used for window sashes, frames, and doors, and in the treament of flooring, furniture, and millwork exposed to fungus and termite attack.

MINERAL SURFACED ASPHALT SHINGLES

The multiplicity of designs and weights of individual and strip shingles that are available, the various methods of laying both types, and the colors offered create almost infinite permutations. In the table on the facing page are listed the combinations of method, grade, and size most generally used.

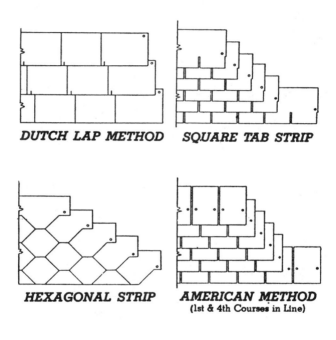

DUTCH LAP METHOD   **SQUARE TAB STRIP**

HEXAGONAL STRIP   **AMERICAN METHOD**
(1st & 4th Courses in Line)

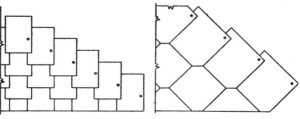

WIDE SPACE METHOD   **FRENCH METHOD**

MINERAL SURFACED ASPHALT SHINGLES

Type	Weight /Square	Size	Bundles /Square	No. /Square	Exposure	Side-Lap	Top-Lap	Head-Lap
American	330 lbs.	12" x 16"	4	226	5"		11"	6"
Dutch Lap	165 lbs.	16" x 12"	2	113	10"	3"	2"	
Strip (Square Butt)	215 lbs. 220 lbs. 225 lbs. 235 lbs. 260 lbs. 265 lbs. 280 lbs. 290 lbs. 300 lbs. 320 lbs. 325 lbs. 330 lbs. 345 lbs.	12" x 36"	3-5	78-84	5"		7"	2"
One-tab Two-tab Three-tab Architectural								
Long Strip	225 lbs. 320 lbs.	13¾" x 39⅜"	3-5	65	5⅝"		7⅜"	2"
Two-Tab Hexagonal	195 lbs.	11⅓" x 36"	3	86	5"		2"	2"
Three-Tab Hexagonal	195 lbs.	11⅓" x 36"	3	86	5"		2"	2"
Staple	145 lbs.	16" x 16"	2	80		2½"		
Lock	150 lbs.	16" x 16"	2	80		2½"		
	180 lbs. 245 lbs. 250 lbs.	19½" x 21½"	2-3	104-108				

Note: Sizes, shapes, styles, weights, and other specifications of asphalt and fiberglass shingles vary widely among manufacturers and change periodically.

MINERAL SURFACED
ASPHALT SHINGLES

DESCRIPTION.   Asphalt shingles are made with a felt base impregnated with asphalt, or with a fiberglass base coated with asphalt. The base is coated with a more viscous asphalt on the surface that is to be exposed to the weather and is surfaced with either slate or rock granules embedded in the asphalt coating. The underside is coated with fine mineral powder, usually talc or mica, to prevent sticking.

Asphalt shingles are variously described as composition shingles, slate-surfaced shingles, or fiberglass shingles, and they are sometimes confusedly called asbestos shingles even though they contain no asbestos fibers. The felt used in the manufacture of asphalt shingles is usually composed of organic (cellulose) fibers derived from wood, paper, and rags, while the fiberglass base is a mat of inorganic (glass) fibers.

The asphalt used is known as asphalt flux; it is a petroleum product derived from the fractional distillation of crude oil. Two grades of asphalt flux are used: a saturant to saturate or coat the base, and a coating asphalt consisting of asphalt mixed with mineral stabilizers. The latter type affords substantially increased resistance to weather and is the material in which the mineral granules are embedded.

Surfacing materials that consist of several types of colored granules are in general use. Natural and fired granules present a wide variety of colors.

FIRE RESISTANCE.   Shingles made and applied according to the specifications of the Underwriters Laboratories are eligible to receive the Class C label that identifies them as being "effective against light fire exposure." That is, they "are not readily flammable and do not readily carry or communicate fire; afford at least a slight degree of fire protection to the roofdeck; do not slip from position; possess no flying brand hazard; and may require occasional repairs or renewals in order to maintain their fire-resisting properties."

INSTALLATION.   The normal roof slope is 4" in 12" or greater. Square butt strip shingles can be applied to slopes as low as 2" in 12". The deck should consist of 1" x 6" T&G No. 3 (or No. 2) boards or No. 2 decking of southern pine, Douglas fir, western larch, hemlock, ponderosa pine, spruce, or white fir, laid across or diagonal to rafters and nailed with 8d nails twice at each bearing point. Plywood decking should be of a thickness appropriate to the rafter span and roof loading factors, of exterior grade C–C wherever edges are exposed to weather and of interior grade with exterior glue within the field. Nailing should conform to APA specifications.

Before shingles are laid in new construction, masonry chimneys should be completed or appliance chimneys installed, chimney flashing should be in place, vent pipes and jacks installed, ventilators should be in place, and gutters should be hung. Valley flashing consists of a lower course of a half-width strip of mineral or smooth-surfaced 90-pound roll roofing (with surfacing down), nailed every 18" along its edges, and an upper course of full-width roll roofing (with surfacing up), not nailed. Alternatively, suitable metal flashing may be used. Eaves and rakes should be flashed with a 1" by 4" angle formed of sheet metal, nailed so that the 1" leg projects ½" beyond the edge of the roof sheathing to serve as a drip edge.

Where roofs abut vertical masonry, metal cap flashing should be in place. Wood cant strips shold be placed where the roof abuts any vertical surface. Redwood or cedar beveled siding (clapboards), with the thick edge butted against the vertical surface, serve nicely for this purpose.

654

MINERAL SURFACED
ASPHALT SHINGLES

WEATHER RESISTANCE. In a survey conducted some years ago by the Bureau of Standards, the average years of service were reported for each state; these are shown on the accompanying map. It should be noted that the averages are composite for all types of asphalt shingles—differing in weight, manufacturing source, and method of laying. Though there have been varying weights, specific compositions, sizes, and minor changes in other similar details through the years (a process that will doubtless continue), these approximate values are valid and probably will remain so in future. It can be safely assumed that using heavier weights, higher quality levels, and methods of laying that increase the lap will produce a lifespan for shingles that exceeds the survey averages. Using lighter weights, lower quality levels, and methods of laying that minimize the lap will be likely to reduce the lifespan of a given installation below that of the survey average. The values given apply to felt-base shingles. The newer fiberglass-base shingles have not yet been in use long enough to build up a use-experience history that would allow a valid comparison of longevity. However, indications are that fiberglass-based mineral surfaced shingles will be somewhat longer-lived than their felt-based counterparts, if other specifications are similar.

COLD WEATHER PRECAUTIONS. Spaces under board-sheathed roofs should be ventilated to prevent warping; this is unnecessary if the sheathing is of plywood. Proper roof ventilation to prevent moisture buildup under the roof deck, however, is essential in either case. It is not advisable to attempt installation at temperatures below 40°F, and if outside temperatures are below 60°F, the roofing should be stored in a warm place for 24 hours before application and removed from storage in small amounts as needed. It is poor practice to walk on asphalt roofing at any time, but this is especially so in very cold weather. Roof service work—particularly snow/ice removal—should be undertaken with caution.

AVAILABILITY. Both felt-base and fiberglass-base asphalt shingles are manufactured at numerous plants around the country. They are available at lumberyards and building supply outlets everywhere, in a wide variety of colors, weights, and sizes.

LOW-TEMPERATURE
INSULANTS

FINISHES. Finishes and overlays applied to low-temperature insulant faces should be of a type and specific material recommended and approved by the manufacturer of the insulant; appropriate finishes vary considerably. Glass block, for example, may be finished with gypsum plaster over metal lath, or with a coating of a special white vinyl emulsion paint. Polyurethane sheets may be covered with a variety of solid finishes. In many cases an interfacing/protective layer of another material must be placed between the insulation and the finish. The drawings below show typical traditional cork insulation installation methods; those for other rigid insulants are generally similar but differ in materials and details.

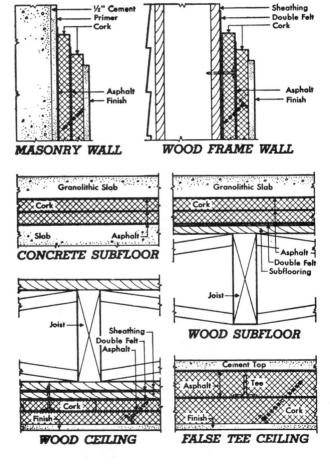

MASONRY WALL

WOOD FRAME WALL

CONCRETE SUBFLOOR

WOOD SUBFLOOR

WOOD CEILING

FALSE TEE CEILING

LOW-TEMPERATURE INSULANTS

DESCRIPTION. The rigid insulants used in commercial/industrial low-temperature applications are often referred to as "low-temperature block insulation." The term does not accurately describe either the forms or the types in which the material is manufactured. Though some insulants are made in forms resembling thin blocks, others are produced in shapes more properly considered as sheets and boards. In addition, some of the same insulants are employed in relatively higher (that is, normal atmospheric) temperature ranges, as well as in residential and general construction applications. Some loose-fill insulants such as expanded perlite or exfoliated vermiculite can also be used in appropriate kinds of low-temperature constructions.

CORK BLOCK is manufactured from ground cork that is molded and baked. The baking melts the natural resinous gums surrounding the cells, binding them together.

GLASS BLOCK consists of true glass that has been cellulated in manufacture so that a section reveals a structure of tiny (5 million per cubic foot) sealed air chambers that are completely impervious to moisture.

MINERAL WOOL BLOCK consists of compressed loose wool with suitable binders to form a rigid material. Mineral wool is a generic term covering a number of similar products differentiated chiefly by the raw materials from which they are made. All are composed of very fine interlaced mineral fibers having the appearance of loose wool or cotton.

RIGID POLYSTYRENE is available in two types—expanded and extruded. Polystyrene is a derivative of styrene that is formed by catalytic polymerization. Expanded polystyrene consists of tiny beads of plastic melted together in board form, and is often called "beadboard." Extruded polystyrene is made in board form and is composed of thousands of tiny closed plastic cells; it is generally of higher density than the expanded type.

RIGID POLYURETHANE AND POLYISOCYANURATE are plastic foam insulants manufactured in board and sheet form. Polyurethane is made as a reactive mix of isocyanurates and alcohols, while polyisocyanurate is catalytically formed from isocyanurates alone. Their physical and insulating characteristics are similar. Both types are generally covered with factory-applied paper or reflective foil "skins."

STRUCTURAL SHELL. Walls, floors, and ceilings should be of solid construction. Monolithic concrete, solid brick, or concrete block is recommended rather than wood frame. Unit masonry joints should be struck flush.

If insulation is to be applied directly to structural shell surfaces, the surfaces must be smooth and true. Unit masonry walls, unless they present an extremely true surface, should be plastered with 1:3 portland cement plaster in two coats, with a minimum total thickness of ¼″. The finish coat should be floated to a true, plane surface to receive the insulation.

Construction with air spaces such as occur in hollow masonry or sheathed wood frame are best avoided if possible; if such construction is used, the spaces may be left open to provide air circulation. Sheathing should be of treated T&G hemlock, pine, spruce, or fir, or of exterior plywood in a suitable thickness.

Self-sustaining partitions and interior walls can be constructed by utilizing temporary studs for alignment.

INSTALLATION. Methods of installation vary between different insulants and sometimes between specific insulation products of the same general type. Glass block, for example, may be installed by first applying an unfilled asphalt cutback to the plastered walls, and then dipping the backs and butting edges of the blocks in hot asphalt and immediately applying them to the primed surface. Alternatively, a special adhesive mastic may be applied to each block with a grooved trowel of proper dimensions, and the block may then be pressed into place against the wall. Another option is to apply gobs of adhesive to the block corners, after which the block can be pressed into place.

Depending upon the installation and the type of insulant, two or more layers may have to be installed to achieve the total thermal efficiency determined by heat loss/gain calculations. Such layering should be done with staggered joints. Vapor barriers, ventilation air spaces, protective coverings, or other additional work may also be needed to complete a satisfactory and guaranteed installation. In all cases the work must follow the manufacturer's application and installation instructions exactly.

LOW-TEMPERATURE INSULANTS

The table below gives typical properties of various forms of insulants used in low-temperature installations. The thermal coefficients given may be used with satisfactory results; however, in the interest of more accurate calculations, specific values may be obtained from the manufacturers of particular insulating materials.

GENERAL PRINCIPLES. Three forces attempt to drive moisture from the warm side to the cold side of a barrier: (1) wind or air current pressures; (2) atmospheric pressures due to difference in the density of air at the different temperatures; and (3) vapor pressure due to the difference in absolute humidity at the different temperatures.

A cold room with one or more walls exposed to extremely low outside temperatures in the winter might have heat, air, and vapor differentials tending to create a flow from the inside of the cold room to the exterior, instead of the other way around, as would occur in the summer.

The basic purpose of low temperature installations utilizing organic or fibrous insulation is to protect the insulating material from moisture permeation's damaging effects on the material itself and on its performance. The joints between rigid blocks of any type must be sealed against the infiltration of air and moisture that would make maintaining the interior temperature uneconomical and that would create a deposit of frost on coils, pipes, or plates.

In addition to a satisfactory thermal coefficient, low temperature insulants should be checked for strength, compressibility/crushing factors, freedom from odors, workability with tools, shrinkage factor, bond strength with adhesives, compatibility with other adjacent materials or finishes, incombustibility, moisture resistance, susceptibility to rot or vermin attack, and of course cost-effectiveness and availability.

TYPICAL THERMAL CONDUCTIVITY (k)
OF LOW TEMPERATURE INSULANTS

Type	Density (lb./cu. ft.)	Temperature (°F)			
		−25°	0°	25°	50°
Fine mineral fiber blanket	1.0	0.23	0.24	0.24	0.27
	1.5	0.21	0.22	0.23	0.25
	2.0	0.20	0.21	0.22	0.23
	3.0	0.19	0.20	0.21	0.22
Semirigid mineral fiber felt	3.0	0.19	0.20	0.21	0.22
Glass fiber board/block	3-10	0.19	0.20	0.22	0.24
Cellular glass	8.5	0.33	0.35	0.36	0.38
Extruded rigid polystyrene	3.5	0.15	0.16	0.16	0.17
	1.8	0.19	0.20	0.21	0.23
Rigid polystyrene — beads	1.0	0.23	0.24	0.25	0.26
Polyurethane	1.5-2.5	0.18	0.18	0.17	0.16

LOW-TEMPERATURE INSULANTS

USES. Cold rooms are used for the cold storage of meat, fruit, vegetables, candy, dairy products, ice, furs, beer, and so forth, and for the processing of foods, ice-cream, beer, and other products. Locker plants and air-conditioning ducts and apparatus require a low-temperature type of insulation. One feature of better-class homes of the future might well be a walk-in cold storage room with a fast-freeze compartment.

THICKNESS OF INSULATION REQUIRED. The correct temperatures of cold rooms for different purposes will vary within quite wide limits. Be wary of manufacturer's literature and various government specifications suggesting minimum thicknesses of insulation based only on such cold room temperatures, without reference to average outside temperatures during the period of peak refrigerating load. This is ridiculous because the total heat leakage through the walls, floor, and ceiling of the cold room is a function of the inside/outside temperature difference—not the interior temperature only. Heat leakage establishes the original cost of the refrigeration plant and maintenance.

By the same token, "rule of thumb" estimates suggesting that 2″ of this kind of insulation or 3″ of that kind will be adequate under assumed approximate operating conditions frequently lead to disappointing and often expensive results. The insulating value of different insulation materials—even of the same type—varies, and outside weather conditions cannot be generalized. The values of specific insulants, as well as highly localized weather conditions in which the cold room will be operating, must be carefully assessed to arrive at the proper and most effective insulation thickness.

The dewpoint temperature of the exterior surface of cold rooms might also be a factor in insulation thickness where spaces adjoining the cold room are subject to high humidity.

A complete analysis of required insulation thickness involves identifying the type of insulant, the insulating value of the specific brand of insulant, the installed cost of the material, the present cost of electric current and projected increases in same, and the interest on the plant investment. These must be compared, often among several alternative insulating/construction arrangements, and balanced against one another; at the same time, any other advantages or disadvantages that might obtain in one alternative or another (availability, speed of installation, relative combustibility, and so forth) must also be taken into consideration so that the most favorable economic balance is reached.

If the supporting construction of the cold room contributes to the overall thermal efficiency of the finished unit, the value of those materials should be included in the final determination of the insulation thickness.

More thickness does not lead to a commensurate increase in thermal efficiency. That is, 4″ of insulation of a given type is not twice as effective as 2″—efficiency is governed by a curve of diminishing returns. After a certain point, added insulation has little or no value, thermally or economically. This point can be readily determined in any given case by running a series of heat loss/gain calculations and plotting the results.

LUMINESCENT PAINTS

DESCRIPTION. Any emission of light not ascribable directly to incandescence, and therefore occurring at low temperatures, is luminescence. (The word *luminous* includes all classes of objects that emit light, whether or not as the result of incandescence, and hence is not as accurate an adjective for cold light–emitting materials as *luminescent*.) Luminescent paints are coatings applied variously by dipping, spraying, or brushing, that will emit light during or after excitation by a light source (these are called photoluminescent) or that will emit light without any form of external excitation (these are called autoluminescent). Commonly used luminescent paints fall into the following three classes:

1. *Phosphorescent paints* are nontoxic and photoluminescent; they exhibit a glow for a considerable time after exposure to an external source of either "near" ultraviolet or visible light. All phosphorescent paints are also fluorescent. However, phosphorescent pigments do not fluoresce as brilliantly as fluorescent paints.

2. *Fluorescent paints* are nontoxic and photoluminescent; for all practical purposes they may be said to emit light only during the period of excitation by an external source such as ultraviolet energy (popularly known as "black light") or some other light source.

3. *Radioactive paints* are autoluminescent and require no excitation from external sources. Radioactive paints are both phosphorescent for brief periods and fluorescent.

LUMINESCENT INTENSITY. Frequently the uninitiated find the intensity of light emitted from luminescent paints disappointing. Optical adaptation to darkness varies widely among people as a result of differences in many complex factors such as the observer's supply of vitamin A, the Purkinje effect, etc.

Immediately after blackout, at dusk, or on a moonlit night, the eyes may have difficulty seeing the light emission of luminescent paints. After complete darkness-adaptation of the eyes, a light intensity of 0.02 microlambert can be distinguished. (This is about equal to 0.000004 times the brightness of an ordinary candle.)

REFERENCES. The information given here is generalized. Before outlining specifications for luminescent paints, up-to-date information on formulations, uses, application techniques, and the like should be obtained. This is best done by contacting specific manufacturers of products required, since specifications (and manufacturers) change with relative frequency.

PHOSPHORESCENT
PAINTS

USES. The use of this material ordinarily should be confined to objects which are to be seen in complete or nearly complete darkness, and for such uses as normally occur after dark-adaptation of the eyes has taken place.

Because the surface to be painted as well as the paint application itself can be controlled better in a factory than at the site, materials such as paper, cardboard, wallboard, adhesive tape, and decalcomanias that are factory-coated with the phosphorescent paint are available. Likewise, markers and other products composed of transparent plastics impregnated with phosphorescent pigments are available.

Murals, decorative designs and ornaments, directional markers and safety warning signs, switch plates, kick plates, doorknobs, furniture trim, and light shades are well-known uses for phosphorescent paints. Phosphorescent materials act as an emergency light source in the case of power failure, to permit movement in a room, place of assembly, or factory. Many new decorative, convenience, and safety applications of phosphorescent materials are possible.

EXCITATION OF PHOSPHORESCENT PAINT. Daylight, visible artificial light, and "near" ultraviolet light (3,200–3,900 AU) will activate phosphorescent paint. Mercury vapor and standard fluorescent lamps are probably the most efficient excitation sources. The greater the intensity of light falling on the pigment, the greater will be its initial emission of afterglow.

COLOR. The daylight color of phosphorescent pigments is generally a light gray or light yellow. Attempts to change the daylight color by the addition of nonluminescent pigments will adversely affect the paint's luminescence by screening out the activating light or absorbing the emitted light. Very small amounts of a transparent synthetic dye may be added to change the daylight color with only slight loss of phosphorescence.

APPLICATION. The painting surface should be clean and dry. It is good practice to use two undercoats prepared with zinc sulfide (regular white pigment, not the luminescent pigment), lithopone, high strength lithopone, titanated lithopone, or titanium dioxide. Lead or other metallic base paints should not be used. The same vehicle used in the luminescent coating should be employed in the undercoat. The white base coat provides a good light-reflecting background and protects the paint from the destructive effects, if any, of the surface to be painted.

After the undercoat is dry, the phosphorescent paint should be applied with an absolutely clean, dry brush; the paint should be stirred with a wooden stick or glass rod just prior to and during application. Being of a coarse, crystalline structure, phosphorescent pigments provide relatively poor brushing and spraying characteristics. Uniform coverage, however, can be obtained by the application of two coats. For maximum phosphorescence these paints should be spread so that a total of 1 gallon of paint is applied to 50 or 60 yards of surface.

The calcium and strontium pigments are particularly susceptible to deterioration by moisture, and if such paint films are to stand up under high humidities or exposure, they must be protected by a coating of protective vehicle. In fact, a protective coating is a desirable precaution for *all* phosphorescent paints.

DURABILITY OF PHOSPHORESCENT PAINT. Phosphorescent pigments will eventually deteriorate, although none of the pigments fail because of continued reexcitation. Zinc and cadmium pigments are quite stable and can remain in continuous use for several years. Calcium and strontium pigments, when properly protected from moisture, can be expected to give service for several months or more under severe outdoor exposure and longer service indoors.

FLUORESCENT
PAINTS

USES. Fluorescent pigments are used in plastics, paints, dyes, printing inks, and paper. Fluorescent pigments have no useful afterglow, and their uses are confined to those applications where it is possible and desirable to have a special black light source which can supply invisible light to the pigments when luminescence is required. Thus, fluorescent pigments are electrically dependent.

Impregnated plastics have been used for luminescent electric lamp shades, costume jewelry, etc.

Fluorescent pigments have been used in the preparation of printing inks and paper for use in airplane instruments, maps, wallpaper, decalcomanias, theater programs, etc.

Fluorescent dyes have been used for draperies, upholstery, wall and floor coverings, theater seats, arm rests, and aisle carpeting.

EXCITATION OF FLUORESCENT PAINT. Fluorescent materials require a light source that contains little or no visible energy if their fluorescent light is to show to best advantage.

An ultraviolet or so-called "black" light serves this purpose well, but any bulb with a suitable nickel oxide glass filter provides a satisfactory (though not always an efficient) activating light source. Ultraviolet sources include argon glow lamps, high pressure mercury arcs, suitably filtered fluorescent lamps, and various "black light" fluorescent tubes and screw-type bulbs made for the purpose.

Some pigments respond immediately to activation, others require several seconds of exposure.

COLOR. The daylight colors are available in a considerable range from pale to fairly vivid colors, none of which corresponds exactly to the fluorescent color. Under activation, the fluorescent pigments display an amazing brilliance and strength of color throughout a wide color range. Attempts to change the daylight color by the addition of non-fluorescent pigments adversely affects the paint's luminescence by screening out the activating light or absorbing the emitted light. Synthetic dyes may be added by the manufacturer in small quantities to alter the daylight color with only slight loss in the fluorescence.

APPLICATION. Applying fluorescent paints raises the same general considerations as does applying phosphorescent paints. The particle sizes of fluorescent pigments correspond to those of ordinary paint pigments, so finished fluorescent paint can be applied readily by brushing or spray gun. To provide satisfactory fluorescence a zinc-cadmium paint should be applied so as to spread 120 yards to the gallon.

DURABILITY. Fluorescent paints vary considerably in their resistance to deterioration from exposure to weather and visible light. Some are relatively unaffected by water, weak acids, or alkalis or by exposure to strong sunlight. Many have been exposed to outdoor weathering for months and even years with little loss of fluorescence. In certain vehicles the pigments are subjected to a photochemical darkening under some conditions of exposure to sunlight in the presence of water.

RADIOACTIVE
PAINTS

GENERAL.   Radioactive paints can be made by combining a minute amount of radium, mesothorium, radiothorium, or thorium in a luminescent base such as zinc sulfide. When such paints were first used, it was thought that elementary precautions sufficed to prevent the radioactive compounds from entering the mouth or lungs of workers and that no detectable injuries would result. Research over the past several years, however, has shown all radioactive materials to be highly dangerous.

USES.   Radioactive pigments were originally used chiefly as a paint for the marking of instruments and dials; they were sometimes placed between two discs of plastic in the form of buttons to be used as guide markers. Because of the dangers involved in handling such materials, the complexities and high cost of doing so safely, the stringent regulations surrounding them, and the availability of other satisfactory materials, however, they are not used today except perhaps in highly specialized applications.

EXCITATION.   Luminescence in radioactive pigments is caused by the internal bombardment by atomic particles of the particles of the phosphorescent-responsive base; no external excitation is required.

COLOR.   The daylight color is a slightly yellowish white and the luminescent color is bluish or greenish white. No other pigment or dyestuff may be added to change the daylight color.

APPLICATION.   Special precautions and equipment are needed for applying radioactive paints. For satisfactory results, 1 gram of pigment should be made to cover an area of not more than 4 square inches. A heavier application will give increased brightness.
The surface to be treated must be clean and free from grease or finger marks. An undercoat of zinc oxide or titanium dioxide white lacquer is recommended.

DURABILITY.   Radioactive compounds can be formulated and applied so as to be stable under outdoor conditions. The amount of radioactive material present should be sufficient to yield optimum brightness; but if used in excess of this amount, it will accelerate the more rapid breakdown of the sensitive base without yielding more light. The luminosity lasts from 6 to 8 years.

STEEL PIPE
FOR ORDINARY USES

DESCRIPTON OF STEEL PIPE FOR SPECIAL USES. A multitude of different types of steel pipe are made for special requirements such as close coiling, bending, high pressure, compression, tension, unusual corrosion resistance, flanging, impact, low temperature, and plating. Standard specifications covering these types of pipe for special uses have been formulated by many organizations, among which are:

> American Society for Testing Materials
> Association of American Railroads
> American National Standards Institute
> American Petroleum Institute
> American Society of Mechanical Engineers
> American Waterworks Association
> Director of Procurement of the United States
> United States Navy

Both physical and chemical tests are usually described in such specifications. Special pipes are normally made by pipe manufacturers on order to conform to the specifications. Pipe meeting such specifications may or may not be regularly found in jobbers' warehouses.

DESCRIPTION OF STEEL PIPE FOR ORDINARY USES. Steel pipe of this description should conform to ASTM Specification A120 which covers black and hot-dipped galvanized, welded and seamless, unalloyed steel pipe from ⅛″ to 12″ nominal inside diameter, purchased mainly from jobbers' warehouse stocks. Hydrostatic pressure tests are the only physical tests made on pipe conforming to ASTM A120 because the pipe is intended for ordinary uses for which special properties are not required. Steel pipe for ordinary uses is manufactured from mild, ductile steel made by the open hearth or Bessemer process. It is available in a range of diameters, wall thicknesses, surface treatments, and methods of manufacture.

If any details or specifications found on this and the following pages are critical to a design or are otherwise of importance, or if further specifics are desired, the American Society for Testing Materials should be contacted for the most up-to-date version of ASTM Specification A120.

LENGTHS. Standard-weight pipe comes in random lengths from 16′ to 22′. Not more than 5% of the total number of lengths may be jointers (two pieces tightly coupled together). Continuously welded pipe comes in 21′ lengths.

Extra-strong and double extra-strong pipe comes in random lengths of 12′ to 22′; 5% may be in lengths of 6′ to 12′.

WELDED AND SEAMLESS TYPES. Steel pipe 3″ and less in diameter is usually butt-welded; sizes 3½″ and over are lap-welded. In the *butt-welding* process, the skelp with square or slightly beveled edges is drawn from the furnace through a funnel-shaped welding die or through welding rolls where it is bent into tubular form and its edges brought together with sufficient pressure to weld them.

In the *lap-welding* process, the skelp with scarfed or beveled edges is heated and bent to tubular form with the edges overlapping; it is then reheated and passed over a mandrel between rolls that compress and weld the lapping edges.

Seamless pipe conforming to A120 is made by piercing solid round steel billets and rolling them.

664

STEEL PIPE
FOR ORDINARY USES

PHYSICAL PROPERTIES. Since warehouse stocks of steel pipe conforming to A120 are not manufactured to meet specific standards established by physical tests, the following figures are approximate.

Tensile strength, minimum pounds per square inch	45,000
Yield point or elastic limit, minimum pounds per square inch	25,000
Coefficient of thermal expansion, ins./in./F°....	.00000674

USES FOR BLACK STEEL PIPE. "Black pipe" is the term commonly applied to uncoated pipe and to pipe that is given an ordinary air-drying lacquer coating for protection against rust during shipment. Applying the lacquer coating is a regular mill practice. Among the many uses for this pipe are to transport natural and manufactured gas for cooking and heating, low-pressure steam for heating systems, air for various purposes, and ammonia.

USES FOR GALVANIZED PIPE. Zinc applied by the hot-dipped galvanizing process is widely used for protection against corrosion. As regular practice, galvanized steel and couplings are hot-galvanized prior to threading. The threading operation, of course, removes the coating from the threaded areas. Regular practice is to furnish galvanized pipe in accordance with the galvanizing requirements given in ASTM Specification A120, and the test procedure given in that specification for determining weight of coating is standard in the industry. Galvanized steel pipe is used for hot and cold water supply lines, plumbing vent lines, and waste lines above ground.

Pipe meeting the dimensional and hydrostatic requirements specified in A120 but having special coatings is available from the mills for underground service.

SPECIAL TREATMENTS. Special treatments are not described in ASTM A120, and specially treated pipe is not normally to be found in jobbers' warehouse stocks. Special treatments include galvanizing after cutting to lengths, galvanizing on outside only, galvanizing on inside only, galvanizing pipe and couplings after threading, producing galvanized coatings heavier than standard, applying tar-base or asphalt-base coatings on inside, outside, or both, adding saturated fabrics over bituminous coatings, coating with primer, lining with cement, and pickling or cleaning mechanically and then oiling. Steel pipe treated with these special finishes is normally available only on special order from the mill and frequently involves an additional cost and a longer time for delivery.

SPECIAL ALLOYS. Although ASTM Specification A120 does not cover wrought iron or alloyed steel pipe, they should be mentioned here. The addition of from 0.20% to 0.35% of copper provides increased resistance to atmospheric and other types of corrosion. The addition of copper and molybdenum increases resistance to various types of corrosion, as well as increasing tensile strength. The corrosion resistance of wrought iron pipe is well established. The first cost of these types of pipe is higher than steel. They are made in standard, extra-heavy, and double extra-heavy weights, in the same dimensions as steel pipe. Fittings for alloyed pipe are usually required to be of the same composition as the pipe.

STEEL PIPE
FOR ORDINARY USES

Size (Nominal Inside Diameter)	O.D. to nearest 1/16th	STANDARD WEIGHT PIPE			EXTRA STRONG PIPE			DOUBLE EXTRA STRONG PIPE		
		Thickness in.	Wt. in lbs. per L.Ft., Threaded and with Couplings	Hydrostatic Test Pressures	Thickness in.	Wt. in lbs. per L.Ft., Plain Ends	Hydrostatic Test Pressures	Thickness in.	Wt. in lbs. per L.Ft., Plain Ends	Hydrostatic Test Pressures
		BUTT WELDED								
1/8	7/16	0.068	0.24		0.095	0.31		—	—	
1/4	9/16	0.088	0.42		0.119	0.54		—	—	
3/8	11/16	0.091	0.57	700	0.126	0.74	850	—	—	1,000
1/2	13/16	0.109	0.85		0.147	1.09		0.294	1.71	
3/4	1 1/16	0.113	1.13		0.154	1.47		0.308	2.44	
1	1 5/16	0.133	1.68		0.179	2.17		0.358	3.66	
		LAP-WELDED, ELECTRIC-WELDED AND SEAMLESS GRADE A								
1 1/4	1 11/16	0.140	2.28		0.191	3.00		0.382	5.21	
1 1/2	1 15/16	0.145	2.73	800	0.200	3.63	1,100	0.400	6.41	1,200
2	2 3/8	0.154	3.68		0.218	5.02		0.436	9.03	
2 1/2	2 7/8	0.203	5.82		0.276	7.66		0.552	13.70	
3	3 1/2	0.216	7.62		0.300	10.25		0.600	18.58	
3 1/2	4	0.226	9.20		0.318	12.51		0.674	27.54	
4	4 1/2	0.237	10.89	1,200	0.337	14.98	1,700	0.750	38.55	2,000
5	5 9/16	0.258	14.81		0.375	20.78		0.864	53.16	
6	6 5/8	0.280	19.18		0.432	28.57		0.875	72.42	2,800
8	8 5/8	0.322	28.81		0.500	43.39		—	—	
10	10 3/4	0.365	41.13	1,000	—	—	—	—	—	—
12	12 3/4	0.375	50.71		—	—	—	—	—	—

STEEL PIPE
FOR ORDINARY USES

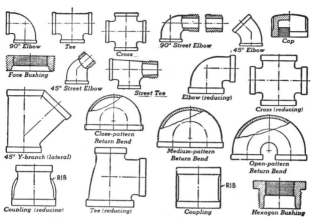

SOME FITTINGS USED WITH STEEL PIPE

FITTINGS FOR STEEL PIPE—Standard weight pipe is furnished with threaded ends and couplings made of wrought iron or steel by the pipe manufacturer. Each length of pipe 2″ and smaller in diameter is furnished with a straight tapped coupling. Each length of pipe 2 1/2″ in diameter and larger is furnished with one taper tapped coupling. All other fittings are made by companies other than the pipe manufacturers.

Any fitting can be *standard* or *heavy* in weight—the heavy weight withstanding pressures greater than the standard weight. Sometimes local building codes require the use of heavy weight fittings with standard weight pipe while others permit the lighter standard fittings to be used.

Fittings are made of either *cast iron* or *malleable iron*. The *cast iron* fittings are less expensive, somewhat more brittle, and slightly heavier in weight of metal. Cast iron fittings are commonly used with steel pipe for steam lines. *Malleable* fittings are less brittle and are somewhat easier to handle because of their lighter weight of metal and are commonly used in water, gas, and air lines.

Fittings may be either *black* or *galvanized,* the choice following type of steel pipe that is used.

INSTALLATION—Joint compound should be used to facilitate assembly and to render pipe connections tight and leakproof. The ends of all threaded pipe should be thoroughly reamed to eliminate burrs. Where water is to be heated for domestic or other uses it is desirable to employ an open or deaerating type of heater which permits dissolved gases in the water to escape, to minimize corrosion. Only ferrous fittings should be used. Pipe covering should be installed so as to prevent air from reaching the pipe surface. Where air gets beneath insulation, condensation can occur on the metal where it becomes acid due to absorption of atmospheric gases.

667

MAGNESIUM OXYCHLORIDE
PLASTIC FLOORING

DESCRIPTION. Plastic magnesium oxychloride cement has been known variously as composition, sorel cement, magnesia cement, and plastic magnesia cement, as well as by various trade names which may or may not suggest the ingredients or appearance of the finished floor. In fact, it is today classed as a terrazzo binder, and full specifications and installation recommendations are found in USA Standard A88.6, *Specifications for Terrazzo Oxychloride Composition Flooring and Its Installation.* Since the advent of resinous terrazzo binders that have greter resistance to chemicals, abrasion, and moisture, the use of this type of flooring has greatly diminished. However, it is still useful in certain applications, particularly as an underlayment for thin-set resinous terrazzo and other finish floor coverings.

These floors may be laid over old or new floors and subbases. Bases, wainscots, and carpet strips can be formed of the same material. Plastic magnesium oxychloride floors are permanent, warm, quiet, resilient, dustproof, relatively non-slip, incombustible, and bacteria- and fungus-inhibiting; they also have good wearing qualities. A wide range of clear, brilliant colors can be obtained.

Because of the inherent resilience of magnesium oxychloride flooring, the material may be applied monolithically—without the need for divider or expansion strips—to areas of large extent and will not crack, provided there is no excessive structural movement. This property allows it to be used in applications where vibration and racking impose difficult demands.

Aggregates may be incorporated to form terrazzo. Metal or plastic strips may be used for design purposes. Ingredients may be controlled to produce a floor that will not spark from friction or static. Floors are sometimes scored to represent tiles or blocks, with the dubious idea that such grooving of the inherently sanitary monolithic surface will improve either its function or its appearance!

COMPOSITION IN GENERAL. The plastic cement is made by adding about a 22° Baumé solution of magnesium chloride hexahydrate—termed magnesium chloride ($MgCl_2 \cdot 6H_2O$)—to magnesium oxide (MgO), which is a fine white powder. Magnesium oxide is also referred to as calcined magnesia and calcined caustic magnesia. The resulting plastic paste is magnesium oxychloride (approximately $3MgO + MgCl_2 + 11H_2O$).

Fillers such as wood flour, hardwood fiber, cork, talc, asbestos, sand, silex, marble flour, and limestone fines, as well as color pigments and aggregates if desired, are mixed into this paste. The paste sets to a hard mass comparable to portland cement in strength. After setting, the cement takes a high wax polish.

RESISTANCE TO LIQUIDS, ETC. The floor is attacked by caustics and sulfuric acid but not by ordinary chemicals in the concentrations in which they would usually be encountered. Alcohol, grease, naphtha, and other solvents do not attack this flooring.

Magnesium oxychloride flooring will not withstand complete immersion over a long period or constant dampness. However, experience in public transit cars and in many industrial and other building installations indicates that, under normal conditions in dry locations, the floors can withstand any wetting incidental to use or maintenance without reducing the normal life expectancy of the installation.

The choice of fillers, aggregates, and aggregate grading affects the solubility and absorption of the flooring. It has been found that the addition of copper powder to compositions that contain no fillers that are themselves affected by water will reduce the solubility of the flooring.

Strong cleaning solutions and water, especially scalding water, will ordinarily attack almost any finish flooring whether it be tile of various sorts, wood, painted surfaces, terrazzo, or even marble—depending upon the strength of the solution. Magnesium chloride is a salt and is water-soluble even after it is crystallized in solid flooring. While under normal circumstances even daily mopping of this flooring is not harmful to it, constant exposure to water without intermittent periods of drying and recovery of the salt is harmful. Therefore, magnesite flooring is not recommended for unprotected subgrade or exterior work.

MAGNESIUM OXYCHLORIDE PLASTIC FLOORING

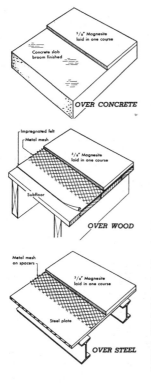

OVER CONCRETE

OVER WOOD

OVER STEEL

INSTALLATION. The usual method is to apply the flooring in one coat to a thickness of approximately ½″ overall. Where floors of 1″ or more in thickness are needed for increased fire resistance, for meeting grade requirements, for greater impact strength under severe service, or for some other reason, two coats may be used with the finish surface coat *not less than* ½″ thick.

The material is brought to the site in two parts. The first consists of magnesium oxide, aggregates, colors, and fillers, which have all been proportioned and dry-premixed by the installer. The second is magnesium chloride, which is furnished in flake form and is reduced to a liquid on the job site by the addition of water to the required specific gravity. These are combined into a mortar, which is spread on the job site, leveled and finished by troweling or grinding.

Flooring can be laid over subfloors of steel, wood, or concrete. Cove bases can be laid against wood, hard cement plaster, brick, concrete, or stone, but should not be laid against glazed brick, gypsum tile, or lime plaster.

Fairly even room temperature must be maintained. In cold weather, a 65° to 70°F temperature should be maintained day and night for 48 hours. A sudden drop of 15° or more may cause cracking by shrinking. A sudden rise in temperature may cause buckling by expansion. Within 24 hours after the final troweling, any roughness can be smoothed by dry rubbing with steel wool. Floors are then given a light coat of wax or a mixture of boiled linseed oil and turpentine or benzene, well rubbed in. New floors should be protected for 72 hours with sawdust or building paper.

WOOD SUBFLOORS. The wood must be clean, sound, and firmly nailed. Expanded metal lath or 1″ wire mesh galvanized after weaving is nailed every 6″ both ways over good quality waterproof building paper or polyethylene sheet.

CONCRETE SUBFLOORS. The concrete must be sound, dry, and not less than 30 days old. No lime fattener should be used in the concrete. Concrete on or below grade should be waterproofed. The concrete surface should be broomed, picked, short-tooth raked, etched, or otherwise roughened, and/or treated with a chemical bonding agent, depending upon circumstances.

STEEL SUBFLOORS. Steel must be clean. The plates may be preformed to provide a key for the magnesite flooring, or a metal mesh can be bolted or spot-welded over the entire surface. Various types of bituminous, paint, or other coatings are used over the steel plate and anchoring mesh before the oxychloride cement is poured.

MAGNESIUM OXYCHLORIDE PLASTIC FLOORING

TYPES OF FLOORING. A wide latitude in properties of the flooring is possible through control of the ingredients. When a flooring possessing certain qualities is desired, the contractor must be allowed to use the combination of ingredients that will produce the stated physical properties. Magnesium oxychloride floors can be grouped into eight basic types as follows:

1. GENERAL PURPOSE TYPE. No coarse aggregates are used. This type may be installed in solid colors, pigmented to match flat paint shades; the colors may have contrasting colors, designs, borders, and base.

Physical characteristics make the general purpose type suitable for school rooms, hospital rooms and wards, ships, corridors and lobbies other than those subjected to extremely heavy traffic, light industrial plants, and retail stores and shops. Flooring is finished by troweling.

2. HEAVY DUTY TYPE. Coarser aggregates are used and a smaller amount of filler than for Type 1. Type 2 can meet more severe service conditions, such as institutional and restaurant kitchens, intermediate industrial plants, and corridors, lobbies, and business establishments having hard usage. It is finished the same way Type 1 is.

3. NONSPARKING AND STATIC DISCHARGING TYPE. There must be no silicious aggregate in this type. The flooring should be laid over a mat formed by lacing bare No. 14 copper wire 12″ o/c both ways. The ends of these wires are soldered to a No. 8 copper wire, which is grounded.

This type of floor prevents either mechanical or static sparking. Nonsparking floors are used in operating rooms, lacquer or paint spray shops, ammunition plants, and wherever explosions, flashes, or fire present a hazard. Flooring is finished by troweling or grinding.

4. NONSLIP ABRASIVE TYPE. Abrasive aggregate may be either sprinkled and troweled into the finish surface or used as an integral aggregate to form a highly nonslip surface suitable for stair treads, ramps, elevator floors, etc. This type may be pigmented but is usually installed in a black color. Flooring is finished by rubbing.

5. TERRAZZO TYPE. Marble chips, colored stone, and pigment provide a practically unlimited range of color combinations. Division strips of brass, white metal, hard rubber, or plastic may be used, but aluminum strips are not suitable. Strips are not necessary for division into small units but are used only for segregating colors, creating design patterns, or localizing cracks over anticipated lines of structural stress. The floor is finished by grinding.

6. INDUSTRIAL TYPE. Crushed granite or trap rock chips are incorporated to create an extremely durable floor comparable to granite. It is installed with a power float of the vibrating type, hard troweled and lightly ground.

Another variety of industrial floor for the most severe use is formed by the installation of a Type 2 bedding—a cast iron or steel grid filled with Type 2 cement. It is finished with steel wool.

7. CORK TERRAZZO TYPE. An extremely resilient and quiet floor can be obtained by using ⅛″ granulated cork. This type is suitable for libraries, hospital wards and corridors, apartment house corridors, schools, and in front of work benches in industrial work. Flooring is finished by grinding.

8. UNDERBED TYPE. This is an uncolored, sometimes fibrous base-coat material for leveling uneven subfloors prior to surfacing with rubber tile, asphalt tile, vinyl sheet, or ½″ magnesite finish. It is also effective as an underlayment for thin-set resinous terrazzo over wood (or other) flooring.

MASTIC SETTING OF
ACOUSTIC TILE

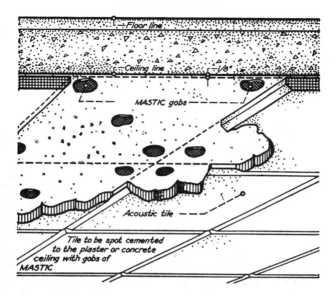

Floor line

Ceiling line — 1/8"

MASTIC gobs

Acoustic tile

Tile to be spot cemented
to the plaster or concrete
ceiling with gobs of
MASTIC

ACOUSTIC TILE MASTIC. Acoustic tile for noise quieting or for acoustic correction is made by several manufacturers in sizes ranging from 9″ x 9″ or 12″ x 12″ to 24″ x 24″ and 24″ x 48″. This can be applied to walls and ceilings with certain mastics or "glues" that are very easy to work with and make for a rapid and trouble-free installation. Special acoustic mastic may be employed, as well as silicone rubber sealant/caulk.

PREPARATION OF SURFACE. The surfaces on which the mastic will be used should be checked for compatibility with the particular mastic. Masonry of any kind, plastered metal lath on frame, thick plasterboard, and other types of "sound wall" constructions possess ample strength and rigidity to support acoustic tile. Wood subsurfaces are often less satisfactory, and proper adherance to vegetable fiberboards is chancy at best. The wall should be thoroughly dry and free of dust, dirt, grease, oil, and loose material. Before acoustic tile is applied, the surface should be cleaned (wire-brushed, if necessary) and coated with an appropriate primer/sealer. If the surface has previously been coated with a very smooth or glossy finish, this should be roughened with sandpaper, steel wool, or by wire-brushing. Allow 48 hours or more after washing or coating before applying the acoustic tile.

APPLICATION OF TILE. The mastic is applied to the back of each tile in daubs of about 2″ in diameter at each corner, as well as at intermediate points for large tiles. If the surface is planar, the daubs should only be of sufficient thickness to leave a small space between the tile back and the mounting surface when the tile is pressed into place. If the mounting surface is uneven, the thickness of the daubs must be regulated accordingly so that the completed job will show a true even finish without irregularity. In most cases the tiles should be placed within a short time of applying the daubs of mastic, before the mastic has a chance to "skin over" and lose some of its adhesive effectiveness.

STRUCTURAL FIBERBOARD
SHEATHING

DESCRIPTION.   Structural fiberboard is made of partially refined vegetable fibers obtained principally from crop plant wastes and wood. The boards are made of fibers from at least five quite different raw materials — wood, bagasse (extracted sugar cane), corn stalks, licorice roots, and waste paper. The general properties of the finished products from these different sources, however, are essentially the same. The basic materials are reduced to fibers by mechanical means or by exploding them with steam, usually after softening them with chemicals or steam. The boards are fabricated from the pulp by a felting or molding process; suitable sizing material is incorporated into the product to render it water-resistant. The drying temperature is such as to destroy rot-producing fungi.

Sheathing is one of a group of five products made by the same process, which are similar in composition and properties:

>   *Class A:* Building Board
>   *Class B:* Lath (for plaster base)
>   *Class C:* Roof Insulating Board
>   *Class D:* Interior Board (factory finished)
>   *Class E:* Sheathing

The use of all of these products has diminished substantially over the past few years as they have been supplanted by newer and better materials, and they are slowly disappearing from the marketplace. Of the five products, sheathing remains the most commonly used and is both popular and widely available, especially in the ½″ thickness.

The sheathing boards are given a surface treatment consisting of one or more coatings of asphalt or a preservative, water-repellent treatment.

THERMAL PROPERTIES.   Fiberboard sheathing has thermal insulating properties that make it suitable for that purpose under mild conditions and for slightly augmenting thermal insulating materials. The thermal values depend upon the density of the material; in the so-called "regular density," the thermal conductivity (k) is approximately 0.38. The thermal conductance (C) of ½″ structural fiberboard is approximately 0.76 (R-1.32); and of the $^{25}/_{32}$″ thickness, 0.49 (R-2.06).

STRENGTH.   Most tests of fiberboard sheathing are inconclusive in comparing strength data with that of either board or plywood sheathing. *However, from the evidence of multitudes of actual installations, it is unquestionable that fiberboard sheathing contributes adequate strength to a complete and properly constructed frame wall.* Certainly, thousands upon thousands of well-built residences and other small buildings have been sheathed with this material. One manufacturer has stated that a 4′ width of $^{25}/_{32}$″ fiberboard sheathing applied vertically has strength comparable to that contributed by diagonal wood sheathing; horizontally applied, it compares with horizontal wood sheathing. Plywood sheathing, however, results in a stronger structure, to a degree dependent upon the thickness of plywood used. Under various accelerated aging tests, samples showed "excellent stability" in all their critical properties.

COST.   The cost of structural fiberboard sheathing fluctuates somewhat vis-a-vis board sheathing and plywood, depending upon market conditions. In general, fiberboard is likely to cost somewhat less than either wood or plywood, but it is also considered to provide a somewhat lower quality level of construction than wood.

STRUCTURAL FIBERBOARD
SHEATHING

INSTALLATION.   Square-edged 4'-wide boards are designed for vertical installation, though they may be installed horizontally if proper edge-blocking is installed, too. They should be nailed 6" o/c on the intermediate supports and 3" o/c on the edges, no less than ⅜" from the edges, using galvanized roofing nails that will ensure not less than 1" to 1¼" penetration into the framing. Structural sheathing may be applied at corners of a conventional stud wall frame, but most building codes will require added corner bracing to meet strength and rigidity requirements. This may be done with let-in diagonal bracing in each corner section. A common construction is to sheath otherwise unbraced corners with plywood, and intermediate areas with structural fiberboard.

Since the maximum coefficient of expansion is 0.5%, it is necessary to allow ⅛" joints between boards.

Building paper is not necessary except under exterior stucco, although it may be required by code in some locales. The presence of building paper, especially at corners, does inhibit air infiltration.

USES.   Fiberboard sheathing is used as structural sheathing with an inherent thermal insulating capacity under wood siding, hardboard siding, shingles, stucco, and masonry veneer. It also can be used as roof sheathing in certain applications and on pitched roofs with wood stripping or solid wood sheathing under various types of roofing. Vertically applied square-edged sheathing 4' wide has been found suitable for the exterior of temporary structures. It can be painted with aluminum paint or with other paint suitable for use over asphalt or water-repellents to reduce absorption of solar heat, to facilitate drying, and to promote subsequent moisture absorption and retention.

SIZES.   Standard width is 4', and lengths generally available are 8' and 9', though the latter length is somewhat less commonly stocked at local lumberyards. Sheets are square-edged, in thicknesses of ½" (most common) and ²⁵/₃₂".

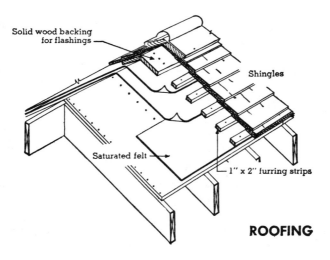

Solid wood backing for flashings

Shingles

Saturated felt

1" x 2" furring strips

ROOFING

STRUCTURAL FIBERBOARD SHEATHING

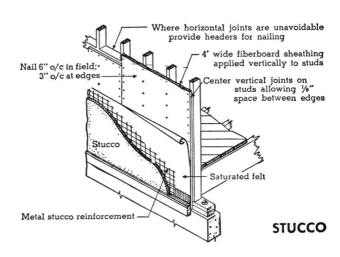

Where horizontal joints are unavoidable provide headers for nailing

4' wide fiberboard sheathing applied vertically to studs

Nail 6" o/c in field; 3" o/c at edges

Center vertical joints on studs allowing ⅛" space between edges

Stucco

Saturated felt

Metal stucco reinforcement

STUCCO

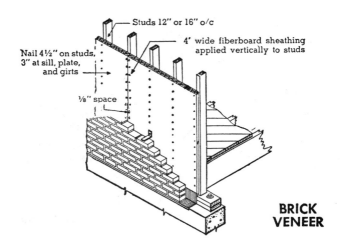

Studs 12" or 16" o/c

4' wide fiberboard sheathing applied vertically to studs

Nail 4½" on studs, 3" at sill, plate, and girts

⅛" space

BRICK VENEER

STRUCTURAL FIBERBOARD
SHEATHING

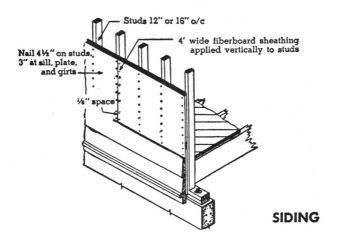

Studs 12" or 16" o/c

4' wide fiberboard sheathing
applied vertically to studs

Nail 4½" on studs;
3" at sill, plate,
and girts

⅛" space

SIDING

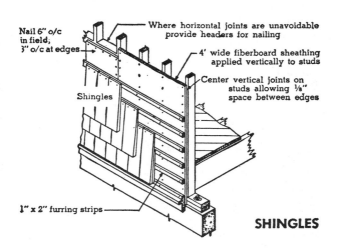

Nail 6" o/c
in field;
3" o/c at edges

Where horizontal joints are unavoidable
provide headers for nailing

4' wide fiberboard sheathing
applied vertically to studs

Center vertical joints on
studs allowing ⅛"
space between edges

Shingles

1" x 2" furring strips

SHINGLES

SPELL IT
RIGHT

The spelling that is recommended in conflicting cases is preferable for use on drawings because of brevity, modernity, or patriotism.

acoustic—adj.; use instead of *acoustical.*
abutment, abutted, abutting—watch your *t*s.
balustrade.
barrel, barreled—shorter than *barrelled*, and preferred.
batt—a short piece of matted cotton, fiberglass, or rock wool insulation. *Bat* is a flying mammal, a club, a small piece of brick. Make the distinction.
beveled, beveling—shorter than *bevelled, bevelling*, and preferred.
bridging.
Btu—not *B.t.u.*, not *B.T.U.*, not *BTU*. Use Btu for both singular and plural.
caulk—(pronounced *kawk*) preferred over *calk.*
cleaned—were you ever *cleansed* in a poker game?
colonnade.
conduit—pronounced *kon-dwit* or *kon-doo-it*, not *kon-dit.*
cupola.
draft, drafting, draftsman/draftsperson—*draughts* is a British checker game.
enameled, enameling.
escalator.
fiber—not *fibre.*
fiberglass—not *fibreglass or fibreglas* (Fiberglas is a trade name).
fluorescent—not *flourescent.*
gauge—seldom spelled *gage*, as a technical term.
grill—a gridiron for cooking.
grille—a grating or screen, especially of decorative intent.
hangars—indoor parking for aircraft.
hangers—supports.
leveled, leveling—no need for extra *l*s.
louver—a slatted opening, pronounced *loo'-ver.*
Louvre—an art museum in Paris, pronounced *loov* (approx.).
mantel—the facing about a fireplace, including the shelf above it. *Mantle* is a cloak, covering, or gas mantle.
marquee—better stick with Webster, who says this is a hood over a door, the word probably originating from the tent or canopy set up to protect a *marquis* from the elements.
miter—not *mitre.*
modillion.
mold, molded, molding—*mould* is archaic or British or both.
movable—no *e* before *a* here.
panel, paneling—shorter than *panelled, panelling.*
parallel, paralleled, paralleling—rather than *parallelled, parallelling*. It's confusing enough as it is.
pavilion—compare *modillion* (for spelling, not meaning).
pinnacle.
precede, preceding.
program—eschew *programme.*
rabbet—pronounced as it appears; means a groove, channel, or recess cut out of the face or edge of a body.
rebate—an old variation of *rabbet* and pronounced the same way when used in the same context. When used to mean *to diminish* or *a return of a part of a payment*, however, it is pronounced *ree'-bate*. Different meaning, different pronunciation.
receptacle—the *a* is the hard part of this word.
removable—not *removeable.*
Renaissance.
stile—of a door; *style* is that of the lady I seen y. w. l. n.
supersede.
template—preferred to *templet*, but both are pronounced *tem'-plit.*
terrazzo—say *ter-rat'-so*, not *ter-rat'-zzo* or *tare-azzo.*
theater—preferable to the affected spelling *theatre.*
thorough—*thoro* is bad.
through—*thru* is generally used only in connection with streets and ways—and even in that limited use, it is not standard English.
transept—modern form of the archaic *transcept.*

LETTERING FOR
WORKING DRAWINGS

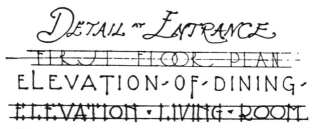

The four examples above were taken from actual published architectural drawings. They all violate one or more of the principles of sane lettering, which are:

1. LEGIBILITY—most important requirement for both direct prints and for reductions, in ink or pencil.
2. SPEED—next most important requirement, necessitating simplicity of letter forms.
3. APPEARANCE—should result from uniformity, not from doo-dads, flourishes, or time-wasting tricks.
4. CHARACTER—should be in keeping with the legal and business nature of the working drawings, which again outlaws freak spacing and letter forms.

Lettering should follow these general principles, in spite of the fact that personal preferences, individual abilities, locally accepted practices, use or nonuse of lettering guides, and similar factors inevitably influence the end result. The architect or draftsperson should be guided in any event by the current recommendations of the American Institute of Architects and by the latest revision of the American National Standards Institute publication *Line Conventions and Lettering* (ANSI Y14.2M), obtainable from the Institute at 1430 Broadway, New York, NY 10018.

ABCDEFGHIJKLMN
OPQRSTUVWXYZ
1234567890 $\frac{3''}{64}$

ABCDEFGHIJKLMN
OPQRSTUVWXYZ
1234567890 $\frac{3''}{64}$

abcdefghijklmnopqrstuvwxyz

FIVE BASIC RULES
OF PERSPECTIVE

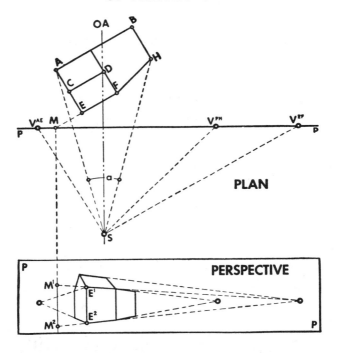

1. ANGLE OF VISION. *The area of the picture embraced by the eye should not represent an angle of greater than 45° in plan.* In the diagram, the angle *a* is the angle of vision. Some authorities set a maximum of 60°, but this often results in distortion at the edge of the picture. The angle *a* should not be much less than 30° if a full perspective effect is to be realized.

2. OPTICAL AXIS. *The optical axis should bisect the angle of vision.* When we look at a picture, we naturally hold it directly in front of us—and perspective drawing should be made as it is going to be looked at. The line *OA* should bisect angle *a*.

3. PICTURE PLANE. *The picture plane is taken perpendicular to the optical axis.* When a drawing is examined, it is held in this position—therefore it should be drawn so that *PP* is normal to *OA*.

4. VANISHING POINTS. *The vanishing point for any system of lines is the point of intersection with the picture plane of a line parallel to the system, running through the observer's eye.* Lines *AB, CD,* and *EF*, together with all other lines parallel to them, constitute a "system." These lines all vanish at V^{EF}.

5. TRUE HEIGHTS. *Projections of points on lines of the object parallel to the picture plane will fall in the picture plane at their true distances apart.* Two points such as E^1 and E^2 (represented by E in plan) may be projected to the picture plane at M. The length M^1M^2 on the picture plane will equal the true height of E^1E^2 on the object.

PERSPECTIVE LAYOUT
MADE EASY

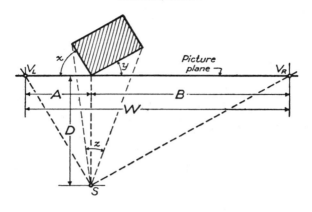

Angle		For W = 41″			For W = Any Measurement		
y	**z**	**A″**	**B″**	**D″**	**A″**	**B″**	**D″**
10°	80°	1¼	39¾	7	.0303xW	.9697xW	.1709xW
15	75	2¾	38¼	10¼	.0669xW	.9330xW	.2500xW
20	70	4¾	36¼	13¼	.1178xW	.8822xW	.3211xW
25	65	7¼	33¾	16	.1796xW	.8211xW	.3826xW
30	60	10¼	30¾	17¾	.2500xW	.7500xW	.4326xW
35	55	13½	27½	19¼	.3292xW	.6708xW	.4695xW
40	50	17	24	20⅛	.4134xW	.5866xW	.4911xW
45	45	20½	20½	20½	.5000xW	.5000xW	.5000xW

The location of the vanishing points V_L, V_R and the station point S must be such that angle z is not greater than 45° in order to prevent distortion of the perspective.

The first A, B, D columns give the location in inches for the vanishing and station points for a standard 42″ drawing board, the total width being 41″.

The last A, B, D columns give the multiplier of any other width that may be used for larger or smaller perspectives, in order to locate the vanishing and station points. For example; *Given* W = 60″, x = 30°. *Solution*, A = .25 × 60″ = 15″, B = .75 × 60″ = 45″, D = .4326 × 60″ = 26″.

679

THREE-CENTERED ARCH

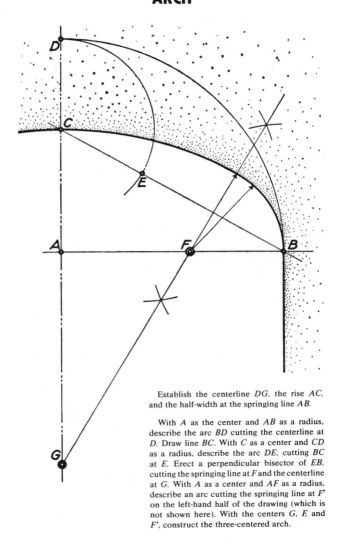

Establish the centerline *DG*, the rise *AC*, and the half-width at the springing line *AB*.

With *A* as the center and *AB* as a radius, describe the arc *BD* cutting the centerline at *D*. Draw line *BC*. With *C* as a center and *CD* as a radius, describe the arc *DE*, cutting *BC* at *E*. Erect a perpendicular bisector of *EB*, cutting the springing line at *F* and the centerline at *G*. With *A* as a center and *AF* as a radius, describe an arc cutting the springing line at *F'* on the left-hand half of the drawing (which is not shown here). With the centers *G*, *F*, and *F'*, construct the three-centered arch.

FOUR-CENTERED
ARCH

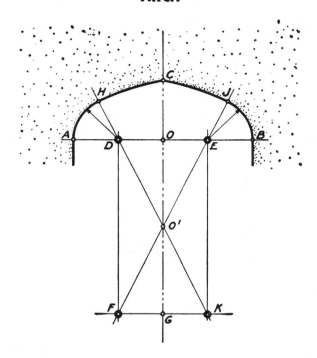

Establish centerline *CG* and springing line *AB*.

Bisect *AO* at *D*.

Bisect *OB* at *E*.

Make *OO'* equal to *DE*.

Drop perpendiculars *DF* and *EK*.

Produce *DO'* to *K*.

Produce *EO'* to *F.*

Using *D* and *E* as centers, describe arcs *AH* and *JB*.

Using *F* and *K* as centers, describe arcs *CJ* and *HC*.

DIVIDING A CIRCUMFERENCE

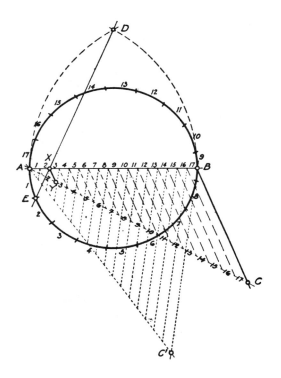

PROBLEM: To divide the circumference of a circle into any number of equal spaces.

SOLUTION: Draw a diameter AB of circle AB. Draw a line from A in any convenient location, as AC or AC'. Connect B and C. Divide AC into a desired number of spaces. Parallel to BC, draw XY through the second division. With B as a center, draw the arc AD. With A as a center, draw the arc BD. From D draw a line through X intersecting the circle at E. AE is the desired spacing.

DIVIDING A CIRCUMFERENCE

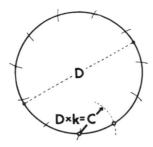

No. of Spaces n	k	No. of Spaces n	k	No. of Spaces n	k
..	.....	35	.0896	68	.0462
3	.8660	36	.0872	69	.0455
4	.7071	37	.0848	70	.0449
5	.5878	38	.0826	71	.0442
6	.5000	39	.0805	72	.0436
7	.4339	40	.0785	73	.0430
8	.3827	41	.0765	74	.0424
9	.3420	42	.0747	75	.0419
10	.3090	43	.0730	76	.0413
11	.2817	44	.0713	77	.0408
12	.2588	45	.0698	78	.0403
13	.2393	46	.0682	79	.0398
14	.2225	47	.0668	80	.0393
15	.2079	48	.0654	81	.0388
16	.1951	49	.0641	82	.0383
17	.1838	50	.0628	83	.0378
18	.1736	51	.0616	84	.0374
19	.1646	52	.0604	85	.0370
20	.1564	53	.0592	86	.0365
21	.1490	54	.0581	87	.0361
22	.1423	55	.0571	88	.0357
23	.1362	56	.0561	89	.0353
24	.1305	57	.0551	90	.0349
25	.1253	58	.0541	91	.0345
26	.1205	59	.0532	92	.0341
27	.1161	60	.0523	93	.0338
28	.1120	61	.0515	94	.0334
29	.1081	62	.0507	95	.0331
30	.1045	63	.0499	96	.0327
31	.1012	64	.0491	97	.0324
32	.0980	65	.0483	98	.0321
33	.0951	66	.0476	99	.0317
34	.0923	67	.0469	100	.0314

To divide a circle into any number of equal spaces, find the proper k factor in the table above. Multiply the diameter times this factor to get the length of the chord, as shown in the illustration.

ELLIPSE AND PARABOLA

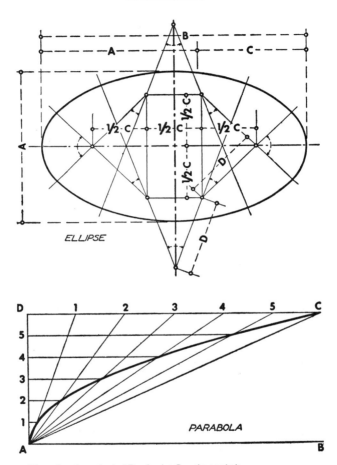

ELLIPSE

PARABOLA

Given: directions of axis *AB* and point *C* on the parabola.

Draw line *AD* through *A* perpendicular to *AB*, and draw line *DC* parallel to *AB* through *C*.

Divide lines *AD* and *DC* into the same number of equal parts.

Connect *A* with the division points on *DC*; draw lines through the division points on *AD* parallel to *AB*.

Intersections of lines of the same number are points on the parabola.

684

HOW TO DRAW
AN ELLIPSE

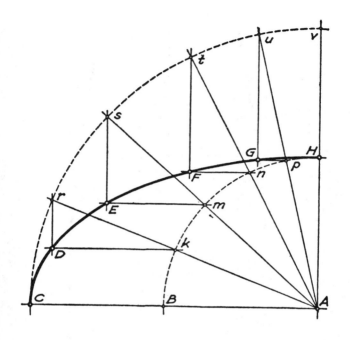

With *A* as the center and with half the minor axis *AH* as the radius, draw the circle *BH*.

With *A* as the center and with half the major axis *AC* as the radius, draw the circle *Cv*.

Draw any number of radii at random, as *Akr, Ams, Ant,* and *Apu.*

Drop a vertical from *r* to intersect a horizontal drawn from its corresponding point *k* at *D*. In a similar manner, find points *E, F* and *G*.

Draw the curve through the points thus found as *CDEFGH*.

Repeat the above steps three times in either a clockwise or counterclockwise series to produce a complete ellipse.

ENTASIS OF COLUMNS

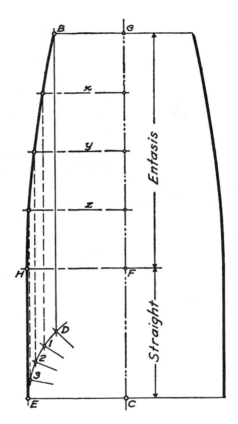

Given:—Radius BG at neck, and radius EC at bottom.

With C as center, draw arc ED to intersect vertical dropped from B at point D.

Divide arc ED into any number of equal parts by points 1, 2, 3.

Divide FG into a corresponding number of equal parts by the lines x, y, z.

The intersections of the lines x, y, z, with the corresponding vertical projections of the points 1, 2, 3, determines the entasis line HB.

SLOPE OF INCLINES

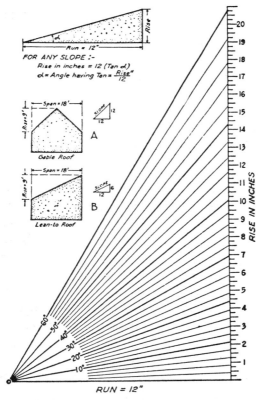

FOR ANY SLOPE :-

Rise in inches = 12 (Tan ∝)

∝ = Angle having Tan = $\frac{Rise''}{12}$

A — Gable Roof — Span = 18', Rise = 9', slope 12/12

B — Lean-to Roof — Span = 18', Rise = 9', slope 12/6

RISE IN INCHES

RUN = 12"

The amount of incline is measured in four ways. The *slope* may be given in inches of rise per 12″ of run. The *pitch* may be given—the rise divided by the span. The *percentage* of the incline may be given—the rise in feet (or inches) over a run of 100 feet (or inches) divided by the run. The *angle* may be given in degrees.

Architectural drawings should always indicate the incline in inches of rise per 12″ of run, using a small triangle as shown at *A* and *B*. Use the above diagram as a template under your tracing to get the slope, or use the trigonometric formulas given above.

The term "pitch" is misleading and ambiguous and should *never* be used except for gable roofs as at *A*. The confusion may be appreciated by considering the two roofs illustrated. Both have the same pitch of "one-half" but they have quite different *slopes* (12 to 12 and 6 to 12).

687

ROMAN NUMERALS

Arabic	Roman	Arabic	Roman
1	I	60	LX
2	II	70	LXX
3	III	80	LXXX
4	IV (IIII)	90	XC
5	V	100	C
6	VI	200	CC
7	VII	300	CCC
8	VIII	400	CCCC
9	IX	500	IƆ or D
10	X	600	DC
11	XI	700	DCC
12	XII	800	DCCC
13	XIII	900	DCCCC or CM
14	XIV	1,000	M or CIƆ
15	XV	2,000	MM
16	XVI	3,000	MMM
17	XVII	4,000	MMMM
18	XVIII	5,000	IƆƆ or $\overline{V}$
19	XIX	10,000	CCIƆƆ or $\overline{X}$
20	XX	50,000	IƆƆƆ or $\overline{L}$
30	XXX	100,000	CCCIƆƆƆ or $\overline{C}$
40	XL	500,000	IƆƆƆƆ or $\overline{D}$
50	L	1,000,000	CCCCIƆƆƆƆ or $\overline{M}$

If the lesser number is placed before the greater, the lesser is to be deducted from the greater: thus IV signifies 1 less than 5, i.e., 4; IX = 9; XC = 90.

If the lesser number be placed after the greater, the lesser is to be added to the greater; thus VI signifies 1 more than 5, i.e., 6; XI = 11; CX = 110.

A horizontal stroke over a numeral denotes 1,000; thus $\overline{V}$ signifies 5,000; $\overline{L}$ = 50,000; $\overline{M}$ = 1,000 × 1,000 = 1,000,000.

INCISED CLASSIC
ALPHABET

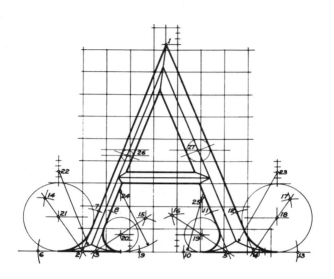

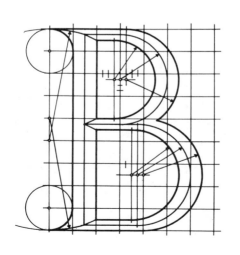

INCISED CLASSIC
ALPHABET

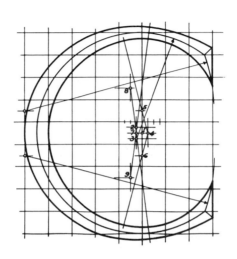

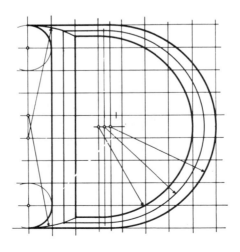

INCISED CLASSIC
ALPHABET

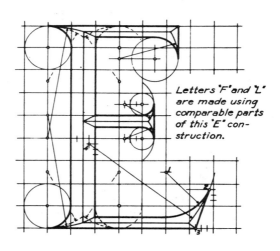

*Letters "F" and "L"
are made using
comparable parts
of this "E" con-
struction.*

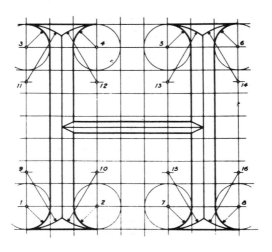

INCISED CLASSIC
ALPHABET

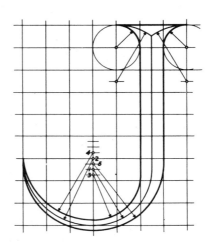

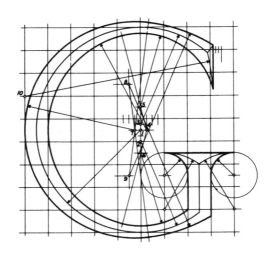

INCISED CLASSIC
ALPHABET

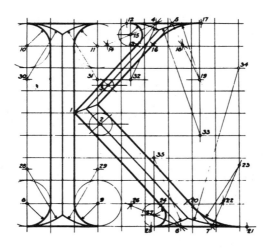

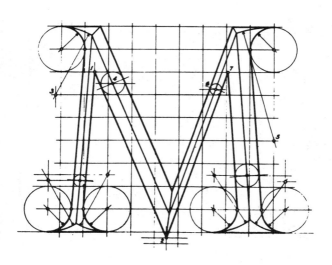

INCISED CLASSIC
ALPHABET

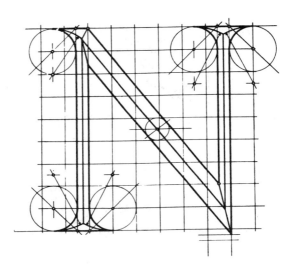

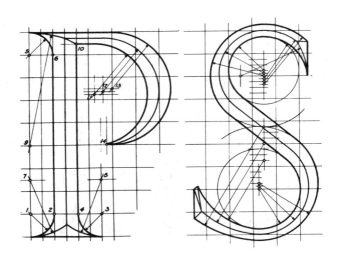

INCISED CLASSIC
ALPHABET

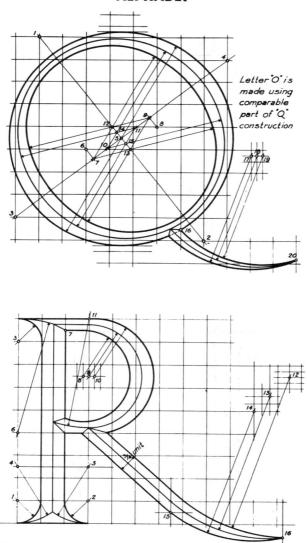

Letter "O" is
made using
comparable
part of "Q"
construction

INCISED CLASSIC
ALPHABET

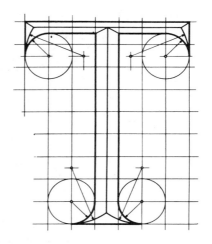

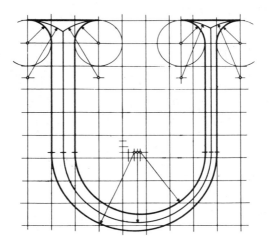

INCISED CLASSIC
ALPHABET

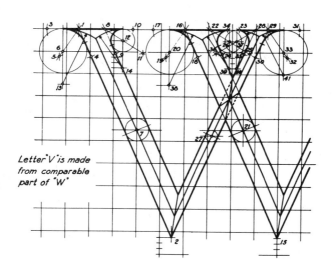

Letter "V" is made
from comparable
part of "W"

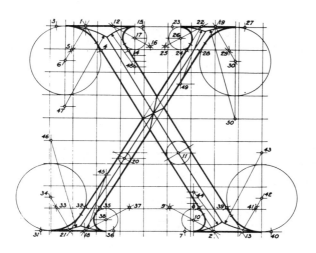

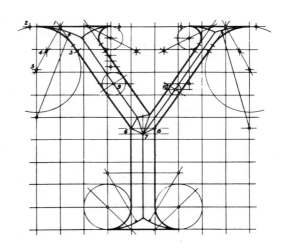

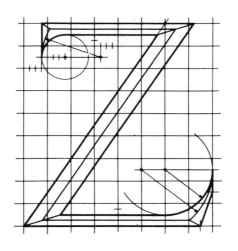

This lettering appears on a slate marker dating from about 1740. From 1725 to 1776 the Colonial lettering reached its greatest development, of which this is a typical example. The use of lowercase, small caps, italics and scratched guide-lines gives the composition its interest and flavor. The letters are incised with a V-cut of about 45° with the plane of the surface.

699

OLD ENGLISH
LETTERING

A B C D E F
G H I J K L
M N O P Q R
S T U V W X Y
Z abcdefgh
ijklmnopqr
stuvwxyy
z ﬀ ﬃ ﬄ ﬁ ﬂ l
2 3 4 5 6 7 8 9 0

JEWISH
ALPHABET

A B C D E
F G H I J K
L M N O P
Q R S T U
V ש א ע
Z מאָרגען
טאָגעבלאַט־

CUBIC COST CALCULATOR

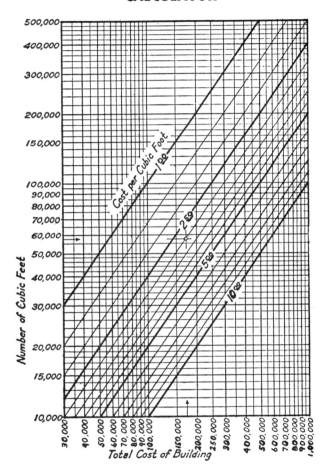

Example: If a building has an interior volume of 57,500 cubic feet and costs $3.00 per cubic foot (follow the arrows), the chart shows the total cost of the building will be $172,500.

Similarly, if the drawings call for a building of 57,500 cubic feet and there is $172,500 available, the chart shows that the cost per cubic foot must not exceed $3.

CHART FOR
FINDING CUBAGE

The approximate cubic contents of a solid can be quickly found with this chart. For example, for a house measuring 32' x 35' in plan by an average height of 33', the volume scale shows the cubic contents to be approximately 37,000 cubic feet.

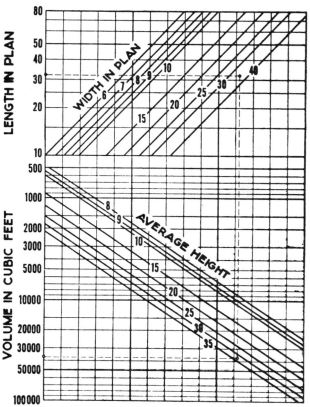

SIZES OF PIANOS

	W	D
Apartment ..	3'- 8"	4'- 0"
Baby	4'-10"	5'- 1"
Baby	4'-11"	5'- 4"
Parlor	5'- 0"	5'-10"
Parlor Concert	5'- 0"	7'- 7"
Concert	5'- 4"	9'- 0"

GRAND

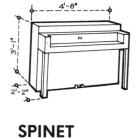

SPINET

BENCH

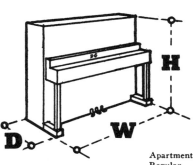

UPRIGHT

	H	D	L
Apartment .	3'-6"	1'-11"	3'-9"
Regular	4'-3"	2'- 0"	5'-0"
Regular	4'-5"	2'- 2"	5'-0"
Regular	4'-7"	2'- 4"	5'-2"

LIVING ROOM
FURNITURE SIZES

Seat 23"×23"   2'-3"   2'-7"

LOVE SEAT

2'-10"   1'-6"   1'-4"

COFFEE TABLE

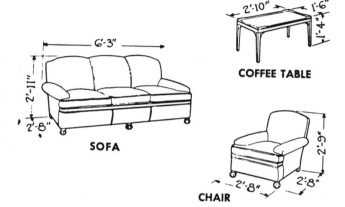

6'-3"

2'-11"

2'-8"

SOFA

2'-9"

2'-8"   2'-8"

CHAIR

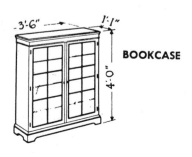

3'-6"   1'-1"

BOOKCASE

4'-0"

1'-7"Dia.

5'-1"

LAMP

SIZES OF
DINING TABLES

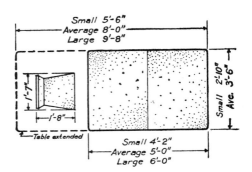

Small 5'-6"
Average 8'-0"
Large 9'-8"

1'-7"

1'-8"

Table extended

Small 2'-10"
Ave. 3'-6"

Small 4'-2"
Average 5'-0"
Large 6'-0"

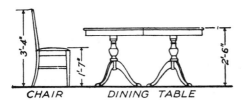

3'-4"

1'-7"

2'-6"

CHAIR   DINING TABLE

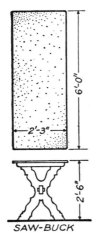

6'-0"

2'-3"

2'-6"

SAW-BUCK

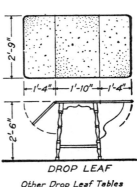

2'-9"

1'-4" 1'-10" 1'-4"

2'-6"

DROP LEAF

Other Drop Leaf Tables
Square 4'-0" x 4'-0"
Oval from —
  1'-8" to 3'-0" wide
  2'-8" to 4'-0" long

DINING
ALCOVE

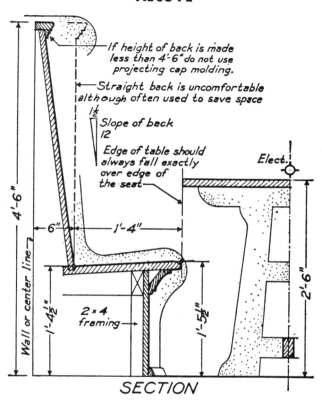

If height of back is made less than 4'-6" do not use projecting cap molding.

Straight back is uncomfortable although often used to save space

1½ Slope of back 12

Edge of table should always fall exactly over edge of the seat

Elect.

Wall or center line

6"

1'-4"

2'-6"

1'-4½"

2 × 4 framing

1'-5½"

SECTION

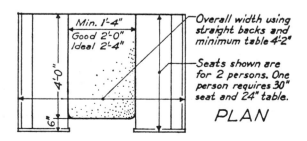

Min. 1'-4"
Good 2'-0"
Ideal 2'-4"

4'-0"

6"

Overall width using straight backs and minimum table 4'-2"

Seats shown are for 2 persons. One person requires 30" seat and 24" table.

PLAN

SIZES OF
BEDROOM FURNITURE

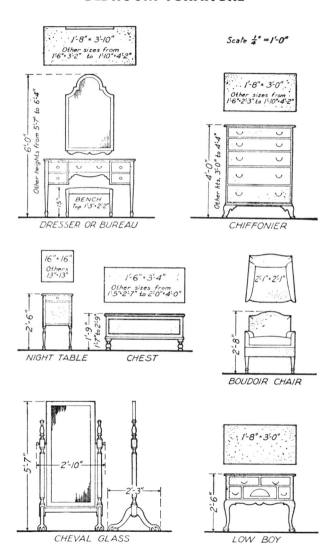

Scale $\frac{1}{4}" = 1'-0"$

1'-8" × 3'-10"
Other sizes from
1'-6"×3'-2" to 1'-10"×4'-2"

6'-0"
Other heights from 5'-7" to 6'-4"

BENCH
Top 1'-3"×2'-2"

15"

DRESSER OR BUREAU

1'-8" × 3'-0"
Other sizes from
1'-6"×2'-3" to 1'-10"×4'-2"

4'-0"
Other hts. 3'-0" to 4'-4"

CHIFFONIER

16"×16"
Others
13"×13"

2'-6"

NIGHT TABLE

1'-6" × 3'-4"
Other sizes from
1'-5"×2'-7" to 2'-0"×4'-0"

1'-9"
1'-7" to 2'-9"

CHEST

2'-1"×2'-1"

2'-8"

BOUDOIR CHAIR

5'-7"

2'-10"

2'-3"

CHEVAL GLASS

1'-8" × 3'-0"

2'-6"

LOW BOY

708

SIZES OF
BEDS

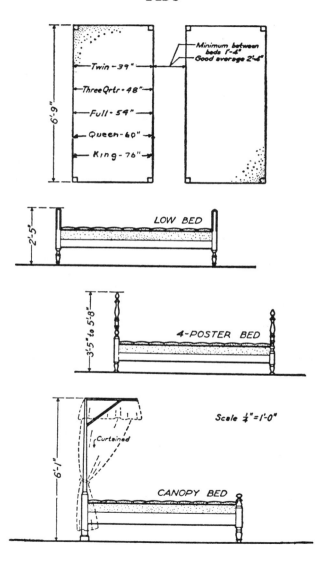

Minimum between beds 1'-4"
Good average 2'-4"

6'-9"

Twin - 39"

Three Qrtr - 48"

Full - 54"

Queen - 60"

King - 76"

LOW BED

2'-5"

4-POSTER BED

3'-5" to 5'-8"

Scale ¼" = 1'-0"

Curtained

CANOPY BED

6'-1"

APPROXIMATE DURATION
OF ARCHITECTURAL PERIODS

© CHAS. N. GEARHART

PERIOD
FURNITURE

Showing the order of the period styles from the beginning of the Renaissance to the nineteenth century.

ENGLAND			FRANCE	
Sovereign	Style		Style	Sovereign
Henry VIII 1509-1557	TUDOR		FRANCOIS PREMIER	Francis I 1515-1547
Elizabeth 1558-1603	ELIZABETHAN		HENRI DEUX	Henry II 1547-1559
James I 1603-1625	JACOBEAN			Francis II 1559-1560
Charles I 1625-1649				
Commonwealth 1649-1660				Charles IX 1560-1574
Charles II 1660-1685				
James II 1685-1688				Henry III 1574-1589
William and Mary 1688-1702	WILLIAM AND MARY		HENRI QUATRE	Henry IV 1589-1610
Anne 1702-1714	QUEEN ANNE		LOUIS TREIZE	Louis XIII 1610-1643
George I 1714-1727	CHIPPENDALE	GEORGIAN PERIOD	ROCOCO LOUIS QUATORZE	Louis XIV 1643-1715
George II 1727-1760	HEPPLEWHITE ADAM		LOUIS QUINZE	Louis XV 1715-1747
George III 1760-1820	SHERATON		CLASSIC REVIVAL LOUIS SEIZE	Louis XVI 1747-1793
			EMPIRE	Napoleon 1793-1814

The duration of the Renaissance in other countries is about as follows: Italy 1443-1546; Germany 1525-1620; Flemish and Dutch 1520-1634; Spain and Portugal 1500-1620; other European countries 1500-1620. The Rococo begins at about the dates given for the end of the Renaissance.

ENGLISH FURNITURE
(1560 - 1690)

TUDOR-ELIZABETHAN, JACOBEAN. Massive, sturdy furniture replaced the stark pieces of feudal days in early England. The Tudor-Elizabethan era was the Renaissance in Britain.

Oak in simple wax finish was carved elaborately in extravagant and forceful forms. Some dining room suites and occasional pieces are reproduced today, but interest in Tudor styles is chiefly because they represent the first swing toward decorative furniture and buildings. When this style is used, it properly belongs in large Gothic rooms.

Early Jacobean furniture, sometimes called Stuart, was particularly sturdy. It utilized the same oak that was employed in Queen Elizabeth's day. It was the style of furniture that inspired early American styles in the colonies. In the middle Jacobean the gateleg table evolved.

Late Jacobean, or Charles 2nd, furniture is increasingly used today; the severity of the Cromwellian morality having been replaced by a merry monarch's love of luxury, the designs reflected this lighter attitude toward life. Both oak and walnut were used in that period.

Gate-leg Table
Jacobean

Chair
Late Jacobean

Armchair
Elizabethan

Draw-top Table
Elizabethan

Armchair
Late Jacobean

FRENCH FURNITURE
(1500 - 1750)

LOUIS 14TH. This period marked the evolution of the straight line toward the curve which was to predominate in the following epoch. The straight line was usual. Proportions were large, massive, dignified and formal. Louis 14th furniture is seldom used today except in large and luxurious quarters. Its purpose was for show—comfort was not considered of great importance.

THE REGENCY. This era marked the beginning of a newer and lighter vein in furniture design. The curved line replaced rectilinear forms.

LOUIS 15TH. Probably the outstanding age of the world in decorative furniture, this period is notable for its rich and luxurious creations. The style is distinctly feminine. Walnut, mahogany and ebony were used effectively. Lacquers and gilding covered much of the woods to good advantage. The cabriolet leg was used almost exclusively and scroll feet were usual. Reproductions are suited to homes where fastidious elegance is desired. Careful selection is necessary to blend Louis 15th furniture with other styles.

Armchair
Louis XIV

Table
French Renaissance

Writing Desk
Louis XV

Armoire
Louis XIV

Commode
Louis XV

Armchair
Louis XV

FRENCH FURNITURE
(1750 - 1815)

LOUIS 16TH. The furniture is a slender, straight line style with a return to classicism. It is a direct and vigorous reaction against the rococo ornamentation and excessive curves of the previous reign. Cherubs, love birds, garlands of flowers and love knots were some of the motifs employed. Round medallions, ovals, heads, busts, human figures, fluting, reeding and beading are features of the style. Mahogany finished either in natural grain or enameled, walnut, and satinwood were much used. Silks, figured satins, brocades, damasks, muslins and velvets in pastoral and floral designs with later extensive use of stripes are all typical. Simple and feminine, the style is used where a marked effect of delicacy and daintiness is desired.

THE DIRECTOIRE. Simple classical forms were substituted for monarchial ornament.

EMPIRE. A militaristic masculine stylistic reaction from the preceding femininity, the furniture was heavy and ponderous. Frequently Empire furniture is adapted to use with modern designs.

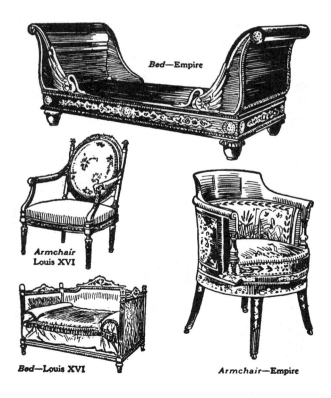

Bed—Empire

Armchair
Louis XVI

Bed—Louis XVI

Armchair—Empire

ITALIAN AND SPANISH
FURNITURE (1453 - 1560)

After the medieval ages, the Renaissance brought renewed interest in furniture as well as art and literature. In Italy, ornate carved pieces were used in formal halls of princes of church and state. The principal wood was walnut; the decoration was classical with fine restraint; rich, colorful dignity was expressed in the upholstery. Italian Renaissance reproductions today are scaled to the large home and would be incongruous in a bungalow.

In Spain, walnut and oak furniture were studded with brass and iron, and metal mounts were freely used. Bright red and rich green velvets were used in the trimmed and fringed upholstering, and decorated leather was also employed. Modern reproductions are well suited to many modern homes, particularly to those of Spanish architecture. Spanish furniture is massive, rugged, masculine, square, and sturdy. It is suited to use with Italian and French Renaissance furniture, as well as some early English designs.

Armchair
Spanish Renaissance

Table
Spanish Renaissance

Vargueno
Spanish Renaissance

Painted Commode
18th Century Venetian

Armchair
Italian Renaissance

"Dante" Chair
Italian Renaissance

Armchair
Italian Renaissance

STANDARD STEEL
LOCKER SIZES

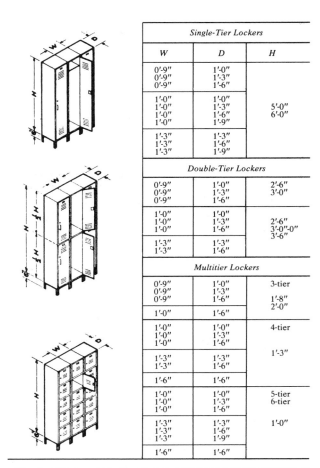

Single-Tier Lockers		
W	D	H
0'-9"	1'-0"	
0'-9"	1'-3"	
0'-9"	1'-6"	
1'-0"	1'-0"	
1'-0"	1'-3"	5'-0"
1'-0"	1'-6"	6'-0"
1'-0"	1'-9"	
1'-3"	1'-3"	
1'-3"	1'-6"	
1'-3"	1'-9"	
Double-Tier Lockers		
0'-9"	1'-0"	2'-6"
0'-9"	1'-3"	3'-0"
0'-9"	1'-6"	
1'-0"	1'-0"	
1'-0"	1'-3"	2'-6"
1'-0"	1'-6"	3'-0"-0"
		3'-6"
1'-3"	1'-3"	
1'-3"	1'-6"	
Multitier Lockers		
0'-9"	1'-0"	3-tier
0'-9"	1'-3"	
0'-9"	1'-6"	1'-8"
1'-0"	1'-6"	2'-0"
1'-0"	1'-0"	4-tier
1'-0"	1'-3"	
1'-0"	1'-6"	
1'-3"	1'-3"	1'-3"
1'-3"	1'-6"	
1'-6"	1'-6"	
1'-0"	1'-0"	5-tier
1'-0"	1'-3"	6-tier
1'-0"	1'-6"	
1'-3"	1'-3"	1'-0"
1'-3"	1'-6"	
1'-3"	1'-9"	
1'-6"	1'-6"	

The sizes listed are generally accepted as "standard," though variations do appear. Frames and doors are usually made of 16-gauge steel, while the remainder varies from 20- to 24-gauge. Units may be of painted carbon steel or plain stainless steel. Doors may be perforated, grilled, louvered, or steel-mesh-paneled. Perforated or steel-mesh sides are available, as are sloped tops, legs of various lengths, closed bases, and interior fittings (hooks, shelves, and so forth). Multiperson, checking, and various combination lockers are also obtainable.

STANDARDS AND SPECIFICATIONS

Standards and specifications covering a host of materials, products, applications, constructions, safety features, installations, certification requirements, and testing procedures now number in the thousands. Several hundred of them directly affect the construction industry. In the normal course of events, the architect, designer, builder, estimator, contractor, and others associated with the industry (and often even do-it-yourself homeowners) must come to grips with at least a few of them. Sometimes it is important to know and understand either a part or all of certain standards or specifications in order to carry on one's work; in other instances, it is sufficient to recognize their import so as to ascertain that work is properly carried out by others or that products or materials conforming to them are employed.

Standards and specifications are promulgated by a great many agencies, organizations, and associations. However, those that originate from six particular sources account for a large percentage of those seen in the building trades.

NBS. Commercial standards from the National Bureau of Standards (NBS), part of the U.S. Department of Commerce, set quality level requirements, establish methods of testing, rating, certifying, and labeling commodities, and provide uniform bases for fair competition. These standards are voluntarily developed by manufacturers, distributors, consumers, and other interested parties through a regular NBS procedure established for just that purpose.

Example: NBS Voluntary Product Standard PS 51-71, *Hardwood and Decorative Plywood.* "51" is the identification number, and "71" is the year of promulgation or revision. Product Standards are gradually replacing the older Commercial Standards (CS). A complete listing of current standards and copies of individual standards may be obtained from NBS in Washington, D.C. Individual copies may also be obtained from whatever trade organization or association promotes the particular product (in this case, the Hardwood Plywood Manufacturers Association).

ASTM. The American Society for Testing Materials (ASTM) conducts tests and analyses of many different materials and constructions and establishes standards for their quality and use. Many of these standards are referred to commonly in building plans and specifications.

Examples: ASTM C549, *Perlite Loose Fill Insulation;* ASTM D1850, *Concrete Joint Sealer, Cold Application Type.* A listing of currently available standards and copies of individual standards may be obtained from ASTM in Philadelphia, Pennsylvania.

ANSI. The American National Standards Institute (ANSI—formerly the American Standards Association (ASA)) also is responsible for the promulgation of a great many standards covering materials, products, testing procedures, installations, and applications. ANSI does not itself develop any of these standards, but rather acts as a coordinating agency through which others may organize and establish them; all such standards are voluntary. Once established and recognized, the standards carry the ANSI designation and are available for voluntary use by others. They are also frequently written into legal codes and regulations, as well as into various kinds of job specifications.

Examples: ANSI A 108.4, *Ceramic Tile Installed with Water-resistant Organic Adhesives;* ANSI/NFPA 80, *Fire Doors and Windows.* A current listing of available standards and copies of individual standards can be obtained from ANSI in New York City, New York.

STANDARDS AND SPECIFICATIONS

FEDERAL SPECIFICATIONS.   There are a great many federal specifications covering a multitude of subjects. A number of them are important to the building trades and to manufacturers of products for the industry. Some of these specifications are developed by particular federal agencies with expertise in the field in question. Others are adoptions, adaptations, and amalgamations of specifications originally developed by recognized industry and technical societies and/or trade associations. They are widely referred to in the construction industry.

Examples: Federal Specification HH-I-521F, *Mineral Fiber Blankets;* Federal Specification TT-F-336E, *Filler, Wood, Paste.* Information on these specifications can be obtained from the Superintendent of Documents, U.S. Government Printing Office, Washington, D.C.

MILITARY SPECIFICATIONS.   Though military specifications are primarily used in conjunction with products and materials being supplied to the various branches of the military, they are also sometimes referred to as general compliance standards for nonmilitary applications.

Examples: MIL S-12158C, *Sealing Compound, Noncuring Polybutene;* DOD-P-15328D, *Primer, Pretreatment.* Information on military specifications can be obtained through the Department of Defense, Washington, D.C.

UL.   The Underwriters' Laboratories, Inc., is responsible for testing materials, products, and systems or constructions of all sorts. Its findings are published, and it develops various standards. UL also operates a certification and labeling program. UL certification and labeling is a common requirement in building specifications, as is compliance with certain of their standards.

Example: Underwriters' Laboratories Standard UL55B, *"Class C" Asphalt Organic-Felt Sheet Roofing and Shingles.* UL standards, as well as lists of labeled products and other publications, can be obtained directly from Underwriters' Laboratories in Chicago, Illinois.

OTHER SOURCES.   In addition to the above standards and specifications, numerous others are established by and may be obtained from a wide variety of specific agencies, organizations, and associations. Among the most important of these are standards developed by the U.S. Department of Housing and Urban Development (HUD), such as the *HUD Minimum Property Standards,* and the American Society of Heating, Refrigerating, and Air Conditioning Engineers (ASHRAE), such as ASHRAE Standard 90, *Energy Conservation in New Building Design.*

In the case of trade associations and the like, a particular name is usually mentioned in connection with their standards and specifications, such as the Red Cedar Shingle and Handsplit Shake Bureau, the Carpet and Rug Institute, the National Particleboard Association, the Steel Window Institute, and so forth. Any such organization is happy to provide pertinent specifications or standards, as well as detailed product information. When an architect or designer wishes to specify materials and/or products that conform to accepted and known quality levels and industry standards, a convenient and time-saving method of successfully doing so is simply to contact the appropriate trade or similar organization for full information.

BUILDING CODES

The principal purpose of building codes is to establish certain minimum requirements with regard to planning, construction, location, and use of public and private buildings of all sorts. When these codes are adopted by local, state, or federal governments they become law, and their provisions must be complied with by all concerned with the construction industry — from the manufacturer of building supplies and products to the worker in the field. At present, construction in all large cities, most small cities and towns, and a large percentage of rural areas is regulated to at least some extent by one or another of the many building codes. In areas where no building codes carry the force of law, the codes are often used as guidelines for construction. There is, on a nationwide scale, little uniformity in the degree of code compliance and enforcement and even less uniformity in the codes themselves; a complex and chaotic code situation therefore exists and will continue to exist for years to come. In addition, code performance standards are frequently inadequate, provisions must be interpreted by individual inspectors and often are largely judgmental, and the extreme slowness of most codes to change and to include modern, up-to-date material and methods leads to a great many problems. In many locales, however, codes are law and must be followed, and they do accomplish their main purpose: to protect the health, safety, and welfare of the general public.

MODEL CODES. There are several nationally recognized and widely accepted building codes, generally known as *model codes,* that have been developed by experts in various fields and made available for adoption by any government agency.

BBC. The *Basic Building Code* (BBC) is a relatively new general building code developed by the Building Officials and Code Administrators, Inc. (BOCA— 17926 South Halstead, Homewood, Illinois 60430). It is in force in many parts of the country. BOCA also publishes auxiliary mechanical codes to be used in conjunction with the BBC, such as the *Basic Mechanical Code,* the *Basic Plumbing Code,* and the *Basic Fire Prevention Code,* among others.

NBC. The *National Building Code* (NBC), originally introduced by the National Board of Fire Underwriters (now the American Insurance Association, 85 John Street, New York, New York 10038) is the oldest of the model building codes, and is widely used throughout the country.

SSBC. The *Southern Standard Building Code* (SSBC) is a product of the Southern Building Code Congress International (SBCCI—900 Montclair Road, Birmingham, Alabama 35213). Other codes published by the SBCCI include the *Southern Standard Plumbing Code,* the *Southern Standard Mechanical Code,* and the *Southern Standard Gas Code.* As the names imply, these codes are primarily in use in the southern states and attempt to address building problems peculiar to that region.

UBC. The *Uniform Building Code* (UBC) is published by the International Conference of Building Officials (ICBO—5360 South Workman Mill Road, Whittier, California 90601). This code is widely accepted across the country. It is a general building code, but the ICBO also publishes additional codes to be used in conjunction with it: the *Uniform Mechanical Code,* the *Uniform Fire Code,* and the *ICBO Plumbing Code.* They offer other publications as well, such as the *Illustrated Mechanical Manual* and the *Concrete Inspection Manual.*

OTHER CODES. In addition to the model building codes noted above and their ancillary mechanical codes, there are several specialized codes that are national in scope and widely used. The National Fire Protection Association (NFPA) publishes the *National Fire Codes,* a part of which is the generally accepted *National Electrical Code.* The *National Plumbing Code* is also in widespread use, and has been adopted as an American National Standard (ANSI A40.8). The *Uniform Plumbing Code* and the *Uniform Heating and Comfort Cooling Code,* both published by the Western Plumbing Officials Association, are also widely recognized. Numerous other even more highly specialized codes, such as the *Boiler and Unfired Pressure Vessel Code* (American Society of Mechanical Engineers), the *Safety Code for Mechanical Refrigeration* (ASHRAE), the *Chimneys, Fireplaces, and Vents Code* (NFPA), and the *Household Fire Warning Equipment Code* (NFPA), are in common use throughout the country.

INDEX

723

724

725